THE ECOLOGY OF THE CAMBRIAN RADIATION

CRITICAL MOMENTS IN PALEOBIOLOGY AND EARTH HISTORY

David J. Bottjer and Richard K. Bambach, editors

The Ecology of the Cambrian Radiation

EDITED BY ANDREY YU. ZHURAVLEV
AND ROBERT RIDING

Columbia University Press ▪ New York

We dedicate this book to David Gravestock and Kirill Seslavinsky.

Columbia University Press
Publishers Since 1893
New York   Chichester, West Sussex

Library of Congress Cataloging-in-Publication Data

The ecology of the Cambrian radiation / edited by
Andrey Yu. Zhuravlev and Robert Riding.
    p.   cm.—(Critical moments in paleobiology and
earth history series)
Includes bibliographical references and index.
ISBN 978-0-231-10613-9 (pbk. : alk. paper)

1. Paleoecology—Cambrian.   2. Paleontology—
Cambrian.   3. Geology, Stratigraphic—Cambrian.
I. Zhuravlev, A. IU. (Andrei IUr'evich).   II. Riding, Robert.
III. Series.
QE720 .E27 2000
560'.1723—dc21   00-063901

Printed in the United States of America

Contents

Acknowledgments

We are indebted to the following specialist reviewers without whose help we could not have accomplished this task: Pierre Adam, Pierre Albrecht, J. Fredrik Bockelie, Gerard C. Bond, Derek E. G. Briggs, Paul Copper, Pierre Courjault-Radé, Mary L. Droser, Richard A. Fortey, Gerd Geyer, Roland Goldring, James W. Hagadorn, Sören Jensen, Viktor E. Khain, Tat'yana N. Kheraskova, Pierre D. Kruse, Ed Landing, John F. Lindsay, Jere H. Lipps, Dorte Mehl, Carl Mendelson, Timothy J. Palmer, Christopher R. C. Paul, John S. Peel, Martin Pickford, Leonid E. Popov, Lars Ramsköld, Robert L. Ripperdan, Philippe Schaeffer, Frederick R. Schram, J. John Sepkoski, Jr., Thomas Servais, Barry D. Webby, Graham L. Williams, Matthew A. Wills, Mark A. Wilson, and Grant M. Young.

We are especially grateful to Françoise Debrenne, Mary Droser, and Alan Smith for help in the preparation of this volume. Françoise Pilard, Max Debrenne, and Henri Lavina assisted greatly in the finalization of many figures. AZ's editing was facilitated by the Muséum National d'Histoire Naturelle, Paris.

We thank our contributors, one and all, for their willingness to join us in this venture and for their forbearance when we acted as editors are only too often prone to do.

Last, but certainly not least, we thank Ed Lugenbeel, Holly Hodder, and Jonathan Slutsky at Columbia University Press, and Mark Smith and his colleagues at G&S Editors, for their expert handling of both the book and us.

THE ECOLOGY OF THE CAMBRIAN RADIATION

CHAPTER ONE

Andrey Yu. Zhuravlev and Robert Riding

Introduction

THE CAMBRIAN RADIATION, which commenced around 550 million years ago, arguably ranks as the single most important episode in the development of Earth's marine biota. Diverse benthic communities with complex tiering, trophic webs, and niche partitioning, together with an elaborate pelagic realm, were established soon after the beginning of the Cambrian period. This key event in the history of life changed the marine biosphere and its associated sediments forever.

At first glance, abiotic factors such us climate change, transgressive-regressive sea level cycles, plate movements, tectonic processes, and the type and intensity of volcanism appear very significant in the shaping of biotic evolution. We can see how rapid rates of subsidence, as expressed in transgressive system tracts on the Australian craton, selectively affected the diversity of organisms such as trace fossil producers, archaeocyath sponges, and trilobites (Gravestock and Shergold—chapter 6); how globally increased rates of subsidence and uplift accompanied dramatic biotic radiation by increasing habitat size and allowing phosphorus- and silica-rich waters to invade platform interiors (Brasier and Lindsay—chapter 4); how climatic effects, coupled with intensive calc-alkaline volcanism, at the end of the Middle Cambrian may have caused a shift from aragonite- to calcite-precipitating seas, providing suitable conditions for development of the hardground biota (Seslavinsky and Maidanskaya—chapter 3; Eerola—chapter 5; Guensburg and Sprinkle—chapter 19); how the reorganization of plate boundaries (Smith—chapter 2; Seslavinsky and Maidanskaya) created conditions for current upwelling, which may in turn have been responsible for the appearance and proliferation of acritarch phytoplankton and many Early Cambrian benthic organisms (Brasier and Lindsay; Ushatinskaya—chapter 16; Moldowan et al.—chapter 21).

However, biotic factors themselves played a remarkable role in the environmental changes that formed the background to the Cambrian radiation. We see how, by means of biomineralization, shell beds and calcite debris contributed to the appearance of hardground communities (Droser and Li—chapter 7; Rozhnov—chapter 11); how the intensification of bioturbation not only obliterated sedimentary structures but also increased aeration of deeper sediments and provided more space for the development of infauna (Brasier and Lindsay; Droser and Li; Crimes—chapter 13); how the Early Cambrian biota changed the quality of seawater, thereby allowing the radiation of diverse phototrophic communities (Zhuravlev—chapter 8; Burzin et al.—chapter 10); how the appearance of framework-building organisms created habitats for diverse reefal communities (Pratt et al.—chapter 12; Debrenne and Reitner—chapter 14; Riding—chapter 20); how the introduction of mesozooplankton in the Eltonian pyramid (in addition to predator and herbivore pressure) produced a cascade of ecologic and evolutionary events in both the pelagic and benthic realms (Butterfield—chapter 9; Zhuravlev); and, finally, how biotic diversity itself, together with community structure, conditioned the intensity of extinction events and the timing and type of abiotic factors that may have caused them (Zhuravlev).

This volume comprises 20 chapters, contributed by 33 authors based in 10 countries. It has three themes: environment; community patterns and dynamics; and radiation of major groups of organisms. The focus is the Cambrian period (tables 1.1 and 1.2), but inevitably discussion of these topics also draws on related events and developments in the adjacent Neoproterozoic and Ordovician time intervals.

ENVIRONMENT

The theme of the environment traces plate tectonic developments, paleogeographic changes, the history of transgressive-regressive cycles, sedimentary patterns, and climate change, as recorded in carbon, strontium, and samarium-neodymium isotope curves, in the context of their influence on biotic development. The records of bioturbation and shell-bed fabrics, which provide links among physical, chemical, and biologic processes, are included, and there are data on biomarkers.

COMMUNITY

The theme of community considers the biotas in their ecologic context, from their diversification to the development of planktonic, level-bottom, reef, hardground, and deep-water communities.

RADIATION

The theme of radiation examines deployment of adaptive abilities by dominant Cambrian groups: brachiopods, cnidarians, coeloscleritophorans, cyanobacteria, algae, echinoderms, hyoliths, lobopods, mollusks, sponges, stenothecoids, trilobites, and other arthropods. Other common groups, such as acritarchs, chaetognaths, hemichordates, conodont-chordates, various worms, and minor problematic animals, are not scrutinized separately, but aspects of their ecology are discussed within analyses of particular communities.

Not all the views expressed in this book are in agreement, nor should they be. We hope that comparison of the facts, arguments, and ideas presented will allow the reader to judge the relative importance of abiotic and biotic factors on the dramatic evolutionary and ecologic expansion that was the Cambrian radiation of marine life.

This volume is a contribution to IGCP Project 366, Ecological Aspects of the Cambrian Radiation. In addition, this work has involved participants from IGCP Projects 303 (Precambrian-Cambrian Event Stratigraphy), 319 (Global Paleogeography of the Late Precambrian and Early Paleozoic), 320 (Neoproterozoic Events and Resources), 368 (Proterozoic Events in East Gondwana Deposits), and 386 (Response of the Ocean/Atmosphere System to Past Global Events).

MUSEUM AND REPOSITORIES ABBREVIATIONS

AGSO (Australian Geological Survey Organisation, Canberra, Australia), GSC (Geological Survey of Canada, Ottawa), HUPC (Harvard University Paleobotanical Collection, Cambridge, USA), IGS (Iranian Geological Survey, Tehran), MNHN (Muséum National d'Histoire Naturelle, Paris, France), PIN (Paleontological Institute, Russian Academy of Sciences, Moscow), SAN (Sansha Collections, J. Reitner, Göttingen, Germany), SMX (Sedgwick Museum, Cambridge University, United Kingdom), UA (University of Alaska, USA), USNM (National Museum of Natural History, Smithsonian Institution, Washington, DC, USA), UW (University of Wisconsin, USA).

REFERENCES

Bowring, S. A., J. P. Grotzinger, C. E. Isachsen, A. H. Knoll, S. M. Pelechaty, and P. Kolosov. 1993. Calibrating rates of Early Cambrian evolution. *Science* 261:1293–1298.

Davidek, K., E. Landing, S. R. Westrop, A. W. A. Rushton, R. A. Fortey, and J. M. Adrain. 1998. New uppermost Cambrian U-Pb date from Avalonian Wales and the age of the Cambrian-Ordovician boundary. *Geological Magazine* 132:305–309.

Jago, J. B. and P. W. Haines. 1998. Recent radiometric dating of some Cambrian rocks in southern Australia: relevance to the Cambrian time scale. *Revista Española de Paleontología,* no. extraordinario, Homenaje al Prof. Gonzalo Vidal, 115–122.

Landing, E., S. A. Bowring, K. Davidek, S. R. Westrop, G. Geyer, and W. Heldmaier. 1998. Duration of the Early Cambrian: U-Pb ages of volcanic ashes from Avalon and Gondwana. *Canadian Journal of Earth Sciences* 35:329–338.

Shergold, J. H. 1995. *Timescales. 1: Cambrian.* Australian Phanerozoic Timescales, Biostratigraphic Charts, and Explanatory Notes, 2d ser. Australian Geological Survey Organisation Record 1995/30.

Zhuravlev, A. Yu. 1995. Preliminary suggestions on the global Early Cambrian zonation. *Beringeria Special Issue* 2:147–160.

Table 1.1 **Correlation Chart for Major Lower Cambrian Regions**

Siberian Platform		Australia		China		Spain
Stages	Trilobite, Archaeocyath, and Small Shelly Fossil Zones	Archaeocyath Zones	Trilobite Zones (Stages)	Stages	Trilobite and Small Shelly Fossil Zones	Stages
Amgan	*Schistocephalus* 1		*Xystridura templetonensis/ Redlichia chinensis*	Maozhuangian	*Yaojiayella*	Leonian
	Anabaraspis splendens 3			Longwangmiaoan	*Redlichia nobilis*	
					Redlichia chinensis	
					Hoffetella	
Toyonian	*Lermontovia grandis/ Irinaecyathus shabanovi- Archaeocyathus okulitchi* beds 2	*Archaeocyathus abacus* beds	(Ordian/ Early Templetonian)	Canglangpuan	*Megapalaeolenus/ Palaeolenus*	Bilbilian
	Bergeroniellus ketemensis 1		*523 Ma			
Botoman	*Bergeroniellus ornata* 4 [EB]	*Syringocnema favus* beds	*Pararaia janeae* *525 Ma		*Drepanuroides*	Marianian
	Bergeroniellus asiaticus 3		*Pararaia bunyerooensis*		*Yunnanaspis/ Yiliangella*	
	Bergeroniellus gurarii [SB] 2	Unnamed beds				
	Bergeroniellus micmacciformis/ Erbiella 1		*Pararaia tatei*	Qiongzhusian [CB]	*Malungia*	
			Abadiella huoi		*Eoredlichia/ Wutingaspis*	
Atdabanian	*Fansycyathus lermontovae* 4	*Jugalicyathus tardus*			"*Parabadiella*"/ *Mianxidiscus*	Ovetian
		Spirillicyathus tenuis			*Lapworthella/ Tannuolina/ Sinosachites*	
	Nochoroicyathus kokoulini 3	*Warriootacyathus wilkawillinensis*				
	Carinacyathus pinus 2			Meishucunian		
	Retecoscinus zegebarti 1					
Tommotian *535 Ma	*Dokidocyathus lenaicus/ Tumuliolynthus primigenius* 4					Cordubian
	Dokidocyathus regularis 2/3					
	Nochoroicyathus sunnaginicus 1				*Siphogonuchites/ Paracarinachites*	
Nemakit- Daldynian *545 Ma	*Purella antiqua* 2					Alcudian
	Anabarites trisulcatus 1				*Anabarites/ Protohertzina/ Arthrochites*	

Note: Approximate correlation of Lower Cambrian stratigraphic subdivisions for different regions, modified from Zhuravlev 1995, and the positions of key Cambrian faunas: CB = Chengjiang fauna, EB = Emu Bay Shale, MC = Mount Cup Formation, SB = Sinsk fauna, SP = Sirius Passet

Morocco		Baltic Platform		Laurentia	Avalonia	
Stages	Trilobite Zones	Trilobite, Small Shelly Fossil, and Ichnofossil Zones	Acritarch Zones	Trilobite Zones	Stages	Trilobite, Small Shelly Fossil, and Ichnofossil Zones
Tissafinian	Ornamentapsis frequens	Eccaparadoxides insularis	"Kibartay"	Albertella	*511 Ma	Protolenus
	Cephalopyge notabilis	Proampyx linnarssoni	Volkovia dentifera/ Liepaina plana	Plagiura/Poliella		
	Hupeolenus			Bonnia/ Olenellus		
Banian	Sectigena	Holmia kjerulfi	Heliosphaeridium dissimilare/ Skiagia ciliosa	[MC]	Branchian	Callavia broeggeri
	Antatlasia guttapluviae					
	Antatlasia hollardi					
Issendalenian	Daguinaspis			"Nevadella"		Camenella baltica
	Choubertella			[SP]		
	Fallotaspis tazemmourtensis	Holmia inusitata		"Fallotaspis"		
	Eofallotaspis	Schmidtiellus mikwitzi	Skiagia ornata/ Fimbriaglomerella membranacea		Placentian	Sunnaginia imbricata
		Rusophycus parallelum				No fauna known
		Platysolenites antiquissimus	Asteridium tornatum/ Comasphaeridium velvetum			Watsonella crosbyi
						"Ladatheca" cylindrica
						No fauna known
						"Phycodes" pedum
		Sabellidites	"Rovno"			Harlaniella podolica

fauna. In addition, in some chapters the Waucoban corresponds to the Early Cambrian, and the Olenellid biomere is used for Atdabanian-Toyonian. Reliable radioisotope ages from Bowring et al. 1993, Jago and Haines 1998, and Landing et al. 1998.

Table 1.2 **Correlation Chart for Major Middle and Late Cambrian**

Kazakhstan & Siberia			Australia		China
Ungurian	*Dikelokephalina*	1	Warendian	*Cordylodus lindstromi*	
Batyrbayan			Datsonian	*Cordylodus prolindstromi*	Xingchanglan
	Euloma limitaris/ Batyraspis			*Hirsutodontus simplex*	
		1		*Cordylodus proavus*	
	Lotagnostus hedini		Payntonian	*Mictosaukia perplexa*	Fengshanian
	Harpidoides/Troedsonia				
	Lophosaukia			*Neoagnostus quasibilobus/ Shergoldia nomas*	
	Trisulcagnostus trisulcus	3		*Sinosaukia impages*	
Aksayan	*Eolotagnostus scrobicularis*	2	Iverian	*Lophosaukia*	Changshanian
				Rhaptagnostus clarki prolatus/ Caznaia sectatrix	
	Neoagnostus quadratiformis			*Rhaptagnostus clarki patulus/ Caznaia squamosa/ Hapsidocare lilyensis*	
	Oncagnostus ovaliformis			*Peichiashania tertia/ Peichiashania quarta*	
	Oncagnostus kazachstanicus			*Peichiashania secunda/ Peichiashania glabella*	
	Pseudagnostus pseudangustilobus	1		*Wentsua iota/ Rhaptagnostus apsis*	
Sakian	*Ivshinagnostus ivshini*	3	Idamean	*Irvingella tropica*	
	Pseudagnostus "curtare"	2		*Stigmatoa diloma*	
	Oncagnostus longifrons			*Erixanium sentum*	
				Proceratopyge cryptica	
	Glyptagnostus reticulatus	1		*Glyptagnostus reticulatus*	
Aysokkanian	*Glyptagnostus stolidotus*	6	Mindyallan	*Glyptagnostus stolidotus*	Kushanian
	Agnostus pisiformis	5		*Acmarhachis quasivespa*	
Mayan	*Leiopyge laevigata/ Aldanaspis truncata*		Boomerangian	*Erediaspis eretis*	Zhangxian
				Damesella torosa/ Ascionepea jantrix	
		4	*495 Ma	*Holteria arepo*	
	Anomocarioides limbataeformis			*Proampyx agra*	
				Ptychagnostus cassis	
		3	Undilian	*Goniagnostus nathorsti*	
	Anopolenus henrici/ Corynexochus perforatus			*Doryagnostus notalibrae*	
				Ptychagnostus punctuosus	
		2		*Euagnostus opimus*	
Amgan	*Pseudanomocarina*	1	Late Templetonian/ Floran	*Acidusus atavus*	Xuzhuangian
		3		*Triplagnostus gibbus*	
	Kounamkites	2		*Xystridura templetonensis/ Redlichia chinensis*	Maozhuangian
	Schistocephalus	1		KF	

Note: Approximate correlation of Middle-Upper Cambrian stratigraphic subdivisions for different regions, modified from Shergold 1995, and the positions of key Cambrian faunas: BS = Burgess Shale, KF = Kaili Formation, MF = Marjum Formation, OR = *orsten,* WF = Wheeler Formation. In addition, in some chapters the Corynexochid,

China (cont.)	Scandinavia		North America (Laurentia)		
Yosimuraspis	Rhabdinopora	Rhabdinopora flabelliforme	Canadian	Symphysurina	Ibexian
Richardsonella/ Platypeltoides	Acerocare	Acerocare ecorne			
		Westergaardia			
		Peltura costata			
		Peltura transiens			
Missisquoia perpetis	Peltura	Peltura scarabaeoides		Missisquoia	
Mictosaukia cf. M. orientalis			Trempealeauan	Eurekia apopsis	
Tsinania/Ptychaspis				Saukiella serotina	Sunwaptan
				Saukiella junia	
	[OR] *492 Ma	Peltura minor		Saukiella pyrene/ Rasettia magna	
Kaolishania pustulosa		Protopeltura praecursor			
	Leptoplastus	Leptoplastus stenotus	Franconian	Ellipsocephaloides	
		Leptoplastus angustatus			
		Leptoplastus ovatus			
		Leptoplastus crassicorne			
		Leptoplastus raphidophorus		Idahoia	
		Leptoplastus paucisegmentatus			
	Parabolina	Parabolina spinulosa			
Maladioidella		Parabolina brevispina		Taenicephalus	
Changshania conica	Olenus	Olenus scanicus		Elvinia	Steptoan
		Olenus dentatus			
		Olenus attenuatus		Dundenbergia	
		Olenus wahlenbergi	Dresbachian		
Chuangia batia	[OR]	Olenus truncatus		Aphelaspis	
		Olenus gibbosus			
Drepanura	Agnostus pisiformis	Agnostus pisiformis		Crepicephalus	
Blackwelderia	[OR]				
Damesella/Yabeia				Cedaria	
Leiopeishania	Paradoxides forchhammeri	Lejopyge laevigata			
Taitzuia/Poshania		Jinsella brachymetopa		Bolaspidella	Marjuman
Amphoton		Goniagnostus nathorsti			
	Paradoxides paradoxissimus	Ptychagnostus punctuosus	Albertian		
Crepicephalina		Hypagnostus parvifrons			
Bailiella/Lioparia		Tomagnostus fissus/ Acidiscus atavus			[MF] [WF]
Poriagraulos Hsuchuangia/Ruichengella		Triplagnostus gibbus		Ehmaniella	
Shantungaspis	Eccaparadoxides oelandicus			Glossopleura	
Yaojiayella		Eccaparadoxides pinus		Albertella	[BS]

Marjumiid, Pterocephaliid, and Ptychaspid biomeres are used for Amgan, Marjuman, Steptoan, and Sunwaptan intervals, respectively. Reliable radioisotope ages from Davidek et al. 1998 and Jago and Haines 1998.

PART I

The Environment

CHAPTER TWO

Alan G. Smith

Paleomagnetically and Tectonically Based Global Maps for Vendian to Mid-Ordovician Time

Recent revisions to the early Paleozoic time scale have been used to recalibrate ages assigned to stratigraphically dated paleomagnetic poles of that era. In particular, a value of 545 Ma has been used for the base of the Cambrian. Selected poles have then been used to derive apparent polar wander paths (APWPs) for the major continents—Laurentia, Baltica, Siberia, and Gondwana—for late Precambrian to Late Ordovician time. The scatter of the paleomagnetic data is high for this interval, and the number of suitable Precambrian poles is very low, with confidence limits (expressed as α_{95}) commonly >20° and occasionally >40°. The scatter is attributed to "noisy" paleomagnetic data rather than to any non-uniformitarian effects such as large-scale "true" polar wander, significant departures from the geocentric axisymmetric dipole field model, very rapid plate motions, and the like. There is a clear need for many more isotopically dated poles of late Precambrian to Cambrian age from all the major continents. The data from Laurentia are considered the most reliable.

Maps have been made for 620–460 Ma at 40 m.y. intervals. For the 460 Ma map the orientation and position of all the major continents have been determined by paleomagnetic data; the longitude separation has been estimated from tectonic considerations. The 500 Ma map has been similarly constructed, except that Baltica's position has been interpolated between a mean pole at 477 Ma and its position on a visually determined reassembly at 580 Ma ("Pannotia"). The 540 Ma map is interpolated between the positions of Gondwana, Baltica, and Siberia at 533 Ma, 477 Ma, and 519 Ma, respectively, and their position in Pannotia. There is a significant difference between the paleomagnetically estimated latitude of Morocco at this time and the latitudes implied by archaeocyaths there. This discrepancy is tentatively attributed to incorrect age assignments to poles of this age, rather than to a period of rapid true polar wander or some such effect. The 580 Ma map represents the time when Pannotia—a late Precambrian Pangea—is considered to have just started to break up. Laurentia's position, interpolated between mean poles at 520 Ma and

589 Ma is used to orient the reassembly. The 620 Ma map is also oriented by inter-polating between Laurentian mean poles at 589 Ma and 719 Ma, with East Gond-wana lying an arbitrary distance from the remainder of Pannotia.

THOSE TECTONIC MODELS that suggest that during late Precambrian and early Pa-leozoic time Baltica and Siberia were close to one another and fringed by more or less laterally continuous island arcs imply that even if the two continents were geographi-cally isolated, faunal interchange between them should have been possible. Other tec-tonic models may not have this requirement.

The maps suggest that nearly all the tillites in the 620–580 Ma interval were de-posited poleward of 40°, rather than reflecting high obliquity or a "snowball Earth." Because of the way in which the maps have been made, some Vendian tillites from Australia lie at much higher latitudes on the maps than the local paleomagnetic data suggest.

Storey (1993) has reviewed significant insights that have recently been made into the likely configurations of Neoproterozoic and early Paleozoic continents. This chap-ter attempts to illustrate some of these developments in five global paleocontinental maps for Vendian to Late Ordovician time, 620–460 Ma, at 40 m.y. intervals. The Ven-dian continents were formed by the breakup of Rodinia, an older "Pangea" that existed at about 750 Ma (McMenamin and McMenamin 1990; Hoffman 1991; Powell et al. 1993; Burrett and Berry 2000). The Rodinian fragments aggregated some time in the later Vendian time to form a possible short-lived second Precambrian "Pangea." This aggregation has been named Pannotia, meaning all the southern continents (Powell 1995), and the term is adopted here despite some controversy (Young 1995). Pan-notia in turn broke up in latest Precambrian time as a result of the opening of the Ia-petus Ocean. Most of the Pannotian fragments eventually came together as Wegener's classic Pangea of Permo-Triassic age. Less detailed maps spanning this interval have been produced by Dalziel (1997), and other maps for shorter intervals are available in the literature (e.g., Scotese and McKerrow 1990; Kirschvink 1992b). The approach adopted here gives primacy to the paleomagnetic and tectonic data. In this it differs somewhat from the approach of some other workers—for example, McKerrow et al. (1992), who use paleoclimate and faunal data as the *primary* constraints and show them to be consistent with some of the paleomagnetic data.

It is assumed that the opening of the Iapetus Ocean began at 580 Ma, causing the breakup of Pannotia. Pannotia's configuration has been found here by visual re-assembly of continents that have been oriented initially by their own paleomagnetic data. Its orientation for the 580 Ma map has been determined by the interpolated mean 580 Ma pole for Laurentia. Most of West Gondwana is assumed to have been joined to Laurentia, Baltica, and Siberia at 620 Ma, with East Gondwana lying some-where offshore. The amount of separation is arbitrary, and Pannotia minus East Gondwana and some pieces of West Gondwana have been oriented by Laurentian paleo-magnetic data to make the 620 Ma map. The 540 Ma map is an interpolation between the 580 Ma reassembly and paleomagnetic data from Laurentia, Baltica, Siberia, and Gondwana. Paleomagnetic poles from these four continents have been used to make the 500 Ma and 460 Ma maps.

The incentives for presenting some new maps for late Precambrian to Late Ordo-vician time include the availability of much new paleomagnetic data; the absence of a series of global maps for this interval based principally on paleomagnetic and tec-tonic data; recent novel suggestions about the relationships between Gondwana and Laurentia during this interval; the substantial revision to the age of the base of the Cambrian period and other early Paleozoic stratigraphic boundaries; and, of course, the great interest in the transition from the late Precambrian to the Cambrian periods as shown by the contributions in this volume.

In principle, it is easy to make pre-Mesozoic paleocontinental reconstructions based on paleomagnetic data: the world is divided into continental fragments that ex-isted at the time (figure 2.1), and the fragments are oriented by paleomagnetic data and repositioned longitudinally by a geologic assessment of their relative positions (Smith et al. 1973). The general geometry of the larger Paleozoic continents is well known: the largest is Gondwana, consisting of South America, Africa, Arabia, Mada-gascar, India, Australia, and Antarctica, together with minor fragments on its periph-ery (such as New Zealand). The northern continents consist of Laurentia, made up of most of North America, Greenland, and northwestern Scotland; and Baltica, essen-tially European Russia and Scandinavia. Laurentia and Baltica united in Early Devo-nian time to form Laurussia (Ziegler 1989). East of Laurussia lay Siberia. In practice, however, the scarcity and scatter of paleomagnetic data make it difficult to reposition even major continents in the interval from Vendian to early Paleozoic. Smaller conti-nental pieces have even less paleomagnetic data, and many other fragments have no paleomagnetic data at all.

An arbitrary method of repositioning such fragments, adopted here, is to "park" them in areas at or not too far from the places where they will eventually reach and where they will not be overlapped. For example, "Kolyma," currently joined to east-ern Siberia (and labeled 53 in figure 2.1), collided with Siberia in earlier Cretaceous time, but its pre-Cretaceous position is unknown (Zonenshain et al. 1990). Seslavin-sky and Maidanskaya (chapter 3 of this volume) consider that in the Vendian to early Paleozoic interval Kolyma lay near its present position relative to Siberia. This view is supported by the presence of very similar Vendian to Cambrian faunas and stratigra-phy on the outer Siberian platform and on Kolyma itself (Zhuravlev, pers. comm.). Kolyma is actually a composite of at least three smaller fragments (Zonenshain et al. 1990), but it is unnecessary to show them on global maps, particularly for the 620–460 Ma interval. Thus Kolyma is simply parked in its present-day position relative to Siberia with its present-day shape throughout the 460–620 Ma interval. However,

Figure 2.1 All fragments. The shaded areas are the outlines of those fragments from which poles have been repositioned by paleomagnetic and tectonic data in the interval ~650–430 Ma. All other fragments have been oriented by miscellaneous tectonic, faunal, and climatically sensitive data. Several fragments have been omitted either because they are small (e.g., Calabria) or they are younger than 460 Ma (e.g., Iceland). Fragments that have been arbitrarily "parked" are in italics. The numbered fragments are as follows: *1, Alaska; 2, Alexander–Wrangellia 1 and 2; 3, Quesnellia; 4, Stikinia; 5, Sonomia; 6,* North America; *7, Baja California; 8,* Mexico; *9,* Yucatan; *10,* Nicaragua-Honduras; *11,* Panama; *12,* Florida; *13,* Carolinas; *14,* Carolina slate belt; *15,* Cuba; *16,* Haiti–Dominican Republic (Hispaniola); *17,* Gander; *18,* west Avalon; *19,* Meguma; *20,* Ellesmere Island; *21,* Greenland; *22,* western, central, and eastern Svalbard; *23,* northwest Scotland; *24,* Grampian; *25,* East Avalonia; *26,* Armorica; *27,* Aquitainia; *28,* South Portuguese terrane; *29,* Cantabria; *30,* Alps; *31,* Italy; *32,* western Greece and Yugoslavia; *33,* Pelagonia; *34,* Silesia; *35,* Pannonia; *36,* Moesia; *37,* Balkans; *38,* Pontides; *39,* Baltica; *40,* Barentsia; *41,* Turkey; *42,* Iran; *43,* Afghanistan; *44,* Taimyr; *45,* Siberia; *46,* North Tibet; *47,* South Tibet; *(46–47, repeated, Greater India); 48,* Indo-Burma; *49,* western Southeast Asia; *50,* Indochina; *51,* South China; *52,* North China; *53,* Kolyma; *54,* Kamchatka; *55,* Chukotka; *56,* Japan; *57, Philippines; 58,* Sulawesi; *59,* Papua New Guinea; *60,* South America; *61, Chilenia; 62, Precordillera (Occidentalia); 63, Patagonia; 64,* Africa; *65,* Arabia; *66,* Somalia; *67,* Madagascar; *68,* India and Sri Lanka; *69,* Australia; *70,* western New Zealand; *71,* eastern New Zealand; *72,* Marie Byrd Land; *73, Thurston Island; 74, Antarctic Peninsula; 75, Ellsworth Mountains; 76,* East Antarctica; *77,* South Tarim; *78,* North Tarim; *79,* Qaidam; *80,* North Korea; *81,* South Korea; *82,* Taiwan; *83,* Pre-Urals. The Altaids (later amalgamated into Kazakhstan) and the Manchurides (later amalgamated into Siberia) are miscellaneous Paleozoic island arcs and related fragments.

Cambro-Ordovician faunas of parts of Kamchatka are typically Laurentian at the species level (Zhuravlev, pers. comm.). Kamchatka has therefore been parked in its present-day position relative to North America for the 620–460 Ma interval.

For ease of recognition, the maps show *present-day* coastlines rather than paleo-coastlines, which are generally unknown. During the plate tectonic cycle, continental crust is, to a first approximation, conserved. Thus, the present-day edges of the continents, taken as the 2,000 m submarine contour, may approximate to the extent at earlier times and is shown on all the maps.

PALEOMAGNETIC DATA

The paleomagnetic data have been taken from the most recent version of the global paleomagnetic database of McElhinny and Lock (1996). This is a Microsoft Access database, giving details of all published paleomagnetic data to 1994.

Time Scale

The time scale used in the paleomagnetic database is that of Harland et al. 1990, which places the base of the Cambrian at 570 Ma, but new high-precision U-Pb zircon dates suggest that it is closer to 545 Ma (Tucker and McKerrow 1995). The problem of relating the two scales is complicated by the fact that the base of the Tommotian was taken as the base of the Cambrian at 570 Ma in Harland et al. 1990. Since then, the Nemakit-Daldynian has been placed in the Cambrian below the Tommotian, with an age of 545 Ma for its base (Tucker and McKerrow 1995), and the base of the Tommotian has been placed at 534 Ma (Tucker and McKerrow 1995). The top of the Early Cambrian is at 536 Ma in Harland et al. 1990 and 518 Ma in Tucker and McKerrow 1995. It is not clear how best to accommodate these changes: the old 536 Ma has been revised to the new 518 Ma, and the old 570 Ma to the new 534 Ma. Clearly, some changes are necessary to poles from rocks with stratigraphic ages just greater than 570 Ma in Harland et al. 1990; here they are assigned to the Nemakit-Daldynian. According to Harland et al. 1990, the base would have been close to 581 Ma. Fortunately, there are very few poles in this age range in the database. The new dates also suggest that significant changes should be made to ages assigned to other Paleozoic stratigraphic boundaries. Thus, all stratigraphically dated poles whose ages lie in the range 386–581 Ma have been changed in accordance with the new scale to lie in the range 391–545 Ma. Isotopically dated poles are unchanged. The changes are similar to those of Gravestock and Shergold (chapter 6 of this volume). No modifications have been made to ages older than 581 Ma, although the time scale will undoubtedly change. Knoll (1996) has reviewed the most recent information and suggests (pers. comm.) that the Varangerian ice age might range from 600 Ma to about 575–580 Ma.

Figure 2.2 Distribution of paleomagnetic sites on the major fragments. Squares are sites of 354 early Paleozoic poles (400–545 Ma). Triangles are sites of 50 late Precambrian poles (>545–640 Ma). Sites in orogenic belts have been included.

Data Selection

The main problem with making maps for 620–440 Ma is obtaining reliable paleo-magnetic data. In particular, if the "quality factor" proposed by Van der Voo (1990), Q_v, is set at 3 or more, virtually all poles measured in former Soviet laboratories would be excluded. For example, a recent list of all Baltica poles considered to be reliable for the Vendian to early Paleozoic time includes only one such pole (Torsvik et al. 1992). This approach would remove most of the data from Siberia. But without paleomag-netic data, it is highly improbable that climatic indicators, faunal distributions, and the like would have led to the conclusion that Siberia was inverted with respect to present-day coordinates for most of the interval discussed here. An alternative qual-ity factor, Q_1, has also been proposed by Li and Powell (1993).

The approach adopted here has been to apply few selection criteria to the pole list, in the belief that some intervals would otherwise be dominated by a few high-quality poles whose magnetization ages may actually be different from the ages assigned to them. One argument in favor of this approach is that there is no significant difference in the mean pole position of high and low Q data for poles of the past 2.5 m.y.: only the scatter of global data increases for lower Q (Smith 1997).

The most important selection criterion used here is that, for the poles selected, the age of the primary magnetization is considered by the authors to be the same as the rock age: all magnetic overprints have been excluded. In addition, only one paleo-magnetic study has been accepted for each rock unit defined in the database. The cri-teria used to select the "best" study from several on the same unit have included the number of sites, the scatter of the data, the magnetic tests, and the pole position rel-ative to other poles of the same age from elsewhere. No attempt has been made to im-pose additional selection criteria, such as whether poles have been subjected to par-ticular field or laboratory tests. Poles from ophiolites or from nappes have been excluded, but other poles from orogenic belts have not been removed, principally be-cause this would commonly significantly reduce the number of poles available. It is assumed that most orogenic poles lie in regions where the necessary tectonic correc-tion—commonly the unfolding of cylindroidal folds—can be reasonably estimated.

Poles with a large age uncertainty have also been eliminated from the pole list, but the size of the acceptable age uncertainty has been varied with age. Thus, the total ac-ceptable age uncertainty for poles whose age is less than 500 Ma is taken as 0.2 pole age, e.g., 400 ± 40 Ma. For poles 500 Ma old or older, the uncertainty has been set at 100 Ma, i.e., 500 ± 50 Ma. The only exception to this age restriction is poles dated as ranging in age from the Neoproterozoic (610 Ma) to Cambrian (495 Ma), with an age range of 115 Ma. The total number of poles on the larger stable fragments in the 650–430 Ma age range is 316, of which only 57 are Precambrian (>545 Ma) in age. Their geographic distribution is shown in figure 2.2.

Poles from Gondwana were repositioned with Africa as the reference frame. The sources of the rotations for reassembling Gondwana are East Antarctica to Madagascar (Fisher and Sclater 1983); Australia to Antarctica (Royer and Sandwell 1989); India to Antarctica (Norton and Sclater 1979 for age, Smith and Hallam 1970 for rotation); Somalia to Africa and Arabia to Somalia (McKenzie et al. 1970; Cochran 1981); Sinai to Arabia (LePichon and Francheteau 1978; Cochran 1981); South America to Africa (Klitgord and Schouten 1986); and Australia to Antarctica (Royer and Sandwell 1989). Laurentia consists of North America, excluding Alaska, Baja California and fragments within the Appalachians (such as the Carolina slate belt, western Avalonia, Meguma, Gander), plus Greenland and NW Scotland. The sources of the rotations for reconstructing Laurentia are Greenland to North America (Roest and Srivastava 1989) and northwest Scotland to North America (Bullard et al. 1965). There are negligible differences between the positions of the paleomagnetic poles on the reassemblies of Gondwana and Laurentia made using the rotations cited above and most others that exist in the literature. The rotations for reassembling the smaller fragments are based on interpretations of the geologic and faunal data, discussed below.

The basic assumption for making global reconstructions from paleomagnetic data is that the continents can be treated as rigid bodies and rotated accordingly. To a very good approximation, Precambrian shields and continental platforms have behaved as rigid bodies since they formed, but younger orogenic belts on their peripheries clearly have not. Paleomagnetic data from foldbelts can be restored reasonably precisely to their original orientation (see above). When orogenic deformation becomes penetrative, as in regional metamorphism, or when plutonism takes place, the repositioning errors become much larger. Areas affected by such deformation have simply been left attached to the platform or cratonic areas of each continent with their present-day shapes. They have not been distinguished on the maps.

In some cases, what was previously regarded as a continental fragment may have been everywhere affected by deformation. For example, Paleozoic Kazakhstan is in reality an amalgam of several island arcs and microcontinents that have collided with one another through Paleozoic time to form the Altaids (Zonenshain et al. 1990; Şengör and Natal'in 1996). It is clearly necessary to show all such areas on global maps. The immensely complex evolution of Kazakhstan, Mongolia, and adjacent areas of Paleozoic Asia has been attributed to an underlying fundamental simplicity by Şengör and Natal'in (1996), but as they acknowledge in the title of their fascinating analysis, for these areas there exists at present only the "fragments of a synthesis." A quite different synthesis for Paleozoic Asia has been proposed by Mossakovsky et al. (1994). The outlines of the Altaid and Manchurid fragments recognized by Şengör and Natal'in (1996) are shown on all the maps. Because there is no agreement on the location of these fragments, they have been "parked" with their present-day shapes and positions unchanged relative to present-day stable Siberia (the Siberian and adjacent platforms). Similar complexities exist elsewhere. For example, Powell et al. (1994: figure 11) suggest that the eastern limit of Precambrian rocks in Australia may have had a rectilinear form, reflecting a ridge-and-transform system created during breakup of the continent in late Precambrian time, but this boundary has not been shown on the maps.

It is also necessary to remove all new areas, like Iceland and Afar, which have been created by mantle plumes and clearly did not exist in Paleozoic time. Apart from these exceptions it is not at the moment practicable to take into account the possible growth in continental area that may have taken place in orogenic belts since 620 Ma. The total volume of new crust might be as much as 80×10^6 km^3 (Howell and Murray 1986), equivalent to an area of about 27×10^6 km^2, or about 15 percent of the present total continental area. The new crust is concentrated in those regions that are in any case difficult to reposition.

APPARENT POLAR WANDER PATHS

The apparent motion of the mean magnetic pole relative to a continent is the apparent polar wander path of that continent, or its APWP. In reality, of course, the continent is wandering relative to the pole. The Mesozoic and Cenozoic motions of large continents are generally smooth for periods of tens of millions of years (figure 2.3). Discontinuities in motion may accompany continental breakup or collision, giving rise to relatively abrupt changes in direction of an APWP. The APWPs assume the geocentric axisymmetric dipole field model—the magnetic field of a centered bar magnet parallel to the earth's spin axis. The present-day, Cenozoic, and Mesozoic fields show relatively small departures from such a model (Livermore et al. 1983, 1984). Such effects undoubtedly existed in late Precambrian and early Paleozoic time, but the errors involved in ignoring them are considered to be much smaller than the likely errors in the late Precambrian and early Paleozoic mean poles.

APWPs for the 620–460 Ma period were calculated at 20 m.y. intervals for Laurentia, Gondwana, Baltica, and Siberia from the 316 poles selected from the database. All poles whose nominal age lay within 30 m.y. of the required age—i.e., in a "window" of 60 m.y. duration—were included. Inspection of the data showed that 40 poles lay more than 60° from their relevant APWP. These deviant poles were removed, and the APWPs were recalculated for the same intervals. In this recalculation all poles lying more than 40° from the new APWPs were excluded from mean pole calculations.

All the resulting APWPs showed segments with features that are absent from Mesozoic and Cenozoic APWPs: they were highly irregular or had very high rates of change of pole position (>100 mm/y) or tracked back on themselves at high rates (figure 2.3). It is assumed that such features reflect aberrations in the Paleozoic and Neoproterozoic paleomagnetic data rather than reflecting some fundamental change in the behavior of the earth for this period, e.g., a nonaxisymmetric field or a field with high nondipole components, very high rates of plate motions (Gurnis and Torsvik 1994), large components of "true" polar wander, or a marked change in obliquity or climate. These uniformitarian assumptions suggest that mean poles that give rise to irregular

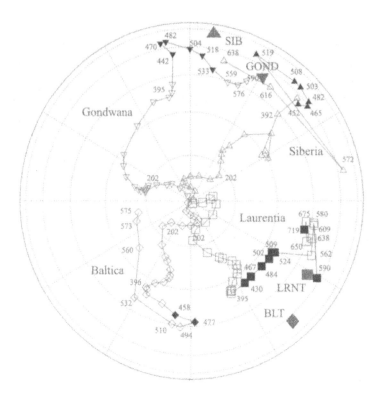

Figure 2.3 North-pole polar wander paths for Laurentia, Siberia, Gondwana, and Baltica on an azimuthal north-polar projection with a latitude-longitude grid at 30° intervals. To avoid overlaps, the APWP for Laurentia has been rotated clockwise by 90°, and the APWP for Baltica has been rotated by 180°. For clarity, the confidence limits have been omitted. The paths start at the present day and have been drawn back to the later Neoproterozoic. The symbols show mean poles at intervals of 20 m.y. for "time windows" of 60 m.y. Ages have been given to the poles closest to 200 Ma and 400 Ma and for poles spanning the 460–620 Ma interval, except for Baltica and Gondwana, which have been truncated at 575 Ma and 590 Ma, respectively. The poles for the 0–250 Ma interval are global mean poles that include data from all the stable continents, which have been repositioned using Euler rotations from the ocean floor. For Siberia and Gondwana the poles that are older than 250 Ma are, respectively, from only Siberia and Gondwana. Laurentia and Baltica poles are combined back to 420 Ma, but older poles are, respectively, from only Laurentia and Baltica.

The backtracking of the Baltica APWP from 477 Ma to 575 Ma is believed to reflect remag-netization. Gondwana was finally assembled at about 550 Ma: older mean poles progressively reflect the mean pole of the fragments from which Gondwana was built rather than poles of a single continent. The Siberian APWP shows a major discontinuity after 519 Ma, possibly reflecting remagnetization. The only Laurentian mean poles older than 509 Ma that are based on 6 or more poles, all of which lie less than 40° from the mean pole, are those for 590 and 719 Ma.

The mean poles used for paleomagnetic interpolation for the maps (figure 2.3) are shown as small filled symbols. All other poles either lie outside the 460–620 Ma interval or are ignored. The larger filled symbols labeled SIB (Siberia), GOND (Gondwana), LRNT (Laurentia), and BLT (Baltica) are the positions of the north pole on the 580 Ma Pannotian reassembly. The mean paleomagnetic pole for Laurentia coincides with its geographic north pole because Laurentia data have been used to position the Pannotian reassembly into its paleo-latitude grid. If the Pannotian reassembly reflects reality, then the poles from Siberia, Gondwana, and Baltica should coincide with the 580 Ma mean paleomagnetic pole for these fragments.

APWP features, to high rates of change of pole position, or to backtracking should be ignored in any reconstructions. The end result is a series of mean poles that produce reasonably smooth APWPs from which the Euler rotations needed to reposition the continents can be calculated.

Laurentia

The best paleomagnetic data are considered to be from Laurentia: it has the most numerous data but the fewest poles that lie more than 60° off the initial APWP. Although data exist for Laurentia for most of the Cambrian and Neoproterozoic periods, the mean pole for 509 Ma is the oldest pole to have more than 30 determinations in the 60 m.y. window. Of the other poles, only the poles for 590 Ma and 719 Ma all lie within 40° of the mean pole and include 6 or more determinations. The positions of the other mean poles form zigzags on the APWP and have been rejected. With such large interpolation intervals, the Cambrian-to-Neoproterozoic motion of Laurentia is inevitably very smooth, possibly misleadingly so. The Pannotian reassembly is placed in a global paleolatitude frame by interpolating between the 509 Ma and 590 Ma mean Laurentian poles rather than using any other paleomagnetic data. Except for West Gondwana (see below), the 620 Ma (figure 2.4e) reconstruction is identical to that for 580 Ma (figure 2.4d) and is oriented by interpolating between the 590 and 719 Ma Laurentian mean poles.

Baltica

Paleomagnetic data from three different areas of Baltica currently offer three distinct solutions to the problem of where Baltica was in earlier Paleozoic and latest Precambrian time. Zonenshain et al. (1990: figure 12) show Baltica at 600 Ma to be lying on its side, with Scandinavia facing west, in the latitude belt 0–30° *south*. Torsvik et al. (1992) show Baltica at 560 Ma to be inverted, with Scandinavia facing east, also in the Southern Hemisphere in the latitude belt 20–50°. By contrast, Êlming et al. (1993: figure 6l) show Baltica at 600 Ma to be in a similar orientation, but they place it in the latitude range 40–70° *north*. The data selected from the database, which include data from all three areas, place Baltica in the latitude range of about 0–30° *north* at 560 Ma, and in the range of 30–60° *north* at 600 Ma. It is not clear how to assess these data, although Torsvik (pers. comm.) considers it likely that the earlier Cambrian Soviet poles from Baltica have been remagnetized. This view is supported by the backtracking of the Baltica APWP for poles older than 477 Ma over the younger part of the same path (see figure 2.3). The remaining non-Soviet poles are too few to give a reliable late Precambrian to Cambrian APWP for Baltica. Acceptable mean poles exist for 458 Ma and 477 Ma. Interpolation gives the 460 Ma position (figure 2.4a). Baltica's position on the 500 Ma and 540 Ma maps (figures 2.4b,c) is obtained by linearly interpolating the difference between the Euler rotation for the 477 Ma pole and the Euler

460 Ma (mid-Ordovician)

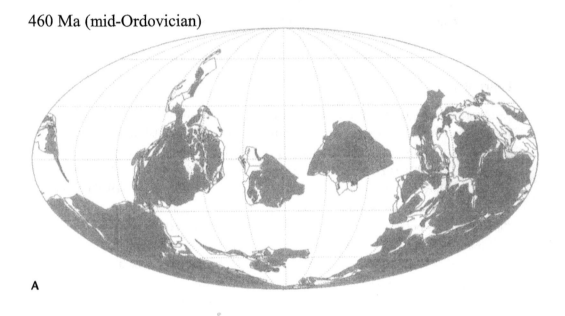

A

500 Ma (latest Cambrian)

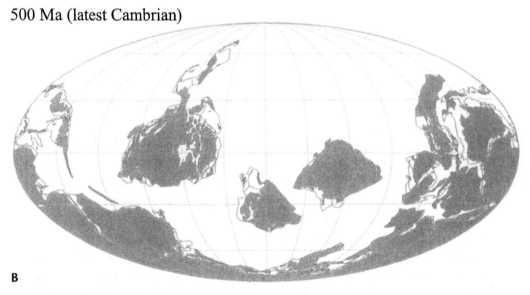

B

Figure 2.4 Global reconstructions for (a) 460 Ma, (b) 500 Ma, (c) 540 Ma, (d) 580 Ma, and (e) 620 Ma. All reconstructions show the present-day coastline (for ease of recognition) and the present-day 2,000-meter submarine contour (to indicate the approximate extent of continental crust). The ages correspond to the time scale used in this chapter and differ slightly from those of Gravestock and Shergold (this volume).

540 Ma (earliest Cambrian)

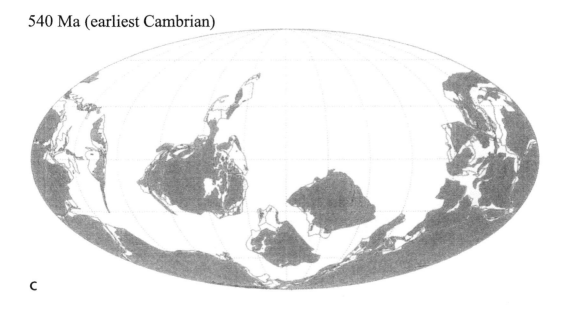

C

580 Ma (late Vendian)

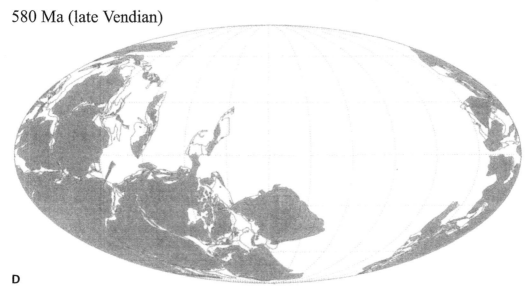

D

Figure 2.4 (Continued)

620 Ma (earliest Vendian)

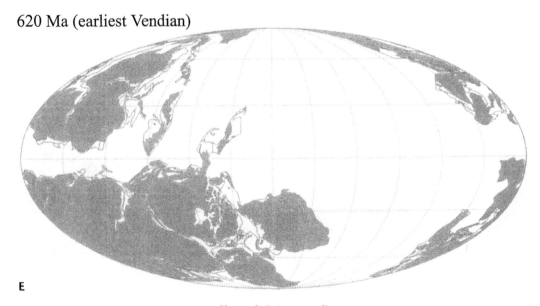

E

Figure 2.4 (*Continued*)

rotation for the visually determined position on Pannotia at 580 Ma (figure 2.4d). Baltica's Cambrian to Neoproterozoic motion is smooth because of this long interpolation interval.

Siberia

Siberia must be repositioned almost entirely by Soviet paleomagnetic data. Removal of all magnetically untested poles from the database does not significantly alter the mean poles. There is no evidence that the poles for 452 Ma to 519 Ma have been remagnetized: this part of the APWP is reasonably smooth. There is an abrupt change in direction and in the rate of change of pole position to the next mean pole at 572 Ma and all others to 638 Ma (see figure 2.3), which may reflect remagnetization. There are 19 or more poles in each 20 m.y. step from 440 Ma to 540 Ma. Only about 10 percent of the poles lie more than 40° from the APWP. The APWP shows an inverted Siberia moving steadily from moderate southerly latitudes in earlier Cambrian time to lower latitudes in Ordovician time. Siberia is positioned by paleomagnetic data for the 460 Ma and 500 Ma maps (figures 2.4a,b). For the 540 Ma map (figure 2.4c), its position has been interpolated between the 519 Ma mean pole and its visually estimated position within Pannotia (see the section "Gondwana" below). The 540–460 Ma positions are similar to those given by Smethurst et al. (1998), based on a more recent analysis of the data.

Gondwana

Reasonable mean poles exist for 442, 470, and 482 Ma. Between one-fifth and one-third of the poles for the 500, 520, 540, 560, and 580 Ma calculations lie more than 40° from the APWP. The ages of the mean poles in these intervals are 504, 518, 533, 559, and 576 Ma, respectively. The 559 and 576 Ma mean poles are considered too close in time to the visually determined 580 Ma position and have been omitted (see figure 2.3). It is interesting to note that the 590 Ma mean pole (from the 600 Ma calculation) is surprisingly close to that implied by the visual reassembly of Pannotia. Gondwana's position on the 460 Ma (figure 2.4a) and 500 Ma (figure 2.4b) maps is given by interpolation between poles that are relatively close in time. The 540 Ma position (figure 2.4c) is an interpolation between the 533 Ma mean pole and the visually estimated 580 Ma position. As noted below (in "Faunal and Climatic Evidence: Archaeocyaths and Gondwana"), there is a significant discrepancy between the paleolatitudes implied by the interpolated mean pole for 540 Ma and the archaeocyath evidence. Eastern Gondwana was still in the process of being joined to western Gondwana until about 550 Ma (Unrug 1997). The 620 Ma map (figure 2.4e) shows eastern Gondwana as a distinct entity, but its position is schematic rather than being based on paleomagnetic data. A summary of the methods used to make the maps is given in figure 2.5.

CONTINENTAL MARGINS

The evolution of the continental margins around each continent is of fundamental importance in estimating longitudinal separations of the continents. In the simplest plate tectonic cycle, a continent splits and separates into two or more continents, each of which eventually collides to form a continent similar to the original continent. In more-complex cases, a continent may split into several fragments, some or all of which might collide with continents different from the one they originally separated from. The age at which two continents separate can be estimated relatively precisely by applying the lithospheric stretching model (McKenzie 1978) to the stratigraphic sequences formed on each margin, even in orogenic belts (Wooler et al. 1992). In the absence of quantitative analyses, the time of separation may be difficult to estimate. Extensional faulting that preceded the formation of ocean floor and the separation of two continents may span some tens of millions of years, as in the present East African rift. The succeeding thermal phase, during which the margin subsides and the post-rift passive margin sequence accumulates, continues until collision takes place. Flexure of the margin prior to actual collision gives rise to a characteristic time-subsidence signature.

460

Gondwana, Laurentia, Baltica & Siberia separated by plausible longitude differences

500

Gondwana, Laurentia, Baltica & Siberia separated by plausible longitude differences

540

Iapetus Ocean spreading

Gondwana, Laurentia, Baltica & Siberia separated by plausible longitude differences

580

E & W Gondwana assumed to have joined by ~580 along collisional belt

Pannotia oriented by Laurentian data

Siberia visually fitted to Laurentia at 580

Baltica fitted visually to Laurentia & S America at 580

620

E & W Gondwana separated by arbitrary amount at 620

No attempt to reposition W Gondwana fragments affected by Pan-African orogenesis

Pannotia minus E Gondwana oriented by Laurentian data

Figure 2.5 Summary of methods used to make figures 2.4a–e. Black areas are large continents oriented by paleomagnetic data (Laurentia, Baltica, Siberia, and Gondwana, to which smaller fragments have been attached using visual, tectonic, and faunal data). Gray areas are large continents and their attached fragments that have been repositioned by interpolation between a paleomagnetically defined orientation and a visually defined fit (Baltica on the 540 Ma map) or by visual estimates alone (most of the fragments on the 580 Ma and 620 Ma maps).

Times of Passive Margin Formation

The time interval of interest here, from Neoproterozoic to Late Ordovician (620–460 Ma) includes two important episodes of passive margin formation. The first is of latest Proterozoic to Early Cambrian age (Bond et al. 1984) and gave rise to passive margin sequences in western North America and Arctic North America (Trettin 1991; Trettin et al. 1991); North Greenland (Higgins et al. 1991); East Greenland (Williams 1995) and eastern North America (Hatcher et al. 1989; Williams 1995); western Baltica (Gee 1975; Gayer and Greiling 1989); northeastern Siberia (Pelechaty 1996); Iran, Turkey, and Pakistan; northwestern Australia; and western South America (Dalla Salda et al. 1992a; Astini et al. 1995). If the Precordillera (Occidentalia) is a fragment of eastern North America (see the section "Positions of Smaller Fragments Around the Larger Continents: Gondwana" below), it probably broke off in Early Cambrian time after North America itself had separated from South America.

The precise time of passive margin formation is uncertain. The Cambro-Ordovician carbonate platforms of Laurentia show that a passive margin existed there in Early Cambrian time, but could it have originated significantly earlier, as suggested by dyke swarms (Bingen et al. 1998)? Clastic sequences conformably underlie the carbonates and are in turn unconformable on significantly older rocks. In the lithospheric stretching model, the thermal phase follows immediately on the stretching phase without a time break. In the model there may be unconformities between the sediments deposited during faulting and those deposited later, but there is no time gap between the cessation of faulting and the onset of the thermal phase. Thus, the Laurentian and other passive margin sequences that lack faulting probably lie outside the zone of stretched continental crust and may correspond to onlapping sequences that are somewhat younger than the age of the oldest ocean floor with a "steer's head" geometry commonly found beyond the margins of zones of continental stretching (White and McKenzie 1988).

Detailed analysis of some of the passive margin sequences of western Baltica suggests that breakup may have been contemporaneous with the deposition of Vendian tillites (Greiling and Smith, n.d.), with ocean-floor spreading beginning at about 580 Ma, a value similar to that adopted by Torsvik et al. (1996).

The second episode of passive margin formation is of Early Ordovician age and created passive margins on the eastern edge of Baltica (Zonenshain et al. 1990) and northwestern Gondwana (Pickering and Smith 1995).

Times of Continental Collision

Collisions took place during the 620–460 Ma interval. East Gondwana (India, East Antarctica, and most of Australia), together with the Arabian-Nubian shield and the Kalahari-Grunehogna cratons of southeastern Africa, had consolidated by 630 Ma and formed a stable nucleus to which the remaining components of West Gondwana were added during the 630–550 Ma interval (Unrug 1997).

Barentsia collided with northern Baltica to cause the Timan orogeny in later Vendian time (Zonenshain et al. 1990: figure 14). Puchkov (1997) summarizes additional evidence for the continuation of the same collisional orogen, with an age of about 630–570 Ma, southward along the Uralian margin of Baltica as the Pre-Uralides; for amphibolites and collisional granites dated at 625–560 Ma in northern Taymir on the Arctic coast of Siberia; and for Late Vendian metamorphic dates ranging from 621 to 556 Ma in Spitsbergen. Eastern Baltica may have collided at this time with a continental fragment (Zonenshain et al. 1990:15).

Western Baltica collided with island arcs in Ordovician time to cause the Finnmarkian orogeny; the contemporaneous Taconic orogeny of eastern Laurentia is regarded here as the result of a collision between eastern Laurentia and other island arcs (Bird and Dewey 1970; Pickering and Smith 1995; Niocaill et al. 1997) rather than as a continent-continent collision between Gondwana and Laurentia (Dalla Salda et al. 1992a,b; Dalziel et al. 1994). The Precordillera, regarded here as a fragment of eastern Laurentia (see the section "Positions of Smaller Fragments Around the Larger Continents: Gondwana" below), may have collided with South America in Early Ordovician time (Astini et al. 1995). Its collision as a fragment may have caused the Famatinan orogeny of Argentina, rather than having been part of a continent-continent collision between Gondwana and Laurentia (Dalziel et al. 1994), a view modified subsequently by Dalziel (1997).

Younger Paleozoic collisions are useful for constructing the maps because they show which continents were approaching one another in earlier Paleozoic time when there may be no other relevant data. For example, in Silurian time the approach of continental fragments originally on the edge of Gondwana and Laurentia caused the early phases of the Acadian orogeny (Williams 1995). Western Baltica and eastern Greenland were probably essentially sutured in Late Silurian time in the late stages of the Caledonian orogeny (Higgins 1995). Arctida (Alaska and Chukotka) first collided with Arctic Canada and North Greenland at about the same time (Zonenshain et al. 1990: figure 191), but in both areas deformation continued until early Carboniferous time (Trettin 1991).

The longitudinal positions of Laurentia, Baltica, Siberia, and Gondwana are shown on the maps. There is no "absolute" reference frame for pre-Mesozoic global reconstructions such as is given by the hot-spots reference frame for Mesozoic and Cenozoic time. Hot spots, or hot mantle areas, have been recognized (Zonenshain et al. 1990) but at present provide only a limited local reference frame.

460 Ma to 580 Ma

North America is kept close to northwestern South America in a position that allows the two continents to join together eventually at 580 Ma without very large strike-slip motions. It must be noted that Kirschvink (1992b) has proposed quite a different set of reconstructions in which Baltica is always in its conventional position east of Laurentia, but Siberia is inserted between eastern Gondwana and western Laurentia.

The width of the Iapetus Ocean between Baltica, Siberia, and Laurentia is arbitrary, but the maps attempt to show the opening and closing of the Iapetus Ocean in a plausible manner. Baltica moves away from Laurentia until about 500 Ma, when the earliest phase of Caledonian deformation (Finnmarkian) began (Andréasson and Albrecht 1995). It is assumed that this phase represents the beginning of the closure of the Iapetus Ocean between Laurentia and Baltica, with eventual collision at 420 Ma.

580 Ma

Laurentian paleomagnetic data have been used to orient Pannotia (figure 2.4d), but its reassembly is visual. The first two pieces that have been joined together in the Pannotia jigsaw are Gondwana and Laurentia. The join is along the southeastern margin of North America and the western margin of South America and is discussed in more detail below. In Cambrian time the East Greenland margin of Laurentia (Higgins 1995) is believed to have been a passive continental margin formed by continental breakup in late Precambrian time (Kumpulainen and Nystuen 1985; Schwab et al. 1988), as was the western Baltica margin (Gee 1975; Gayer and Greiling 1989).

The Neoproterozoic successions of East Greenland appear significantly different from those in western Baltica (Kumpulainen and Nystuen 1985). These authors suggest that the Grenville "front" in North America can be correlated with the Sveconorwegian "front" in southern Scandinavia. This front provides a line on each continent at a high angle to the continental margins. When matched, Baltica has a more southerly position relative to Laurentia than it had in Devonian time. A more recent fit of Archean and early Proterozoic provinces gives a similar position (Condie and Rosen 1994). There are as yet no agreed precise "piercing points" such as might be provided by giant dyke swarms (Ernst and Buchan 1997). Baltica has been repositioned here by visually fitting it to an aggregate of northern South America and Laurentia independently of, but in agreement with, the geologic evidence.

Figure 2.4c shows Chukotka and Alaska against the present-day continental margin of northern North America and Ellesmere Island. This reassembly is based on a possible Mesozoic reconstruction of the Arctic Ocean. It is assumed that the Chukotka-Alaska fragment underwent earlier periods of collision and separation and includes the late Vendian to early Paleozoic conjugate margin of northern North America. There is still no conjugate margin for the more northerly half of East Greenland. It is assumed here that Siberia, with Kolyma attached, fulfilled that role (but see below for an alternative view). It too has been visually fitted against Laurentia.

There are several similar suggestions for Siberia's position in Pannotia and Rodinia. Condie and Rosen (1994) suggest that at ~800 Ma the Verkhoyansk margin of Siberia—its eastern margin—was joined to the Franklin margin of North America and North Greenland. By contrast, Hoffman (1991) places the northern margin of Siberia against these two continents at ~700 Ma, a reconstruction supported by Pelechaty (1996) for the whole of the 700–550 Ma interval.

Clearly, Siberia's position in the Pannotian reassembly is not well established. For example, the conjugate margin for northeastern Greenland may be along the margin of another fragment such as Barentsia, as sketched by Condie and Rosen (1994). Were this the case, then the longitudinal uncertainty of paleomagnetic data permits entirely different reconstructions. For example, the Cryogenian/Vendian to Cambrian/Ordovician maps of Kirschvink (1992b: figures 12.6–12.11) show Siberia lying between the *eastern* margin of East Gondwana and the western margin of Laurentia.

That Siberia may not have been attached to *any part* of Laurentia at 580 Ma is suggested by Jaccards coefficients of similarity for Vendian–Early Cambrian faunas, which show no similarity before late Early Cambrian between Siberia and Laurentia but a progressive increase in similarity during this and subsequent periods (Zhuravlev, pers. comm.). In addition, Siberian faunas appear to be more similar to some faunas from China than to coeval faunas from Laurentia and Baltica. These data suggest that alternative 540–620 Ma reconstructions are ones in which Siberia is closer to China than shown in figures 2.4c–e. There is no conflict with the paleomagnetic data, but the tectonic data seem to require the creation of passive margins in Siberia that have no obvious conjugate counterpart and appear inconsistent with convergence between Siberia and Laurentia for this interval.

620 Ma

Gondwana was still being assembled at 620 Ma (Unrug 1997). East Gondwana (East Antarctica, India, and most of Australia) had been joined to Madagascar and to the Kalahari-Grunehogna craton of southeastern Africa since 1000 Ma. The Arabian-Nubian shield was joined to it by 630 Ma, but parts of West Gondwana were probably undergoing the final stages of consolidation with, for example, subduction still continuing in the western Tuareg shield of the central Sahara (Black et al. 1994).

The remainder of West Gondwana, including the Amazonian, West African, Congo, and smaller shields probably formed a continental mosaic set in relatively small oceans that was colliding with Laurentia and the East Gondwana unit.

The 620 Ma map (figure 2.4e) shows Laurentia, Siberia, and Baltica, to which the components of the continental mosaic of West Gondwana have been added as a unit, rather than being separated. The result has been oriented by Laurentian paleomag-

Table 2.1 **Rotations for Major Continents**

	LATITUDE	LONGITUDE	ANGLE
460 Ma			
North America	0.0	38.1	67.2
Africa	2.6	105.1	122.7
Siberia	23.1	81.5	133.0
Baltica	−15.4	69.5	84.9
500 Ma			
North America	0.0	53.0	67.9
Africa	11.9	112.6	124.0
Siberia	18.0	85.7	141.4
Baltica	−13.0	87.1	97.5
540 Ma			
North America	0.0	59.1	87.4
Africa	27.1	114.7	118.4
Siberia	6.2	83.4	137.7
Baltica	−14.1	93.6	116.3
580 Ma			
North America	0.0	58.8	110.8
Africa	34.7	127.5	155.5
Siberia	−3.1	72.3	144.7
Baltica	−14.8	98.8	135.7
620 Ma			
North America	0.0	62.1	108.6
Africa	33.3	129.4	159.6
Siberia	−4.1	75.0	143.1
Baltica	−15.8	101.8	134.8

netic data. The East Gondwana–Arabia–southeastern Africa unit has been given an arbitrary separation from the first major unit.

The finite rotations used to reposition the major continents are listed in table 2.1. For brevity the pole lists used and the resulting APWPs have not been included.

POSITIONS OF SMALLER FRAGMENTS
AROUND THE LARGER CONTINENTS

Laurentia

The western margin of Laurentia has had several terranes added to it in Mesozoic time, mostly in Canada and Alaska (Gabrielse et al. 1992: figure 2.6; Saleeby and Busby-Spera 1992: plate 5). Five of the largest are schematically shown on the maps: Quesnellia, Stikinia, Alexander-Wrangellia 1 and 2, and Sonomia. Carter et al. (1992: figure 2.10) postulate that in early Carboniferous time some of these fragments may have lain quite close to Australia and migrated several thousand kilometers east by

Late Permian time. Their positions on the maps simulate the suggestions of Gabrielse et al. (1992: figure 2.6) and Miller et al. (1992: plate 6) for later Mesozoic time, except that the terranes have been arbitrarily parked in these positions against the western North American margin for the whole of the 620–460 Ma interval.

What was joined to the western Laurentian margin in latest Precambrian time is unclear. Bell and Jefferson (1987) suggested that ~1200 Ma ago the eastern Australian margin was joined to the western Canadian margin. A logical extension of this hypothesis was made by Dalziel (1991) and Moores (1991), who speculated that parts of East Gondwana had been joined to the western United States. Breakup is suggested to have occurred at ~750 Ma, long before the beginning of Vendian time. If breakup did occur at 750 Ma, then the second breakup episode, which began at about 625 Ma near the beginning of the Vendian (Bond et al. 1984), created the early Paleozoic passive margin of the western United States. Possible conjugates include Siberia (Kirschvink 1992b) and South China (Keppie et al. 1996). The model of Hoffman (1991) for Laurentian margins is more wide-ranging but is not discussed in detail here.

Baja California, Mexico, Yucatan, and Nicaragua have been kept attached in the early Mesozoic positions. The Appalachian fragments (see figure 2.1) form part of western Avalonia (see the section "Gondwana" below).

Zonenshain et al. (1990) have sketched a possible early Paleozoic evolution of the Arctic region. It includes the two fragments "Arctida" and "Barentsia." Arctida is the name of the amalgamated Alaskan and Chukotkan fragments. Barentsia is the name for the continental fragment north of the Timan orogenic belt, active in Vendian to Early Cambrian time. Where Barentsia was located prior to the Timan orogeny is not known: it has arbitrarily been separated from Baltica along with the Pre-Uralides on the 620 Ma map. Similarly, the position of Arctida prior to its Late Silurian collision with Arctic North America and North Greenland is not well constrained. On figures 2.4a–e it has been joined to northern North America throughout the 620–460 Ma interval.

Baltica and Siberia

Except possibly for Spitsbergen, Timan, and the Pre-Uralide margins, Baltica was bordered by passive margins in earlier Vendian time or joined to another continent (Zonenshain et al. 1990). The Uralian margin of Baltica was deformed and metamorphosed in latest Vendian or Cambrian time (Zonenshain et al. 1990: figures 36 and 45; Puchkov 1997). The deformation is attributed to collisions with island arcs and microcontinents, probably originally forming part of Siberia (Zonenshain et al. 1990:41). Here it is assumed that the Pre-Uralides and Barentsia were joined at about 620 Ma and had collided (or were in the process of colliding) with Baltica (see figure 2.4e). By 580 Ma it is assumed that Pre-Uralides–Barentsia had collided with Baltica, giving rise to the orogenic belt that affected the Pre-Uralides and Barentsia and somehow involved Spitsbergen and Taimyr, i.e., extending along the length of the Uralian margin of Baltica and possibly beyond to Arctida. The detailed kinematics of this orogenic belt are presently obscure. An alternative view is that there was a collision with some fragments of Avalonia (Puchkov 1997).

During the 620–460 Ma interval, eastern Baltica and all except northern Siberia (present coordinates) were bordered by back-arc basins, island arcs, and microcontinents (Zonenshain et al. 1990; Şengör et al. 1993; Mossakovsky et al. 1994; Şengör and Natal'in, 1996) that resembled the present-day western Pacific. Figures 2.4a–e show the tectonic elements recognized by Şengör et al. (1993) and Şengör and Natal'in (1996) attached in their present-day positions relative to Siberia but do not attempt to reproduce their evolution.

Gondwana

South America today includes several "terranes" that were probably elsewhere in early Paleozoic time. Their positions on figures 2.4a–e are intended only to suggest a possible evolution. Other models suggesting differing possible extents, shapes, and evolution are reviewed by Dalziel (1997). Three terranes are shown on the maps: Patagonia, the Precordillera terrane (or Occidentalia), and Chilenia (Ramos et al. 1986; Ramos 1988). Patagonia is considered here to have collided with the rest of South America in Carboniferous time, but because there are no ophiolites and no obvious suture, this orogenic belt may not reflect a collision (Dalziel, pers. comm.). The Precordillera may have been an elongate fragment that started life on the edge of southeastern Laurentia (Dalla Salda et al. 1992a,b; Dalziel et al. 1994; Astini et al. 1995; Dalziel and Dalla Salda 1996; Thomas and Astini 1997; Keller 1999) and may have collided with South America during Arenig-Llanvirn time (Astini et al. 1995). However, alternative models, reviewed by Dalziel (1997), exist. Chilenia is a second elongate fragment that collided with the Precordillera in the later Carboniferous (Astini et al. 1995; Pankhurst and Rapela 1998). Apart from the geometric requirement of avoiding overlap, there are few constraints on the relative positions of Patagonia and Chilenia prior to collision. For simplicity it is assumed that they were joined together as a single fragment that lay somewhere off Antarctica, where they have been parked for most of the Vendian period.

Several large continental fragments that now lie in eastern North America were originally attached to northwestern Gondwana (mostly Africa). They constitute the Avalon zone, named after the Avalon Peninsula in southeastern Newfoundland (Williams et al. 1974; Williams 1995). Here, late Precambrian sediments and volcanics are overlain by Cambro-Ordovician shales and sandstones, rather than the platform carbonates of Laurentia. Similar rocks can be recognized in the Appalachians as far south as the Carolina slate belt and across the Atlantic in Wales, Brittany, Iberia, northwestern Africa, and parts of central Europe (Pickering and Smith 1995). Avalonia is used here as an informal name for the entire belt, which is up to 750 km wide in its type area and several thousand kilometers long. In reality it is made up of a number of fragments, as indicated on figures 2.4a–e, several of which behaved independently (Nance and Thompson 1996). Its treatment here as a single unit is *simply a convenience.*

Avalonia's present tectonic position against the Laurentian and Baltica margins is the result of a tectonic process that appears to be common when an ocean basin closes. The available evidence suggests that the Avalonian fragments originally bordered one continent (here northwestern Gondwana), were split from it, and migrated across the ocean (Iapetus), though not necessarily as the single fragment shown in figure 2.4a, colliding with the opposing continent (eastern North America and southern Baltica) before the collision of the two major continents. Eastern Avalonia may have separated from Gondwana some time in the Late Cambrian to Early Ordovician interval (Prigmore et al. 1997), although Landing (1996) argues that Avalon was a unified, isolated continent by latest Precambrian time that had no relationship to Gondwana or Baltica at that time. It resembles a one-way tectonic windshield wiper, which sweeps a pre-existing ocean away (the Iapetus), creating a new one behind it (part of which was the Rheic Ocean), and leaving the "wiper" on the opposite continent (the Avalonian fragments). The process has no formal name nor is the underlying physics of it well understood. A simplified model that conveys the gist of the evolution of Avalonia is shown in figures 2.4a–e.

The kinematic evolution of Avalonia is analogous to that of Sibumasu (Metcalfe 1992), discussed below, and its continuation westward as far as Greece, as the Cimmerian continent (Şengör et al. 1984): a long, narrow, more or less continuous continental sliver that migrated across the Mesozoic Tethys, sweeping away Paleotethys and inaugurating Neotethys behind it. "Avalonias" are important paleontologically because they may act as rafts that transport floras and faunas across an ocean in a time span during which significant evolution can take place en route. The Precordillera may provide another possible example of the splitting off of a fragment and its migration across a small ocean.

Late Precambrian and early Paleozoic poles determined for fragments such as North China, South China, and elsewhere show that the paleolatitudes of all three fragments are more or less in the latitude range one might expect, i.e., close to the latitude range of the edge of East Gondwana. However, in a discussion of the Cambro-Ordovician poles from these three fragments, Kirschvink (1992b) concludes that although all three were close to one another and joined to East Gondwana, North and South China were *both inverted* at the time.

There is good stratigraphic evidence, reviewed by Burrett et al. (1990) and Metcalfe (1992), suggesting the Cambrian–Early Ordovician evolution of these and other Southeast Asian fragments. Although most of the boundaries of the fragments in central and Southeast Asia on figures 2.4a–e are close to those suggested by Metcalfe (1992: figure 1) and Yin and Nie (1996: figure 20.1), some differences exist. In particular, Metcalfe subdivides what is western Southeast Asia on figures 2.4a–e into two fragments: Sibumasu (northwestern Burma + Burma + Malaysia + Sunda) and East Malaya. In essence, the following fragments were probably contiguous with Australia and India in Cambrian to Early Ordovician time: Indo-Burma, western Southeast Asia, South China, Indochina, and North China, together with smaller fragments to the north such as the North and South Tarim fragments in China. South China is believed to have been close to Pakistan at the time. The post–460 Ma evolution is complex and not relevant to the maps.

Alternative views exist about the placing of North and South China in the period 620–460 Ma. For example, Yin and Nie (1996: figure 20.18) regard the North Tarim and North China fragments as having been joined together from about 630–438 Ma and to have been bordered on all sides by passive margins, rather like present-day Madagascar. Li et al. (1996) show possible earlier Cambrian to Vendian positions for an isolated North China and also for an isolated South China.

Because of the considerable uncertainties involved, it is assumed for simplicity that all these fragments remained attached to Gondwana in their likely Cambrian positions until at least 620 Ma (figures 2.4a–e), which are essentially those of Kirschvink (1992b), slightly modified.

Still further east there are many continental and igneous fragments lying off northern and eastern Australia, but it is difficult to assess what proportion of them consists of material that existed in the 620–460 Ma interval. Some, like the Tonga-Kermadec arc, may be entirely Cenozoic in age. For completeness, they are shown on the maps.

Finally, although Antarctica has been divided into several fragments, the only ones that have been moved relative to the others on the maps are the Antarctic Peninsula and Thurston Island. These have been visually repositioned to conform to the Jurassic reconstruction (not shown) of Lawver et al. (1992), and then moved with Patagonia and Chilenia to avoid overlap. There has been relative motion between East Antarctica and other Antarctic fragments, but on a global scale this can be (and has been) neglected. The configuration of West Antarctica and its position relative to other continents in pre-Mesozoic time is not known.

FAUNAL AND CLIMATIC EVIDENCE

Tillites

There are several exposures of middle Vendian glacigenic sediments in Baltica and Laurentia. They form two main horizons that are believed to be correlative (Fairchild and Hambrey 1995). The base of the older horizon and the top of the younger define the Varangerian (Harland et al. 1990), which has an age estimated at 620–610 Ma for the base and 590–575 Ma for the top. In North America the till left behind by the Pleistocene ice sheets was locally within 40° of the equator. By analogy, the Varangerian tillites are likely to have been deposited poleward of 40° latitude. Outcrops of the later Varangerian sequence all lie poleward of 40° south (not illustrated), but the earlier Varangerian tillites, assumed to be about 610 Ma old, lie close to 40° south, though there are considerable uncertainties in the chronostratigraphic and paleomagnetic data for early Varangerian time. Figures 2.4d–e show that any other Varangerian tillites on Baltica will be poleward of 40° south. Similar conclusions have previously been reached by Meert and Van der Voo (1994) and Torsvik et al. (1996).

Tillites occur in the 620–460 Ma interval at several horizons in Gondwana. The mid-Cambrian Tamale Group tillites of northwestern Africa, dated at about 520 Ma (Villeneuve and Cornée 1994), are high-latitude tillites on figures 2.4d and 2.4e. The Mali Group tillites of northwestern Africa dated at about 620–580 Ma (Villeneuve and Cornée 1994) all lie poleward of 45°, but the uncertainties in the reconstructions are large.

Figures 2.4d,e show that any Australian tillites in the age range 620–580 Ma lie poleward of 40° north. Meert and Van der Voo (1994:11) reached a similar conclusion. Both conclusions depend on rejecting some of the highest-quality poles from Australia (e.g., from the Elatina Formation) (Meert and Van der Voo 1994:8–9). Acceptance of these and similar poles has led to some novel hypotheses such as the "snowball Earth" (Kirschvink 1992a) or dramatic changes in the earth's obliquity (Williams 1993). However, the distribution of ice during the Ordovician and Permo-Carboniferous glaciations of Gondwana appears analogous to the distribution of ice sheets during Pleistocene glacial maxima (Smith 1997). In this writer's view, the latitudinal distribution of ice during the late Precambrian glaciations, which took place only some 100–200 m.y. earlier, is likely to have been similar to that in the subsequent 440 m.y.

Archaeocyaths and Gondwana

The most poleward occurrences of Cambrian archaeocyaths are in southern Morocco at about 70° south, according to Gondwana paleomagnetic data. The selection of Gondwana data by Dalziel (1997: figure 12) for his 545 Ma map independently supports this view. Archaeocyaths are regarded as indicative of subtropical latitudes— 30° south–30° north—(Courjault-Radé et al. 1992; Debrenne and Courjault-Radé 1994). The Moroccan archaeocyaths are associated with red beds, oolitic limestones, and sediments with halite pseudomorphs (Geyer and Landing 1995), all of which indicate warm, nonpolar conditions and are likely to have formed at lower latitudes than those indicated by the paleomagnetic data.

Kirschvink et al. (1997) recognize that unrealistically high plate velocities are implied by the currently accepted high-quality Vendian-to-Cambrian paleomagnetic data (which include the Elatina pole). They resolve this dilemma by proposing that not only do these poles record plate motions in the range inferred for Mesozoic and Cenozoic plate motions, but they also include a large component of true polar wander in which the mantle as a whole is rotating rapidly. The net result of this ingenious model is to place the archaeocyaths in the low-latitude range suggested for them.

The present author's view is more conservative and uniformitarian: most of the discrepancies between the inferred paleolatitudes could readily stem from "noisy" paleomagnetic data. Many of the latest Precambrian poles are dated as simply "Vendian to Early Cambrian," i.e., with a mean age of about 565 Ma and an age range of some 90 m.y. Other poles may have been assigned an unlikely age, as may be the case for the pole from the Elatina Formation, discussed above. It is also possible that archaeocyaths did span a wider range of latitude, particularly if parts of the Cambrian were warm.

Faunal Provinces of Baltica and Siberia

Distinct Early Ordovician trilobite provinces of Baltica and Siberia have long been recognized in the literature (Cocks and Fortey 1990), commonly attributed to geographic separation. In a recent review, Torsvik et al. (1996: figure 11) show these provinces on a reconstruction at 490 Ma in which Baltica is an isolated continent south of Siberia. The 500 Ma map (see figure 2.4b) also shows Baltica and Siberia as isolated entities, but with Siberia to the east of Baltica and with Baltica in lower latitudes. It is not clear which is the better fit to the paleontological evidence, though if Baltica was in a more southerly position on figure 2.4b, it would be more geographically isolated. By contrast, models in Şengör et al. 1993 and Şengör and Natal'in 1996 require Siberia and Baltica to have been linked by island arc systems for most of early Paleozoic time. If such systems did exist, they would have allowed trilobite interchange between Baltica and Siberia, providing that the faunas could tolerate the differing environments.

OTHER MAPS

Dalziel 1997

The time periods that are relevant are 465, 475, 515, and 545 Ma. In Dalziel's view, Pannotia existed until 545 Ma (Dalziel 1997: figure 12) and the Iapetus Ocean started to form at this time. The present author also accepted a similar interpretation until recently, but the difficulties of fitting the passive margin sequences into such a model (discussed in the section "Times of Passive Margin Formation" above), suggest that the Iapetus Ocean began to spread some time before 545 Ma. The geometry of Pannotia differs significantly, with Amazonia placed opposite southern Greenland, and Baltica in a more northerly position than on figure 2.4e. Most of the differences between Dalziel's maps and the one shown here stem from this very different initial starting reassembly and age: the 420 Ma map (Dalziel 1997: figure 17) is similar to an extrapolated figure 2.4a.

Torsvik et al. 1996

Torsvik et al. (1996) include a series of maps showing the positions of Baltica, Laurentia, Siberia, and some other continents for 580, 550, 490, and 440 Ma. Their partial reassembly of Pannotia at 580 Ma (Torsvik et al. 1996: figure 9A) is similar to Dalziel's except that Baltica is rotated by 180°, with the western Urals placed against eastern Greenland and Amazonia abutting western Baltica. Pannotia breaks up at about this time, as in figure 2.4d. Its evolution to 440 Ma differs from that in figure 2.4, principally in relation to the orientation and paleolatitude of Baltica.

Seslavinsky and Maidanskaya (Chapter 3)

The seven paleogeographic maps by Seslavinsky and Maidanskaya in chapter 3 of this volume were made completely independently of figures 2,4a–e, but in many respects are similar. The main differences between figures 2.4c–e and their maps for 540, 580, and 620 Ma are in the orientations of the reassemblies, principally because Seslavinsky and Maidanskaya have also used paleontological and paleoclimatic evidence in their construction.

CONCLUSIONS

1. Apart from longitudinal uncertainties, Laurentia, Siberia, and Gondwana can be repositioned by paleomagnetic data for most of mid-Ordovician to Cambrian time. Cambrian paleomagnetic data from different parts of Baltica are inconsistent.

2. Baltica appears to have been joined to Laurentia in late Precambrian time in a more southerly position than it had after the closure of the Iapetus Ocean in Devonian time.

3. Tectonic and paleomagnetic evidence suggests that South America was also probably joined to Laurentia in latest Precambrian time. However, in early Vendian time, East Gondwana (East Antarctica, India, and most of Australia) was joined to the Arabian-Nubian shield and the Kalahari-Grunehogna craton of southeastern Africa. This assemblage was probably distinct from the rest of West Gondwana, which was in the process of colliding with it and with Laurentia.

4. Baltica, Siberia, and Gondwana can be visually joined to Laurentia to form a latest Precambrian Pangea (Pannotia) at 580 Ma. Pannotia can be oriented using Laurentian paleomagnetic data. After breakup at 580 Ma, the latest Neoproterozoic latitudes and orientation of Baltica, Siberia, and Gondwana can be estimated by interpolation between their position on Pannotia and the oldest paleomagnetic data that are considered reliable. Baltica's position in Cambrian

time has been estimated by interpolating between its 580 Ma position and a paleomagnetic pole at 477 Ma.

5. The relative longitudinal separations of Laurentia and Baltica can be plausibly estimated from the tectonic history of their continental margins. Siberia's longitude is determined by placing it as an isolated continent between Baltica and East Gondwana for the 580–460 Ma interval. Gondwana's separation from Laurentia is determined by keeping southeastern North America close to northwestern South America throughout the same interval.

6. Many other continental fragments, such as North and South China, can be plausibly repositioned by using a combination of tectonic, stratigraphic, and faunal data, though other fragments, such as Kolyma, have to be arbitrarily parked next to areas where they are today.

7. The only significant misfit between the paleolatitudes of the continents and paleolatitudes inferred from paleobiogeographic indicators are the paleolatitudes of early Cambrian archaeocyaths in Morocco. The misfit is tentatively attributed to incorrect age assignments to some (unspecified) paleomagnetic poles.

8. There is no misfit between the paleolatitudes of the repositioned continents and the paleolatitudes inferred from most sedimentary paleoclimatic indicators, particularly tillites. Varangerian tillites in Baltica and Laurentia lie poleward of 40° south, as do Mali Group tillites in Africa. Late Precambrian Australian tillites could all lie poleward of 40° north, in direct contrast to subequatorial latitudes estimated from high-quality paleomagnetic poles from Australia. However, in all cases the uncertainties in the paleolatitudes are large.

9. In contrast to the assumption that Baltica and Siberia were joined to Laurentia at 580 Ma and retreated from it during 580–500 Ma, similarity coefficients show very low values between the Vendian faunas of Siberia and Laurentia before late Early Cambrian time. The progressive increase in similarity during this and subsequent periods suggests that Siberia was approaching Laurentia during this interval rather than retreating from it as shown on figures 2.4c–e.

10. The longitude of Siberia relative to all other continents and the Cambrian-to-late-Neoproterozoic latitude and orientation of Baltica are two of the major uncertainties of Vendian to early Paleozoic reconstructions.

Acknowledgments. I thank Irina Maidanskaya and Andrei Maslov for their help and interest during the conference and field trip for IGCP Project 319 (Global Paleogeography of Latest Precambrian and Early Paleozoic) in Russia in 1996. I also thank David Gee for the invitation to participate in the Europrobe conference in St. Petersburg in 1997, during which there were many useful discussions, and Pierre Courjault-Radé,

Ian Dalziel, Richard Fortey, Zheng-Xiang Li, Chris Powell, Reinhard Greiling, and two anonymous reviewers for constructive criticism. Andrey Zhuravlev provided an excellent, wide-ranging critical review that has substantially improved the paper. Robert Riding is thanked for his editorial work and for the invitation to submit this paper. St. John's College, Cambridge, is thanked for a travel grant. This paper is a contribution to IGCP Project 366 and is Contribution no. 5156, Department of Earth Sciences, Cambridge.

REFERENCES

Andréasson, P. and L. Albrecht. 1995. Derivation of 500 Ma eclogites from the passive margin of Baltica and a note on the tectonometamorphic heterogeneity of eclogite-bearing crust. *Geological Magazine* 132:729–738.

Astini, R. A., J. L. Benedetto, and N. E. Vaccari. 1995. The early Paleozoic evolution of the Argentine Precordillera as a Laurentian rifted, drifted, and collided terrane: A geodynamic model. *Bulletin of the Geological Society of America* 107:253–273.

Astini, R. A., J. L. Benedetto, and N. E. Vaccari. 1996. The early Paleozoic evolution of the Argentine Precordillera as a Laurentian rifted, drifted, and collided terrane: A geodynamic model: Discussion and reply. Reply. *Bulletin of the Geological Society of America* 108:373–375.

Bell, R. T. and C. W. Jefferson. 1987. An hypothesis for an Australian-Canadian connection in the Neoproterozoic and the birth of the Pacific Ocean. In *Proceedings, Pacific Rim Congress 1987, Parkville, Australia,* pp. 39–50. Parkville, Victoria: Australian Institute of Mining and Metallurgy.

Bingen B., D. Demaiffe, and O. van Breemen. 1998. The 616 Ma old Egersund basaltic dyke swarm, SW Norway, and late Neoproterozoic opening of the Iapetus Ocean. *Journal of Geology* 106:565–574.

Bird, J. M. and J. F. Dewey. 1970. Lithosphere plate: Continental margin tectonics and the evolution of the Appalachian orogen. *Bulletin of the Geological Society of America* 81:1031–1059.

Black, R., L. Latouche, J. P. Liegeois, R. Caby, and J. M. Bertrand. 1994. Pan-African displaced terranes in the Tuareg shield (central Sahara). *Geology* 22:641–644.

Bond, G. C., P. A. Nickeson, and M. A. Kominz. 1984. Breakup of a supercontinent between 625 Ma and 555 Ma: New evidence and implications for continental histories. *Earth and Planetary Science Letters* 70:325–345.

Bullard, E., J. E. Everett, and A. G. Smith. 1965. The fit of the continents around the Atlantic. *Philosophical Transactions of the Royal Society of London* A 258:41–51.

Burrett, C. and R. Berry. 2000. Proterozoic Australia–western United States (AUSWUS) fit between Laurentia and Australia. *Geology* 28:103–106.

Burrett, C. F., J. Long, and B. Stait. 1990. Early-middle Palaeozoic biogeography of Asian terranes derived from Gondwana. *Geological Society of London, Memoir* 12:163–164.

Carter, E. S., M. J. Orchard, C. A. Ross, J. R. P. Ross, P. L. Smith, and H. W. Tipper. 1992. Paleontological signatures of terranes. In H. Gabrielse and C. J. Yorath, eds., *Geology of the Cordilleran Orogen in Canada,* vol. 4 of *Geology of Canada,* pp. 28–38. Ottawa: Geological Survey of Canada.

Cochran, J. R. 1981. The Gulf of Aden: Structure and evolution of a young ocean basin and continental margin. *Journal of Geophysical Research* 86:263–287.

Cocks, L. R. M. and R. A. Fortey. 1990. Biogeography of Ordovician and Silurian faunas. In W. S. McKerrow and C. R. Scotese, eds., *Palaeozoic Palaeogeography and Biogeography,* Geological Society Memoir no. 12, pp. 97–104. London: Geological Society.

Condie, K. C. and O. M. Rosen. 1994. Laurentia-Siberia connection revisited. *Geology* 22:168–170.

Courjault-Radé, P., F. Debrenne, and A. Gandin. 1992. Palaeogeographic and geodynamic evolution of the Gondwana continental margins during the Cambrian. *Terra Nova* 4:657–667.

Dalla Salda, L., C. Cingolani, and R. Varela. 1992a. Early Paleozoic orogenic belt of the Andes in southwestern South America: Result of Laurentia-Gondwana collision. *Geology* 20:617–620.

Dalla Salda, L. H., I. W. D. Dalziel, C. A. Cingolani, and R. Varela. 1992b. Did the Taconic Appalachians continue into South America? *Geology* 20:1059–1062.

Dalziel, I. W. D. 1991. Pacific margins of Laurentia and East Antarctica–Australia as a conjugate rift-pair: Evidence and implications for an Eocambrian supercontinent. *Geology* 19:598–602.

Dalziel, I. W. D. 1997. Neoproterozoic-Paleozoic geography and tectonics: Review, hypothesis, environmental speculation. *Geological Society of America Bulletin* 109:16–42.

Dalziel, I. W. D. and L. H. Dalla Salda. 1996. The early Paleozoic evolution of the Argentine Precordillera as a Laurentian rifted, drifted, and collided terrane: A geodynamic model: Discussion and reply. Discussion. *Bulletin of the Geological Society of America* 108:372–373.

Dalziel, I. W. D., L. H. Dalla Salda, and L. M. Gahagan. 1994. Paleozoic Laurentia-Gondwana interaction and the origin of the Appalachian-Andean mountain system. *Bulletin of the Geological Society of America* 106:243–252.

Debrenne, F. and P. Courjault-Radé. 1994. Répartition paléogéographique des archéocyathes et délimitation des zones intertropicales au Cambrian inférieur. *Bulletin de la Société géologique de France* 165:459–467.

Elming, S.-Å., L. J. Pesonen, M. A. H. Leino, A. N. Khramov, N. P. Mikhailova, A. F. Krasnova, S. Mertanen, G. Bylund, and M. Terho. 1993. The drift of Fennoscandian and Ukrainian shields during the Precambrian: A paleomagnetic analysis. *Tectonophysics* 223:177–198.

Ernst, R. E. and K. L. Buchan. 1997. Giant radiating dyke swarms: Their use in identifying pre-Mesozoic large igneous provinces and mantle plumes. In J. Mahoney and M. Coffin, eds., *Large Igneous Provinces: Continental, Oceanic, and Planetary Flood Volcanism.* Geophysical Monograph 100, pp. 297–333. Washington, D.C.: American Geophysical Union.

Fairchild, I. J. and M. J. Hambrey. 1995. Vendian basin evolution of East Greenland and NE Svalbard. *Precambrian Research* 73:217–233.

Fisher, R. L. and J. G. Sclater. 1983. Tectonic evolution of the southwest Indian Ocean since the mid-Cretaceous: Plate motions and stability of the pole of Antarctica/Africa for at least 80 myr. *Geophysical Journal of the Royal Astronomical Society* 73:553–576.

Gabrielse, H., J. W. H. Monger, J. O. Wheeler, and C. J. Yorath. 1992. Morphogeological belts, tectonic assemblages, and terranes.

In H. Gabrielse and C. J. Yorath, eds., *Geology of the Cordilleran Orogen in Canada,* vol. 4 of *Geology of Canada,* pp. 15–28. Ottawa: Geological Survey of Canada.

Gayer, R. A. and R. O. Greiling. 1989. Caledonian nappe geometry in north-central Sweden and basin evolution on the Baltoscandian margin. *Geological Magazine* 126:499–513.

Gee, D. G. 1975. A tectonic model for the central part of the Scandinavian Caledonides. *American Journal of Science* 275A:468–515.

Geyer, G. and E. Landing. 1995. The Cambrian of Moroccan Atlas regions. *Beringeria Special Issue* 2:7–46.

Greiling, R. O. and A. G. Smith. n.d. The Dalradian of Scotland: missing link between the Vendian of northern and southern Scandinavia? *Physics and Chemistry of the Earth.*

Gurnis, M. and T. H. Torsvik. 1994. Rapid drift of large continents during the late Precambrian and Paleozoic. *Geology* 22:1023–1026.

Harland, W. B., R. L. Armstrong, A. V. Cox, L. E. Craig, A. G. Smith, and D. G. Smith. 1990. *A geologic time scale 1989.* Cambridge: Cambridge University Press.

Hatcher, R. D., Jr., W. A. Thomas, and G. W. Viele, eds. 1989. *The Appalachian-Ouachita Orogen in the United States.* F2 of *The Geology of North America.* Boulder, Colo.: Geological Society of America.

Higgins, A. K. 1995. Caledonides of East Greenland. In H. Williams, ed., *Geology of the Appalachian-Caledonide Orogen in Canada and Greenland,* vol. 6 of *Geology of Canada,* pp. 891–921. Ottawa: Geological Survey of Canada.

Higgins, A., J. R. Ineson, J. S. Peel, F. Surlyk, and M. Sønderholm. 1991. Cambrian to Silurian basin development and sedimentation, North Greenland. In H. P. Trettin,

ed., *Geology of the Innuitian Orogen and Arctic Platform of Canada and Greenland,* vol. 3 of *Geology of Canada,* pp. 111–161. Ottawa: Geological Survey of Canada.

Hoffman, P. 1991. Did the breakout of Laurentia turn Gondwanaland inside out? *Science* 252:1409–1412.

Howell, D. G. and R. W. Murray. 1986. A budget for continental growth and denudation. *Science* 233:446–449.

Keller, M. 1999. Argentine Precordillera: Sedimentary and plate tectonic history of a Laurentian continental fragment in South America. *Geological Society of America Special Paper* 341:140.

Keppie, J. D., J. Dostal, J. B. Murphy, and R. D. Nance. 1996. Terrane transfer between Eastern Laurentia and Western Gondwana in Early Paleozoic: Constraints on global reconstructions. In R. D. Nance and M. D. Thompson, eds., *Avalonian and Related Peri-Gondwanan Terranes of the Circum-North Atlantic,* pp. 369–380. Geological Society of America Special Paper 304.

Kirschvink, J. L. 1992a. Late Proterozoic low-latitude global glaciation: The snowball Earth. In J. W. Schopf and C. Klein, eds., *The Proterozoic Biosphere: A Multidisciplinary Study,* pp. 51–52. Cambridge: Cambridge University Press.

Kirschvink, J. L. 1992b. A paleogeographic model for Vendian and Cambrian time. In J. W. Schopf and C. Klein, eds., *The Proterozoic Biosphere: A Multidisciplinary Study,* pp. 569–581. Cambridge: Cambridge University Press.

Kirschvink, J. L., R. L. Ripperdan, and D. A. Evans. 1997. Evidence for a large-scale reorganization of Early Cambrian continental masses by inertial interchange true polar wander. *Science* 277:541–545.

Klitgord, K. D. and H. Schouten. 1986. Plate kinematics of the central Atlantic. In P. R. Vogt and B. E. Tucholke, eds., *The West-*

ern North Atlantic Region, M of *The Geology of North America,* pp. 351–378. Boulder, Colo.: Geological Society of America.

Knoll, A. H. 1996. Daughter of time. *Paleobiology* 22:1–7.

Kumpulainen, R. and J. P. Nystuen. 1985. Late Proterozoic basin evolution and sedimentation in the westernmost part of Baltoscandia. In D. G. Gee and B. A. Sturt, eds., *The Caledonian Orogen—Scandinavia and Related Areas,* pp. 213–232. New York: John Wiley and Sons.

Landing, E. 1996. Avalon: Insular continent by the latest Precambrian. In R. D. Nance and M. D. Thompson, eds., Avalonian and related peri-Gondwanan terranes of the circum–North Atlantic, *Geological Society of America Special Paper* 304:29–63.

Lawver, L. A., L. M. Gahagan, and M. C. Coffin. 1992. The development of paleoseaways around Antarctica. In J. P. Kennett and D. A. Warnke, eds., *The Antarctic Paleoenvironment: A Perspective on Global Change,* Antarctic Research Series, no. 56, pp. 7–30. Washington, D.C.: American Geophysical Union.

LePichon, X. and J. Francheteau. 1978. A plate-tectonic analysis of the Red Sea—Gulf of Aden area. *Tectonophysics* 46:369–406.

Li, Z. X. and C. M. Powell. 1993. Late Proterozoic to early Paleozoic paleomagnetism and the formation of Gondwanaland. In R. H. Findlay, R. Unrug, M. R. Banks, and J. J. Veevers, eds., *Gondwana Eight: Assembly, Evolution, and Dispersal,* pp. 9–21. Rotterdam: Balkema.

Li, Z. X., L. Zhang, and C. M. Powell. 1996. Positions of the East Asian craton in the Neoproterozoic supercontinent Rodinia. *Australian Journal of Earth Sciences* 43:593–604.

Livermore, R. A., F. J. Vine, and A. G. Smith. 1983. Plate motions and the geomagnetic field. 1: Quaternary and Late Tertiary.

Geophysical Journal of the Royal Astronomical Society 73:153–171.

Livermore, R. A., F. J. Vine, and A. G. Smith. 1984. Plate motions and the geomagnetic field. 2: Jurassic to Tertiary. *Geophysical Journal of the Royal Astronomical Society* 79:939–961.

McElhinny, M. W. and J. Lock. 1996. IAGA paleomagnetic databases with Access. *Surveys in Geophysics* 17:575–591.

McKenzie, D. P. 1978. Some remarks on the development of sedimentary basins. *Earth and Planetary Science Letters* 40:25–32.

McKenzie, D. P., D. Davies, and P. Molnar. 1970. Plate tectonics of the Red Sea and East Africa. *Nature* 226:243–248.

McKerrow, W. S., C. R. Scotese, and M. D. Brasier. 1992. Early Cambrian continental reconstructions. *Journal of the Geological Society, London* 149:599–606.

McMenamin, M. A. S. and D. L. S. McMenamin. 1990. *The Emergence of the Animals—The Cambrian Breakthrough.* New York: Columbia University Press.

Meert, J. G. and R. Van der Voo. 1994. The Neoproterozoic (1000–540 Ma) glacial intervals: No more snowball Earth? *Earth and Planetary Science Letters* 123:1–13.

Meert, J. G., R. Van der Voo, C. M. Powell, Z-X. Li, M. W. McElhinny, Z. Chen, and D. T. A. Symonds. 1993. A plate-tectonic speed limit? *Nature* 363:216–217.

Metcalfe, I. 1992. Ordovician to Permian evolution of Southeast Asian terranes: NW Australian Gondwana connections. In B. D. Webby and J. R. Laurie, eds., *Global Perspectives on Ordovician Geology,* pp. 293–305. Rotterdam: Balkema.

Miller, D. M., T. H. Nilsen, and W. L. Bilodeau. 1992. Late Cretaceous to early Eocene geology of the United States Cordillera. In B. C. Burchfiel, P. W. Lipman, and M. L. Zoback, eds., *The Cordilleran Orogen:*

Conterminous U.S., G-3, plate 6. Boulder, Colo.: Geological Society of America.

Moores, E. M. 1991. Southwest U.S.–East Antarctic (SWEAT) connection: A hypothesis. *Geology* 19:425–428.

Mossakovsky, A. A., S. V. Ruzhentsev, S. G. Samygin, and T. N. Kherasova. 1994. Central Asian Fold Belt: Geodynamic evolution and formation history. *Geotectonics* 27:445–474.

Nance, R. D. and M. D. Thompson, eds. 1996. *Avalonian and Related Peri-Gondwana Terranes of the Circum–North Atlantic.* Geological Society of America Special Paper 304.

Niocaill, C. M., B. A. van der Pluijm, and R. Van der Voo. 1997. Ordovician paleogeography and the evolution of the Iapetus Ocean. *Geology* 25:159–162.

Norton, I. O. and J. G. Sclater. 1979. A model for the evolution of the Indian Ocean and the breakup of Gondwanaland. *Journal of Geophysical Research* 84:6803–6830.

Pankhurst, R. J. and C. W. Rapela. 1998. The Proto-Andean margin of Gondwana. *Geological Society of London Special Publication* 142:369.

Pelechaty, S. M. 1996. Stratigraphic evidence for the Siberia-Laurentia connection and Early Cambrian rifting. *Geology* 24:719–722.

Pickering, K. T. and A. G. Smith. 1995. Arcs and backarc basins in the Early Paleozoic Iapetus Ocean. *Island Arc* 4:1–67.

Powell, C. M. 1995. Are Neoproterozoic glacial deposits preserved on the margins of Laurentia related to the fragmentation of two supercontinents? Comment. *Geology* 23:1053–1054.

Powell, C. M., Z. X. Li, M. W. McElhinny, J. G. Meert, and J. K. Park. 1993. Paleomagnetic constraints on timing of the Neoproterozoic breakup of Rodinia and the Cambrian formation of Gondwana. *Geology* 21:889–892.

Powell, C. M., W. V. Preiss, C. G. Gatehouse, B. Krapez, and Z. X. Li. 1994. South Australian record of a Rodinian epicontinental basin and its mid-Neoproterozoic breakup (~700 Ma) to form the Palaeo-Pacific Ocean. *Tectonophysics* 237:113–140.

Prigmore, J. K., A. J. Butler, and N. H. Woodcock. 1997. Rifting during separation of eastern Avalonia from Gondwana: Evidence from subsidence analysis. *Geology* 25:203–206.

Puchkov, V. N. 1997. Structure and geodynamics of the Uralian orogen. *Geological Society of London Special Publication* 121:201–236.

Ramos, V. 1988. Late Proterozoic–Early Paleozoic of South America: A collisional history. *Episodes* 19:425–428.

Ramos, V., T. Jordan, R. Allmedinger, C. Mpodozis, S. M. Kay, J. Cortés, and M. Palma. 1986. Paleozoic terranes of the central Argentine-Chilean Andes. *Tectonics* 5:855–880.

Roest, W. R. and S. P. Srivastava. 1989. Sea floor spreading in the Labrador Sea: A new reconstruction. *Geology* 17:1000–1003.

Royer, J.-Y. and D. T. Sandwell. 1989. Evolution of the eastern Indian Ocean since the late Cretaceous: Constraints from Geosat altimetry. *Journal of Geophysical Research* 94:13755–13782.

Saleeby, J. and C. Busby-Spera. 1992. Early Mesozoic tectonic evolution of the western U.S. Cordillera. In B. C. Burchfiel, P. W. Lipman, and M. L. Zoback, eds., *The Cordilleran Orogen: Conterminous U.S.,* G-3, plate 5. Boulder, Colo.: Geological Society of America.

Schwab, F. L., J. P. Nystuen, and I. Gunderson. 1988. Pre-Arenig evolution of the Appalachian-Caledonide orogen: Sedimentation and stratigraphy. *Geological Society of London Special Publication* 38:75–91.

Scotese, C. R. and W. S. McKerrow. 1990. Revised world maps and introduction. In W. S. McKerrow and C. R. Scotese, eds., *Palaeogeography and Biogeography,* Geological Society Memoir no. 12, pp. 1–21. London: Geological Society.

Şengör, A. M. C. and B. A. Natal'in. 1996. Paleotectonics of Asia: Fragments of a synthesis. In A. Yin and M. Harrison, eds., *The Tectonic Evolution of Asia,* pp. 486–640. Cambridge: Cambridge University Press.

Şengör, A. M. C., Y. Yilmaz, and O. Sungurlu. 1984. Tectonics of the Mediterranean Cimmerides: Nature and evolution of the western termination of the Palaeo-Tethys. *Geological Society of London Special Publication* 17:77–112.

Şengör, A. M. C., B. A. Natal'in, and V. S. Burtman. 1993. Evolution of the Altaid tectonic collage and Palaeozoic crustal growth in Eurasia. *Nature* 364:299–307.

Smethurst, M. A., A. N. Khramov, and T. H. Torsvik. 1998. The Neoproterozoic and Palaeozoic palaeomagnetic data for the Siberian Platform: From Rodinia to Pangea. *Earth-Science Reviews* 43:1–24.

Smith, A. G. 1997. Estimates of the earth's spin (geographic) axis relative to Gondwana from glacial sediments and paleomagnetism. *Earth-Science Reviews* 42:161–170.

Smith, A. G. and A. Hallam. 1970. The fit of the southern continents. *Nature* 225:139–144.

Smith, A. G., J. C. Briden, and G. E. Drewry. 1973. Phanerozoic world maps. In N. F. Hughes, ed., Organisms and continents through time, *Special Papers in Palaeontology* 12:1–42. London: Palaeontological Association.

Storey, B. C. 1993. The changing face of late Precambrian and early Palaeozoic recon-

structions. *Journal of the Geological Society, London* 150:665–668.

Thomas, W. A. and R. A. Astini. 1997. The Argentine Precordillera: A traveler from the Ouachita embayment of North America Laurentia. *Science* 273:752–756.

Torsvik, T. H., M. A. Smethurst, R. Van der Voo, A. Trench, N. Abrahams, and E. Halvorsen. 1992. Baltica: A synopsis of Vendian-Permian palaeomagnetic data and their palaeotectonic implications. *Earth-Science Reviews* 33:133–152.

Torsvik, T. H., M. A. Smethurst, J. G. Meert, R. Van der Voo, W. S. McKerrow, M. D. Brasier, B. A. Sturt, and H. J. Walderhaug. 1996. Continental break-up and collision in the Neoproterozoic and Palaeozoic: A tale of Baltica and Laurentia. *Earth-Science Reviews* 40:229–258.

Trettin, H. P. 1991. The Proterozoic to Late Silurian record of Pearya. In H. P. Trettin, ed., *Geology of the Innuitian Orogen and Arctic Platform of Canada and Greenland,* vol. 3 of *Geology of Canada,* pp. 241–259. Ottawa: Geological Survey of Canada.

Trettin, H. P., U. Mayr, G. D. F. Long, and J. J. Packard. 1991. Cambrian to Early Devonian basin development, sedimentation, and volcanism, Arctic Islands. In H. P. Trettin, ed., *Geology of the Innuitian Orogen and Arctic Platform of Canada and Greenland,* vol. 3 of *Geology of Canada,* pp. 165–238. Ottawa: Geological Survey of Canada.

Tucker, R. D. and W. S. McKerrow. 1995. Early Paleozoic chronology: A review in light of new U-Pb zircon ages from Newfoundland and Britain. *Canadian Journal of Earth Sciences* 32:368–379.

Unrug, R. 1997. Rodinia to Gondwana: The geodynamic map of Gondwana Supercontinent reassembly. *GSA Today* 7:1–6.

Van der Voo, R. 1990. Phanerozoic paleomagnetic poles from Europe and North

America and comparisons with continental reconstructions. *Reviews of Geophysics* 28:167–206.

Villeneuve, M. and J. J. Cornée. 1994. Structure, evolution, and palaeogeography of the West African craton and bordering belts during the Neoproterozoic. *Precambrian Research* 69:307–326.

White, N. and D. P. McKenzie. 1988. Formation of the "steer's head" geometry of sedimentary basins by differential stretching of the crust and mantle. *Geology* 16:250–253.

Williams, G. E. 1993. History of the earth's obliquity. *Earth-Science Reviews* 34:1–45.

Williams, H. 1995. *Geology of the Appalachian-Caledonian Orogen in Canada and Greenland*, vol. 6 of *Geology of Canada*. Ottawa: Geological Survey of Canada.

Williams, H., M. J. Kennedy, and E. R. W. Neale. 1974. The northeastward termination of the Appalachian Orogen. In A. E. M. Nairn and F. G. Stehli, eds., *The Ocean Basins and Margins*, vol. 2, *The North Atlantic*, pp. 79–123. New York: Plenum Press.

Wooler, D., A. G. Smith, and N. White. 1992. Measuring lithospheric stretching on Tethyan margins. *Journal of the Geological Society, London* 149:517–532.

Yin, A. and S. Nie. 1996. A Phanerozoic palinspastic reconstruction of China and its neighbouring regions. In A. Yin and M. Harrison, eds., *The Tectonic Evolution of China*, pp. 442–485. Cambridge: Cambridge University Press.

Young, G. M. 1995. Are Neoproterozoic glacial deposits preserved on the margins of Laurentia related to the fragmentation of two supercontinents? Reply. *Geology* 23:1054–1055.

Ziegler, P. A. 1989. *Evolution of Laurussia*. London: Kluwer Academic Publishers.

Zonenshain, L. P., M. I. Kuzmin, and L. M. Natapov. 1990. *Geology of the USSR: A Plate-Tectonic Synthesis*. Geodynamics Series, no. 21. Washington, D.C.: American Geophysical Union.

CHAPTER THREE

Kirill B. Seslavinsky and Irina D. Maidanskaya

Global Facies Distributions
from Late Vendian to Mid-Ordovician

Global paleogeographic world maps compiled for the late Vendian, Cambrian, and Early to Middle Ordovician bring together, possibly for the first time, a systematic and uniform overview of paleogeographic and facies distribution patterns for this interval. This 150 Ma period of Earth history was a cycle of oceanic opening and closing. These processes were accompanied by formation of spreading centers and subduction zones, and systems of island arcs and orogenic belts replaced one another successively in time and space. The main features of our planet during this period were the vast Panthalassa Ocean and several smaller oceanic basins (Iapetus, Rheic, Paleoasian).

A VARIETY OF plate tectonic reconstructions has been proposed for the Neoproterozoic and early Paleozoic (e.g., Zonenshain et al. 1985; Courjault-Radé et al. 1992; Kirschvink 1992; Storey 1993; Dalziel et al. 1994; Kirschvink et al. 1997; Debrenne et al. 1999). Some of these are reproduced elsewhere in this volume (Brasier and Lindsay: figure 4.2; Eerola: figure 5.4). However, none of them wholly satisfies current data on paleobiogeography, facies distributions, and metamorphic, magmatic, and tectonic events. Pure paleomagnetic reconstructions often ignore paleontologic data and contain large errors in pole position restrictions. Paleobiogeographic subdivisions developed for single groups, mainly trilobites and archaeocyaths, do not fit either each other or paleomagnetic data, and they ignore the possibility that Cambrian endemism may have been a result of high speciation rates rather than basin isolation (e.g., Cowie 1971; Sdzuy 1972; Jell 1974; Repina 1985; Zhuravlev 1986; Shergold 1988; Pillola 1990; Palmer and Rowell 1995; Gubanov 1998). Furthermore, terrane theory suggests even more-complex tectonic models due to inclusion of multiple "suspect" terranes and drifting microcontinents (Coney et al. 1980). Such terranes are now recognized in a large number of Cordilleran and Appalachian zones of North America (Van der Voo 1988; Samson et al. 1990; Gabrielse and Yorath 1991; Pratt and

Waldron 1991), Kazakhstan, Altay Sayan Foldbelt, Transbaikalia, Mongolia, the Russian Far East (Khanchuk and Belyaeva 1993; Mossakovsky et al. 1993), and western and central Europe (Buschmann and Linnemann 1996).

Paleomagnetic data, which form the basis for the present reconstructions, were obtained from Paleomap Project Edition 6 of Scotese (1994). These reconstructions differ in some details from the earlier reconstructions of Scotese and McKerrow (1990) and McKerrow et al. (1992). The present edition has been chosen only as a working model and, inevitably, does not escape inconsistencies. Certainly, there are problems, such as the position of some blocks or the evolution of the Innuitian Belt in the Canadian Arctic, which still await solution. There is no general agreement on the paleogeographic boundaries of Siberia for the Vendian and Cambrian. The southern and southwestern boundaries of the ancient Siberian craton (in contemporary coordinates) are now formed by large sutures. For example, the Baikal-Patom terrane, where numerous sedimentation and tectonic events occurred during the Cambrian, is separated from Siberia by one such suture, and it is now difficult to determine its original paleogeographic position in the Cambrian. The Vendian-Cambrian succession of the Kolyma Uplift is characterized by species and facies typical of the Yudoma-Olenek Basin of the Siberian Platform (Tkachenko et al. 1987). Thus, Kolyma was probably a part of Siberia, at least during the Vendian-Cambrian, and was displaced much later. In contrast, the Central Asian belt is a complex fold structure now located between the Siberian Platform and Cathaysia (North China and Tarim platforms), which united the Riphean, Salairian, Caledonian, Variscan, and Indo-Sinian zones. Structurally it is a very complicated region that includes accretionary (Altay, Sayan, Transbaikalia, Mongolia, Kazakhstan) and collision (North China, South Mongolia, Dzhungaria, South Tien Shan, northern Pamir) structures, the formation of which was closely related to numerous Precambrian microcontinents. The appearance of the belt was a result of the tectonic development of several oceans (Paleoasian, Paleothetis I and Paleothetis II) (Ruzhentsev and Mossakovsky 1995). The width of the Paleoasian Ocean itself is conventional on the maps, and probably this ocean was never so wide. The position of the northern Taimyr in this and all later reconstructions seems inappropriate. At that time this terrane was not yet part of Siberia, and it was separated from the Siberian craton by an oceanic basin of unknown width (Khain and Seslavinsky 1995).

In the present work, we initially attempted to determine the exact spatial and temporal location of glacial deposits, transgressions and regressions, orogenic belts, volcanic complexes, granitization, regional metamorphism, large tectonic deformations, and some lithologic assemblages, which are indicators of past paleogeographic conditions. It is of particular importance to determine real boundaries (i.e., established by reliable geologic data) of island arcs and subduction zones. Such data may in future be used to constrain plate tectonic reconstructions.

The values of absolute ages shown on the maps refer to the time slices for which reconstructions based on the software Paleomap Project Edition 6 were obtained. However, the maps accumulate all available geologic information for an entire epoch and do not reflect events at any particular moment. For instance, the paleogeographic map of the Early Ordovician (490 Ma) shows all geologic events that took place during the entire Early Ordovician, and the coastlines shown indicate their maximum extent.

PREAMBLE

The Vendian–early Paleozoic includes the Caledonian tectonic cycle of Earth evolution. It was a part of a megacycle (the Wilson cycle), which lasted approximately 600 Ma from late Riphean to the end of the Paleozoic. This megacycle covered the time span from the breakup of a supercontinent until the moment when its fragments, with newly accreted continental crust, joined to form a new supercontinent. This megacycle included three subcycles: Baikalian (oldest), Caledonian, and Hercynian (youngest).

Recent investigations of strontium and carbon isotope variations suggest that the Vendian–early Cambrian interval was, on the whole, a time of extremely high erosion rates that were probably greater than in any other period of Earth history (Kaufman et al. 1993; Derry et al. 1994). Moreover, during the latest Proterozoic these high rates of erosion were accompanied by high organic productivity and anoxic bottom-water conditions (Kaufman and Knoll 1995). Abundant ophiolites formed during the initial stage of the cycle (late Vendian), whereas mountain building, granitization, and the first Phanerozoic generation of volcano-plutonic marginal belts, were characteristic of its later part. The middle-late Ordovician peak of island-arc volcanic activity was confined to the Caledonian cycle (Khain and Seslavinsky 1994).

Accretion of Gondwana ended in the early Vendian. This process lasted for about 200 Ma. The Congo, Parana, Amazonia, Sahara, and other microcontinents became closer together, and manifestations of island arc volcanism in the Atakora, Red Sea, and the central Arabia zones were associated with this accretion. It is likely that rifting between South America sensu lato and Laurentia (North America, excluding Avalonian and other terranes, but including northwestern Scotland, northern Ireland, and western Svalbard) started at the end of the early Vendian. This process influenced the development of the South Oklahoma rift and ophiolite complexes of the southern Appalachians. At this time, the largest epicratonic sedimentary basin covered Siberia, which separated from Laurentia probably at the beginning or just before the early Vendian (Condie and Rosen 1994).

Glaciation was an important paleogeographic event in the early Vendian. At that time, most of the Gondwana fragments were located in polar latitudes and distribution of tillite horizons in modern North America and Europe (Varangerian Horizon) is in good agreement with such a reconstruction. When the reconstruction by Dalziel

Legend for Figures 3.1–3.6 The areas numbered in circles are as follows: *1*, Qilianshan zone; *2*, Shara-Moron zone of North China; *3*, Yunnan-Malaya zone; *4*, Cathaysian zone; *5*, southern Queensland–New South Wales; *6*, Thomson zone; *7*, Bowers Trough, Marie Byrd Land; *8*, Lachlan zone; *9*, Adelaide zone; *10*, West Antarctic zone; *11*, Argentinian and Chilean Cordillera; *12*, Patagonian Massif; *13*, Argentinian Precordillera; *14*, Argentinian Andes; *15*, Pampeanos Massif; *16*, Colombian Andes; *17*, Bolivian Andes; *18*, southern Ireland and Wales; *19*, Anti-Atlas; *20*, Iberia (West Asturias-León zone); *21*, Armorica-Massif Central (France); *22*, southern Balkans; *23*, Scandinavia; *24*, Finnmark Zone; *25*, southern Carpathians; *26*, North Caucasus zone; *27*, Urals; *28*, northern Scotland; *29*, East Greenland zone; *30*, northern Canada; *31*, western Koryak zone; *32*, Innuitian Belt; *33*, northwestern Alaska; *34*, southern Cordillera zone; *35*, South Oklahoma zone; *36*, Appalachian zone (*36a*, northern Appalachian zone; *36b*, southern and central Appalachian zones); *37*, southern margin of Siberia; *38*, Dzhida-Vitim zone; *39*, Mongolian-Amurian zone; *40*, Chingiz-Tarbagatay zone of Kazakhstan; *41*, eastern Tuva; *42*, Kuznetsky Alatau, Gorny Altay, western Tuva (Altay Sayan Foldbelt); *43*, Great and Little Hinggan zone; *44*, Taimyr; *45*, Saxo-Thuringian zone.

LEGEND

⌊LLLLL⌋	*Ophiolite Complexes*
⌊ᵛᵛᵛ⌋	*Island Arc Complexes*
⌊↗↗⌋	*Tectonic Deformation*
⌊⌐●⌐⌐⌐●⌐⌐⌐●⌐⌋	*Granitization*
⌊✗⌋	*Flysch (Turbidite) Complexes*
⌊°₀°₀°⌋	*Molasse*
⌊⁺₊⁺₊⁺⌋	*Regional Metamorphism*
⌊⌒ₙ⌒ₙ⌋	*Tillites*
⌊▼▼▼⌋	*Subduction Zones*
⌊░░░⌋	*Land*
⌊▬▬▬▬⌋	*Epicontinental Seas*
⌊⸜⸍⸍⌋	*Sutures*

et al. (1994) of the Neoproterozoic supercontinent Rodinia is considered, the glacigene deposits at ~600 Ma form a continuous belt from Scandinavia to Namibia, passing through Greenland, Scotland, eastern North America, Paraguay, Bolivia, western and southern Brazil, Uruguay, and Argentina (Eerola, this volume: figure 5.3). This zone could also probably be extended to Antarctica (Nimrod) and Australia (Marino Group, Kanmantoo Trough). The second region where tillites of the Varangerian glaciation are known (Australia and South China) is located in mid-latitudes on paleoreconstructions.

LATE VENDIAN (EDIACARIAN-KOTLIN)

The formation of Gondwana ended in the late Vendian (figure 3.1). Long mountain belts appeared at the sites of plate collisions in North America, Arabia, and the eastern part of South America. Molasse formed in intramontane depressions, and analysis of molasse distribution reveals that the late Vendian was an epoch of global orogeny (Khain and Seslavinsky 1995). During this time, rifting between South America sensu lato and Laurentia reached the middle and northern Appalachians (Keppie 1993). As in the early Vendian, the Siberian basin was the largest sedimentary shelf basin. In addition, extensive transgressions developed in Baltica and Arabia. By contrast, regressions commenced in northwestern and western Africa, and in North

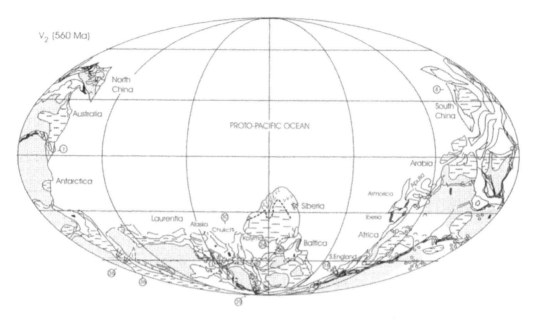

Figure 3.1 Paleogeography of the late Vendian (560 Ma). See legend on page 50.

China. The prevalence of passive continental margins in the peripheral parts of Gondwana should be noted.

New glaciations developed in circumpolar areas of Gondwana. The late Sinian (Baykonurian) glaciation covered Kazakhstan, Mongolia, and North China (Chumakov 1985), and the Fersiga glaciation expanded in West Africa and Brazil (Bertrand-Sarfati et al. 1995; Eerola 1995; Eerola, this volume). Except for Avalonia, Baltica, and Australia, where siliciclastic deposits accumulated, Late Vendian sedimentation was dominated by carbonates, commonly stromatolitic and oolitic dolostones. These were widespread in Siberia (Mel'nikov et al. 1989a; Astashkin et al. 1991), on the microcontinents of the Altay Sayan Foldbelt, Transbaikalia, Mongolia, Russian Far East (Astashkin et al. 1995), North and South China (Liu and Zhang 1993), Somalia, Near and Middle East (Gorin et al. 1982; Wolfart 1983; Hamdi 1995), Morocco (Geyer and Landing 1995), and the Canadian Cordillera (Fritz et al. 1991).

EARLY CAMBRIAN

The most important paleogeographic events of the Early Cambrian were the opening and relatively rapid widening of Iapetus (Bond et al. 1988; Harris and Fettes 1988), and the breakup of Laurasia into three large fragments—Laurentia, Siberia, and Baltica (Condie and Rosen 1994; Torsvik et al. 1996) (figure 3.2). Intense volcanic and tectonic processes occurred at this time, as well as in the late Vendian, along the northwestern periphery of Gondwana where rift-to-drift transition involved a num-

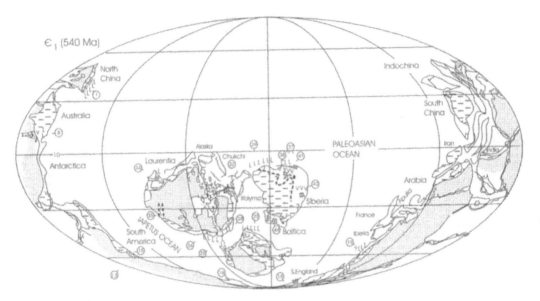

Figure 3.2 Early Cambrian paleogeography (540 Ma). See legend on page 50.

ber of central Asian microcontinents (Zavkhan, Tuva-Mongolia, South Gobi, North Tien Shan, etc.) (Mossakovsky et al. 1993). The total combination of tectonic, facies, paleomagnetic, and paleontologic data allow suggestion, contrary to the view of Şengör et al. (1993), that these blocks drifted from northwestern Gondwana to Siberia during this time interval (Didenko et al. 1994; Kheraskova 1995; Ruzhentsev and Mossakovsky 1995; Chuyko 1996; Evans et al. 1996; Svyazhina and Kopteva 1996). The results of a number of studies (stratigraphic, structural, isotope, and so on) combine to show that some segments (microcontinents or terranes) of the East Antarctic margin were also tectonically active, and there were allochthonous movements of such segments relative to each other and to the East Antarctic craton (Dalziel 1997).

In the Early Cambrian, there were no high mountain ranges in the central parts of Gondwana such as were present in the late Vendian; high hills and uplands dominated (Khain and Seslavinsky 1995). As for the Vendian, continental margins of Gondwana were passive, with the exception of small mobile belts in Australia, Antarctica, and North China (Qinlianshan zone) (Courjault-Radé et al. 1992; Kheraskova 1995). Siberia was the largest sedimentary inland basin; the second in size was South China. Carbonate sedimentation dominated on both of them, as well as in Morocco.

The Early Cambrian epoch is broadly subdivided into four phases, each of which was dominated by a characteristic type of sedimentation. During the Nemakit-Daldynian–Tommotian phase, phosphate-rich sediments occurred on a global scale. Areas of phosphate enrichment are now the sites of many prominent and even economically important phosphate deposits. Such areas were restricted to the northwestern (South China; Mongolian and Kazakhstan terranes) and southwestern (West

Africa and Iberia) regions of Gondwana (Parrish et al. 1986; Vidal et al. 1994; Culver et al. 1996). This pattern of phosphorite distribution in high-mid latitudes, and the restriction of phosphorites to the probable narrow rift zone of an incipient Paleoasian Ocean, closely match the model of upwelling of nutrient-rich and isotopically heavy brines onto continental margins (Donnelly et al. 1990). At the same time, very extensive evaporite basins occurred on subequatorial parts of Siberia (Turukhansk-Irkutsk-Olekma Basin) (Astashkin et al. 1991) and Gondwana (Oman–southern Iran–Saudi Arabia; northern Pakistan) (Wolfart 1983; Mattes and Conway Morris 1990). Other epicontinental seas were sites of siliciclastic accumulation, mainly fluviatile and deltaic. These were in Laurentia, including Svalbard (Holland 1971; Knoll and Swett 1987; Fritz et al. 1991), South America (Bordonaro 1992), Scandinavia and Baltica (Holland 1974; Rozanov and Łydka 1987), Avalonia (Landing et al. 1988), Iberia-Armorica (Pillola et al. 1994), the Montagne Noire–Sardinia fragment (Gandin et al. 1987), Turkey (Dean et al. 1993), and Australia (Shergold et al. 1985; Cook 1988).

During the next phase (Atdabanian), reddish carbonates became widespread on Siberia and some microcontinents of the Altay Sayan Foldbelt and Mongolia (Astashkin et al. 1991, 1995); Iberia, Germany, and Morocco (Moreno-Eiris 1987; Elicki 1995; Geyer and Landing 1995); Australia (Shergold et al. 1985); and Avalonia (Landing et al. 1988). On the whole, the Atdabanian-Botoman interval was the time of most widespread carbonate development in the Early Cambrian, mainly due to intense calcimicrobial-archaeocyath reef building within a belt extending on either side of the paleoequator from 30° north to 30° south (Debrenne and Courjault-Radé 1994).

In the early-middle Botoman, the Cambrian transgression reached its maximum extent (Gravestock and Shergold, this volume). This was marked by extensive accumulation of black shales and black finely bedded limestones in low latitudes (Siberia, some microcontinents of the Altay Sayan Foldbelt, Transbaikalia, Mongolia, the Russian Far East [Astashkin et al. 1991, 1995], Kazakhstan [Kholodov 1968], Iran [Hamdi 1995], Turkey [Dean et al. 1993], South Australia [Shergold et al. 1985], South China [Chen et al. 1982]) and by pyritiferous green shales or oolitic ironstones, commonly strongly pyritized, in temperate regions of Avalonia (Brasier 1995) and Baltica (Brangulis et al. 1986; Pirrus 1986), respectively. Features characteristic of transgression are observed in the sedimentary record of Iberia (Liñan and Gámez-Vintaned 1993), Germany (Elicki 1995), the Montagne Noir-Sardinia fragment (Gandin et al. 1987), Morocco (Geyer and Landing 1995), Tarim (Chang 1988), and Laurentia, including Greenland (Mansy et al. 1993; Vidal and Peel 1993).

The fourth phase, the late Botoman-Toyonian, was probably the time of major regression, variously known as the Hawke Bay, Daroka, or Toyonian regression. The Toyonian sedimentary record is characterized by widespread *Skolithos* pipe rock in Iberia (Gámez et al. 1991), Morocco (Geyer and Landing 1995), and eastern Laurentia (Palmer and James 1979) and by other intertidal siliciclastic rocks on Baltica

(Bergström and Ahlberg 1981; Brangulis et al. 1986) and in Iran (Hamdi 1995), Laurentia (Fritz et al. 1991; McCollum and Miller 1991; Mansy et al. 1993), and South China (Atlas 1985; Belyaeva et al. 1994). Sabkha conditions affected large areas of Siberia and Australia (Cook 1988; Mel'nikov et al. 1989b; Astashkin et al. 1991). Bimodal and acid volcanism occurred in Ossa-Morena, Normandy, and southern France (Pillola et al. 1994), as well as in the island arcs of central Asia (Kheraskova 1995).

MIDDLE CAMBRIAN

In the Middle Cambrian, Laurentia continued to drift toward the equator, while Iapetus became wider (figure 3.3). Ophiolites, reflecting the spreading of Iapetus, are found in the Appalachians, Scandinavia, and perhaps the North Caucasus (Belov 1981; Harris and Fettes 1988). By Middle Cambrian time, mountain ridges of collisional orogenic systems in Africa and South America had been eroded, and Gondwana became a vast plateau (Khain and Seslavinsky 1995). Subsequently, island arc systems developed widely on the Gondwana margins facing the Panthalassa Ocean. They include the volcanic arcs of North China, southeastern Australia, Antarctica, the Cordillera of Chile and Argentina, and possibly the Cordillera of Colombia (Aceñolaza and Miller 1982; *Atlas* 1985; Rowell et al. 1992; Gravestock and Shergold, this volume), where submarine andesite and basalt and marine sedimentary-volcanogenic complexes formed. Carbonate sedimentation, however, was still widely developed in marginal basins of Gondwana, and conditions of almost exclusively carbonate sedimentation existed in inland shelf basins such as those in Siberia and in North and South China (Courjault-Radé et al. 1992).

At the beginning of the Middle Cambrian (Amgan stage), a general sinking of carbonate ramps is expressed by the accumulation of black and other deeper-water shales in Siberia, northern Mongolia-Transbaikalia, the Russian Far East (Astashkin et al. 1991, 1995), the Baykonur-Karatau province of Kazakhstan (Kheraskova 1995), Pakistan and Turkey (Wolfart 1983; Dean et al. 1993), Iberia (Liñan and Quesada 1990), Scandinavia (Holland 1974), Novaya Zemlya (Andreeva and Bondarev 1983), and Avalonia (Thickpenny and Leggett 1987). Distinct deepening was typical of large parts of southwestern Gondwana, including the Montagne Noire–Sardinia fragment and large parts of Iberia and Morocco (Bechstädt and Boni 1994; Geyer and Landing 1995), as well as of South America (Bordonaro 1992). On Laurentia, transgression of the western margin and subsequent reduction of terrigenous input to the shelf led to the development of extensive carbonate platforms (Bond et al. 1989; Mansy et al. 1993). The largest, although extremely shallow-water, basin occupied Baltica (Dmitrovskaya 1988). The last major phase of Cambrian phosphogenesis was related to this globally recognizable sea level rise (Freeman et al. 1990).

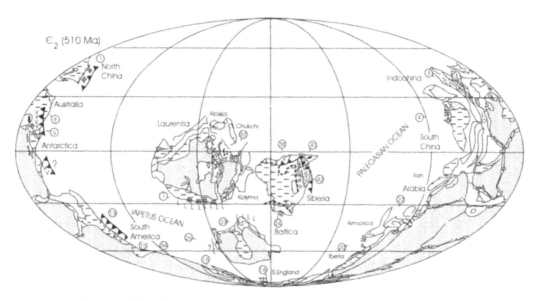

Figure 3.3 Middle Cambrian paleogeography (510 Ma). See legend on page 50.

On the whole, the late Middle Cambrian (Marjuman stage) was the time of the most intense tectonic activity in the Cambrian, coinciding with the peak of the Salairian orogeny (Seslavinsky 1995). Island arc calc-alkali, mostly subaerial volcanism was especially abundant. During the Marjuman stage, collapse of carbonate platforms and establishment of mixed-sediment shelves occurred in Iberia (Gámez et al. 1991; Sdzuy and Liñan 1993), Morocco (Geyer and Landing 1995), and Siberia. By that time, accretion of the Altay Sayan, Mongolia, and Baikal-Patom region had mostly ended, and the constituent fragments were added to Siberia. This collision led to formation of an elongate semicircular orogenic belt around Siberia, from Salair to Transbaikalia. Rugged mountain relief was formed here, and extensive molasse developed (Kremenetskiy and Dalmatov 1988; Astashkin et al. 1995; Kheraskova 1995). Sedimentation in the remainder of these regions was characterized by shallow-water sandstones, arkoses, and conglomerates. The regression on Siberia, a large part of which was covered by a subaerial plain (Budnikov et al. 1995), was one of the consequences of collision. Similar sedimentary features are observed in Novaya Zemlya (Andreeva and Bondarev 1983) and Turkey (Dean et al. 1993). Major hiatuses are typical of Avalonia and Baltica (Rushton 1978; Brangulis et al. 1986), and in Scandinavia the monotonous Alum Shale temporarily gave way to formation of the Andrarum Limestone (Harris and Fettes 1988). Quartzites and dolostones are the principal lithologies of the Canadian Cordillera (Fritz et al. 1991). Reef building was restricted to the Anabar-Sinsk Basin of Siberia and some parts of Laurentia, North China, and Iran, but the reefs were mainly thrombolitic (Hamdi et al. 1995).

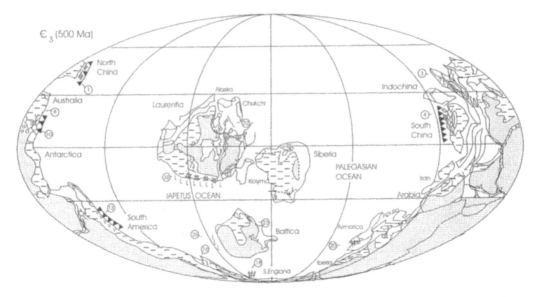

Figure 3.4 Late Cambrian paleogeography (500 Ma). See legend on page 50.

LATE CAMBRIAN

In the Late Cambrian, Laurentia and Siberia continued to drift toward the equator (figure 3.4). The Iapetus Ocean increased in area and had its maximum width at this time (Harris and Fettes 1988; Khain and Seslavinsky 1995). Two additional oceans, the Panthalassa and the Paleoasian, also existed. The imbrication of oceanic crust on the periphery of Siberia and the central Kazakhstan terranes continued during the Late Cambrian. It was probably related to marked spreading of the ocean floor between the North China and Bureya-Khanka, South Gobi, and Central Mongolia terranes, where new oceanic crust continued to grow and where accumulation of siliciclastics, often ore-bearing formations, and, on local uplifts, reef limestones occurred. During this time, the Bureya-Khanka terrane appears to have amalgamated into the single large Amur Massif (Amuria), where coarse molasse developed and orogenic acid volcanism occurred (Kheraskova 1995).

Widespread transgression on Laurentia was an important Late Cambrian event, and shelf basins covered vast areas of the midcontinent from the Cordillera to the Appalachians (Link 1995; Long 1995). Marine conditions were reestablished over the whole of Siberia (Budnikov et al. 1995), but the proportions of siliciclastic sediments increased in marginal sedimentary basins (Markov 1979). Scarcity of marine Late Cambrian deposits, and the onset of bimodal subaerial volcanism, are indicative of uplift of the peri-African shelf of Gondwana (Liñan and Gámez-Vintaned 1993; Geyer and Landing 1995; Buschmann and Linnemann 1996). A general restriction of marine basins occurred in Australia and in North and South China (*Atlas* 1985; Cook 1988).

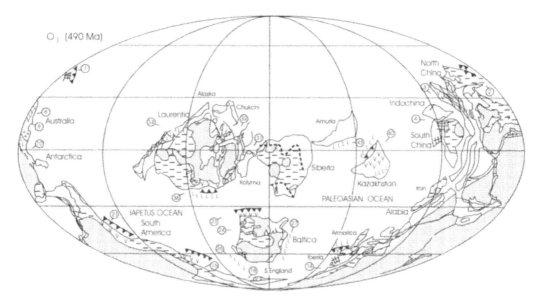

Figure 3.5 Early Ordovician paleogeography (490 Ma). See legend on page 50.

EARLY ORDOVICIAN

In the Early Ordovician, Laurentia was located at the same position in the equatorial zone, Siberia drifted slightly northward, and Gondwana started drifting toward Laurentia (figure 3.5). Opening of Iapetus either ended or continued very slowly. Subduction zones and island arc systems of the northern Appalachians, Scandinavia, Armorica, and the Andes formed in marginal parts of the approaching plates. It is likely that opening of the Rheic Ocean started in the Early Ordovician as a result of rifting and, later, spreading between new volcanic arcs of Avalonia and Gondwana (Keppie 1993). Baltica rotated counterclockwise, accompanied by opening of the Uralian mobile belt, where intense rifting and accumulation of graben facies occurred followed by the initiation of the Ural Ocean. Early Ordovician graben facies comprise shallow quartzose sandstones and arkoses with subordinate subalkali basalts, which later were replaced by deeper-water siliceous passive continental margin sediments (Samygin 1980; Maslov et al. 1996; Maslov and Ivanov 1998). Kazakhstan can be identified for the first time on Early Ordovician reconstructions as an entity of several microcontinents (Appollonov 1995).

About half of Gondwana's continental margins developed as active margins. The evidence for this is complexes of island arc volcanism, granitization, metamorphism, and tectonic deformation. Siliciclastics dominated the sedimentary deposits of the marginal basins, and the proportion of turbidites increased. Terrigenous sedimentation totally replaced carbonates in inland shelf basins of Siberia, China, Australia, and West Gondwana (*Atlas* 1985; Cook 1988; Wensink 1991; Budnikov et al. 1995), and carbonate sedimentation continued only in Laurentia (Holland 1971; Long 1995).

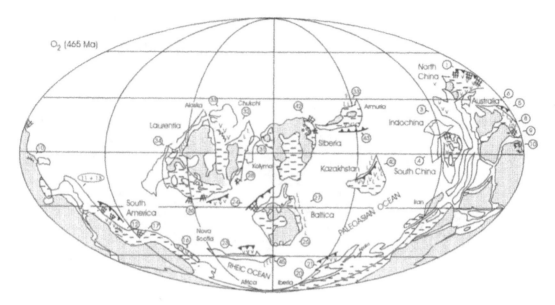

Figure 3.6 Middle Ordovician paleogeography (465 Ma). See legend on page 50.

MIDDLE ORDOVICIAN

Iapetus became narrower in the Middle Ordovician (figure 3.6), and drift directions changed for all plates. This was the epoch of global activation of tectonic processes (Khain and Seslavinsky 1994). Laurentia was approaching Gondwana, while Baltica continued its counterclockwise rotation and moved from Gondwana to the equator. Siberia drifted northward significantly. By the Middle Ordovician, Avalonia had compensated for the closure of the Iapetus Ocean. Complicated processes of interaction between the Appalachian mobile belt and the Iapetus lithosphere resulted in formation of the volcanic Popologan arc (Van Staal 1994), where the subduction zone dipped southeast toward the oceanic plate.

The Middle Ordovician is characterized by sea level rise and the largest global transgression in the Paleozoic (Algeo and Seslavinsky 1995). Seas covered vast areas along the margins of Gondwana. Prominent carbonate platforms developed in Siberia and Baltica, in the Canadian Cordillera, South America, Spitsbergen, and West Gondwana (Dmitrovskaya 1989; Fritz et al. 1991; Wensink 1991; Aceñolaza 1992; Harland et al. 1992; Khain and Seslavinsky 1995). This was a time of extremely widespread turbidite deposition.

POSTSCRIPT

Narrowing of Iapetus was compensated for in the Late Ordovician by opening of the Rheic Ocean, while the mutual positions of Laurentia, Baltica, and Gondwana did not change. This is an interesting feature of Earth evolution in that there were no significant changes in relative plate positions for 25 Ma. This standstill did not occur during the entire history described above. Probably it reflects some reduction in the total intensity of endogenic processes. However, spreading and subduction processes were not completely halted. Island arcs formed in the east, west, and north of Laurentia, in the Appalachian, Cordilleran, and Innuitian belts, respectively, and in some regions of the Gondwana continental margin; and the evolution of southeastern Australia and the Qinlianshan zone of North China continued, which by this time was of about 100 Ma duration.

The Iapetus Ocean was still open at the end of the Ordovician. It subsequently closed at the end of the Silurian. It is obvious that this 150 Ma period of Earth history was a cycle of opening of new oceans and their subsequent closing. These processes of the Caledonian cycle were accompanied by formation of spreading and subduction zones, and systems of island arcs and orogenic belts replaced each other successively in time and space. These features combined to determine global paleogeography during this period. The principal features were the very extensive Panthalassa Ocean and the smaller Iapetus, Rheic, and Paleoasian oceans. Numerous global tectonic, magmatic, and sedimentary events occurred within, and at the margins of, these oceans.

Acknowledgments. Plate tectonic reconstructions were kindly provided by C. R. Scotese. We very much appreciate the help of all colleagues in IGCP Project 319: data were prepared by D. Long for Canada, Zhang Qi Rui for China, N. V. Mel'nikov and S. S. Sukhov for the Siberian Platform, Yu. E. Dmitrovskaya for the eastern Baltic Platform, K. Mens and L. Hints for the Baltic states and Scandinavia, and N. M. Chumakov for Vendian tillites. New information on geologic events in mobile belts was obtained by A. V. Maslov and K. S. Ivanov for the Urals, by T. N. Kheraskova for central Asia, by P. Link for the Cordillera of the United States, by J. A. Gámez-Vintaned for Spain, by M. D. Brasier for Great Britain, and by other researchers. Many thanks go to Prof. V. E. Khain for long-term collaboration and for sharing ideas and data. We are very grateful to T. N. Kheraskova, A. G. Smith, and A. Yu. Zhuravlev for helpful comments and review of the manuscript. We thank N. B. Kuznetsov and I. N. Makarova for technical assistance. Research funding was provided by International Science Foundation Project MLW 300 and Russian Foundation for Basic Research Project 95-05-14545. This paper is a contribution to IGCP Projects 319 and 366.

REFERENCES

Aceñolaza, F. G. 1992. El Sistema ordovícico de Latinoamérica. In J. G. Gutiérrez Marco, J. Saavedra, and I. Rábano, eds., *Paleozoico Inferior de Ibero-América*, pp. 85–118. Universidad de Extremadura, Spain.

Aceñolaza, F. G. and H. Miller. 1982. Early Paleozoic orogeny in southern South America. *Precambrian Research* 17:133–146.

Algeo, T. J. and K. B. Seslavinsky. 1995. The Paleozoic world: Continental flooding, hypsometry, and sea level. *American Journal of Science* 295:520–541.

Andreeva, I. A. and V. I. Bondarev. 1983. Nizhniy-sredniy paleozoy tsentral'noy chasti Novoy Zemli [Lower-middle Paleozoic in the central part of Novaya Zemlya]. In V. I. Bondarev, ed., *Paleontologicheskoe obosnovanie raschleneniya paleozoya i mezozoya arkticheskikh rayonov SSSR* [Paleontological grounds for the subdivision of the Paleozoic and Mesozoic in the Arctic USSR], pp. 5–15. Leningrad: PGO Sevmorgeologiya.

Appollonov, M. K. 1995. The evolution of tectonic structure, environments, and communities of fauna in the Ordovician of Kazakhstan. In J. D. Cooper, M. L. Droser, and S. C. Finney, eds., *Ordovician Odyssey: Short Papers for the Seventh International Symposium on the Ordovician System (Las Vegas, Nevada, USA, June 1995)*, pp. 456–466. Fullerton, Calif.: Pacific Section Society for Sedimentary Geology.

Astashkin, V. A., T. V. Pegel', L. N. Repina, A. Yu. Rozanov, Yu. Ya. Shabanov, A. Yu. Zhuravlev, S. S. Sukhov, and V. M. Sundukov. 1991. *The Cambrian System on the Siberian Platform*. International Union of Geological Sciences, Publication 27.

Astashkin, V. A., G. V. Belyaeva, N. V. Esakova, D. V. Osadchaya, N. N. Pakhomov, T. V. Pegel', L. N. Repina, A. Yu. Rozanov, and A. Yu. Zhuravlev. 1995. *The Cambrian System of the Foldbelts of Russia and Mongolia*. International Union of Geological Sciences, Publication 32.

Atlas of the Paleogeography of China. 1985. Institute of Geology, Chinese Academy of Geological Sciences, Wuhan College of Geology. Beijing: Cartographic Publishing House.

Bechstädt, T. and M. Boni, eds. 1994. *Sedimentological, Stratigraphical, and Ore Deposits Field Guide of the Autochthonous Cambro-Ordovician of Southwestern Sar-*

dinia. Memorie Descrittive della Carta Geologica d'Italia, no. 48.

Belov, A. A. 1981. *Tektonicheskoe razvitie Al'piyskoy skladchatoy oblasti v paleozoe* [Tectonic development of the Alpine Foldbelt in the Paleozoic]. Trudy, Geologicheskiy institut, Akademiya nauk SSSR, no. 347.

Belyaeva, G. V., Y. Tian, K.-X. Yuan, and A.-D. Xu. 1994. Nizhniy kembriy severa platformy Yantsy: Raschlenenie i korrelyatsiya po arkheotsiatam s razrezami yugovostochnoy Rossii [Lower Cambrian on the north of the Yangtze Platform: Subdivision and correlation according to archaeocyaths with sections on southeastern Russia]. *Tikhookeanskaya geologiya* 1994 (5):48–59.

Bergström, J. and P. Ahlberg. 1981. Uppermost Lower Cambrian biostratigraphy in Scania, Sweden. *Geologiska Föreningens i Stockholm Förhandlingar* 103:193–214.

Bertrand-Sarfati, J., A. Moussine-Pouchkine, B. Amard, and A. Ait Kaci Ahmed. 1995. First Ediacaran fauna found in western Africa and evidence for an early Cambrian glaciation. *Geology* 23:133–136.

Bond, G. C., M. A. Kominz, and J. P. Grotzinger. 1988. Cambro-Ordovician eustasy: Evidence from geophysical modeling of subsidence in Cordilleran and Appalachian passive margins. In K. L. Kleinspehn and C. Paola, eds., *New Perspectives in Basin Analysis*, pp. 129–159. New York: Springer Verlag.

Bond, G. C., M. A. Kominz, M. S. Steckler, and J. P. Grotzinger. 1989. Role of thermal subsidence, flexure, and eustasy in the evolution of early Paleozoic passive-margin carbonate platforms. In P. D. Crevello, J. L. Wilson, J. F. Sarg, and J. F. Read, eds., Controls on carbonate platform and basin development, *Society of Economic Paleontologists and Mineralogists Special Publication* 44:39–61.

Bordonaro, O. L. 1992. El Cámbrico de Sudamérica. In J. G. Gutiérrez Marco, J. Saavedra, and I. Rábano, eds., *Paleozoico Inferior de Ibero-América,* pp. 69–84. Universidad de Extremadura, Spain.

Brangulis, A., A. Murnieks, A. Natle, and A. Fridrihsone. 1986. Srednebaltiyskiy fatsial'nyy profil' venda i kembriya [Middle Baltic facies profile of the Vendian and Cambrian]. In E. A. Pirrus, ed., *Fatsii i stratigrafiya venda i kembriya zapada Vostochno-Evropeyskoy platformy* [Facies and stratigraphy of the Vendian and Cambrian of the western East-European Platform], pp. 24–33. Tallin: Academy of Sciences of Estonian SSR.

Brasier, M. D. 1995. The basal Cambrian transition and Cambrian bio-events (from terminal Proterozoic extinctions to Cambrian biomeres. In O. H. Walliser, ed., *Global Events and Event Stratigraphy in the Phanerozoic,* pp. 113–118. Berlin: Springer Verlag.

Budnikov, I. V., S. S. Sukhov, V. P. Devyatov, T. V. Lopushinskaya, B. B. Shishkin, A. G. Yadrenkina, and A. M. Kasakov. 1995. Stratigraphy and lithology of sedimentary cover from the western Yakutia. In *Sixth International Kimberlite Conference, Kimberlites of Yakutia, Field Guide Book,* pp. 27–34. Novosibirsk, Russia.

Buschmann, B. and U. Linnemann. 1996. Geotectonic aspects of Vendian–early Paleozoic paleogeography in the European Variscides. In *Paleogeografiya venda-rannego paleozoya* [Vendian–early Paleozoic paleogeography], pp. 173–175. Ekaterinburg: Uralian Branch, Russian Academy of Sciences.

Chang, W. T. 1988. *The Cambrian System in Eastern Asia.* International Union of Geological Sciences, Publication 24.

Chen, N.-S., X.-Z. Yang, D.-H. Liu, X.-J. Xiao, D.-L. Fan, and L.-F. Wang. 1982. Lower Cambrian black argillaceous and arenaceous rock series in South China and its associated stratiform deposits. *Mineral Deposits* 1 : 39–51.

Chumakov, N. M. 1985. Laplandskiy lednikovyy gorizont i ego analogy [Laplandian glacial horizon and its analogies]. In B. S. Sokolov and M. A. Fedonkin, eds., *Vendskaya sistema. Istoriko-geologicheskoe i paleontologicheskoe obosnovanie,* vol. 2, *Stratigrafiya i geologicheskie protsessy,* pp. 167–198. Moscow: Nauka [English translation: B. S. Sokolov and A. B. Iwanowski, eds. 1990. *Regional Geology,* vol. 2 of *The Vendian System,* pp. 191–225. Berlin: Springer Verlag].

Chuyko, V. S. 1996. Paleomagnitnye shiroty nekotorykh blokov kontinental'noy kory vostochnoy Azii v vende-kembrii [Paleomagnetic latitudes of some continental crust blocks from eastern Asia in the Vendian-Cambrian]. *Paleogeografiya venda-rannego paleozoya* [Vendian–early Paleozoic paleogeography], pp. 164–166. Ekaterinburg: Uralian Branch, Russian Academy of Sciences.

Condie, K. C. and O. M. Rosen. 1994. Laurentia-Siberia connection revisited. *Geology* 22 : 168–170.

Coney, P. J., D. L. Jones, and J. W. H. Monger. 1980. Cordilleran suspect terranes. *Nature* 288 : 329–333.

Cook, P. J., ed. 1988. *Cambrian,* vol. 1 of *Palaeogeographic Atlas of Australia.* Canberra: Australian Government Publishing Survey.

Courjault-Radé, P., F. Debrenne, and A. Gandin. 1992. Paleogeographic and geodynamic evolution of the Gondwana continental margins during the Cambrian. *Terra Nova* 4 : 657–667.

Cowie, J. W. 1971. Lower Cambrian faunal provinces. In F. A. Middlemiss, P. F. Rawson, and G. Newall, eds., *Faunal Provinces in Space and Time,* pp. 31–46. *Geological Journal, Special Issue, London* 4.

Culver, S. J., J. E. Repetski, J. Pojeta, Jr., and D. Hunt. 1996. Early and Middle (?)Cambrian metazoan and protistan fossils from West Africa. *Journal of Paleontology* 70:1–6.

Dalziel, I. W. D. 1994. Precambrian Scotland as a Laurentia-Gondwana link: Origin and significance of cratonic promontories. *Geology* 22:589–592.

Dalziel, I. W. D. 1997. Neoproterozoic-Paleozoic geography and tectonics: Review, hypothesis, environmental perturbations. *Geological Society of America Bulletin* 109:16–42.

Dalziel, I. W. D., L. H. Dalla Salda, and L. M. Gahagan. 1994. Paleozoic Laurentia-Gondwana interaction and the origin of the Appalachian-Andean mountain system. *Geological Society of America Bulletin* 106:243–252.

Dean, W. T., F. Martin, O. Monod, M. A. Gül, N. Bozdogan, and N. Özgül. 1993. Early Palaeozoic evolution of the Gondwanaland margin in the western and central Taurids, Turkey. In S. Turgut, ed., *Tectonics and Hydrocarbon Potential of Anatolia and Surrounding Regions*, Ozan Sungurlu Symposium Proceedings, November 1991, pp. 262–273. Ankara.

Debrenne, F. and P. Courjault-Radé. 1994. Répartition paléogéographique des archéocyathes et délimitation des zones intertropicales au Cambrien inférieur. *Bulletin de la Société géologique de France* 165:459–467.

Debrenne, F., I. D. Maidanskaya, and A. Yu. Zhuravlev. 1999. Faunal migrations of archaeocyaths and Early Cambrian plate dynamics. *Bulletin de la Société géologique de France* 170:189–194.

Derry, L. A., M. D. Brasier, R. M. Corfield, A. Yu. Rozanov, and A. Yu. Zhuravlev. 1994. Sr and C isotopes in Lower Cambrian carbonates from the Siberian craton: A paleoenvironmental record during the "Cambrian explosion." *Earth and Planetary Science Letters* 128:671–681.

Didenko, A. N., A. A. Mossakovsky, D. M., Pecherskiy, S. V. Ruzhentsev, S. G. Samygin, and T. N. Kheraskova. 1994. Geodinamika paleozoyskikh okeanov Tsentral'noy Azii [Geodynamics of the Paleozoic oceans of the central Asia]. *Geologiya i geofizika* 35 (7–8):29–40.

Dmitrovskaya, Yu. E. 1988. Novye dannye po stratigrafii nizhnego paleozoya Moskovskoy sineklizy. Stat'ya 1: Kembriy [New data on the Lower Paleozoic stratigraphy of the Moscow syneclise. Paper 1: The Cambrian]. *Byulleten' Moskovskogo obshchestva ispytateley prirody, Otdel geologicheskiy* 63 (2):47–54.

Dmitrovskaya, Yu. E. 1989. Novye dannye po stratigrafii nizhnego paleozoya Moskovskoy sineklizy. Stat'ya 2: Ordovik [New data on the Lower Paleozoic stratigraphy of the Moscow syneclise. Paper 2: The Ordovician]. *Byulleten' Moskovskogo obshchestva ispytateley prirody, Otdel geologicheskiy* 64 (2):82–93.

Donnelly, T. H., J. H. Shergold, P. N. Southgate, and C. J. Barnes. 1990. Events leading to global phosphogenesis around the Proterozoic/Cambrian boundary. In A. J. G. Notholt and I. Jarvis, eds., *Phosphorite Research and Development*, pp. 273–287. Geological Society Special Publication 52.

Eerola, T. T. 1995. From ophiolites to glaciers? Review on geology of the Neoproterozoic-Cambrian Lavras Do Sul region, Southern Brazil. *Geological Survey of Finland, Special Paper* 20:5–16.

Elicki, O. 1995. Fazie und stratigraphische Stellung des deutschen Unterkambriums. *Zentralblatt für Geologie und Paläontologie*, Teil 1 (1/2):245–255.

Evans, D. A., A. Yu. Zhuravlev, C. J. Budney, and J. L. Kirschvink. 1996. Palaeomagnetism of the Bayan Gol Formation, western

Mongolia. *Geological Magazine* 133:487–496.

Freeman, M. J., J. H. Shergold, D. J. Morris, and M. R. Walter. 1990. Late Proterozoic and Palaeozoic basins of central and northern Australia—regional geology and mineralization. In F. E. Hughes, ed., *Geology of the Mineral Deposits of Australia and Papua New Guinea,* pp. 1125–1133. Melbourne: Australasian Institute of Mining and Metallurgy.

Fritz, W. H., M. P. Cecile, B. S. Norford, D. Morrow, and H. H. J. Geldsetzer. 1991. Cambrian to Middle Devonian assemblages. In H. Gabrielse and C. J. Yorath, eds., *Geology of the Cordilleran Orogen of Canada,* vol. 4 of *Geology of Canada,* pp. 151–218. Ottawa: Geological Survey of Canada.

Gabrielse, H. and C. J. Yorath, eds. 1991. *Geology of the Cordilleran Orogen of Canada,* vol. 4 of *Geology of Canada.* Ottawa: Geological Survey of Canada.

Gámez, J. A., C. Fernandez-Nieto, R. Gozalo, E. Liñan, J. Mandado, and T. Palacios. 1991. Biostratigrafía y evolución ambiental del Cámbrico de Borobia (Provincia de Soria, Cadena Ibérica Oriental). *Cuadernos do Laboratorio Xeolóxico de Laxe, Coruña* 16:251–271.

Gandin, A., N. Minzoni, and P. Courjault-Radé. 1987. Shelf to basin transition in the Cambrian–Lower Ordovician of Sardinia (Italy). *Geologische Rundschau* 76:827–836.

Geyer, G. and E. Landing. 1995. The Cambrian of the Moroccan Atlas regions. *Beringeria Special Issue* 2:7–46.

Gorin, G. E., L. G. Racz, and M. R. Walter. 1982. Late Precambrian–Cambrian sediments of Huqf Group, Sultanate of Oman. *American Association of Petroleum Geologists, Bulletin* 66:2602–2627.

Gubanov, A. 1998. The Early Cambrian mol-luscs and their palaeogeographic implications. *Schriften des Staatlichen Museums für Mineralogie und Geologie zu Dresden* 9:139.

Hamdi, B. 1995. Precambrian-Cambrian deposits in Iran. *Treatise on the Geology of Iran* 20.

Hamdi, B., A. Yu. Rozanov, and A. Yu. Zhuravlev. 1995. Latest Middle Cambrian metazoan reef from northern Iran. *Geological Magazine* 132:367–373.

Harland, W. B., R. A. Scott, K. A. Auckland, and I. Snape. 1992. The Ny Friesland Orogen, Spitsbergen. *Geological Magazine* 129:679–708.

Harris, A. L. and D. J. Fettes, eds. 1988. *The Caledonian-Appalachian Orogen.* Geological Survey Special Publication 38.

Holland, C. H., ed. 1971. *Cambrian of the New World.* London: John Wiley and Sons.

Holland, C. H., ed. 1974. *Cambrian of the British Isles, Norden, and Spitsbergen.* London: John Wiley and Sons.

Jell, P. A. 1974. Faunal provinces and possible planetary reconstruction of the Middle Cambrian. *Journal of Geology* 82:319–350.

Kaufman, A. J. and A. H. Knoll. 1995. Neoproterozoic variations in the C-isotopic composition of seawater: Stratigraphic and biogeochemical implications. *Precambrian Research* 73:27–49.

Kaufman, A. J., S. J. Jacobsen, and A. H. Knoll. 1993. The Vendian record of Sr and C isotopic variations in seawater: Implications for tectonic and paleoclimate. *Earth and Planetary Science Letters* 120:409–430.

Keppie, J. D. 1993. Synthesis of Paleozoic deformational events and terranes accretion in the Canadian Appalachians. *Geologische Rundschau* 82:381–431.

Khain, V. E. and K. B. Seslavinsky. 1994. Global rhythms of the Phanerozoic endogenic activity of the Earth. *Stratigraphy and Geological Correlation* 2:520–541.

Khain, V. E. and K. B. Seslavinsky. 1995. *Historical Geotectonics: Paleozoic.* New Delhi: Oxford and IBH Publishing Co.

Khanchuk, A. and G. Belyaeva. 1993. Relationship between the terranes of Paleoasian and Paleopacific oceans in the Far East, Russia. In N. L. Dobretsov and N. A. Berzin, eds., *Fourth International Symposium on the Geodynamic Evolution of the Paleoasian Ocean, Abstracts, 15–24 June 1993, Novosibirsk, Russia,* pp. 84–86. IGCP Project 283, Report 4. Novosibirsk: United Institute of Geology, Geophysics, and Mineralogy, Siberian Branch, Russian Academy of Sciences.

Kheraskova, T. N. 1995. Paleogeography of Central Asia paleoocean in Vendian and Cambrian. In M. D. Rodríguez Alonso and J. C. Gonzalo Coral, eds., *XIII Reunión de Geología del Oeste Peninsula: Caracterización y evolución de la cuenca Neoproterozoico-Cámbrico en la Península Ibérica, 19–30 de Septiembre de 1995,* pp. 77–80. Salamanca-Coimbra.

Kholodov, V. N. 1968. *Obrazovanie rud v osadkakh i metallogeniya vanadiya* [The genesis of ores in sediments and metallogeny of vanadium]. Moscow: Nauka.

Kirschvink, J. L. 1992. A paleogeographic model for Vendian and Cambrian time. In J. W. Schopf and C. Klein, eds., *The Proterozoic Biosphere: A Multidisciplinary Study,* pp. 569–581. Cambridge: Cambridge University Press.

Kirschvink, J. L., R. L. Ripperdan, and D. A. Evans. 1997. Evidence for a large-scale reorganization of Early Cambrian continental masses by inertial interchange true polar wander. *Science* 277:541–545.

Knoll, A. H. and K. Swett. 1987. Micropaleontology across the Precambrian-Cambrian boundary in Spitsbergen. *Journal of Paleontology* 61:898–926.

Kremenetskiy, I. G. and B. A. Dalmatov. 1988. Novye dannye po stratigrafii Vostochnogo Pribaykal'ya: Kembriy, podstilayushchie i perekryvayushchie ego otlozheniya [New data on the stratigraphy of eastern Cisbaikalia: Cambrian, its underlying and overlying strata]. *Trudy, Institut geologii i geofiziki, Sibirskoe otdelenie, Akademiya nauk SSSR* 720:83–97.

Landing, E., G. M. Narbonne, P. Myrow, A. P. Benus, and M. M. Anderson. 1988. Faunas and depositional environments of the upper Precambrian through Lower Cambrian, southeastern Newfoundland. In E. Landing, G. M. Narbonne, and P. Myrow, eds., *Trace Fossils, Small Shelly Fossils, and the Precambrian-Cambrian Boundary,* pp. 18–52. *New York State Museum Bulletin* 463.

Liñan, E. and J. A. Gámez-Vintaned. 1993. Lower Cambrian palaeogeography of the Iberian Peninsula and its relations with some neighbouring European areas. *Bulletin de la Société géologique de France* 164:831–842.

Liñan, E. and C. Quesada. 1990. Rift phase (Cambrian). In R. D. Dallmeyer and E. Martinez Garcia, eds., *Pre-Mesozoic Geology of Iberia,* pp. 257–266. Berlin: Springer Verlag.

Link, P. K. 1995. Vendian and Cambrian paleogeography of the western United States and adjacent Sonora, Mexico. In M. D. Rodríguez Alonso and J. C. Gonzalo Coral, eds., *XIII Reunión de Geología del Oeste Peninsula: Caracterización y evolución de la cuenca Neoproterozoico-Cámbrico en la Península Ibérica, 19–30 de Septiembre de 1995,* pp. 91–95. Salamanca-Coimbra.

Liu, H. and Q. R. Zhang. 1993. The Sinian System in China. *Advances in Sciences of China, Earth Sciences* 3:1–24.

Long, D. J. F. 1995. Late Neoproterozoic to Early Ordovician paleogeography of Canada and Greenland. In M. D. Rodríguez

Alonso and J. C. Gonzalo Coral, eds., *XIII Reunión de Geología del Oeste Península: Caracterización y evolución de la cuenca Neoproterozoico-Cámbrico en la Península Ibérica, 19–30 de Septiembre de 1995*, pp. 96–101. Salamanca-Coimbra.

Mansy, J.-L., F. Debrenne, and A. Yu. Zhuravlev. 1993. Calcaires à archéocyathes du Cambrien inférieur du Nord de la Colombie britannique (Canada): Implications paléogéographiques et précisions sur l'extension du continent Américano-Koryakien. *Géobios* 26:643–683.

Markov, E. P. 1979. Paleogeografiya rannego ordovika Sibirskoy platformy [Early Ordovician paleogeography of the Siberian Platform]. In N. V. Mel'nikov and B. B. Grebenyuk, eds., *Zakonomernosti razmeshcheniya skopleniy nefti i gaza na Sibirskoy platforme* [*Regularities in the distribution of the oil and gas accumulations on the Siberian Platform*], pp. 32–42. Novosibirsk: Siberian Scientific-Research Institute of Geology, Geophysics, and Mineral Resources.

Maslov, A. V. and K. S. Ivanov. 1998. Paleogeografiya i osnovnye tektonicheskie sobytiya rifeya–rannego paleozoya na Yuzhnom Urale [Riphean–early Paleozoic paleogeography and principal tectonic events on the South Urals]. In V. A. Koroteev and A. V. Maslov, eds., *Paleogeografiya venda–rannego paleozoya Severnoy Evrazii* [*Vendian–early Paleozoic paleogeography of Northern Eurasia*], pp. 8–24. Ekaterinburg: Uralian Branch, Russian Academy of Sciences.

Maslov, V. A., O. V. Artyushkova, and R. R. Yakupov. 1996. Paleogeografiya ordovica Yuzhnogo Urala [Ordovician paleogeography of the South Urals]. In *Paleogeografiya venda-rannego paleozoya* [*Vendian–early Paleozoic paleogeography*], pp. 102–103. Ekaterinburg: Uralian Branch, Russian Academy of Sciences.

Mattes, B. W. and S. Conway Morris. 1990. Carbonate/evaporite deposition in the late

Precambrian–Early Cambrian Ara Formation of Southern Oman. In A. H. F. Robertson, M. P. Searl, and A. C. Ries, eds., *The Geology and Tectonics of the Oman Region*, pp. 617–636. Geological Society Special Publication 49.

McCollum, L. B. and D. M. Miller. 1991. Cambrian stratigraphy of the Wendover Area, Utah and Nevada. *United States Geological Survey, Bulletin* 1948:1–43.

McKerrow, W. S., C. R. Scotese, and M. D. Brasier. 1992. Early Cambrian continental reconstructions. *Journal of the Geological Society, London* 149:599–606.

Mel'nikov, N. V., G. G. Shemin, and A. O. Efimov. 1989a. Paleogeografiya Sibirskoy platformy v vende [Vendian paleogeography of the Siberian Platform]. In R. G. Matukhin, ed., *Paleogeografiya fanerozoya Sibiri* [Phanerozoic paleogeography of Siberia], pp. 3–10, 4 figs. Novosibirsk: Siberian Scientific-Research Institute of Geology, Geophysics, and Mineral Resources.

Mel'nikov, N. V., V. A. Astashkin, L. I. Kilina, and B. B. Shishkin. 1989b. Paleogeografiya Sibirskoy platformy v rannem kembrii [Early Cambrian paleogeography of the Siberian Platform]. In R. G. Matukhin, ed., *Paleogeografiya fanerozoya Sibiri* [Phanerozoic paleogeography of Siberia], pp. 10–17, 4 figs. Novosibirsk: Siberian Scientific-Research Institute of Geology, Geophysics, and Mineral Resources.

Moreno-Eiris, E. 1987. Los montículos arrecifales de Algas y Arqueociatos del Cámbrico Inferior de Sierra Morena. 1: Estratigrafía y facies. *Boletín Geológico y Minero* 98:295–317.

Mossakovsky, A. A., S. V. Ruzhentsev, S. G. Samygin, and T. N. Kheraskova. 1993. Central Asian Fold Belt: Geodynamic evolution and formation history. *Geotectonics* 27 (6):3–32.

Palmer, A. and N. P. James. 1979. The Hawke Bay event: A circum-Iapetus regression near the lower Middle Cambrian boundary. In D. R. Wones, ed., *The Caledonides in the USA: Caledonide Orogen,* pp. 15–18. IGCP Project 27. Blacksburg, Va.: Department of Geological Sciences, Virginia Polytechnic Institute and State University.

Palmer, A. R. and A. J. Rowell. 1995. Early Cambrian trilobites from the Shackleton limestones of the central Transantarctic Mountains. *Paleontological Society Memoir* 45:1–28.

Parrish, J. T., A. M. Zeigler, C. R. Scotese, R. G. Humphreville, and J. L. Kirschvink. 1986. Proterozoic and Cambrian phosphorites —special studies: Early Cambrian palaeogeography, palaeoceanography, and phosphorites. In P. J. Cook and J. H. Shergold, eds., *Proterozoic and Cambrian Phosphorites,* vol. 1 of *Phosphorite Deposits of the World,* pp. 280–294. Cambridge: Cambridge University Press.

Pillola, G. L. 1990. Lithologie et trilobites du Cambrien inférieur du SW de la Sardaigne (Italie): Implications paléobiogéographiques. *Comptes rendus sommaires de l'Académie des Sciences, Paris,* 2d ser., 310:321–328.

Pillola, G. L., J. A. Gámez Vintaned, M. P. Dabard, F. Leone, E. Liñan, and J.-J. Chauvel. 1994. The Lower Cambrian ichnofossils *Astropolichnus hispanicus:* Palaeoenvironmental and palaeogeographic significance. *Bollettino della Società Paleontologia Italiana, Special Volume* 2:253–267.

Pirrus, E. A. 1986. Fatsial'nye osobennosti stroeniya vergal'sko-rausveskogo zhelezorudnogo urovnya Baltiyskoy sineklizy [Facies features in the structure of the Vergale-Rausve iron-ore level in the Baltic Syneclise]. In E. A. Pirrus, ed., *Fatsii i stratigrafiya venda i kembriya zapada Vostochno-Evropeyskoy platformy* [Facies and stratigraphy of the Vendian and Cambrian of the western East-European Platform], pp. 99–109. Tallin: Academy of Sciences of Estonian SSR.

Pratt, B. R. and J. W. F. Waldron. 1991. A Middle Cambrian trilobite faunule from the Meguma Group of Nova Scotia. *Canadian Journal of Earth Sciences* 28:1843–1853.

Repina, L. N. 1985. Rannekembriyskie morya zemnogo shara i paleobiogeograficheskie podrazdeleniya po trilobitam [Early Cambrian seas of the earth and paleobiogeographic subdivisions according to trilobites]. *Trudy, Institut geologii i geofiziki, Sibirskoe otdelenie, Akademiya nauk SSSR* 628:5–17.

Rowell, A. J., M. N. Rees, and K. R. Evans. 1992. Evidence of major Middle Cambrian deformation in the Ross orogen, Antarctica. *Geology* 20:31–34.

Rozanov, A. Yu. and K. Łydka, eds. 1987. *Palaeogeography and Lithology of the Vendian and Cambrian of the Western East-European Platform.* Warsaw: Wydawnictwa Geologiczne.

Rushton, A. W. A. 1978. Fossils from the Middle-Upper Cambrian transition in the Nuneaton district. *Palaeontology* 21:245–283.

Ruzhentsev, S. V. and A. A. Mossakovsky. 1995. Geodinamika i tektonicheskoe razvitie paleozoid Tsentral'noy Azii kak rezul'tat vzaimodeystviya tikhookeanskogo i indo-atlanticheskogo segmentov Zemli [Geodynamics and tectonic development of the Paleozoids in Central Asia as a result of interactions of the Pacific and Indian-Atlantic segments of the earth]. *Geotektonika* 1995 (4):29–47.

Samson, S., A. R. Palmer, R. A. Robison, and D. T. Secor. 1990. Biogeographical significance of Cambrian trilobites from the Carolina Slate Belt. *Geological Society of America Bulletin* 102:1459–1470.

Samygin, S. G. 1980. Differentsirovannoe smeshchenie obolochek litosfery i evolyutsiya formatsionnykh kompleksov (Ural) [Differentiated shift of the lithospheric covers and the evolution of formation complexes (Urals)]. In A. V. Peyve, ed., *Tektonicheskaya rassloennost' litosfery* [Tectonic stratification of the lithosphere], pp. 29–63. Moscow: Nauka.

Scotese, C. R. 1994. *Continental Drift. Edition 6, Paleomap Project*. Department of Geology, University of Texas at Arlington.

Scotese, C. R. and W. S. McKerrow. 1990. Revised world maps and introduction. In W. S. McKerrow and C. R. Scotese, eds., *Paleozoic Paleogeography and Biogeography*, pp. 1–21. *Geological Society of London, Memoir* 12.

Sdzuy, K. 1972. Das Kambrium der acadobaltischen Faunenprovinz. *Zentralblatt für Geologie und Paläontologie*, Teil 2 (1/2):1–91.

Sdzuy, K. and E. Liñan. 1993. Rasgos paleogeográficos del Cámbrico Inferior y Medio del norte de España. *Cuadernos do Laboratorio Xeolóxico de Laxe, Coruña* 18:189–215.

Şengör, A. M. C., B. A. Natal'in, and V. S. Burtman. 1993. Evolution of the Altaid tectonic collage and Palaeozoic crustal growth in Eurasia. *Nature* 364:299–307.

Seslavinsky, K. B. 1995. Glavnye vulkanicheskie, sedimentatsionnye, i tektonicheskie sobytiya paleozoya i ikh svyaz' s izmeneniyami klimata Zemli [Main volcanic, sedimentary, and tectonic events of the Paleozoic and their correlation with the earth climate changes]. *Byulleten' Moskovskogo obshchestva ispytateley prirody, Otdel geologicheskiy* 70 (2):3–13.

Shergold, J. H. 1988. Review of trilobite biofacies distributions at the Cambrian-Ordovician boundary. *Geological Magazine* 125:363–380.

Shergold, J. H., J. Jago, R. Cooper, and J. Laurie. 1985. *The Cambrian System in Australia, Antarctica, and New Zealand*. International Union of Geological Sciences, Publication 19.

Storey, B. C. 1993. The changing face of late Precambrian and early Palaeozoic reconstructions. *Journal of the Geological Society, London* 150:665–668.

Svyazhina, I. A. and R. A. Kopteva. 1996. The paleogeography of the southern Urals in early and late Paleozoic according to paleomagnetic data. *Paleogeografiya vendarannego paleozoya* [Vendian–early Paleozoic paleogeography], pp. 144–146. Ekaterinburg: Uralian Branch, Russian Academy of Sciences.

Thickpenny, A. and J. K. Leggett. 1987. Stratigraphic distribution and palaeooceanographic significance of European early Palaeozoic organic-rich sediments. In J. Brooks and A. Fleet, eds., *Marine Petroleum Source Rocks*, pp. 231–248. *Geological Society Special Publication* 26.

Tkachenko, V. I., Ushatinskaya, G. T., Zhuravlev, A. Yu. and L. N. Repina. 1987. Kembriyskie otlozheniya Prikolymskogo podnyatiya [Cambrian sediments of the Kolyma Uplift]. *Izvestiya Akademii nauk SSSR, Seriya geologicheskaya* 1987 (8):55–62.

Torsvik, T. H., M. A. Smethurst, R. Van der Voo, W. S. McKerrow, M. D. Brasier, B. A. Sturt, and H. J. Walderhaug. 1996. Continental break-up and collision in the Neoproterozoic and Palaeozoic—a tale of Baltica and Laurentia. *Earth-Science Reviews* 40:229–258.

Van der Voo, R. 1988. Paleozoic paleogeography of North America, Gondwana, and intervening displaced terranes: Comparisons of paleomagnetism with paleoclimatology and biogeographical patterns. *Geological Society of America Bulletin* 100:311–324.

Van Staal, C. R. 1994. Brunswick subduction complex in the Canadian Appalachians: Record of the Late Ordovician to Late Silurian collision between Laurentia and the Gander margin of Avalon. *Tectonics* 13:946–962.

Vidal, G. and J. S. Peel. 1993. Acritarchs from the Lower Cambrian Buen Formation in North Greenland. *Grønlands Geologiske Undersøgelse, Bulletin* 164:1–35.

Vidal, G., T. Palacios, J. A. Gámez-Vintaned, M. A. Díez Balda, and S. W. F. Grant. 1994. Neoproterozoic–Early Cambrian geology and palaeontology of Iberia. *Geological Magazine* 131:729–765.

Wensink, H. 1991. Late Precambrian and Paleozoic rocks of Iran and Afghanistan. In M. Moullade and A. E. M. Nairn, eds., *The Phanerozoic Geology of the World. I: The Palaeozoic,* pp. 147–218. Amsterdam: Elsevier Science Publishers.

Wolfart, R. 1983. *The Cambrian System in the Near and Middle East.* International Union of Geological Sciences, Publication 15.

Zhuravlev, A. Yu. 1986. Evolution of archaeocyaths and palaeobiogeography of the Early Cambrian. *Geological Magazine* 123:377–385.

Zonenshain, L. P., M. I. Kuz'min, and M. N. Kononov. 1985. Absolute reconstructions of the Paleozoic oceans. *Earth and Planetary Sciences Letters* 74:103–116.

CHAPTER FOUR

Martin D. Brasier and John F. Lindsay

Did Supercontinental Amalgamation Trigger the "Cambrian Explosion"?

A global overview of sediment patterns and accumulation rates, and carbon, stron-
tium, and neodymium isotopes confirms that increasing rates of subsidence and up-
lift accompanied the dramatic radiation of animal life through the Neoproterozoic-
Cambrian interval (ca. 600 to 500 Ma). Peritidal carbonate platforms were
drowned, to be followed in places by phosphorites and black shales, while thick evap-
orites accumulated in interior basins. This drowning of cratons during the latest
Neoproterozoic-Cambrian could have brought about major taphonomic changes.
The shoreward spread of oxygen-depleted and nutrient-enriched waters favored the
preservation of thin skeletons by secondary phosphate and chert in peritidal carbon-
ates and, later, the occurrence of Burgess Shale–type preservation in deeper-water
shales. The burial of event sands in rapidly subsiding basins also allowed the para-
doxical preservation of deep-water Nereites *ichnofacies in shallow-water sediments.*

THIS CHAPTER ATTEMPTS to put the "Cambrian explosion" into the wider context of events in the lithosphere. The formation and later rapid extensional subsidence of supercontinents in the Neoproterozoic have recently become apparent from a wide range of disciplines, including paleomagnetism, facies and fossil distributions, sub-sidence curves, and isotopic studies (e.g., Bond et al. 1984; Lindsay et al. 1987; Dalziel 1991; McKerrow et al. 1992; Derry et al. 1992, 1994). At some time before ca. 900 Ma B.P., Antarctica, Australia, Laurentia, Baltica, and Siberia appear to have been united in a Neoproterozoic supercontinent called Rodinia or Kanatia (Torsvik et al. 1996). It is possible that this may have begun to rift apart as early as 800 Ma (e.g., Lindsay and Korsch 1991; Lindsay and Leven 1996); certainly early rift suc-cessions can preserve deposits of the older, Rapitan-Sturtian glaciations (ca. 750–700 Ma; Young 1995). At some point after 725 Ma, the western margins of Laurentia and Antarctica-Australia were certainly separated and moving apart (Dalziel 1992a,b; Powell et al. 1993). By ca. 600–550 Ma, Laurentia, Baltica, and Siberia were also in

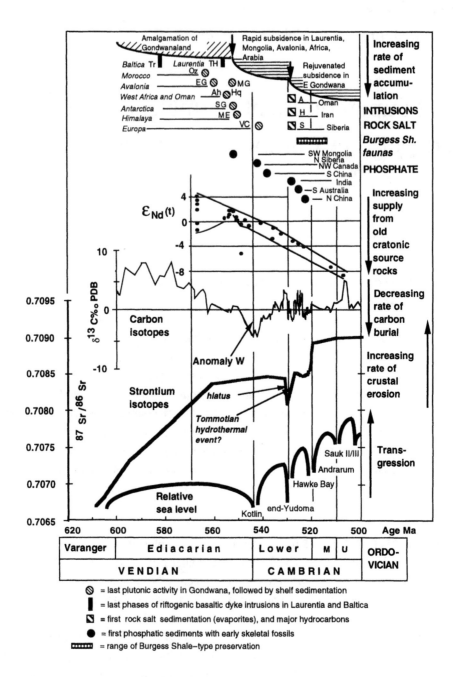

the process of rifting apart (McKerrow et al. 1992; Torsvik et al. 1996), and here the rift sequences may preserve deposits of the younger, Varangerian (or Marinoan) glaciations (ca. 620–590 Ma; e.g., Young 1995).

The assembly of another supercontinent, Gondwana, also took place during the Ediacarian to Early Cambrian interval. (*Ediacarian* is here used to indicate that period of the Late Neoproterozoic between the Marinoan glaciation at ca. 600 Ma and the base of the Cambrian at ca. 543 Ma). This involved the amalgamation of the separate

Figure 4.1 Isotopes, sea level, fossil taphonomy, and global tectonic changes during the Vendian-Cambrian interval. Basic dykes in Baltica and Laurentia indicate a final phase of rifting: Tr = Troms, Norway (582 ± 30 Ma; Torsvik et al. 1996); TH = Tibbit Hill, Quebec (554 Ma; Kumarapeli et al. 1989). Latest Pan-African plutonic events may indicate the final phases of amalgamation in West Gondwana: EG = Ercall Granophyre, England (560 ± 1 Ma, U/Pb zircon; Tucker and Pharaoh 1991); Ah = Ahaggar plutons, West Africa (556 ± 12 Ma, U/Pb zircon; Betrand-Sarfati et al. 1995); Hq = granite and ignimbrite below Huqf Group, Oman (556 ± 10 Ma, Rb/Sr; Burns et al. 1994); ME = granites from the Mount Everest region, Nepal, Himalaya (550 ± 16 Ma, Rb/Sr; Ferrara et al. 1983); MG = Marystown Group volcanics, southeastern Newfoundland (552 ± 3 Ma, U/Pb zircon; Myrow and Hiscott 1993); Oz = Ourzazate volcanics, Morocco (563 ± 2.5 Ma, U/Pb zircon; Odin et al. 1983); SG = postorogenic quartz syenite, Skelton Group, Antarctica (551 ± 4 Ma, U/Pb zircon; Rowell et al. 1993); VC = Vires-Carolles granite, Brioverian France (540 ± 10 Ma, U/Pb monazite; Dupret et al. 1990). Thick rock salt accumulated during rapid subsidence of extensional basins: A = Ara Salt Formation, Oman (Burns and Matter 1993; Loosveld et al. 1996); H = Hormuz Salt Formation, Iran (Brasier et al. 1990; Husseini and Husseini 1990). Burgess Shale–type faunas are confined to the medial Lower to Middle Cambrian (Butterfield 1996). Phosphatic sediments with early skeletal fossils first appear in the transition to more rapid subsidence and/or flooding of the platforms (sources cited in figures 4.2 and 4.3). $\varepsilon_{Nd}(t)$ data recalculated from Thorogood 1990, using revised ages. The carbon isotope curve is composite, compiled from the Vendian of southwestern Mongolia (Brasier et al. 1996), Early to Middle Cambrian of Siberian Platform (Brasier et al. 1994), and Middle to Upper Cambrian of the Great Basin, USA (Brasier 1992b). The strontium isotope curve is based on least-altered samples (compiled from Burke et al. 1982; Keto and Jacobsen 1987; Donnelly et al. 1988, 1990; Derry et al. 1989, 1992, 1994; Narbonne et al. 1994; Nicholas 1994, 1996; Smith et al. 1994; Brasier et al. 1996). The sea level curve is based on data in Brasier 1980, 1982, and 1995; Notholt and Brasier 1986; Palmer 1981; and Bond et al. 1988.

crustal blocks of Avalonia, Europa, Arabia, Africa, Madagascar, South America, and Antarctica (together forming West Gondwana) and resulted in the compressional Pan-African orogeny, which culminated between ca. 560 and 530 Ma. Orogenic closure of the Pan-African compressional basins was accompanied in many places by igneous intrusions. In figure 4.1, we have plotted some of the youngest dated phases of igneous activity, as well as the riftogenic dyke swarms of Laurentia. Although geologic evidence indicates that East Gondwana (India, South China, North China, Australia) collided with West Gondwana along the Mozambique suture between ca. 600 and 550 Ma, recent paleomagnetic evidence has also suggested that final amalgamation did not take place until the Early Cambrian (Kirschvink 1992; Powell et al. 1993).

Pan-African amalgamation of Gondwana appears to have been accompanied by the widespread development of subsiding foreland basins, as documented in figures 4.1–4.3. Sediments of "rift cycle 1" (sensu Loosveld et al. 1996) begin with the Sturtian Ghadir Mangil glaciation in Arabia, dated to ca. 723 Ma (Brasier et al. 2000). The development of thick salts in the Ara Formation, once thought to be rift deposits of Tommotian age (Loosveld et al. 1996; Brasier et al. 1997), now appear to be foreland basin deposits of late Ediacarian age (Millson et al. 1996; Brasier et al. 2000).

Subductive margins were also developed along the borders of eastern Australia and Antarctica (e.g., Millar and Storey 1995; Chen and Liu 1996) and Mongolia (e.g., Şengör et al. 1993; but see also Ruzhentsev and Mossakovsky 1995) in the Early to

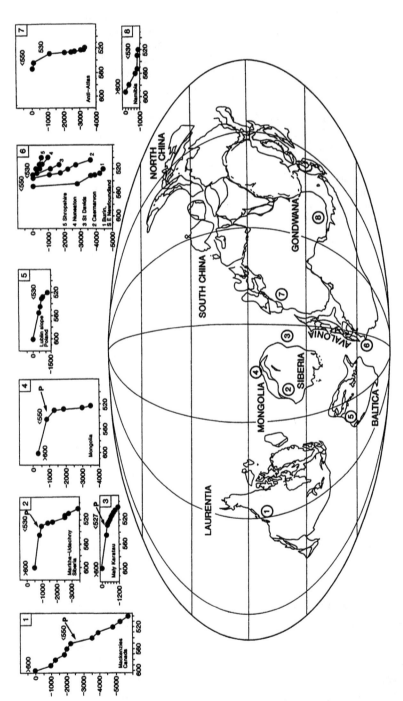

Figure 4.2 Sediment accumulation rates for the Ediacarian-Cambrian interval placed against latest Vendian to Nemakit-Daldynian paleogeography of McKerrow et al. (1992). Based on data in the following sources: Mackenzies, Canada (Narbonne and Aitken 1990); Markha-Udachny area of Siberian Platform (Astashkin et al. 1991); Maly Karatau in Kazakhstan (Cook et al. 1991); southwestern Mongolia (Brasier et al. 1996); Lublin Slope, Poland, Baltica (Moczydłowska 1991); Avalonian, Burin Peninsula, Newfoundland (Landing 1992): Caernarvon, North Wales; St. Davids, South Wales, Nuneaton, English Midlands; Shropshire, Welsh borderlands (Rushton 1974; Brasier 1989); Anti-Atlas, Morocco (Sdzuy and Geyer 1988); Namibia, southwestern Africa (Kaufman et al. 1994). The time scale is adapted from sources cited in Bowring et al. 1993, Tucker and McKerrow 1995, and Brasier 1995: base of Vendian = 610 Ma; Varangerian glacials = 610–600 Ma; base of Ediacaran = 600 Ma; main Ediacaran faunal interval = 580–555 Ma; late Ediacaran (Kotlin) interval = 555–545 Ma; base of Nemakit-Daldynian = base of Cambrian herein = 545 Ma; base of Tommotian = 530 Ma; base of Atdabanian = 528 Ma; base of Botoman = 526 Ma; base of Toyonian = 523 Ma; base of Middle Cambrian = 520 Ma; base of Upper Cambrian = 510 Ma; base of Ordovician = 500 Ma. P marks the first phosphatic sediments with early skeletal fossils. The numbers (e.g., <550) give the suggested timing of renewed rift/drift in millions of years ago (Ma).

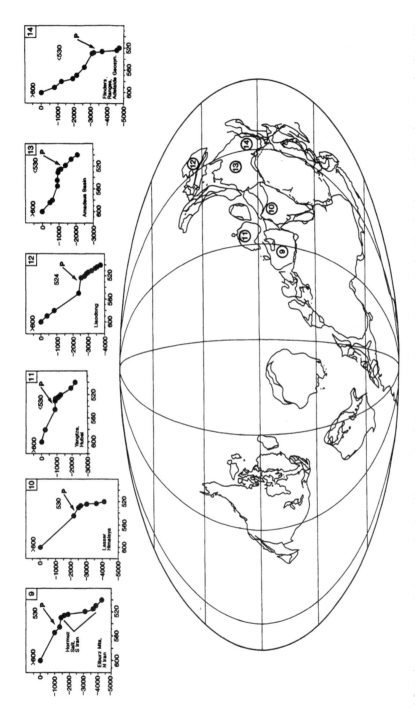

Figure 4.3 Sediment accumulation rates for the Ediacarian–Cambrian placed against the Atdabanian–Toyonian paleogeography of McKerrow et al. (1992). Based on data in the following sources: Elburz Mountains, Iran (Hamdi et al. 1989); Lesser Himalaya, India (Shanker and Mathur 1992); Yangtze Gorges, Hubei; South China, and Liaodong, North China (Wang 1986; Chang 1988; Lindsay 1993); Amadeus Basin, central Australia, and Flinders Ranges, Adelaide Geosyncline, southern Australia (Lindsay et al. 1987; Jenkins et al. 1993; Lindsay 1993). Key as for figure 4.2.

Middle Cambrian. Below, we explore the possibility that the amalgamation of Gond-wana between ca. 555 and 510 Ma helped to bring about dramatic changes in the rate of sediment accumulation and in the biosphere over the Precambrian-Cambrian transition.

SEDIMENT ACCUMULATION RATES

Plots of sediment thickness against time can give an impression of the changing rate of sediment accumulation (figures 4.2 and 4.3). Such curves may, however, be skewed by the effects of compaction, which is greatest in siliciclastic sediments (especially argillites) and least in early-cemented carbonates. Rather than make assumptions about the degree of compaction and cementation, we here plot the raw data. Sediment accumulation rates are therefore likely to be underestimates in the case of finer clay-rich clastic lithologies. Inspection of the data, however, suggests that changes in sediment accumulation rate cannot be explained by changes in lithology and compaction alone.

In order to portray the tectonic component, data on the sediment accumulation rate should be "backstripped" by making corrections not only for the assumed effects of cementation and compaction but also for the isostatic effects of sediment loading; and further corrections should be made for the effects of water depth, the isostatic effect of seawater loading, and the stretch factor due to crustal extension (e.g., Watts 1982). If the sediments are mainly shallow-water deposits, as in this study, then back-stripping tends to change the amplitude but not the general shape of the curves. In this study, we have found that selection of different time scales has relatively little effect on the shapes of the curves.

Backstripped tectonic subsidence curves have been used to track the thermal relaxation of the crust following rifting events, such as those during the Neoproterozoic-Cambrian. As rift turned to drift and ocean basins widened, extension on the margins of cratons is believed to have encouraged rapid rates of subsidence that diminished with time, in general accordance with geophysical models (e.g., Bond et al. 1985, 1988; Lindsay et al. 1987). The latter authors, by backtracking post-rift tectonic subsidence curves from the Middle-Late Cambrian, have estimated that a major phase of continental breakup took place in the Neoproterozoic–Early Cambrian (then dated at 625–555 Ma).

In figures 4.2 and 4.3 we have plotted sediment accumulation data against a time scale adapted from sources in Bowring et al. 1993, Tucker and McKerrow 1995, and Brasier 1995. We note that the rifting cratons of "Rodinia" are widely believed to have resulted from the breakup of Rodinia before ca. 720 Ma (Laurentia, Baltica, Siberia; figure 4.2), and show relatively low average rates of sediment accumulation during the early Ediacarian (ca. 600–550 Ma), followed by more rapid rates in the latest Ediacarian (after ca. 550 Ma, Mackenzies, Mongolia) to Early Cambrian (after ca. 530 Ma, Siberia, Kazakhstan, Baltica). These patterns may be attributed to a progressive attenuation in the thermal relaxation of the crust following the initial rifting of Rodinia in the Riphean, followed either by renewed phases of rifting (Laurentia, Baltica) or by the development of foreland basins (Siberia, Mongolia) across the Precambrian-Cambrian transition.

A similar pattern is seen in East Gondwana (Iran to Australia; figure 4.3), where an initial phase of rifting also seems to have been Riphean-Varangerian (ca. 725–600 Ma). There the rates of sediment accumulation in the Ediacarian interval (ca. 600–543 Ma) appear to have been relatively low, with some evidence for condensation and hiatus in the earliest Cambrian. A sharp change in the estimated rate of sediment accumulation coincides with major facies changes that suggest a renewed phase of subsidence close to the Precambrian-Cambrian boundary (ca. 545–530 Ma).

In West Gondwana (e.g., Avalonia, Morocco), the Ediacarian was characterized by rapid rates of sediment accumulation in compressive settings, which concluded with igneous intrusions, uplift, and cratonic amalgamation by ca. 550 Ma (figures 4.1 and 4.2). This phase was rapidly followed by the formation of extensional strike-slip basins that began to accumulate thick volumes of sediment.

LITHOFACIES CHANGES

Lithofacies changes provide further evidence for the rapid flooding of carbonate platforms between ca. 550 and 530 Ma B.P. The replacement of peritidal carbonates, especially "primary" dolomite, by neritic limestones and/or siliciclastic units above the Precambrian-Cambrian boundary (Tucker 1992; Brasier 1992a) broadly coincides in places (e.g., Mongolia; Lindsay et al. 1996) with the change from slower to more rapid rates of sediment accumulation. Hence, the mineralogic shift from dolomite to calcite/aragonite can be explained, in part, by the "drowning" of peritidal platforms, brought about by increased subsidence and relative sea level rise.

The widespread occurrence of phosphorites and cherts across the Precambrian-Cambrian boundary interval has for many years been related to the explosion of skeletal fossils in the Early Cambrian (e.g., Brasier 1980; Cook and Shergold 1984), but the connection has remained somewhat enigmatic. Brasier (1989, 1990, 1992a,b) has summarized evidence for the widespread development of "nutrient-enriched waters" during this interval and has argued that their incursion dramatically enhanced the preservation potential of early, thin-shelled skeletal fossils that herald the Cambrian period. These phosphatic sediments typically lie within the upper parts of dolomitic facies or rapidly succeed them. In figures 4.1–4.3 it can be seen that the first appearance of phosphatic beds with early skeletal fossils tends to coincide with the switch from slow to more rapid sediment accumulation. This may be explained by the interaction between phosphorus-rich oceanic waters and calcium-rich platformal waters under relatively low rates of sediment accumulation. Such conditions appear to have been widespread in the late Ediacarian to Tommotian (ca. 555–530 Ma). At first, the peritidal carbonate banks discussed above may have acted as barriers.

Later drowning of these barriers allowed incursions of nutrient-enriched water masses from the outer shelf and open sea. This drowning of barriers was made possible by the interrelated factors of increased subsidence and relative sea level rise.

Many Asiatic successions also show abrupt transitions from a restricted carbonate platform to organic-rich black shales over this interval, as, for example, in the latest Ediacarian of southwestern Mongolia (ca. 550–543 Ma, Brasier et al. 1996; Lindsay et al. 1996), and between the latest Ediacarian and mid-Atdabanian of southern Kazakhstan, Oman, Iran, Pakistan, India, and South China (ca. 545–527 Ma). These laminated black shales have many distinctive features: (1) they are basin-wide; (2) they follow a well-defined sequence boundary indicated by a major break in deposition, often with evidence for karstic solution of underlying peritidal carbonates; (3) they coexist with or overlie phosphatic dolostone beds and bedded cherts; (4) they contain high levels of organic matter with distinctively negative $\delta^{13}C$ values and positive $\delta^{34}S$ values; (5) they are highly metalliferous, with high concentrations of vanadium, molybdenum, cobalt, and barium; (6) in India, Oman, and China, they are accompanied by carbonates yielding a large negative carbon isotope anomaly (e.g., Hsu et al. 1985; Brasier et al. 1990, 2000), which is consistent with the turnover of aged, nutrient-enriched, and poorly oxygenated bottom waters (Brasier 1992a).

These anoxic marker events appear to lie in the interval between slower and more rapid rates of sediment accumulation. Drowning of the platform is indicated by the abrupt change in facies, from dolomites and peritidal phosphorites beneath. It therefore appears that a change in sedimentary regime took place, from one in which sediment accumulation rates were "space limited" (in the carbonate platform) to one in which they were "supply limited" (in the black shales).

Although gypsum, anhydrite, and evaporitic fabrics are not uncommon within the peritidal dolomite facies discussed above, thick layers of rock salt (halite) became widespread in the latest Ediacarian to the Early Cambrian. Indeed, some of the world's thickest successions of rock salt were laid down from ca. 545 Ma onward (e.g., figures 4.1 and 4.3). These include the Hormuz Salt of Iran, the Ara Salt of Oman (both thought to be latest Ediacarian), the Salt Range salt of Pakistan (Atdabanian-Botoman), and the Usolka and contemporaneous salts of Siberia (Tommotian-Atdabanian; see Husseini and Husseini 1990; Kontorovitch et al. 1990; Burns and Matter 1993; Brasier et al. 2000). The preservation of thick halite implies interior basins with low siliciclastic supply, restricted by major barriers. The Hormuz and Oman salt horizons are also associated with volcanic rocks (e.g., Husseini and Husseini 1990; Brasier et al. 2000), which are taken to indicate an extensional tectonic setting. These salt deposits are therefore thought to have accumulated within interior barred basins formed by renewed subsidence of the basement (e.g., Loosveld et al. 1996). Poor bottom-water ventilation also led to anoxic conditions, so that associated sediments can be important as hydrocarbon source rocks (e.g., Gurova and Chernova 1988; Husseini and Husseini 1990; Mattes and Conway Morris 1990; Korsch et al. 1991).

THE EDIACARIAN-CAMBRIAN Sr AND Nd ISOTOPE RECORD

Figure 4.1 shows that least-altered values of $^{87}Sr/^{86}Sr$ rose almost continuously from ca. 0.7072 in the Varangerian to 0.7090 in the Late Cambrian, punctuated by a fall in values during the Tommotian (Derry et al. 1994; Brasier et al. 1996; Nicholas 1996). The low Riphean-Varangerian values have been attributed to the influence of hydrothermal flux on new ocean floors during rifting of the Rodinia (e.g., Veizer et al. 1983; Asmerom et al. 1991). The rise in Vendian $^{87}Sr/^{86}Sr$ ratios has been explained by accelerating rates of uplift and erosion associated with the Pan-African orogeny (e.g., Derry et al. 1989, 1994; Asmerom et al. 1991; Kaufman et al. 1994) and late Precambrian glaciations (Burns et al. 1994). The decline in seawater $^{87}Sr/^{86}Sr$ values in the Tommotian perhaps reflects a drop in the rate of erosion and subsidence, a decrease in silicate weathering rate, and/or the influence of rift-related hydrothermal activity (Derry et al. 1994; Nicholas 1996). It is interesting to note that this $^{87}Sr/^{86}Sr$ shift and the preceding hiatus found across much of the Siberian Platform and possibly beyond (Corsetti and Kaufman 1994; Ripperdan 1994; Knoll et al. 1995; Brasier et al. 1996) (figure 4.1) are both broadly coincident with the inferred shift from slower to more rapid sediment accumulation on many separate cratons (figures 4.2 and 4.3).

High crustal erosion rates have been inferred from late Tommotian to Late Cambrian $^{87}Sr/^{86}Sr$ values (Derry et al. 1994). This suggests that uplift and erosion of Pan-African orogenic belts (Avalonia and the Damara-Gariep belt of Namibia, for example) may have provided a source of radiogenic ^{87}Sr through the Cambrian. This interpretation is supported by studies of $\varepsilon_{Nd}(t)$ values in Ediacarian to Cambrian clastics from the Avalonian terranes of England and Wales. These sediments show a progressive reduction in the signal left by juvenile igneous rocks and an increase in the radiogenic component, between ca. 563 and 500 Ma (Thorogood 1990). Such a change in sediment supply suggests that younger accretionary margins became progressively submerged while older, interior crystalline rocks of the craton were uplifted and eroded, presumably as bulging of the crust and transgression of the platform proceeded. Comparison of the $^{87}Sr/^{86}Sr$ record of the Ediacarian-Cambrian with that of the Cenozoic (Derry et al. 1994) suggests that the inferred uplifted regions of Gondwana could even have experienced major montane glaciations through the latest Ediacarian-Cambrian interval.

THE EDIACARIAN-CAMBRIAN CARBON ISOTOPIC RECORD

Carbon isotopes show a long-term trend of falling values, from maxima of $+11\%_0$ $\delta^{13}C_{PDB}$ in the post-Sturtian interval (ca. 730–600 Ma B.P.) to $+8$ in the Ediacarian and $+5.5$ in the Cambrian (Brasier et al. 1996, 2000). On this broad-scale trend are superimposed a series of second-order cycles, which in the Cambrian appear to have been about 1 to 5 m.y. long, some of which can be correlated globally (e.g., Brasier et al. 1990; Kirschvink et al. 1991; Ripperdan 1994; Brasier et al. 1996; Calver and Lindsay 1998).

Above, we have argued for increasing rates of sediment accumulation through this time interval, which might be expected to have increased the global rates of carbon burial (cf. Berner and Canfield 1989). The long-term trend for carbon burial, however, is for falling values through the Neoproterozoic-Cambrian (figure 4.1). This means that increases in carbon burial due to raised rates of sediment accumulation must have been offset by raised rates of organic carbon oxidation. Such oxidation could have been brought about by a range of factors, including uplift and erosion of sedimentary carbon, greater ocean-atmosphere mixing (e.g., glacial climates, Knoll et al. 1996) and innovations in the biosphere (e.g., fecal pellets, Logan et al. 1995; bioturbation, Bottjer and Droser 1994, Brasier and McIlroy 1998).

The second order, 1–5 m.y. cycles in $\delta^{13}C$ may contain signals that relate to subsidence and sea level. Such a connection has been argued at higher levels in the geological column, as, for example, in the Late Cambrian (Ripperdan et al. 1992) and in the Jurassic-Cretaceous (e.g., Jenkyns et al. 1994). This has led to the suggestion that positive $\delta^{13}C$ excursions may record an increase in the burial of organic matter connected with the rapid areal expansion of marine depositional basins during "transgressions." Conversely, the negative $\delta^{13}C$ excursions may record reduced rates of carbon burial and increased rates of carbon oxidation during "regressions."

It is difficult to test for a connection between $\delta^{13}C$ and sea level in the Ediacarian-Cambrian interval without access to a set of rigorously derived sea level curves. Figure 4.1 shows a notional global sea level curve that depicts the major Cambrian transgression divided into major transgressive pulses. It is notable that several of the carbon isotopic maxima can be traced to these pulses; e.g., the appearance of laminated black limestones of the Sinsk Formation in Siberia coincided with the Botoman $\delta^{13}C$ maximum (Brasier et al. 1994; Zhuravlev and Wood 1996), and the influx of flaggy, phosphatic "outer detrital belt" carbonates of the Candland Shales in the Great Basin coincided with the Upper Cambrian sea level maximum (Bond et al. 1988; Brasier 1992c). Negative excursions can also, in several cases, be connected with evidence for emergence and omission surfaces. These are named in figure 4.1 and include the *Kotlin regression* prior to negative anomaly "W"; the *end-Yudoma regression* at the top of the Nemakit-Daldynian in Siberia (e.g., Khomentovsky and Karlova 1993; correlated with the top of the Dahai Member in South China, according to Brasier et al. 1990); the *Hawke Bay regression* across the Lower-Middle Cambrian boundary interval (i.e., the Sauk I-II boundary of Laurentia, according to Palmer 1981; with similar breaks in Baltica and Avalonia, according to Notholt and Brasier 1986); the *Andrarum regression* associated with the *Lejopyge laevigata* Zone of the Middle Cambrian in Scandinavia (correlated into Avalonia by Notholt and Brasier [1986] and possibly into Laurentia); and the Sauk II-III regression of Laurentia (Sauk II-III boundary of Palmer 1981 and Bond et al. 1985).

Of particular interest is the negative $\delta^{13}C$ interval of anomaly W, here taken to correlate the Precambrian-Cambrian boundary. Major sedimentary breaks occur close to this anomaly across the globe, which could be taken to indicate a synchronous regression during which sediments were removed by erosion (Brasier et al. 1997). At earlier times in the Neoproterozoic, negative $\delta^{13}C$ anomalies of this amplitude are associated with glacial/deglacial carbonates (Knoll et al. 1996). Although no glacial sediments are known from the Precambrian-Cambrian boundary interval, a glacially driven overturn of a stratified water column provides a possible explanation for the advection of light ^{12}C into surface waters (cf. Aharon and Liew 1992; Knoll et al. 1996). A similar explanation has also been put forward to explain falling $\delta^{13}C$ in the Tommotian (e.g., Ripperdan 1994).

Some positive $\delta^{13}C$ anomalies in shallow-water carbonates are accompanied by very positive $\delta^{34}S$ values in anhydrite (e.g., Mattes and Conway Morris 1990). Oxygen released by the burial of organic carbon may not, therefore, have been counterbalanced by the oxidation of sulfides to sulfates in the oceans (cf. Veizer et al. 1980). This has raised the possibility that the partial pressure of atmospheric oxygen could have risen over this interval, favoring the radiation of large, oxygen-hungry metazoans (cf. Knoll 1992; Sochava 1992). More-detailed $\delta^{34}S$ records are, however, needed to test this hypothesis.

IMPLICATIONS FOR THE BIOSPHERE

In figure 4.4 the ways in which raised rates of sediment accumulation/subsidence could have affected the hydrosphere, lithosphere, biosphere, and fossil record across the Precambrian-Cambrian transition are summarized. Submergence of shallow shelves inevitably led to an expansion of habitat area and, as we have argued, also caused phosphorus- and silica-rich waters to invade platform interiors. It may be argued that these environmental changes had a major ecologic impact upon the biota, encouraging blooms of eutrophic plankton, which in turn may have favored the development of a wide range of suspension feeders and the migration of pandemic phosphatic and siliceous taxa (figure 4.4). The reciprocal uplift of hinterland margins, indicated by the strontium isotope curve and by the thick succession of siliciclastic sediments, may well have delivered yet more phosphorus and iron into the oceans, thereby sustaining or raising its productivity (Derry et al. 1994).

This evidence for drowning of platforms also helps to explain some peculiar aspects of Cambrian fossil preservation. The development of secondary phosphatization of thin $CaCO_3$ or organic-walled skeletons during the latest Ediacarian to Atdabanian (e.g., Brasier 1980) is closely related to the timing of subsidence of carbonate platforms (figures 4.1–4.3). In Mongolia, for example, Cambrian-type siliceous sponge spicules and phosphatized early skeletal fossils first appear in the latest Ediacarian (Brasier et al. 1996, 1997) (figure 4.1). In India and North China, flood-

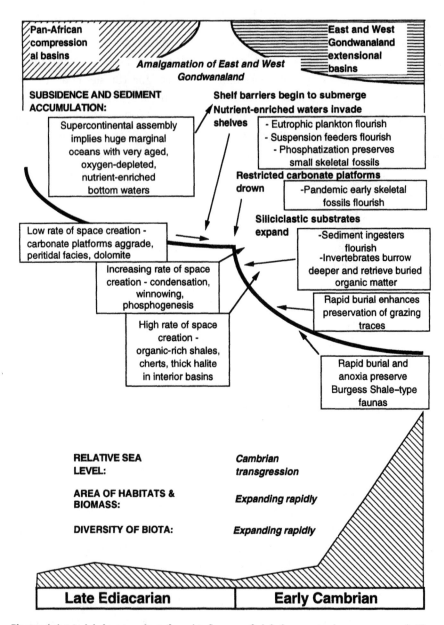

Figure 4.4 Model showing the inferred influence of global tectonic changes upon subsidence, sediment accumulation rate, sea level, nutrients, fossil preservation, and the adaptive radiation of the Cambrian fauna.

ing of the carbonate platforms brought phosphatic sediments with early skeletal fossils that were a little younger (figure 4.1).

In clastic sediments, the first main indications of the Cambrian radiation are given by trace fossils. Here, one of the main puzzles has been the preservation of deep-water *Nereites* ichnofacies in shallow waters during the Cambrian (e.g., Crimes 1994). At

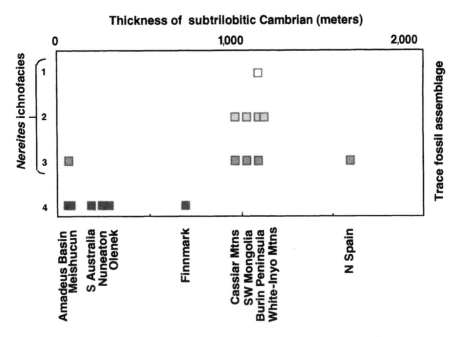

Figure 4.5 The paradox of deeper-water *Nereites* ichnofacies traces (assemblages 1, 2, and 3) in shallow water subtrilobitic Cambrian facies, which may be related to rapid rates of sediment accumulation. *1* = graphoglyptids (e.g., *Palaeodictyon* isp.), *2* = *Helminthopsis* and *Helminthoida* isp., *3* = *Taphrhelminthopsis* isp., *4* = other ichnogenera. Thickness and ichnotaxa from sources in Crimes (1989).

higher levels in the stratigraphic column, the distribution of these grazing traces has been related to the incidence of event sands, such as turbidites, which cast and preserve the delicate top tier of the ichnofauna (Bromley 1990). A review of the literature suggests that these grazing traces tend to be best represented in subtrilobitic Cambrian successions that are relatively thick (figure 4.5). Hence, the paradox of deeper-water *Nereites* ichnofacies traces in shallow-water sediments may well have been enhanced by conditions of rapid deposition, which led to the preservation of a greater number of sand-mud interfaces.

A further paradox of Cambrian fossil preservation concerns the restriction of Burgess Shale–type Lagerstätten to the Early and Middle Cambrian (Conway Morris 1992) (figure 4.1), despite the presence of suitable, anoxic, and poorly bioturbated facies at other times. Butterfield (1995, 1996) has suggested that this paradox could be explained by the restricted temporal distribution of volcanogenic clay minerals with antienzymatic and/or stabilizing effects. Here, we wonder whether Burgess Shale–type preservation was enabled by frequent pulses of fine-grained sedimentation along rapidly subsiding margins, leading to rapid burial and early diagenesis.

One of the most dramatic effects of sea floor subsidence on the Cambrian fossil record was arguably that of sudden "explosive phases" in diversification. The explosion in diversity at the base of the Tommotian certainly coincides with a rapid change in ^{87}Sr/^{86}Sr and lies above a major karstic surface (Brasier et al. 1996). In southwestern Mongolia, where rejuvenation of subsidence began in the latest Ediacarian, there is no clear Tommotian explosion in diversity (Brasier et al. 1996). This "Tommotian explosion" can therefore be regarded as an artefact brought about by missing time followed by abrupt facies changes (Lindsay et al. 1996), together caused by a rejuvenation of subsidence along the margin of the Mongolian arc.

CONCLUSION

Evidence is given above for increasing rates of subsidence and sediment accumulation during the Cambrian. This is no longer consistent, however, with the simple hypothesis that a rift-to-drift transition took place over the Precambrian-Cambrian boundary interval (Bond et al. 1984, 1985, 1988); foreland basins related to the amalgamation of Gondwana were also forming at this time. Such subsidence is also in line with the evidence for a major rise in relative sea level, from a low point during the Varangerian glaciation to a high point somewhere in the Late Cambrian (see figure 4.1; e.g., Brasier 1980, 1982, 1995; Bond et al. 1988).

A rise in the rate of sediment accumulation between ca. 550 and 530 Ma suggests that rapid subsidence took place in cratonic margins and interior basins around the globe. The large supplies of clastic sediment that flooded into these basins imply high rates of uplift and erosion of the basin hinterlands, which in turn can provide a plausible explanation for the progressive rise in ^{87}Sr/^{86}Sr and change in $\varepsilon_{Nd}(t)$ through the Cambrian.

Carbon isotopic maxima in the Early and Late Cambrian appear to coincide with transgressive pulses, and several carbon isotopic minima (e.g., anomaly W close to the Precambrian-Cambrian boundary, and that at the base of the Tommotian) are associated with widespread breaks in deposition. Further work is needed to test the hypothesis that these negative excursions relate to episodes of cooler (glacial?) climate and oceanic overturn. Recent work on the very complex section preserved in the Nama Basin in Namibia emphasizes the possibility of breaks in sedimentation in other basins (Grotzinger et al. 1995).

A picture emerges of Neoproterozoic to Early Cambrian oceans that were well fed with biolimiting nutrients, derived perhaps from high rates of erosion and runoff and enhanced by montane glaciations. As the platforms extended and subsided across the Neoproterozoic-Cambrian transition, new kinds of fossil preservation became possible: phosphatization and silicification of early skeletal fossils; rapid burial of delicate, grazing trace fossils; and rapid burial and diagenesis of Burgess Shale–type faunas. These changes have amplified and distorted our view of the evolutionary radiation, making the fossil record appear more stepwise, with explosive phases in diversity that we suspect are largely illusory.

Acknowledgments. We thank Simon Conway Morris, Gerard Bond, Rob Ripperdan, and Andrey Zhuravlev for helpful critical comments at various stages in the preparation of this paper.

REFERENCES

Aharon, P. and T. C. Lieu. 1992. An assessment of the Precambrian-Cambrian transition events on the basis of carbon isotope records. In M. Schidlowski, S. Golubic, M. M. Kimberley, D. M. McKirdly, and P. A. Trudinger, eds., *Early Organic Evolution,* pp. 212–223. Berlin: Springer Verlag.

Asmerom, Y. A., S. B. Jacobsen, N. J. Butterfield, and A. H. Knoll. 1991. Implications for crustal evolution. *Geochimica et Cosmochimica Acta* 55:2883–2894.

Astashkin, V. A., T. V. Pegel', L. N. Repina, A. Yu. Rozanov, Yu. Ya. Shabanov, A. Yu. Zhuravlev, S. S. Sukhov, and V. M. Sundukov. 1991. *The Cambrian System on the Siberian Platform.* International Union of Geological Sciences, Publication 27.

Berner, R. A. and D. Canfield. 1989. A model for atmospheric oxygen over Phanerozoic time. *American Journal of Science* 289:333–361.

Bond, G. C., P. A. Nickeson, and K. A. Kominz. 1984. Breakup of a supercontinent between 625 Ma and 555 Ma: New evidence and implications for continental histories. *Earth and Planetary Science Letters* 70:325–345.

Bond, G. C., N. Christie-Blick, M. A. Kominz, and W. J. Devlin. 1985. An early Cambrian rift to post-rift transition in the Cordillera of western North America. *Nature* 316:742–745.

Bond, G. C., M. A. Kominz, and J. P. Grotzinger. 1988. Cambro-Ordovician eustasy: Evidence from geophysical modelling of subsidence in Cordilleran and Appalachian passive margins. In K. L. Kleinspehn and C. Paola, eds., *New Perspectives in Basin Analysis,* pp. 129–159. New York: Springer Verlag.

Bottjer, D. J. and M. L. Droser. 1994. The Phanerozoic history of bioturbation. In S. K. Donovan, ed., *The Palaeobiology of Trace Fossils,* pp. 155–176. Chichester, England: John Wiley and Sons.

Bowring, S. A., J. P. Grotzinger, C. E. Isachsen, A. H. Knoll, S. Pelechaty, and P. Kolosov. 1993. Calibrating Cambrian evolution. *Science* 261:1293–1298.

Brasier, M. D. 1980. The Lower Cambrian transgression and glauconite-phosphate facies in western Europe. *Journal of the Geological Society, London* 137:695–703.

Brasier, M. D. 1982. Sea-level changes, facies changes, and the lower Precambrian–Early Cambrian evolutionary explosion. *Precambrian Research* 17:105–123.

Brasier, M. D. 1989. China and the Palaeotethyan Belt (India, Pakistan, Iran, Kazakhstan, and Mongolia). In J. W. Cowie and M. D. Brasier, eds., *The Precambrian-Cambrian Boundary,* pp. 40–74. Oxford: Clarendon Press.

Brasier, M. D. 1990. Phosphogenic events and skeletal preservation across the Precambrian-Cambrian boundary interval. In A. J. G. Notholt and I. Jarvis, eds., *Phosphorite Research and Development,* pp. 289–303. *Geological Society of London, Special Publication* 52.

Brasier, M. D. 1992a. Nutrient-enriched waters and the early skeletal fossil record.

Journal of the Geological Society, London 149:621–629.

Brasier, M. D. 1992b. Palaeoceanography and changes in the biological cycling of phosphorus across the Precambrian-Cambrian boundary. In J. H. Lipps and P. W. Signor, eds., *Origin and Early Evolution of the Metazoa*, pp. 483–523. New York: Plenum Press.

Brasier, M. D. 1992c. Towards a carbon isotope stratigraphy of the Cambrian System: Potential of the Great Basin succession. In E. A. Hailwood and R. B. Kidd, eds., *High Resolution Stratigraphy*, pp. 341–350. *Geological Society of London, Special Publication* 70.

Brasier, M. D. 1995. The basal Cambrian transition and Cambrian bio-events (from terminal Proterozoic extinctions to Cambrian biomeres)—A contribution from IGCP Project 303. In O. H. Walliser, ed., *Global Events and Event Stratigraphy*, pp. 113–138. Berlin: Springer Verlag.

Brasier, M. D. and D. McIlroy. 1998. *Neonereites uniserialis* from c. 600 Ma year old rocks in western Scotland and the emergence of animals. *Journal of the Geological Society, London* 155:5–12.

Brasier, M. D., M. Magaritz, R. Corfield, H. Luo, X. Wu, L. Ouyang, Z. Jiang, B. Hamdi, T. He, and A. G. Fraser. 1990. The carbon- and oxygen-isotope record of the Precambrian-Cambrian boundary interval in China and Iran and their correlation. *Geological Magazine* 127:319–332.

Brasier, M. D., R. M. Corfield, L. A. Derry, A. Yu. Rozanov, and A. Yu. Zhuravlev. 1994. Multiple $\delta^{13}C$ excursions spanning the Cambrian explosion to the Botomian crisis in Siberia. *Geology* 22:455–458.

Brasier, M. D., G. Shields, V. N. Kuleshov, and E. A. Zhegallo. 1996. Carbon and oxygen isotope stratigraphy of the terminal Proterozoic to early Cambrian: West Mongolia. *Geological Magazine* 133:445–485.

Brasier, M. D., O. Green, and G. Shields. 1997. Ediacarian sponge spicule clusters from southwestern Mongolia and the origins of the Cambrian fauna. *Geology* 25:303–306.

Brasier, M., G. McCarron, R. Tucker, J. Leather, P. Allen, and G. Shields. 2000. New U-Pb zircon dates for the Neoproterozoic Ghubrah glaciation and for the top of the Huqf Supergroup, Oman. *Geology* 28:175–178.

Bromley, R. G. 1990. *Trace Fossils: Biology and Taphonomy*. London: Unwin Hyman.

Burke, W. M., R. E. Denison, E. A. Hetherington, R. B. Koepnik, M. F. Nelson, and J. B. Omo. 1982. Variations of seawater 87Sr/86Sr throughout Phanerozoic time. *Geology* 10:516–519.

Burns, S. J., U. Haudenschild, and A. Matter. 1994. The strontium isotopic composition of carbonates from the late Precambrian (ca. 560–540 Ma) Huqf Group of Oman. *Chemical Geology* 111:269–282.

Burns, S. J. and A. Matter. 1993. Carbon isotopic record of the latest Proterozoic from Oman. *Eclogae Geologa Helvetiae* 86:595–607.

Butterfield, N. J. 1995. Secular distribution of Burgess-Shale–type preservation. *Lethaia* 28:1–13.

Butterfield, N. J. 1996. Fossil preservation in the Burgess Shale: Reply. *Lethaia* 29:109–112.

Calver, C. R. and J. F. Lindsay. 1998. Ediacarian sequence and isotope stratigraphy of the Officer Basin, South Australia. *Australian Journal of Earth Sciences* 45:513–532.

Chang, W. T. 1988. *The Cambrian System in Eastern Asia*. International Union of Geological Sciences, Publication 24.

Chen, Y. D. and S. F. Liu. 1996. Precise U-Pb zircon dating of a post-D2 meta-dolerite:

Constraints for rapid development of the southern Adelaide Foldbelt during the Cambrian. *Journal of the Geological Society, London* 153:83–90.

Conway Morris, S. 1992. Burgess Shale–type faunas in the context of the "Cambrian explosion": A review. *Journal of the Geological Society, London* 149:631–636.

Cook, H. E., M. E. Taylor, M. K. Apollonov, G. K. Ergaliev, Z. S. Sargaskaev, S. V. Dubinina, and L. M. Mel'nikova. 1991. Comparison of two early Paleozoic carbonate submarine fans—western United States and southern Kazakhstan, Soviet Union. In J. D. Cooper and C. H. Stevens, eds., *Paleozoic Paleogeography of the Western United States,* vol. 2, pp. 847–872. *Pacific Section of the Society for Economic Mineralogists and Petrologists* 67.

Cook, P. J. and J. H. Shergold. 1984. Phosphorus, phosphorites, and skeletal evolution at the Precambrian-Cambrian boundary. *Nature* 308:231–236.

Corsetti, F. A. and A. J. Kaufman. 1994. Chemostratigraphy of Neoproterozoic-Cambrian units, White-Inyo region, eastern California, and western Nevada: Implications for global correlations and faunal distributions. *Palaios* 9:211–219.

Crimes, T. P. 1989. Trace fossils. In J. W. Cowie and M. D. Brasier, eds., *The Precambrian-Cambrian Boundary,* pp. 166–185. Oxford: Clarendon Press.

Crimes, T. P. 1994. The period of evolutionary failure and the dawn of evolutionary success: The record of biotic changes across the Precambrian-Cambrian boundary. In S. K. Donovan, ed., *The Palaeobiology of Trace Fossils,* pp. 105–133. Chichester, England: John Wiley and Sons.

Dalziel, I. W. D. 1991. Pacific margins of Laurentia and East Antarctica–Australia as a conjugate rift pair: Evidence and implications for an Eocambrian supercontinent. *Geology* 19:598–601.

Dalziel, I. W. D. 1992a. Antarctica: A tale of two supercontinents? *Annual Review of Earth and Planetary Sciences* 20:501–526.

Dalziel, I. W. D. 1992b. On the organization of American plates in the Neoproterozoic and the breakout of Laurentia. *GSA Today* 2:237, 240–241.

Derry, L. A., L. S. Keto, S. B. Jacobsen, A. H. Knoll, and K. Swett. 1989. Sr isotopic variations in late Proterozoic carbonates from Svalbard and East Greenland. *Geochimica et Cosmochimica Acta* 54:2331–2339.

Derry, L. A., A. J. Kaufman, and S. B. Jacobsen. 1992. Sedimentary cycling and environmental change in the late Proterozoic: Evidence from stable and radiogenic isotopes. *Geochimica et Cosmochimica Acta* 56:1317–1329.

Derry, L. A., M. D. Brasier, R. M. Corfield, A. Yu. Rozanov, and A. Yu. Zhuravlev. 1994. Sr isotopes in Lower Cambrian carbonates from the Siberian craton: A palaeoenvironmental record during the "Cambrian explosion." *Earth and Planetary Science Letters* 128:671–681.

Donnelly, T. H., J. H. Shergold, and P. N. Southgate. 1988. Anomalous geochemical signals from phosphatic Middle Cambrian rocks in the southern Georgina Basin. *Sedimentology* 35:549–570.

Donnelly, T. H., J. H. Shergold, P. N. Southgate, and C. J. Barnes. 1990. Events leading to global phosphogenesis around the Proterozoic/Cambrian boundary. In A. J. G. Notholt and I. Jarvis, eds., *Phosphorite Research and Development,* pp. 273–287. *Geological Society of London Special Publication* 52.

Dupret, L., E. Dissler, F. Doré, F. Gresselin, and J. Le Gall. 1990. Cadomian geodynamic evolution of the northeastern Armorican Massif (Normandy and Maine). In R. S. D'Lemos, R. A. Strachan, and C. G. Topley, eds., *The Cadomian Orogeny,* pp. 115–131. *Geological Society of London Special Publication* 51.

Ferrara, G. B., F. Lombardo, and S. Tonarini. 1983. Rb/Sr geochronology of granites and gneisses from the Mount Everest region, Nepal Himalaya. *Geologische Rundschau* 72:119–136.

Grotzinger, J. P., S. A. Bowring, B. Z. Saylor, and A. J. Kaufman. 1995. Biostratigraphic and geochronologic constraints on early animal evolution. *Science* 270:598–604.

Gurova, T. I. and L. S. Chernova, eds. 1988. *Litologiya i usloviya formirovaniya rezervuarov nefti i gaza Sibirskoy platformy* [Lithology and conditions of the formation of oil and gas reservoirs on the Siberian Platform]. Moscow: Nedra.

Hamdi, B., M. D. Brasier, and Z. Jiang. 1989. Earliest skeletal fossils from Precambrian-Cambrian boundary strata, Elburz Mountains, Iran. *Geological Magazine* 126:283–289.

Hsu, K. J., H. Oberhansli, J. Y. Gao, S. Sun, H. Chen, and U. Krahenbuhl. 1985. "Strangelove ocean" before the Cambrian explosion. *Nature* 316:809–811.

Husseini, M. I. and S. I. Husseini. 1990. Origin of the Infracambrian Salt Basins of the Middle East. In J. Brooks, ed., *Classic Petroleum Provinces*, pp. 279–292. *Geological Society of London Special Publication* 50.

Jenkins, R. J. F., J. F. Lindsay, and M. R. Walter. 1993. *Field Guide to the Adelaide Geosyncline and Amadeus Basin, Australia.* Australian Geological Survey Organisation Record 1993/35.

Jenkyns, H. C., A. S. Gale, and R. M. Corfield. 1994. Carbon- and oxygen-isotope stratigraphy of the English Chalk and Italian Scaglia and its palaeoclimatic significance. *Geological Magazine* 131:1–34.

Kaufman, A. J., S. B. Jacobsen, and A. H. Knoll. 1994. The Vendian record of Sr- and C-isotopic variations in seawater: Implications for tectonics and paleoclimate. *Earth and Planetary Science Letters* 84:27–41.

Keto, L. S. and S. B. Jacobsen. 1987. Nd and Sr isotopic variations of Early Paleozoic oceans. *Earth and Planetary Science Letters* 84:27–41.

Khomentovsky, V. V. and G. A. Karlova. 1993. Biostratigraphy of the Vendian-Cambrian beds and Lower Cambrian boundary in Siberia. *Geological Magazine* 130:29–45.

Kirschvink, J. L. 1992. A paleogeographic model for Vendian and Cambrian time. In J. W. Schopf and C. Klein, eds., *The Proterozoic Biosphere: A Multidisciplinary Study,* pp. 569–581. Cambridge: Cambridge University Press.

Kirschvink, J. L., M. Magaritz, R. L. Ripperdan, A. Yu. Zhuravlev, and A. Yu. Rozanov. 1991. The Precambrian-Cambrian boundary: Magnetostratigraphy and carbon isotopes resolve correlation problems between Siberia, Morocco, and South China. *GSA Today* 1:69–71, 87, 91.

Knoll, A. H. 1992. Biological and biogeochemical preludes to the Ediacaran radiation. In J. H. Lipps and P. W. Signor, eds., *Origin and Early Evolution of the Metazoa,* pp. 53–84. New York: Plenum Press.

Knoll, A. H., A. J. Kaufman, M. A. Semikhatov, J. P. Grotzinger, and W. Adams. 1995. Sizing up the sub-Tommotian unconformity in Siberia. *Geology* 23:1139–1143.

Knoll, A. H., R. K. Bambach, D. E. Canfield, and J. P. Grotzinger. 1996. Comparative Earth history and Late Permian mass extinction. *Science* 273:452–457.

Kontorovitch, A. E., M. M. Mandel'baum, V. S. Surkov, A. A. Trofimuk, and A. N. Zolotov. 1990. Lena-Tunguska Upper Proterozoic-Palaeozoic petroleum superprovince. In J. Brooks, ed., *Classic Petroleum Provinces,* pp. 473–489. Geological Society Special Publication 50.

Korsch, R. J., H. Mai, Z. Sun, and J. D. Gorter. 1991. The Sichuan Basin, southwest China: A late Proterozoic (Sinian) petroleum province. *Precambrian Research* 54: 45–63.

Kumarapeli, S., H. Pintson, and G. R. Dunning. 1989. Age of Tibbit Hill Formation and its implications for the timing of Iapetan rifting. *Geological Association of Canada, Annual Meeting, Program with Abstracts* 14: A125.

Landing, E. 1992. Lower Cambrian of southeastern Newfoundland: Epeirogeny and Lazarus faunas, lithofacies-biofacies linkages, and the myth of a global chronostratigraphy. In J. H. Lipps and P. W. Signor, eds., *Origin and Early Evolution of the Metazoa*, pp. 283–310. New York: Plenum Press.

Lindsay, J. F. 1993. Preliminary sequence stratigraphic comparison of the Neoproterozoic and Cambrian sections of the Yangtze Platform, China, and the Amadeus Basin, Australia. *Professional Opinion of the Australian Geological Survey Organization* 1993/002: 1–16.

Lindsay, J. F. and R. J. Korsch. 1991. The evolution of the Amadeus Basin, central Australia. *Bureau of Mineral Resources, Australia, Bulletin* 236: 7–32.

Lindsay, J. F. and J. F. Leven. 1996. Evolution of a Neoproterozoic to Palaeozoic intracratonic setting, Officer Basin, South Australia. *Basin Research* 8: 403–424.

Lindsay, J. F., R. J. Korsch, and J. R. Wilford. 1987. Timing the breakup of a Proterozoic supercontinent: Evidence from Australian intracratonic basins. *Geology* 15: 1061–1064.

Lindsay, J. F., M. D. Brasier, D. Dornjamjaa, P. Kruse, R. Goldring, and R. A. Wood. 1996. Facies and sequence controls on the appearance of the Cambrian biota in southwestern Mongolia: Implications for the Precambrian-Cambrian boundary. *Geological Magazine* 133: 417–428.

Logan, G. A., J. M. Hayes, G. B. Hieshima, and R. E. Summons. 1995. Terminal Proterozoic reorganization of biogeochemical cycles. *Nature* 376: 53–56.

Loosveld, R., A. Bell, and J. Terken. 1996. A concise tectonic history of Oman. In A. Wood, ed., *Exploration in Oman: New Ideas from Old Basins*, pp. 13–26. Muscat: Petroleum Development Oman.

Mattes, B. W. and S. Conway Morris. 1990. Carbonate/evaporite deposition in the late Precambrian–Early Cambrian Ara Formation of southern Oman. In A. H. F. Robertson, M. P. Searle, and A. C. Ries, eds., *The Geology and Tectonics of the Oman Region. Geological Society of London Special Publication* 49: 617–636.

McKerrow, W. S., C. R. Scotese, and M. D. Brasier. 1992. Early Cambrian continental reconstructions. *Journal of the Geological Society, London* 149: 599–606.

Millar, I. L. and B. C. Storey. 1995. Early Palaeozoic rather than Neoproterozoic volcanism and rifting within the Transantarctic Mountains. *Journal of the Geological Society, London* 152: 417–422.

Millson, J. A., C. G. L. Mercadier, S. E. Livera, and J. M. Peters. 1996. The Lower Palaeozoic of Oman and its context in the evolution of a Gondwana continental margin. *Journal of the Geological Society, London* 153: 213–230.

Moczydłowska, M. 1991. Acritarch biostratigraphy of the Lower Cambrian and the Precambrian-Cambrian boundary in southeastern Poland. *Fossils and Strata* 29: 1–127.

Myrow, P. M. and R. Hiscott. 1993. Depositional history and sequence stratigraphy of the Precambrian-Cambrian boundary stratotype section, Chapel Island Formation, Southeast Newfoundland. *Palaeogeography, Palaeoclimatology, Palaeoecology* 104:13–35.

Narbonne, G. M. and J. D. Aitken. 1990. Ediacaran fossils from the Sekwi Brook area, Mackenzie Mountains, northwestern Canada. *Palaeontology* 33:945–980.

Narbonne, G. M., A. H. Kaufman, and A. H. Knoll. 1994. Integrated chemostratigraphy and biostratigraphy of the Windermere Supergroup, northwestern Canada: Implications for Neoproterozoic correlations and early evolution of animals. *Geological Society of America Bulletin* 106:1281–1292.

Nicholas, C. J. 1994. New stratigraphical constraints on the Durness Group of NW Scotland. *Scottish Journal of Geology* 30:73–85.

Nicholas, C. J. 1996. The Sr isotope evolution of the oceans during the "Cambrian Explosion." *Journal of the Geological Society, London* 153:243–254.

Notholt, A. G. and M. D. Brasier. 1986. Regional review: Europe. In P. J. Cook and J. H. Shergold, eds., *Proterozoic and Cambrian Phosphorites*, pp. 91–100. Cambridge: Cambridge University Press.

Odin, G. S., N. H. Gale, B. Auvray, M. Bielski, F. Doré, J. R. Lancelot, and P. Pateels. 1983. Numerical dating of Precambrian-Cambrian boundary. *Nature* 301:21–23.

Palmer, A. R. 1981. Subdivision of the Sauk sequence. In M. E. Taylor, ed., *Short Papers for the Second International Symposium on the Cambrian System*, pp. 160–162. U.S. Geological Survey Open-File Report 81-743.

Powell, C. M., Z. X. Li, M. W. McElhinny, J. G. Meert, and J. K. Park. 1993. Paleomagnetic constraints on timing of the Neoproterozoic breakup of Rodinia and the Cambrian formation of Gondwana. *Geology* 21:880–892.

Ripperdan, R. 1994. Global variations in carbon isotope composition during the latest Neoproterozoic and earliest Cambrian. *Annual Review of Earth and Planetary Sciences* 22:385–417.

Ripperdan, R. L., M. Magaritz, R. S. Nicoll, and J. H. Shergold. 1992. Simultaneous changes in carbon isotopes, sea level, and conodont biozones within the Cambrian-Ordovician boundary interval at Black Mountain, Australia. *Geology* 20:1039–1042.

Rowell, A. J., M. N. Rees, E. M. Duebendorfer, E. T. Wallin, W. R. Schmus, and E. I. Smith. 1993. An active Neoproterozoic margin: Evidence from the Skelton Glacier area, Transantarctic Mountains. *Journal of the Geological Society, London* 150:677–82.

Rushton, A. W. A. 1974. The Cambrian of Wales and England. In C. H. Holland, ed., *Cambrian of the British Isles, Norden, and Spitsbergen*, pp. 43–122. London: John Wiley and Sons.

Ruzhentsev, S. V. and A. A. Mossakovsky. 1995. Geodinamika i tektonikoe razvitie paleozoid Tsentral'noy Azii kak rezul'tat vzaimodeystviya Tikhookeanskogo i Indo-Atlanticheskogo segmentov Zemli [Geodynamics and tectonic evolution of the Central Asian paleozoids as a result of interaction between the Pacific and Indo-Atlantic segments of Earth]. *Geotektonika* 1995 (4):29–47.

Sdzuy, K. and G. Geyer. 1988. The base of the Cambrian in Morocco. In V. H. Jacobshagen, ed., *The Atlas System of Morocco: Studies on Its Geodynamic Evolution*, pp. 91–106. *Lecture Notes in Earth Sciences* 15. Berlin: Springer Verlag.

Şengör, A. M. C., B. A. Natal'in, and V. S. Burtman. 1993. Evolution of the Altaid tectonic collage and Palaeozoic crustal growth in Eurasia. *Nature* 364:299–307.

Shanker, R. and V. K. Mathur. 1992. Precambrian-Cambrian sequence in Krol belt and additional Ediacaran fossils. *Geophytology* 22:27–39.

Smith, L. H., A. J. Kaufman, A. H. Knoll, and P. K. Link. 1994. Chemostratigraphy of predominantly siliciclastic Neoproterozoic successions: A case study of the Pocatello Formation and Lower Brigham Group, Idaho, USA. *Geological Magazine* 131:301–314.

Sochava, A. V. 1992. Kvazistatsionarnaya model' geokhimicheskikh tsiklov i evolyutsionnykh sobytiy na rubezhe dokembriya-fanerozoya [Quasi-steady model of geochemical cycles and evolutionary events at the Precambrian-Phanerozoic boundary]. *Izvestiya Rossiyskoy Akademii nauk, Seriya geologicheskaya* 6:41–56.

Thorogood, E. J. 1990. Provenance of pre-Devonian sediments of England and Wales: Sm-Nd isotopic evidence. *Journal of the Geological Society, London* 147:591–594.

Torsvik, T. H., M. A. Smethurst, J. G. Meert, R. Van der Voo, W. S. McKerrow, M. D. Brasier, B. A. Sturt, and H. J. Walderhaug. 1996. Continental break-up and collision in the Neoproterozoic and Palaeozoic: A tale of Baltica and Laurentia. *Earth-Science Reviews* 40:229–258.

Tucker, M. E. 1992. The Precambrian-Cambrian boundary: Seawater chemistry, ocean circulation, and nutrient supply in metazoan evolution, extinction, and biomineralization. *Journal of the Geological Society, London* 149:655–668.

Tucker, R. D. and W. S. McKerrow. 1995. Early Paleozoic chronology: A review in light of new U-Pb zircon ages from Newfoundland and Britain. *Canadian Journal of Earth Sciences* 32:368–379.

Tucker, R. D. and T. C. Pharaoh. 1991. U-Pb ages for Late Precambrian igneous rocks in southern Britain. *Journal of the Geological Society, London* 148:435–443.

Veizer, J., W. T. Holser, and V. P. Wilgus. 1980. Correlation of $^{13}C/^{12}C$ and $^{34}S/^{32}S$ secular variations. *Geochimica et Cosmochimica Acta* 44:579–587.

Veizer, J., W. Compston, N. Clauer, and M. Schidlowski. 1983. $^{87}Sr/^{86}Sr$ in Late Proterozoic carbonates: Evidence for a "mantle event" at 900 Ma ago. *Geochimica et Cosmochimica Acta* 47:295–302.

Young, G. M. 1995. Are Neoproterozoic glacial deposits preserved on the margins of Laurentia related to the fragmentation of two supercontinents? *Geology* 23:153–156.

Wang, H. 1986. The Sinian System. In Z. Yang, Y. Cheng, and H. Wang, eds., *The Geology of China*, pp. 50–63. Oxford: Clarendon Press.

Watts, A. B. 1982. Tectonic subsidence, flexure, and global changes of sea level. *Nature* 297:469–474.

Zhuravlev, A. Yu. and R. A. Wood. 1996. Anoxia as the cause of the mid-Early Cambrian (Botomian) extinction event. *Geology* 24:311–314.

CHAPTER FIVE

Toni T. Eerola

Climate Change at the Neoproterozoic-Cambrian Transition

Varangerian and lower Sinian glacial deposits are found in Argentina, Uruguay, Mato Grosso (Brazil), Namibia, Laurentia, and probably southern Brazil, which were all situated close together during Neoproterozoic-Cambrian times. According to continental paleoreconstructions, glacial deposits of these regions, together with those of Scotland, Scandinavia, Greenland, Russia, Antarctica, and Australia, formed the Varangerian-Sinian Glacial Zone of the supercontinent Rodinia. Tectonic activity associated with the amalgamation of Rodinia and Gondwana was probably related to the origin of these deposits, as in the case of mountain glaciers that formed in uplifted areas of fragmenting or colliding parts of this supercontinent. In such circumstances, the Pan-African and Brasiliano orogenies and the site of opening of the Iapetus Ocean would have been in key positions. However, some paleomagnetic reconstructions locate these regions near the South Pole, where glaciers could have formed even in the absence of tectonic events. In this case, the change to warm climate and the evolutionary explosion of the Cambrian could have been due to rapid shift of continents to equatorial latitudes, although these changes might also have been triggered by supercontinent breakup. These events are reflected in the isotopic records of strontium and carbon, which provide some of the best available indicators of the climatic and environmental changes that occurred during the Neoproterozoic-Cambrian transition. They also appear to reveal the occurrence of a discrete cold period in the Cambrian: the disputed lower Sinian glaciation.

INTRODUCTION

The Neoproterozoic-Cambrian transition was characterized by ophiolite formation (Yakubchuk et al. 1994), the formation and breakup of supercontinents (e.g., Bond et al. 1984), the Cambrian evolutionary explosion (Moores 1993; Knoll 1994), and intense climatic changes, among which the most important might be considered glacia-

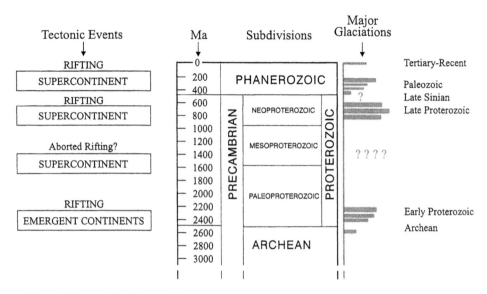

Figure 5.1 Time distribution of glaciogenic sedimentary rocks, showing their sporadic nature and possible relationship with supercontinentality. *Source:* Modified from Young 1991.

tions (e.g., Hambrey and Harland 1985) and the shift from Neoproterozoic icehouse to Cambrian greenhouse conditions (Veevers 1990; Tucker 1992).

At least 10 major glacial periods have been recorded prior to the Pleistocene (Young 1991; Eyles 1993) (figure 5.1). Probably the most extensive and enigmatic of these occurred during the Neoproterozoic and at the beginning of the Cambrian, at ~900–540 Ma (Hambrey and Harland 1985; Young 1991; Eyles 1993; Meert and Van der Voo 1994). There are signs of four Neoproterozoic-Cambrian glacial periods (figures 5.1–5.2): the Lower Congo (~900 Ma), the Sturtian (~750–700 Ma), the Varangerian (~650–600 Ma), and the lower Sinian (~600–540 Ma) (Hambrey and Harland 1985; Eyles 1993; Meert and Van der Voo 1994). There are, however, also proposals for only two (Kennedy et al. 1998) or even five (Hoffman et al. 1998a; Saylor et al. 1998).

This chapter presents a brief overview of Neoproterozoic-Cambrian climate changes and events, with emphasis on the Varangerian and lower Sinian glacial periods and the subsequent global warming in the Cambrian (see also chapters in this volume by Brasier and Lindsay; Seslavinsky and Maidanskaya; Smith; and Zhuravlev).

PALEOMAGNETIC RECONSTRUCTIONS AND GLACIERS

The application of paleomagnetic investigations to research into the Neoproterozoic has yielded important findings. It is now recognized that continental drift may have been faster than at present (Gurnis and Torsvik 1994) and that glaciers might have formed at sea level even in low latitudes (e.g., Hambrey and Harland 1985; Schmidt

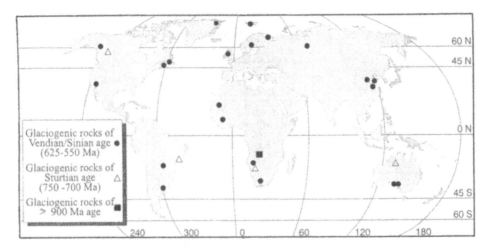

Figure 5.2 Locations of some glaciogenic deposits formed during the 1000–540 Ma interval.
Source: Modified from Meert and Van der Voo 1994.

and Williams 1995), implying a significant climatic paradox (Chumakov and Elston 1989).

The glacial interpretation of many Neoproterozoic deposits was questioned by Schemerhorn (1974). Many factors have been presented to explain the generation of glaciers at low latitudes (see Meert and Van der Voo 1994), such as the incorrect interpretation of paleolatitudes due to remagnetization (e.g., Gurnis and Torsvik 1994); global glaciation, i.e., "the snow-ball Earth" (Kasting 1992; Kirschvink 1992; Hoffman et al. 1998b); astronomical causes, such as modification of the obliquity of the earth's rotation (Williams 1975; Schmidt and Williams 1995); and tectonic causes, such as the formation of mountain glaciers in rift and collisional zones of supercontinents (Eyles 1993; Eyles and Young 1994; Young 1995).

According to Dalziel et al. (1994) and Gurnis and Torsvik (1994), continents were situated close to the southern pole during the Vendian (figure 5.3), in which case continental glaciation would be expected. Meert and Van der Voo (1994) argued, however, that continents occupied middle latitude position at that time.

SUPERCONTINENTS, CORRELATIONS, AND THE VARANGERIAN–LOWER SINIAN GLACIAL ZONE

Glacial horizons are often treated as the best markers for stratigraphic correlation (e.g., Hambrey and Harland 1985; Christie-Blick et al. 1995), although this has been contested by Chumakov (1981). Varangerian glacial deposits, ~600 Ma (figure 5.3), seem to be correlative in Namibia (Numees Formation, Gariep Group), in Laurentia (e.g., Gaskiers and Ice Brook formations; Eyles and Eyles 1989; Young 1995), and

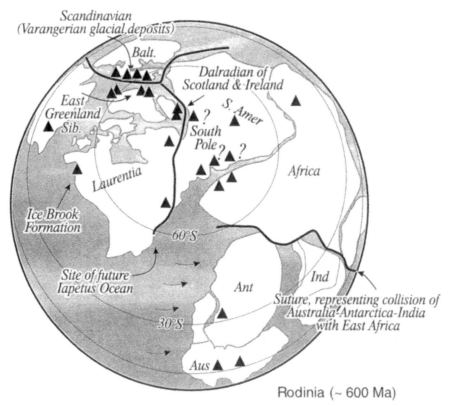

Rodinia (~ 600 Ma)

Figure 5.3 Reconstruction of the Neopro-terozoic supercontinent Rodinia, at ~600 Ma (modified from Dalziel et al. 1994) and its coeval glaciogenic record: the Varangerian–Lower Sinian Glacial Zone (cf. Eerola and Reis 1995; Young 1995). Deposits of Antarctica (Stump et al. 1988) and Australia (Schmidt and Williams 1995) are also included (cf. Eerola 1996).

possibly also in the Santa Bárbara Basin, Rio Grande do Sul State, southern Brazil (Eerola 1995, 1997; Eerola and Reis 1995). Coeval glacial deposits in the present-day North Atlantic region have also been related to these (e.g., Hambrey 1983). Glacial formations of similar age are also found in Mato Grosso and Minas Gerais, Brazil (Uh-lein et al. 1999), western Brazil (Alvarenga and Trompette 1992), Argentina (Spalletti and Del Valle 1984), and possibly Uruguay (F. Preciozzi, pers. comm., 1994) (see figures 5.2 and 5.3). Evidences for lower Sinian cold climate are found in West Gondwana (Schwarzrand Subgroup, Nama Group in Namibia [Germs 1995]; and the Taoudenni Basin in West Africa [Bertrand-Sarfati et al. 1995; Trompette 1996]) and in China and Kazakhstan (Hambrey and Harland 1985). A glacial deposit of Cambrian age has been tentatively identified in the Itajaí Basin, Santa Catarina State, southern Brazil (P. Paim, pers. comm., 1996), but the origin and age have still to been confirmed.

Given that Laurentia and Fennoscandia were situated close to South America in Neoproterozoic-Cambrian times, forming the supercontinent Rodinia (e.g., Bond et al. 1984; Dalziel et al. 1994; Young 1995) (figure 5.3), extensive glaciation is possible (Meert and Van der Voo 1994). Such connections may play an important role in paleogeographic reconstructions.

According to the paleogeography of Dalziel et al (1994), the glacial formations at 600 Ma constituted a continuous zone that can be traced from Svalbard, through Scandinavia, Greenland, and Scotland, to eastern Laurentia and western South America (Young 1995) (figure 5.3). Eerola and Reis (1995) and Eerola (1996) called this zone the Varangerian-Sinian Glacial Zone, on the basis of the ages of the glacial deposits, and suggested that the zone appears to continue to Mato Grosso, Argentina, probably to Uruguay, southern Brazil, Namibia, Antarctica (Nimrod area, Stump et al. 1988), and Australia (Marinoan glacial deposits; e.g., Schmidt and Williams 1995). The tectonics of Rodinia probably had a strong influence on the generation and distribution of these glacial deposits (Eyles 1993; Moores 1993; Young 1995).

DEBATE ON THE SEDIMENTARY RECORD
OF NEOPROTEROZOIC GLACIATIONS

Although the existence of Neoproterozoic glaciations is widely accepted, there have been authors who have questioned the concept with reference to some particular deposits, for instance, the Bigganjargga tillite in northern Norway (figure 5.4) (Crowell 1964; Jensen and Wulff-Pedersen 1996), some parts of the basal Windermere Group in Canada (Mustard 1991), and the Schwarzrand Subgroup of the Nama Group in Namibia (P. Crimes, pers. comm., 1995; Saylor et al. 1995). The whole concept of the Neoproterozoic glaciation was put in doubt by Schemerhorn (1974) and recently criticized by P. Jensen (pers. comm., 1996). The problem is that in the case of some Neoproterozoic deposits, the simple presence of diamictites has been considered sufficient proof of glacial origin (Schemerhorn 1974; Eyles 1993; Jensen and Wulff-Pedersen 1996).

Distinguishing between the results of glacial and other processes is a difficult task, both in ancient sequences (Chumakov 1981) and in more recent deposits—for instance, in alluvial fan facies (Carraro 1987; Kumar et al. 1994; Marker 1994; Owen 1994; Hewitt 1999), especially in volcanic settings (Ui 1989; Eyles 1993), and even when glacial influence is evident (Vinogradov 1981; Clapperton 1990; LeMasurier et al. 1994). The problem is that a variety of processes can generate deposits that may easily be confused with those of glaciation (e.g., Crowell 1957; Eyles 1993; Bennett et al. 1994). This is especially true in relation to diamictites (figure 5.5), which could also result from mud flows, debris flows, lahars, debris-avalanches, or meteorite impacts in many different environments (Crowell 1957, 1964; Ui 1989; Rampino 1994) and are not, in themselves, climatic indicators (Crowell 1957, 1964; Heezen and Hol-

Figure 5.4 The Neoproterozoic diamictite and striated pavement, Bigganjargga Tillite, Smalfjord Formation, northern Norway.

lister 1964; Schemerhorn 1974, 1983; Eyles 1993; Jensen and Wulff-Pedersen 1996). Dropstones (figure 5.6) may be generated by many processes other than glacial rafting—for instance, by turbidites (Crowell 1964; Heezen and Hollister 1964; Donovan and Pickerill 1997) or volcanic bombs (Bennett et al. 1994). Consequently, they too have been queried as climatic indicators (e.g., Crowell 1964; Bennett et al. 1994; Bennett and Doyle 1996; Donovan and Pickerill 1997). Although varves indicate climatic seasonality, they can also occur under warm climatic conditions, such as in the

Figure 5.5 Clast of boulder size in Neoproterozoic diamictite, Passo da Arcia
sequence. Lavras do Sul, southern Brazil. Note rhythmic shales above.

Santa Barbara Basin of present-day California (Thunell et al. 1995). Even striated and
faceted clasts and pavements are not exclusive to glaciated terrains, because these can
also be generated by mud flows and lahars (e.g., Crowell 1964; Eyles 1993; Jensen
and Wulff-Pedersen 1996). There has been much debate, for example, in relation to
striations below the Smalfjord Formation (see figure 5.4) (Crowell 1964; Jensen and
Wulff-Pedersen 1996; Edwards 1997). Interpretation of the shape and surface textures of sediment grains as possible indicators of ancient glacial deposits has been

Figure 5.6 Supposed dropstone of Neoproterozoic age, Passo da Arcia sequence, Lavras do Sul, southern Brazil.

contested by Mazzullo and Ritter (1991). A great variety of clast lithology is also not sufficient to establish the glacial origin of deposits (e.g., Jensen and Wulff-Pedersen 1996). Similarly, clast similarity does not necessarily reflect absence of glacial influence but çan merely reflect provenance. Jensen and Wulff-Pedersen (1996) suggested that in relation to the Bigganjargga Tillite (Smalfjord Formation), local provenance is evidence against glacial transport. This view is contested by H. Hirvas and K. Nenonen (pers. comm., 1996) and Edwards (1997), because in southern Finland, for example, boulders and other clasts in Quaternary till deposits were transported only about 3–6 km, so that they strongly reflect the local geology (Perttunen 1992).

Troughlike sandstone downfolds and dykes, similar to those caused by cryoturbation (subaerial frost churning), have been considered as proof of cold climate (e.g., Spencer 1971). These features could, however, also be produced by subaqueous gravitational loading (e.g., Eyles and Clark 1985).

In this context, there seems to be no diagnostic evidence that consistently proves glacial influence, either in the Neoproterozoic (P. Jensen, pers. comm., 1996) or at any other time prior to the present day. Perhaps the only reliable evidence is provided by indications of large transport distance on stable platforms and by the occurrence of "bullet" boulders (Eyles 1993).

Ojakangas (1985) argued that the occurrence of the diamictite-dropstone association is sufficient to characterize Proterozoic glacial deposits. This, however, is not the case, especially in volcanic sequences, as discussed above. There are many supposed Neoproterozoic glacial sequences related to volcanism (Schemerhorn 1983; Eyles and Eyles 1989; Eyles 1993; Eyles and Young 1994; Eerola 1995, 1997) in which the recognition of a glacial contribution could be ambiguous but in which the absence or dearth of volcanic clasts would support a glacial interpretation.

Crowell (1964) and P. Jensen (pers. comm., 1996) argued that in environments lacking vegetation, and with intense tectonism, as during the Neoproterozoic, the generation of diamictites by debris flows is to be expected. However, no proposals for glaciation, for instance, in the Mesozoic, have been made based on the simple occurrence of diamictites (P. Jensen, pers. comm., 1996). Mesozoic glaciation has, however, been proposed on the basis of supposed dropstones (see references in Bennett and Doyle 1996). Intense and worldwide tectonic activity (rifting and/or orogeny) in the Neoproterozoic could have produced extensive debris flows due to uplift (Crowell 1964; Schemerhorn 1974, 1983). These are the same tectonic zones that are argued to have generated mountain glaciers by Eyles (1993), Eyles and Young (1994), and Young (1995). The lack of vegetation, however, does not provide an explanation for coeval dropstones and extensive marine diamictites.

The other problem is that supposed glacial deposits do not occur in all coeval Neoproterozoic basins and sequences, notably on stable platforms (Schemerhorn 1983). If, however, the glacial interpretation for most of the inferred Neoproterozoic deposits is correct, then their localized preservation seems to be evidence against the worldwide glaciation of Hambrey and Harland (1985) and in favor of the occurrence of local glaciers in uplifted areas, as argued by Schemerhorn (1983), Eyles (1993), Eyles and Young (1994), and Young (1995).

ISOTOPIC RECORD

Probably the strongest evidence for environmental change in the Neoproterozoic-Cambrian transition is provided by the stable isotopic records of strontium and carbon, a subject that has been extensively studied in recent years (e.g., Asmerom et al. 1991; Tucker 1992; Kaufman et al. 1993; Derry et al. 1994; Kaufman and Knoll 1995; Nicholas 1996; Hoffman et al. 1998a,b; Saylor et al. 1998; Myrow and Kaufman 1999; Prave 1999; Brasier and Lindsay, this volume). The strontium isotopic record demonstrates the influence of weathering and erosion rates, and variations in the hydrothermal flux from the mid-ocean ridges, on seawater composition, i.e., the relationships among climatic, oceanographic, and tectonic events. Variations in the carbon isotopic record show the influence of burial of organic matter and the relationships between oceanography and climate (Donnelly et al. 1990; Kaufman et al. 1993).

Negative $\delta^{13}C$ excursions coincide with the Neoproterozoic Sturtian, Varangerian, and lower Sinian glaciations, while $^{87}Sr/^{86}Sr$ values rise almost continuously (e.g.,

Donnelly et al. 1990; Kaufman et al. 1993; Kaufman and Knoll 1995; Saylor et al. 1998) (see figure 5.1). These indicate variations in weathering rate, hydrothermal flux, and organic matter burial, reflecting climatic change and tectonic events. Weathering rate and the production and burial of organic matter both decline in cold climates, and there is significant oceanic overturning (e.g., Kaufman and Knoll 1995; Kimura et al. 1997; Myrow and Kaufman 1999).

The Neoproterozoic-Cambrian strontium isotopic record seems to provide evidence in favor of glaciations and may also relate to tectonic uplift (Asmerom et al. 1991; Derry et al. 1994), linking these events and supporting the views of Schemerhorn (1983), Eyles (1993), Eyles and Young (1994), Young (1995), Prave (1999), and Uhlein et al. (1999) on possible tectonic influence in the generation of glaciers. The Pan-African and Avalonian orogenies have been cited as possible uplifted sources of abundant ^{87}Sr to seawater (Asmerom et al. 1991; Derry et al. 1994). In this sense, the coeval and related Brasiliano orogeny, which affected a large part of the Brazilian shield, causing vigorous uplift and the generation of numerous molasse basins (e.g., Chemale 1993), should also be considered.

LATE SINIAN GLACIATION?

While the ^{87}Sr/^{86}Sr isotopic ratio continued to rise at the beginning of the Cambrian (Asmerom et al. 1991; Tucker 1992; Kaufman et al. 1993; Derry et al. 1994; Kaufman and Knoll 1995; Nicholas 1996), the δ^{13}C value declined near the Precambrian-Cambrian boundary and at the beginning of the Cambrian (Donnelly et al. 1990; Kaufman and Knoll 1995; Saylor et al. 1998), coinciding with the proposed late Sinian glaciation (Hambrey and Harland 1985). This cold period was probably of short duration and low intensity, as argued by Hambrey and Harland (1985), Meert and Van der Voo (1994), Germs (1995), and Saylor et al. (1998). It was probably related to tectonic uplift and erosion of the Brasiliano–Pan-African orogenies, which were in a post- to late-orogenic stage during the Cambrian (Chemale 1993; Derry et al. 1994; Germs 1995). The cold period at the Neoproterozoic-Cambrian transition was of very limited extent (Hambrey and Harland 1985; Meert and Van der Voo 1994; Germs 1995; Saylor et al. 1998), being recorded only in West Africa (Bertrand-Sarfati et al. 1995; Trompette 1996), probably in the Nama Group of Namibia (e.g. Germs 1995; Saylor et al. 1998), and in China, Kazakhstan, and Poland (see Hambrey and Harland 1985). The evidence for the occurrence of the Cambrian glaciation has, however, been contested (Derry et al. 1994; Saylor et al. 1995; Kennedy et al. 1998). Derry et al. (1994) noted a fall in the ^{87}Sr/^{86}Sr isotopic ratio at the Neoproterozoic-Cambrian boundary and in the beginning of the Early Cambrian, which they attributed to one or more of (1) reduced rates of tectonic uplift or climate change and decreased weathering, (2) changes in the type of crust undergoing erosion, (3) rift-associated volcanic activity, and (4) worldwide marine transgression.

The link between uplift, ^{87}Sr/^{86}Sr rise, and glaciation has, however, been contested by Kaufman et al. (1993). They suggest that Cenozoic uplift of the Himalayas and Andes corresponds with both $^{87}Sr/^{86}Sr$ rise and glaciation and argue that detailed comparison of the Cenozoic and Vendian suggests more complex relationships among climate, continental erosion rates, glaciation, and changes in atmospheric CO_2 than previously envisioned. Raymo (1991) also refers to problems arising from such links in the Mesozoic and Cenozoic.

Although study of Neoproterozoic-Cambrian isotopic curves has revealed clear trends, many uncertainties surround them, and their correct interpretation will probably continue to be a matter of controversy and debate for some time.

NEOPROTEROZOIC-CAMBRIAN TRANSITION AND CAMBRIAN CLIMATES

Despite profound disagreement concerning many aspects of Cambrian climates, there is consensus that the most significant events relating to the Neoproterozoic-Cambrian transition (544 Ma; Brasier et al. 1994) were global warming, sea level rise, extensive phosphogenesis, and marine biodiversification at ~570–540 Ma (e.g., Cook 1992; Kaufman et al. 1993; Moores 1993; Knoll 1994; Zhuravlev, this volume). These changes, which are also registered as strontium and carbon isotopic variations in marine sediments (e.g., Asmerom et al. 1991; Tucker 1992; Kaufman et al. 1993; Derry et al. 1994; Kaufman and Knoll 1995; Nicholas 1996; Brasier and Lindsay, this volume), were probably related to the breakup of Rodinia, which resulted in the formation of new oceans and shallow seas, and affected seawater and atmospheric composition and circulation patterns (e.g., Bond et al. 1984; Donnelly et al. 1990; Kaufman et al. 1993; Moores 1993; Knoll 1994; Kimura et al. 1997). Almost without exception, wherever there are signs of Neoproterozoic-Cambrian glacial influence, these occur with, or are followed by, warm climate indicators such as red beds, phosphate and evaporite deposits, or carbonates (e.g., Hambrey and Harland 1985; Chumakov and Elston 1989; Eyles 1993). In the Rio Grande do Sul State of southern Brazil, an inferred Neoproterozoic glacial deposit (Eerola 1995) is succeeded by the Cambrian Guaritas Formation, which is composed of red beds and aeolian dunes thought to be formed in a warm desert (e.g., De Ros et al. 1994; Eerola and Reis 1995; Paim 1995; Eerola 1997). According to Gurnis and Torsvik (1994), this climate change was due to the rapid drift of continents from the South Pole to equatorial latitudes. However, Schemerhorn (1983), Veevers (1990), Raymo (1991), Tucker (1992), Kaufman et al. (1993), and Saylor et al. (1998) argued that variations in atmospheric CO_2, controlled by the episodic uplift, volcanic activity, and erosion of major mountain ranges, should have an important (if not the most important) influence on global temperatures; i.e., lowering of atmospheric CO_2 levels in the Neoproterozoic could have produced glaciation, and the reverse could have led to climate warming in the Cambrian. The combined evidence is, although indirectly, against application of the greenhouse model to the Early Cambrian earth (A. Zhuravlev, pers. comm., 1996). Increasing evidence of Early Cambrian cold climate (e.g., Trompette 1996; Saylor et al. 1998) makes the coldhouse model of Tucker (1992) more applicable to this period. This assumes transitional conditions existing after the melting of polar icecaps until high-latitude temperatures exceeded 5°C, when greenhouse conditions would take over. In these circumstances, there should be intermediate sea-level, vigorous oceanic circulation due to the temperature gradient (which also promotes nutrient supply), increased CO_2 supply from the atmosphere to the oceans, and rise in the carbonate compensation depth (which is one of the requirements for aragonite-sea conditions in the Early Cambrian) (A. Zhuravlev, pers. comm.). In addition, the coldhouse model obviates the need for unusually rapid continental drift proposed by Gurnis and Torsvik (1994).

CONCLUSIONS

Despite the lack of consensus regarding many aspects of Neoproterozoic-Cambrian climates and events, this interval was characterized by intense environmental change, as is evident from the sedimentary and isotopic records. The sedimentary evidence for Neoproterozoic glaciations has been questioned by some authors. Schemerhorn (1974), Eyles and Young (1994), and Jensen and Wulff-Pedersen (1996) have called for objective studies to determine the proportion of glacial components and the origin and extent of glacial activity in diamictite deposits, and for improved definition of the tectonic and depositional setting, paleoclimate modeling, and geochronologic and paleomagnetic control of the glacial record. Because of difficulties and uncertainties of interpreting sedimentary deposits, isotopic records appear to provide some of the best available indicators of climatic and environmental change for the Neoproterozoic-Cambrian transition. The isotopic record seems to indicate glaciations in Sturtian and Varangerian times. It also appears to link periods of uplift, weathering, erosion, atmospheric CO_2, and glaciation, supporting the evidence for local mountain glaciers in uplifted areas in the Varangerian with a probable extension to the lower Sinian.

Together with increasing evidence for a renewed cold period in the Cambrian, the isotopic record suggests the unique nature of the period. On the basis of strontium and carbon isotopic studies, the tectonism of Rodinia, especially that related to the Brasiliano–Pan-African belts and the opening of Iapetus, seems to be a key factor affecting climate change during the Neoproterozoic-Cambrian transition. In this sense, investigations in South America and surrounding areas are important. South America might play an important role in future discussions concerning Laurentia-Gondwana interactions and Neoproterozoic-Cambrian climate change.

Acknowledgments. I wish to thank the editors, A. Zhuravlev and R. Riding for inviting me to participate in this volume. Discussions with R. da Cunha Lopes, M. Eronen, A. Garcia, H. Hirvas, P. Jensen, J. Kohonen, J. Marmo, K. Nenonen, P. Paim, P. Crimes, L. Pesonen, and G. Young were very helpful in preparation of this article. J. Kohonen,

L. Pesonen, and J. Rantataro revised the first version of the manuscript, and the final review was made by G. Young, A. Zhuravlev, and J. Karhu, whom I thank for their criticism and comments, which greatly improved the work. The English was checked by G. Häkli. This work was supported by the Geological Survey of Finland and the Department of Geology of the University of Helsinki and is a contribution to IGCP Projects 319, 320, 366, 368, and 440. It is dedicated to the memory of K. Rankama, a Finnish geologist and researcher who, among his numerous activities, studied also Neoproterozoic glacial deposits in Namibia and Australia and who gave me the impetus to start the work related to southern Brazil.

REFERENCES

Alvarenga, C. J. S and R. Trompette. 1992. Glacially influenced sedimentation in the Later Proterozoic of the Paraguay belt (Mato Grosso, Brazil). *Palaeogeography, Palaeoclimatology, Palaeoecology* 92:85–105.

Asmerom, Y., S. B. Jacobsen, A. H. Knoll, N. J. Butterfield, and K. Swett. 1991. Strontium isotopic variations of Neoproterozoic seawater: Implications for crustal evolution. *Geochimica et Cosmochimica Acta* 55: 2883–2894.

Bennett, M. R. and P. Doyle. 1996. Global cooling inferred from dropstones in the Cretaceous: Fact or wishful thinking? *Terra Nova* 8:182–185.

Bennett, M. R., P. Doyle, A. E. Mather, and J. L. Woodfin. 1994. Testing the climatic significance of dropstones: An example from southeast Spain. *Geological Magazine* 131:845–848.

Bertrand-Sarfati, J., A. Moussine-Pouchkini, B. Amard, and A. Aït Kaci Ahmed. 1995. First Ediacaran fauna found in western Africa and evidence for an early Cambrian glaciation. *Geology* 23:133–136.

Bond, G. C., P. A. Nickeson, and M. A. Kominz. 1984. Breakup of a supercontinent between 625 Ma and 555 Ma: New evidence and implications for continental histories. *Earth and Planetary Science Letters* 70:325–345.

Brasier, M., J. Cowie, and M. Taylor. 1994. Decision on the Precambrian-Cambrian boundary stratotype. *Episodes* 17:3–8.

Carraro, F. 1987. Remodelling and reworking as causes of error in distinguishing between glacial and non-glacial deposits and landforms. In R. Kujansuu and M. Saarnisto, eds., *INQUA Till Symposium*, pp. 39–48. Geological Survey of Finland, Special Paper 3.

Chemale, F., Jr. 1993. Bacias molássicas Brasilianas. *Acta Geologica Leopoldensia* 16: 109–118.

Christie-Blick, N., I. A. Dyson, and C. C. Von der Borch. 1995. Sequence stratigraphy and the interpretation of Neoproterozoic Earth history. *Precambrian Research* 73:3–26.

Chumakov, N. M. 1981. Upper Proterozoic glaciogenic rocks and their stratigraphic significance. *Precambrian Research* 15: 373–195.

Chumakov, N. M. and D. P. Elston. 1989. The paradox of late Proterozoic glaciations at low latitudes. *Episodes* 12:120.

Clapperton, C. M. 1990. Glacial and volcanic geomorphology of the Chimborazo-Carihuairazo Massif, Ecuadorian Andes. *Transactions of the Royal Society of Edinburgh (Earth Sciences)* 81:91–116.

Cook, P. J. 1992. Phosphogenesis around the Proterozoic-Cambrian transition. *Journal of the Geological Society, London* 149:615–620.

Crowell, J. C. 1957. Origin of pebbly mudstones. *Geological Society of America Bulletin* 68:993–1010.

Crowell, J. C. 1964. Climatic significance of sedimentary deposits containing dispersed megaclasts. In A. E. M. Nairn, ed., *Problems in Paleoclimatology*, pp. 87–99. Proceedings, NATO Paleoclimates Conference, Newcastle upon Tyne, January 1963. London: John Wiley and Sons Interscience.

Dalziel, I. W. D., L. H. Dalla Salda, and L. M. Gahagan. 1994. Paleozoic Laurentia-Gondwana interaction and the origin of the Appalachian-Andean mountain system. *Geological Society of America Bulletin* 106:243–252.

De Ros, L. F., S. Morad, and P. S. G. Paim. 1994. The role of detrital composition and climate on the diagenetic evolution of continental molasses: Evidence from the Cambro-Ordovician Guaritas Sequence, southern Brazil. *Sedimentary Geology* 92:197–228.

Derry, L. A., M. D. Brasier, R. M. Corfield, A. Yu. Rozanov, and A. Yu. Zhuravlev. 1994. Sr and C isotopes in Lower Cambrian carbonates from the Siberian craton: A paleoenvironmental record during the "Cambrian explosion." *Earth and Planetary Science Letters* 128:671–681.

Donnelly, T. H., J. H. Shergold, P. N. Southgate, and C. J. Barnes. 1990. Events leading to global phosphogenesis around the Proterozoic/Cambrian boundary. In A. J. Notholt and I. Jarvis, eds., *Phosphorite Research and Development*, pp. 273–87. Geological Society Special Publication 52.

Donovan, S. K. and R. K. Pickerill. 1997. Dropstones: Their origin and significance: A comment. *Palaeogeography, Palaeoclimatology, Palaeoecology* 131:175–178.

Edwards, M. B. 1997. Comments: Discussion of glacial or non-glacial origin for the Bigganjargga Tillite, Finnmark, northern Norway. *Geological Magazine* 134:873–874.

Eerola, T. T. 1995. From ophiolites to glaciers? Review on geology of the Neoproterozoic-Cambrian Lavras do Sul region, southern Brazil. In S. Autio, ed., *Geological Survey of Finland, Current Research 1993–1994*, pp. 5–16. Geological Survey of Finland, Special Paper 20.

Eerola, T. T. 1997. Neoproterozoic glaciation in southern Brazil? *Gondwana Newsletter* 8:6.

Eerola, T. T. and M. R. Reis. 1995. The Neoproterozoic glacial record and the Passo da Areia Sequence in the Lavras do Sul region, southern Brazil. In P. Heikinheimo, ed., *International Conference on Past, Present, and Future Climate*, pp. 52–55. Proceedings, SILMU Conference, Helsinki, Finland, 22–25 August 1995. *Publications of the Academy of Finland* 6.

Eyles, N. 1993. Earth's glacial record and its tectonic setting. *Earth Science Reviews* 35:1–248.

Eyles, N. and B. M. Clark. 1985. Gravity-induced soft-sediment deformation in glaciomarine sequences of the Upper Proterozoic Port Askaig Formation, Scotland. *Sedimentology* 32:789–814.

Eyles, N. and C. H. Eyles. 1989. Glacially influenced deep-marine sedimentation of the Late Precambrian Gaskiers Formation, Newfoundland, Canada. *Sedimentology* 36:601–620.

Eyles, N. and G. M. Young. 1994. Geodynamic controls on glaciation in Earth history. In M. Deynoux, J. G. M. Miller, E. W. Domack, N. Eyles, I. J. Fairchild, and G. M. Young, eds., *Earth's Glacial Record*, pp. 1–27. London: Cambridge University Press.

Germs, G. J. B. 1995. The Neoproterozoic of southwestern Africa, with emphasis on platform stratigraphy and paleontology. *Precambrian Research* 73:137–151.

Gurnis, M. and T. H. Torsvik. 1994. Rapid drift of large continents during the late Precambrian and Paleozoic: Paleomagnetic constraints and dynamic models. *Geology* 22:1023–1026.

Hambrey, M. J. 1983. Correlation of Late Precambrian tillites in the North Atlantic region and Europe. *Geological Magazine* 120:209–320.

Hambrey, M. J. and W. B. Harland. 1985. The Late Proterozoic glacial era. *Palaeogeography, Palaeoclimatology, Palaeoecology* 51:255–272.

Heezen, B. C. and C. Hollister. 1964. Turbidity currents and glaciation. In *Problems in Paleoclimatology*, pp. 99–112. *Proceedings, NATO Paleoclimates Conference, Newcastle upon Tyne, January 1963*. London: John Wiley and Sons Interscience.

Hewitt, K. 1999. Quaternary moraines vs. catastrophic rock avalanches in the Karakoram Himalaya, northern Pakistan. *Quaternary Research* 51:220–237.

Hoffman, P. F., A. J. Kaufman, and G. P. Halverson. 1998a. Comings and goings of global glaciation on a Neoproterozoic tropical platform in Namibia. *GSA Today* 8:1342–1346.

Hoffman, P. F., A. J. Kaufman, G. P. Halverson, and D. P. Schrag. 1998b. A Neoproterozoic snowball Earth. *Science* 281:1342–1346.

Jensen, P. A. and E. Wulff-Pedersen. 1996. Glacial or non-glacial origin for the Bigganjargga Tillite, Finnmark, northern Norway. *Geological Magazine* 133:137–145.

Kasting, J. F. 1992. Proterozoic climates: The effect of changing atmospheric carbon dioxide concentrations. In J. W. Schopf and C. Klein, eds., *The Proterozoic Biosphere: A Multidisciplinary Study*, pp. 165–168. Cambridge: Cambridge University Press.

Kaufman, A. J. and A. H. Knoll. 1995. Neoproterozoic variations in the C-isotopic composition of seawater: Stratigraphic and biogeochemical implications. *Precambrian Research* 73:27–49.

Kaufman, A. J., S. B. Jacobsen, and A. H. Knoll. 1993. The Vendian record of Sr and C isotopic variations in seawater: Implications for tectonics and paleoclimate. *Earth and Planetary Science Letters* 120:409–430.

Kennedy, M. J., B. Runnegar, A. R. Prave, K.-H. Hoffman, and M. A. Arthur. 1998. Two or four Neoproterozoic glaciations? *Geology* 26:1059–1063.

Kimura, H., R. Matsumoto, Y. Kakuwa, B. Hamdi, and H. Zibaseresht. 1997. The Vendian-Cambrian $\delta^{13}C$ record, north Iran: Evidence for overturning the ocean before the Cambrian explosion. *Earth and Planetary Science Letters* 147: E1–E7.

Kirschvink, J. L. 1992. Late Proterozoic low-latitude global glaciation: The snowball Earth. In J. W. Schopf and C. Klein, eds., *The Proterozoic Biosphere: A Multidisciplinary Study*, pp. 51–52. Cambridge: Cambridge University Press.

Knoll, A. H. 1994. Neoproterozoic evolution and environmental change. In S. Bengtson, ed., *Early Life on Earth*, pp. 439–449. Nobel Symposium no. 84. New York: Columbia University Press.

Kumar, T. N., R. K. Bagati, and R. K. Mazari. 1994. Uplifted late Quaternary debris fan in the Upper Spiti Valley (H.P.) and its environmental significance. *Journal of the Geological Society of India* 43:603–611.

LeMasurier, W. E., D. M. Harwood, and D. C. Rex. 1994. Geology of Mount Murphy Volcano: An 8-m.y. history of interaction between a rift volcano and the West Antarctic ice sheet. *Geological Society of America Bulletin* 106:265–280.

Marker, M. E. 1994. Sedimentary sequences

at Sani Top, Lesotho highlands, southern Africa. *Holocene* 4:406–412.

Mazzullo, J. and C. Ritter. 1991. Influence of sediment source on the shapes and surface textures of glacial quartz and sand grains. *Geology* 19:384–388.

Meert, J. G. and R. Van der Voo. 1994. The Neoproterozoic (1000–540 Ma) glacial intervals: No more snowball Earth? *Earth and Planetary Science Letters* 123:1–13.

Moores, E. M. 1993. Neoproterozoic oceanic crustal thinning, emergence of continents, and origin of the Phanerozoic ecosystem: A model. *Geology* 21:5–8.

Mustard, P. S. 1991. Normal faulting and alluvial-fan deposition, basal Windermere Tectonic Assemblage, Yukon, Canada. *Geological Society of America Bulletin* 103:1346–1364.

Myrow, P. M. and A. J. Kaufman. 1999. A newly discovered cap carbonate above Varanger-age glacial deposits in Newfoundland, Canada. *Journal of Sedimentary Research* 69:789–793.

Nicholas, C. J. 1996. The isotopic evolution of the oceans during the "Cambrian Explosion." *Journal of the Geological Society, London* 153:243–254.

Ojakangas, R. W. 1985. Evidence for early Proterozoic glaciation: The dropstone unit-diamictite association. In K. Laajoki and J. Paakkola, eds., *Proterozoic Exogenic Processes and Related Metallogeny*, pp. 51–72. *Proceedings, Symposium, Oulu, Finland, 15–16 August 1983*. Geological Survey of Finland, Bulletin 331.

Owen, L. A. 1994. Glacial and non-glacial diamictons in the Karakoram Mountains and western Himalayas. In W. P. Warren and D. G. Croot, eds., *Formation and Deformation of Glacial Deposits*, pp. 9–28. Proceedings, Meeting of the Commission on the Formation and Deformation of Glacial Deposits, Dublin, Ireland, May 1991. Rotterdam: Balkema.

Paim, P. S. G. 1995. Alluvial palaeogeography of the Guaritas depositional sequence of southern Brazil. *International Association of Sedimentologists, Special Publication* 22:2–16.

Perttunen, M. 1992. The transport of till in southern Finland. In E. Pulkkinen, ed., *Environmental Geochemistry in Northern Europe*, pp. 79–86. Geological Survey of Finland, Special Paper 9.

Prave, A. R. 1999. Two diamictites, two cap carbonates, two $\delta^{13}C$ excursions, two rifts: The Neoproterozoic Kingston Peak Formation, Death Valley, California. *Geology* 27:339–342.

Rampino, M. R. 1994. Tillites, diamictites, and ballistic ejecta of large impacts. *Journal of Geology* 102:439–456.

Raymo, M. E. 1991. Geochemical evidence supporting T. C. Chamberlin's theory of glaciation. *Geology* 19:344–347.

Saylor, B. Z., J. P. Grotzinger, and G. J. B. Germs. 1995. Sequence stratigraphy and sedimentology of the Neoproterozoic Kuibis and Schwarzrand Subgroups (Nama Group), southwestern Namibia. *Precambrian Research* 73:153–171.

Saylor, B. Z., A. J. Kaufman, J. P. Grotzinger, and F. Urban. 1998. A composite reference section for terminal Proterozoic strata of southern Namibia. *Journal of Sedimentary Research* 68:1223–1235.

Schemerhorn, L. J. G. 1974. Late Precambrian mixtites: Glacial and/or nonglacial? *American Journal of Science* 274:673–824.

Schemerhorn, L. J. G. 1983. Late Proterozoic glaciation in the light of CO_2 depletion in the atmosphere. In L. G. Medaris, C. W. Byers, D. M. Mickelson, and W. C. Shanks, eds., Proterozoic geology: Selected papers from an International Proterozoic Symposium, *Geological Society of America, Memoir* 161:309–315.

Schmidt, P. W. and G. E. Williams. 1995. The

Neoproterozoic climatic paradox: Equatorial paleolatitude for Marinoan glaciation near sea level in South Australia. *Earth and Planetary Science Letters* 134:107–124.

Spalletti, L. A. and A. Del Valle. 1984. Las diamictitas del sector oriental de Tandilia: Caracteres sedimentologicas y origen. *Revista da Associación Geológica Argentina* 39: 188–206.

Spencer, A. M. 1971. *Late Pre-Cambrian Glaciation in Scotland.* Geological Society of London, Memoir 6.

Stump, E., J. M. G. Miller, R. J. Korsch, and D. G. Edgerton. 1988. Diamictite from Nimrod Glacier area, Antarctica: Possible Proterozoic glaciation on the seventh continent. *Geology* 16:225–228.

Thunell, R. C., R. C. Tappa, and D. M. Anderson. 1995. Sediment fluxes and varve formation in Santa Barbara Basin, offshore California. *Geology* 23:1083–1086.

Trompette, R. 1996. Temporal relationship between cratonization and glaciation: The Vendian–Early Cambrian glaciation in western Gondwana. *Palaeogeography, Palaeoclimatology, Palaeoecology* 123:373–383.

Tucker, M. E. 1992. The Precambrian-Cambrian boundary: Seawater chemistry, ocean circulation, and nutrient supply in metazoan evolution, extinction, and biomineralization. *Journal of the Geological Society, London* 149:655–668.

Uhlein, A., R. R. Trompette, and C. J. S. Alvarenga. 1999. Neoproterozoic glacial and gravitational sedimentation on a continental rifted margin: The Jequitaí-Macaúbas

sequence (Minas Gerais, Brazil). *Journal of South American Earth Sciences* 12:435–451.

Ui, T. 1989. Discrimination between debris avalanches and other volcaniclastic deposits. In J. H. Latter, ed., *Volcanic Hazards: Assessment and Monitoring,* pp. 201–229. IAVCEI, Proceedings in Volcanology 1. Berlin: Springer Verlag.

Veevers, J. J. 1990. Tectonic-climatic supercycle in the billion-year plate-tectonic eon: Permian Pangean icehouse alternates with Cretaceous dispersed-continents greenhouse. *Sedimentary Geology* 68:1–16.

Vinogradov, V. N. 1981. Glacier erosion and sedimentation in the volcanic regions of Kamchatka. *Annals of Glaciology* 2:164–169.

Williams, G. E. 1975. Late Precambrian glacial climate and the earth's obliquity. *Geological Magazine* 112:441–465.

Yakubchuk, A. S., A. M. Nikishin, and A. Ishiwatari. 1994. A late Proterozoic ophiolite pulse. In A. Ishiwatari, J. Malpas, and H. Ishizuka, eds., *Proceedings of the 29th International Geological Congress, Part D: Circum Pacific Ophiolites,* pp. 273–286. Utrecht: VNU Science Press.

Young, G. 1991. The geologic record of glaciation: Relevance to the climatic history of Earth. *Geoscience Canada* 18:100–206.

Young, G. M. 1995. Are Neoproterozoic glacial deposits preserved on the margins of Laurentia related to the fragmentation of two supercontinents? *Geology* 23:153–156.

CHAPTER SIX

David I. Gravestock and John H. Shergold

Australian Early and Middle Cambrian Sequence Biostratigraphy with Implications for Species Diversity and Correlation

This description of Lower and Middle Cambrian strata from the Stansbury, Arrowie, Amadeus, and Georgina basins combines elements of biostratigraphy and sequence stratigraphy. The record of some South Australian Lower Cambrian sequences is missing, or has not been recognized, in central Australia. Deposition in the Middle Cambrian of the central Australian basins and the Stansbury Basin reflects subsidence-induced transgression, but these sequences cannot be differentiated in the almost unfossiliferous clastic deposits of the Arrowie Basin. Trace fossil assemblages in basal siliciclastic rocks are most diverse in lowstand half-cycles of relative sea level. Archaeocyath species diversity is highest in transgressive tracts, whereas lowstands are accompanied by extinction on shallow to emergent carbonate shelves. Trilobite species diversity is likewise highest in transgressive tracts but is seemingly unaffected by lowstand conditions. Duration of the Early and Middle Cambrian is 25–35 m.y. and 10–15 m.y., respectively, indicating very high rates of trilobite speciation in successive transgressive systems tracts.

AUSTRALIAN LOWER AND Middle Cambrian sedimentary rocks contain rich assemblages of fossil marine invertebrates, calcified and organic-walled microbial fossils, and traces of organic activity. Knowledge of the taxonomy and affinities of Australian Cambrian invertebrate fossils has increased significantly in the past decade, but at present only the archaeocyaths and trilobites have been studied in detailed stratigraphic successions. Progress is being made in the further study of mollusks and other small skeletal fossils, superbly described by Bengtson et al. (1990).

In this chapter we document the species distribution of archaeocyaths in the Lower Cambrian and trilobites in the Middle Cambrian of the Stansbury and Arrowie basins in South Australia and the Amadeus and Georgina basins in the Northern Territory and western Queensland (figure 6.1). Upper Cambrian trilobite faunas are well preserved

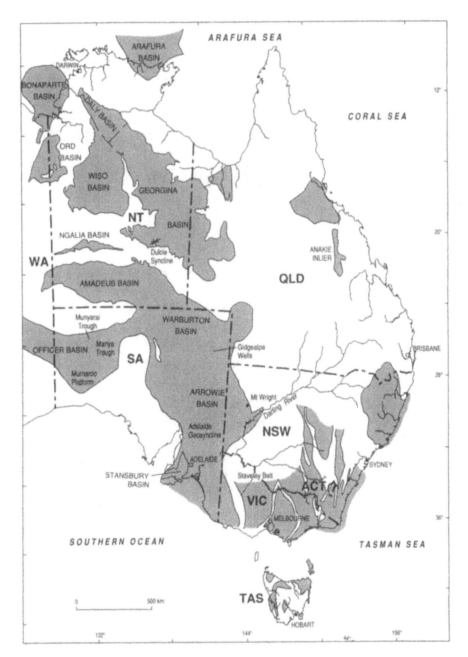

Figure 6.1 Cambrian and undifferentiated Cambrian-Ordovician sedimentary basins
of central and eastern Australia. *Source:* Modified after Cook 1988.

in the Georgina and Warburton basins, but are beyond the scope of this study because correlative strata in the Stansbury, Arrowie, and Amadeus basins have yielded
few fossils.

Trace fossils occur in basal Cambrian siliciclastic rocks beneath archaeocyath-
bearing carbonates in all of these basins (Daily 1972). For completeness the occur-

rences of trace fossils are investigated, together with archaeocyaths and trilobites, in a sequence stratigraphic context (sensu Vail et al. 1977; van Wagoner et al. 1988). On the basis of our analysis, we discuss three key attributes of the Cambrian radiation in Australia: species diversity and relative sea level change; correlation of sequences between basins; and rates of speciation, assisted by the increasing number and accuracy of radiometric ages of Cambrian successions.

SEQUENCE BIOSTRATIGRAPHY

A number of sequence stratigraphic frameworks have been proposed for the Early and Middle Cambrian of Australia (Amadeus Basin: Lindsay 1987; Kennard and Lindsay 1991; Lindsay et al. 1993; Arrowie Basin: Gravestock and Hibburt 1991; Mount and McDonald 1992; Stansbury Basin: Gravestock et al. 1990; Jago et al. 1994; Gravestock 1995; Dyson et al. 1996).

Sequence stratigraphy relates patterns of sediment accumulation at various scales to recurring cycles of marine transgression and regression, as well as to rates of sediment supply and subsidence. The depositional components of a sequence are systems tracts (Brown and Fisher 1977), which describe the associations of shelf-to-basin facies at low relative sea level (lowstand systems tracts), rising relative sea level (transgressive systems tracts), and falling relative sea level (highstand, or forced regressive, systems tracts).

Systems tracts or entire sequences may be condensed or incomplete, and hiatuses occur close to basin margins in regions undergoing slow relative subsidence and in structural belts where tectonic uplift opposes regional subsidence. Sequence biostratigraphy permits the interpretation of depositional sequences within biozonal frameworks, which often represent a wide sample of paleoenvironments. Without a detailed faunal succession, it is difficult to determine whether all sequences have been preserved. In this work, archaeocyath and trilobite biostratigraphic schemes correlate sequences and determine which are missing. Within a sequence, facies analysis of systems tracts helps explain why a particular species assemblage occurs at a given place and time relative to a cycle of sea level change.

Sequence nomenclature in the Stansbury and Arrowie basins is shown in figure 6.2. Four third-order sequences (Uratanna sequence, €1.1, €1.2, €1.3) span much of the Early Cambrian. The late Early to Middle Cambrian sequences €2.1–€3.2 rely principally on data from the Stansbury Basin, with the Middle Cambrian being placed at the base of the Coobowie Limestone on Yorke Peninsula (see the section "Stansbury Basin" below).

A relative sea level curve illustrated in figure 6.2 indicates the positions of lowstands and highstands in the stratigraphic succession. Based on the ideas of Zhuravlev (1986) and Rowland and Gangloff (1988), the dashed envelope that connects high sea level culminations corresponds to the Botoman transgression and Toyonian re-

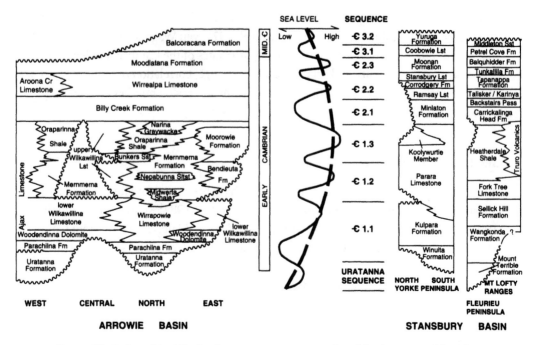

Figure 6.2 Early and Middle Cambrian sequence stratigraphy of the Arrowie and Stansbury basins. Third-order high sea level culminations are linked by a dashed curve to depict Botoman transgression and Toyonian regression.

gression. These are considered to be global phenomena. The third-order sequences illustrated in figure 6.2 operated in all basins under review where a rock record is preserved.

URATANNA SEQUENCE BIOSTRATIGRAPHY

The Uratanna sequence (Mount and McDonald 1992) is represented by the Uratanna Formation in the Arrowie Basin and the Mount Terrible Formation in the Stansbury Basin. Mount (1993) has reported a new occurrence of *Sabellidites* cf. *cambriensis* from the Uratanna Formation interpreted here to be at or just beneath the level of Daily's (1976a) Mount Terrible skeletal fauna, and well below his first reported occurrence of *Saarina*.

Arrowie Basin

The Uratanna Formation (Daily 1973) contains three informal members that indicate lowstand, abrupt upward deepening, then gradual shoaling of the succession (McDonald 1992; Mount and McDonald 1992; Mount 1993). A relative sea level curve, its component systems tracts, and a composite stratigraphic column (from Mount 1993) are illustrated in figure 6.3.

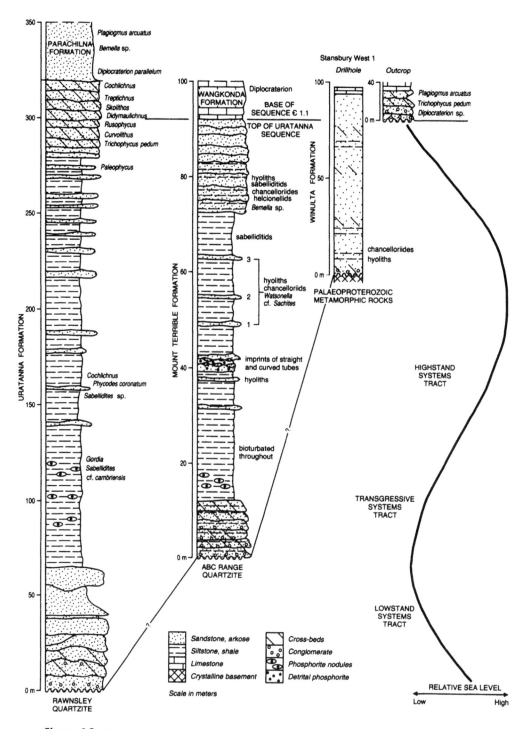

Figure 6.3 Uratanna sequence stratigraphy. Sections are drawn at different scales to illustrate their location within systems tracts.

Incised channels at the lower sequence boundary contain massive, amalgamated sandstone beds that have locally eroded to a level bearing the Ediacara fauna (Daily 1973). The beds lack fossils and are interpreted to represent the lowstand systems tract (Mount 1993) (figure 6.3). The transgressive systems tract is represented by laminated siltstone and shale with phosphorite nodules at lower levels. Rare, but upwardly increasing, interbeds of fine-grained sandstone mark the incoming highstand tract. The first recorded specimens of *Sabellidites* cf. *cambriensis* occur within the transgressive tract, and the trace fossil *Phycodes coronatum* occurs about 60 m above in the highstand tract. Upper parts of the highstand tract are recorded by passage into fine-grained, cross-bedded quartz sandstone deposited in upward-shallowing cycles. Within these, Mount (1993) lists 10 ichnotaxa including *Treptichnus pedum* (referred to as *Phycodes pedum* in figure 6.3), *Treptichnus*, and *Rusophycus*. *Diplocraterion parallelum, Plagiogmus arcuatus,* and the mollusk *Bemella* sp. occur in the overlying Parachilna Formation (Daily 1976a), which we interpret with Mount (1993) to be in the lowstand tract of the overlying sequence.

On present evidence, the first organic-walled fossils (sabelliditids) are preserved in the transgressive systems tract, the first Cambrian trace (*P. coronatum*) is found in the lower part of the highstand tract, and abundant traces occur in its upper part.

Stansbury Basin

The Mount Terrible Formation is composed of three informal members exposed on Fleurieu Peninsula (Daily 1976a). In outcrop, the lowest member disconformably overlies the Neoproterozoic ABC Range Quartzite and comprises thin, planar-tabular bed sets with scoured bases. Each bed consists of fine-grained arkosic sandstone with a pebbly, phosphatized base and a bioturbated pyritic siltstone top. Low-angle cross-beds and streaming lineations indicate high-energy conditions. We interpret these beds to be transgressive marine deposits, because a lowstand tract is not preserved. The middle member comprises 60 m of bioturbated siltstone with phosphorite concretions at lower levels and rare, thin interbeds of fine-grained feldspathic sandstone. Two beds bearing large discoidal clasts of fine-grained sandstone occur at midlevels. The upper member comprises 20 m of bioturbated feldspathic, fine-grained sandstone with pyritic and argillaceous siltstone interbeds.

The first shelly fossils, hyoliths (cf. *Turcutheca*), occur immediately beneath the clast-bearing beds. Daily (1976a) also recorded shelly fossils from three overlying levels (labeled 1–3 in figure 6.3), comprising hyoliths, chancelloriids, cf. *Sachites* and *Watsonella* (=*Heraultia*). The first sabelliditids (*Saarina*) were recorded above the third fossiliferous level of the middle member and in the lower part of the upper member. In the latter, hyoliths, chancelloriids, helcionelloid mollusks, and *Bemella* sp. are recorded. Imprints of tubular fossils were noted in the sandstone clasts of the middle member. We interpret the lower, phosphorite-enriched level of the middle member to contain the maximum flooding surface, and hence the organic-walled and shelly fossils found to date occur in the highstand tract.

The suggested position of the Winulta Formation on Yorke Peninsula is also shown in figure 6.3 (note differing scale). Daily (1972, 1976a, 1990) has recorded hyoliths and chancelloriids from near the base of the Winulta Formation in drill cores, where the formation approaches 100 m in thickness. Drill cores are composed of glauconitic and pyritic sandstone and arkose with siltstone interbeds and dolomitic cement. Outcrops comprise cross-bedded conglomeratic to fine-grained sandstones, which yield *Treptichnus pedum, Plagiogmus arcuatus,* and *Diplocraterion* sp. On northern Yorke Peninsula (e.g., outcrops at Winulta and Kulpara), the Winulta Formation is represented by a basal conglomerate and flaggy trace-bearing sandstones, whereas on southern Yorke Peninsula it is thicker and fine-grained and contains shelly fossils.

The sequence biostratigraphic scheme in figure 6.3 illustrates the observations of Daily (1976a), McDonald (1992), Mount and McDonald (1992), and Mount (1993). The base of the Uratanna Formation represents the base of the Uratanna sequence in the Arrowie Basin. In the Stansbury Basin, depending on location, the base of the Mount Terrible Formation is in the transgressive systems tract of the Uratanna sequence (Sellick Hill), and the base of the Winulta Formation is in the highstand tract of the Uratanna sequence (southern Yorke Peninsula drillholes). The Uratanna-Є1.1 sequence boundary is placed either within the trace fossil–bearing sandstones of the upper Uratanna Formation or at the base of the Parachilna Formation in the Arrowie Basin (Mount and McDonald 1992). The boundary is placed at the base of the Wangkonda Formation and at the base of the trace fossil–bearing sandstones of the Winulta Formation in the Stansbury Basin.

The Precambrian-Cambrian boundary in South Australia is the base of the Uratanna sequence, and the most complete representative section is in the Arrowie Basin. It is unlikely that the first appearance of *Phycodes coronatum* in the Uratanna Formation correlates with the GSSP (Global Stratotype Section and Point) in Newfoundland. *Treptichnus pedum* appears at Fortune Head, Newfoundland, in the transgressive tract of a sequence that comprises Member 1 and part of Member 2 of the Chapel Island Formation. Skeletal fossils are preserved about 400 m higher in a second sequence, which comprises the remainder of Member 2, as well as Members 3 and 4 of the Chapel Island Formation (Myrow and Hiscott 1993). This latter succession may correlate with the Uratanna sequence in the Stansbury Basin, which also contains skeletal fossils, although as Myrow and Hiscott have pointed out, it is by no means certain that the Newfoundland sequences have global correlation potential either.

Amadeus and Georgina Basins

The facies succession of the Uratanna Formation (Mount 1993) resembles Arumbera Sandstone units 3 and 4 in the Amadeus Basin (Lindsay 1987; Kennard and Lindsay

1991; Lindsay et al. 1993). Unit 3 overlies the Ediacaran metazoan-bearing unit 2 with a conformable to disconformable contact. Arumbera unit 3 comprises siltstone with interbeds of laminated and rippled sandstone. Arumbera unit 4 comprises thick sandstone beds with climbing ripples and hummocky cross-stratification, followed by bioturbated and channel-filling, cross-bedded sandstone that passes conformably into tidal deposits of the Todd River Dolomite.

Arumbera Sandstone units 3 and 4 record upward transition from prodelta or basinal muddy deposits at the base through delta front to coastal delta plain deposits at the top. This succession was placed in the highstand systems tract by Lindsay (1987) and in the lowstand tract by Kennard and Lindsay (1991) and Lindsay et al. (1993), as shown in figure 6.5.

Trace fossils are abundant in Arumbera Sandstone units 3 and 4, with 36 taxa noted by Walter et al. (1989). The first records of *Treptichnus pedum*, *Diplichnites* sp., and *Rusophycus* sp. occur in the delta slope facies 20 m above the base of Arumbera Sandstone unit 3 (Arumbera II of Daily 1972), and *Plagiogmus* sp. occurs in Arumbera 4 (Daily's Arumbera III), 2 m above the first occurrence of hyoliths (Haines 1991). We follow Mount and McDonald (1992) in correlating Arumbera Sandstone unit 3 with the upper Uratanna and upper Mount Terrible formations, and Arumbera Sandstone unit 4 with the uppermost Uratanna, uppermost Winulta and Parachilna formations. These occurrences span the Uratanna-€1.1 sequence boundary. Trace fossils in the Namatjira Formation are placed here in the lowstand of sequence €1.1. It is likely on present evidence that the Precambrian-Cambrian boundary in the Amadeus Basin occurs in upper Arumbera 2, which lacks trace fossils (Walter et al. 1989).

Trace fossils in the Huckitta region of the Georgina Basin are diverse and well preserved in the 300 m-thick quartzose Mount Baldwin Formation (Walter et al. 1989). They include ?*Bergaueria* sp., *Treptichnus* sp., *Helminthopsis* sp., and *Diplocraterion parallelum*. Although the stratigraphic context of the traces is not reported, they also appear to span the Uratanna-€1.1 sequence boundary, and they occur in the thickest accumulation of sandstone at this level in Australia.

ARCHAEOCYATH SEQUENCE BIOSTRATIGRAPHY

Stratigraphic studies of South Australian archaeocyaths (Gravestock 1984; Debrenne and Gravestock 1990; Lafuste et al. 1991; Zhuravlev and Gravestock 1994) and taxonomic revision of the whole class (Debrenne et al. 1990; Debrenne and Zhuravlev 1992) provide sufficient information to assess the distribution of archaeocyath species within a sequence stratigraphic framework.

The four sequences are depicted in figure 6.4 with a relative sea level curve for the Arrowie and Stansbury basins. Archaeocyath assemblage zones (Zhuravlev and Gravestock 1994) are shown at the base of the figure, and the number of species within each zone is depicted in columns. Older trace and shelly fossil occurrences are also shown. Horizontal scales are arbitrary, as is the relative sea level curve, although de-

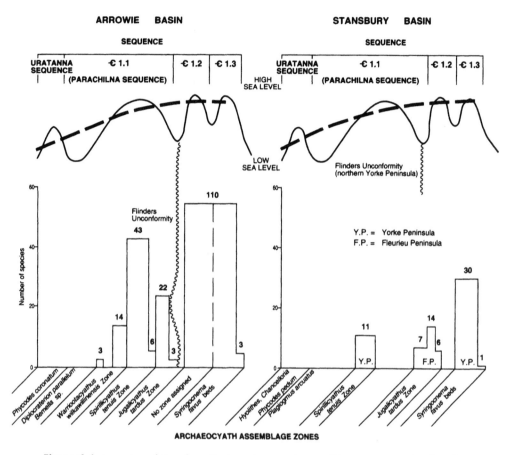

Figure 6.4 Arrowie and Stansbury Basin archaeocyath assemblage zones, species diversity, sequences, and relative sea level curve.

piction of increasing water depth through the Early Cambrian (dashed envelope in figure 6.4) is in accord with a generally transgressive setting. This envelope represents a second-order cycle of sea level change from the terminal Proterozoic to late Botoman, an estimated 20–25 m.y.; thus each third-order sequence spanned about 5 m.y.

Arrowie Basin

Sandstone of the Parachilna Formation, interpreted as a lowstand deposit near the base of sequence €1.1, lacks archaeocyaths but bears in its lowermost part abundant burrows of *Diplocraterion parallelum.* The first shelly fossil, *Bemella* sp., appears at a higher level (Daily 1976a). The conformably overlying Woodendinna Dolomite (Haslett 1975) represents a lowstand tidal flat composed of stromatolitic and oolitic carbonates.

The first archaeocyaths, together with *Epiphyton* and *Renalcis,* formed small bioherms 38–50 m above the base of the Wilkawillina Limestone at Wilkawillina Gorge. (Calcimicrobes in South Australia referred to as *Epiphyton* [cf. James and Gravestock

1990] are more likely *Gordonophyton* [A. Zhuravlev, pers. comm., 1995]). Initially there were 14 species (*Warriootacyathus wilkawillinensis* Zone). Submarine erosion surfaces within this zone at Wilkawillina Gorge are interpreted as marine flooding surfaces. With continued transgression, the pioneer species were replaced by 43 new species, which formed the *Spirillicyathus tenuis* Zone (figure 6.4). A deep-water bioherm in the Mount Scott Range is overlain by small bioherms composed mostly of "*Epiphyton*" with only six archaeocyath species, suggesting agitated, shoaling marine conditions. Continued sea level fall and moderate-to-high energy conditions are evidenced by cross-bedded fossil packstone with scarce, small bioherms. These beds contain 22 species of archaeocyaths of the *Jugalicyathus tardus* Zone and are interpreted to be late highstand deposits of sequence €1.1.

At Wilkawillina Gorge, species diversity remained moderately high to the base of the Flinders Unconformity, a distinctive exposure surface capped by red microstromatolites (Daily 1976b; James and Gravestock 1990). The *tardus* zone was truncated, with no preservation of regressive facies. The excursion of the Flinders Unconformity to the left in figure 6.4 depicts this truncation. In contrast, abundance and diversity dropped markedly in the Mount Scott Range, where thinly laminated limestones rich in other skeletal fossils yielded only three archaeocyath species. Thus the top of the *tardus* zone in the Arrowie Basin is defined by disconformity and facies change depending on locality, both resulting from the interplay of relative sea level fall and subsidence.

A lowstand wedge of Bunkers Sandstone intervenes between the Mernmerna Formation and Oraparinna Shale, separating sequences €1.2 and €1.3. Archaeocyaths are scarce in slope deposits between the Flinders Unconformity and Bunkers Sandstone, and species in adjacent shelfal facies are poorly studied. The informal name "*Syringocnema favus* beds" applies only to upper shelf carbonates of the Ajax and Wilkawillina limestones and the Moorowie Formation (Zhuravlev and Gravestock 1994). The 110 or so species from these younger limestones are arbitrarily shared equally between the unzoned interval and the *favus* beds (figure 6.4). The first appearance of *S. favus* above the Bunkers Sandstone suggests that the *favus* beds are entirely within sequence €1.3.

Botoman time (sequences €1.2 and €1.3) in the Arrowie and Stansbury basins was characterized by the appearance of distinct shelves with abrupt margins and of slopes with mass flow deposits and basin plains; the last contains mainly shales variably enriched in organic matter, pyrite, and phosphorite. Examples are the Midwerta Shale, Nepabunna Siltstone, Mernmerna Formation, and Oraparinna Shale in the Arrowie Basin and the Heatherdale Shale in the Stansbury Basin (see figure 6.2). [Note that the name "Mernmerna Formation" (Dalgarno and Johnson 1962) is now applied to the formation previously mapped as Parara Limestone in the Arrowie Basin. Usage of "Parara Limestone" is restricted to the Stansbury Basin–type area.] Growth of submarine topography was accompanied by rift-related volcanic activity in the Stansbury

Basin (Truro Volcanics), with eruptive phases recorded as tuff beds within sequences €1.2 and €1.3 in both basins. Submarine volcanism was more marked in western New South Wales, where correlative archaeocyaths accumulated in lenticular lime-stones enclosed in the Mount Wright Volcanics and in tuffs and cherts of the Cymbric Vale Formation (Kruse 1982).

During Botoman time, shelf and shelf-margin settings in the Arrowie Basin were favored sites of reef growth, the principal constructors being archaeocyaths and calci-microbes (James and Gravestock 1990). In the Moorowie Formation, reefs are inter-preted as having developed in a sea-marginal fan setting. Competition for space is re-flected in complex growth interactions between reef builders (Savarese et al. 1993), which include archaeocyaths, "sphinctozoans," calcimicrobes, coral-like cnidarians, and possibly true tabulates (Lafuste et al. 1991; Fuller and Jenkins 1994; Sorauf and Savarese 1995). There is a distinct tendency among some of these archaeocyaths to-ward modular growth, a habit that increased through the Early Cambrian (Wood et al. 1992).

Archaeocyath diversity was high, as witnessed by the 110 species shown in fig-ure 6.4, and coincided with the second-order high sea level curve, considered global in extent (Zhuravlev 1986). In the uppermost 20–50 m of the Andamooka Limestone on the Stuart Shelf and upper levels of the Ajax, Wilkawillina and Moorowie lime-stones and Oraparinna Shale in the Flinders Ranges, there is evidence of widespread regression (Daily 1976b). Oolitic, fenestral, stromatolitic, and evaporitic units, which typify this final phase of carbonate deposition, represent the late highstand systems tract of sequence €1.3. Immediately beneath these deposits in the upper Andamooka and Wilkawillina limestones is a distinctive bioherm type composed of thrombolite-like intergrowths of *Renalcis* and *Botomaella* (type 1 calcimicrobe boundstones; James and Gravestock 1990). In such bioherms, there are no more than three species of dwarfed archaeocyaths, which are assigned to the upper *favus* beds.

Archaeocyaths and corals disappeared from the Arrowie Basin principally because of tectonic adjustments and attendant shifts of facies belts. Only *Archaeocyathus aba-cus, Ajacicyathus* sp., and the radiocyath *Girphanovella gondwana* are recorded in the Wirrealpa Limestone (Kruse 1991). Correlation with the *Redlichia chinensis* zone of the Chinese Longwangmiaoan stage is evident from trilobites at this stratigraphic level (Jell in Bengtson et al. 1990).

Stansbury Basin

Archaeocyaths in the Stansbury Basin (Debrenne and Gravestock 1990; Zhuravlev and Gravestock 1994) occur in the Kulpara Formation and Parara Limestone on Yorke Peninsula and in the Sellick Hill Formation and Fork Tree Limestone on Fleurieu Pen-insula (see figure 6.2). Figure 6.4 depicts archaeocyath species diversity as it is pres-ently known. The same sequences and sea level curve as used for the Arrowie Basin are shown at the top of the figure, but the curve is "generic" and intended only as a guide. Variations in subsidence, sediment supply, and the position of studied sections relative to the paleoshoreline lead to different local sea level curves. The studied outcrops are on opposite sides of the present Gulf St. Vincent, which necessitates switching from Yorke Peninsula to Fleurieu Peninsula (designated Y.P. and F.P., respectively, in figure 6.4) in the following account.

Peritidal oolite, stromatolites, and fenestral carbonates in the Wangkonda Formation and through most of the Kulpara Formation indicate that conditions unsuited to archaeocyaths prevailed longer in the Stansbury Basin than elsewhere (Daily 1972). The *wilkawillinensis* zone is thus not recorded, but the *tenuis* zone on southern Yorke Peninsula is represented by 11 species in the upper Kulpara Formation and basal Parara Limestone where these units are conformable.

Deposition on Yorke Peninsula was controlled by a tectonically active hinge, south of which the Kulpara Formation and Parara Limestone are conformable and north of which they are disconformable (Zhuravlev and Gravestock 1994). On southern Yorke Peninsula (e.g., at Curramulka Quarry), the Parara Limestone contains a rich invertebrate fauna (Bengtson et al. 1990) in dark, micritic, and nodular phosphorite-enriched limestone, indicating upwardly deepening marine conditions. Archaeocyaths of the *tenuis* zone are rapidly lost, and the *tardus* zone is not represented. On northern Yorke Peninsula, at Horse Gully, the Flinders Unconformity surface overlies a condensed section in the upper 2 m of the Kulpara Formation, which displays evidence of subaerial exposure (Wallace et al. 1991; Zhuravlev and Gravestock 1994). Archaeocyaths of the *tenuis* zone in this section are overlain by skeletal fossils found elsewhere in the *tardus* zone (e.g., *Microdictyon depressum*).

Archaeocyaths on Fleurieu Peninsula occur near the top of the Sellick Hill Formation, which Daily (1972) correlated with the top levels of his Faunal Assemblage 2 (= *tardus* zone) or Faunal Assemblage 3 on Yorke Peninsula. Alexander and Gravestock (1990) interpreted the Sellick Hill Formation to comprise outer shelf and ramp sediments deposited during marine transgression. Lower levels contain hyoliths, mollusks, and a rich ichnofauna of predominantly horizontal traces (including *T. pedum*). Middle levels show evidence of slope instability and intense storm activity (Mount and Kidder 1993), and upper levels contain archaeocyath framestone bioherms.

The 14 species of regular archaeocyaths (including the Botoman genus *?Inacyathella*) are assigned to the *tardus* zone (Debrenne and Gravestock 1990; Zhuravlev and Gravestock 1994). Two of the 14 species also occur in the *tenuis* zone in the Arrowie Basin, but not on Yorke Peninsula, and 5 species are restricted to Fleurieu Peninsula. There is no evidence of subaerial exposure as found at the top of the Kulpara Formation on northern Yorke Peninsula, but Alexander and Gravestock (1990) recorded a thin, laterally persistent bioclastic packstone containing 7 species of abraded archaeocyaths and other fossil debris. They suggested that this bed was the reworked product of eroded bioherms. It overlies multiple corroded and phosphatized surfaces and is interpreted here as the culmination of a series of high-energy events on the carbonate ramp during low sea level at the top of sequence €1.1 (see figure 6.4). The impact of the fall in relative sea level that gave rise to the Flinders Unconformity was not great, because unlike those on the shelf, these ramp carbonates were not exposed.

The overlying bioherms thus grew in the transgressive systems tract of sequence €1.2. Six archaeocyath species persisted into the conformably overlying Fork Tree Limestone. The postulated outer ramp setting may explain the oligotypic archaeocyath faunas in these bioherms, within which exocyathoid outgrowths, rather than calcimicrobes, bound the cups together (Debrenne and Gravestock 1990).

Continued marine transgression is evidenced by deposition of the conformably overlying Heatherdale Shale, which contains a bivalved arthropod and rare conocoryphid trilobite fauna (Jago et al. 1984; Jenkins and Hasenohr 1989). There is no evidence, either on Fleurieu Peninsula or on Yorke Peninsula (where Parara Limestone continued to be deposited), of the lowstand that marked the boundary between sequences €1.2 and €1.3, except perhaps immediately beneath the mottled upper member of the Fork Tree Limestone, where small calcimicrobe-archaeocyath bioherms indicate shallow marine conditions. The most likely explanation for a cryptic boundary is tectonic subsidence, which exceeded sea level fall as rifting and volcanism commenced only a few tens of kilometers to the east.

There is outcrop, drill core, and seismic evidence that a Botoman reef complex extended from Horse Gully to Edithburgh on Yorke Peninsula and probably to Kangaroo Island, a distance of 120 km. Pale pink, massive limestone of the Koolywurtie Member of the Parara Limestone (Daily 1990) is composed of calcimicrobe-archaeocyath boundstone, *Girvanella* crust boundstone, and oncolitic and bioclastic packstone, capped by peritidal fenestral limestone. Bioherms are overlain by mud-cracked red beds or fissile micrite and shale interpreted as coastal lagoon deposits. The Emu Bay Shale on Kangaroo Island with its Lagerstätte of *Hsuaspis bilobata, Redlichia takooensis,* anomalocaridids, and *Isoxys* may be a contemporaneous lagoonal deposit (Nedin 1995). The underlying White Point Conglomerate contains reworked boulders of reef rock resulting from tectonic activity (Kangarooian Movements; Daily and Forbes 1969).

Twenty-eight species of archaeocyath (plus *Acanthinocyathus* and a radiocyath) in the Koolywurtie Member are assigned to the *favus* beds (Zhuravlev and Gravestock 1994) (see figure 6.4). These species occur in the Flinders Ranges, western New South Wales (Kruse 1982), or Antarctica (Hill 1965; Debrenne and Kruse 1986, 1989). Syringocnemidids also occur in eastern Tuva and western Sayan in Russia. The Koolywurtie reefs are interpreted to have formed in the highstand systems tract of sequence €1.3, and the wide dispersal of archaeocyath species testifies to high global sea level in middle to late Botoman time.

A single species, *Archaeopharetra irregularis,* is interpreted to have survived sea level fall prior to the onset of red bed deposition represented by the Minlaton Formation (see figure 6.2). The overlying Ramsay and Stansbury limestones contain brachio-pods and small skeletal fossils (Brock and Cooper 1993), but archaeocyaths have not been found in these units.

Amadeus and Georgina Basins

The Todd River Dolomite in the northeastern Amadeus Basin is composed of six facies described in detail by Kennard (1991). Three siliciclastic-carbonate units are overlain by high-energy reef shoals, low-energy shelf deposits with patch reefs, and stromato-litic mudrocks. Six archaeocyath taxa and a radiocyath were described by Kruse (in Kruse and West 1980) as predominantly from the reef-shoal facies at Ross River. Most are restricted to the Amadeus and Georgina basins, but *Beltanacyathus* sp. at the base of the reef-shoal facies, an indeterminate trilobite, and the brachiopod *Edreja* aff. *distincta* (Laurie and Shergold 1985; Laurie 1986) higher in the section indicate that both the upper *tenuis* and *tardus* zones may be represented. Rare archaeocyaths in microbial bioherms in the underlying barrier bar facies have not been described.

In their sequence stratigraphic study of the Amadeus Basin, Lindsay et al. (1993) concluded that the barrier-bar, reef, and stromatolitic mudflat facies were deposited in transgressive and highstand systems tracts. In the Arrowie and Stansbury basins the *tenuis* and *tardus* zones occur in these systems tracts in sequence €1.1, confirming that the same sequence is represented in all three basins. Subaerial exposure and dissolution at the top of the Todd River Dolomite (Kennard 1991) are complex and may be related not only to the Flinders Unconformity but also to lowstand at the top of sequence €1.3. The long hiatus in figure 6.5 between the Todd River Dolomite and overlying units reflects these lowstand events.

The disconformably overlying Chandler Formation is considered by Lindsay et al. (1993) to be of Botoman age. Like Shergold (1995), we favor an Ordian–early Templetonian age because that is the age of fossils in the laterally equivalent Chandler Formation limestone and the lower Giles Creek Dolomite ("Giles Creek Dolomite" in figure 6.5). The Chandler Formation is composed primarily of halite with a medial unit of fetid limestone devoid of fossils. Bradshaw (1991) envisages a deep desiccated basin with two stages of drawdown and an intervening flooding event. It is overlain by the late Templetonian–Floran Tempe Formation, Hugh River Shale, or Giles Creek Dolomite. Major changes in coastline configuration wrought by late Botoman tectonic activity in the Arrowie and Stansbury basins may also have resulted in epeirogenic uplift of the Amadeus Basin region (Chandler Movement; Oaks et al. 1991).

The Chandler Formation salt may result from alternating lowstand and transgression in sequences €2.1 to €2.3, a time of global fall in sea level (Toyonian regression of Rowland and Gangloff 1988). If the salt indeed marks this Early Cambrian regression, an age discrepancy arises because of the Ordian–Early Templetonian fossils, which seemingly correlate with the South Australian sequence €3.1 (cf. figures 6.2

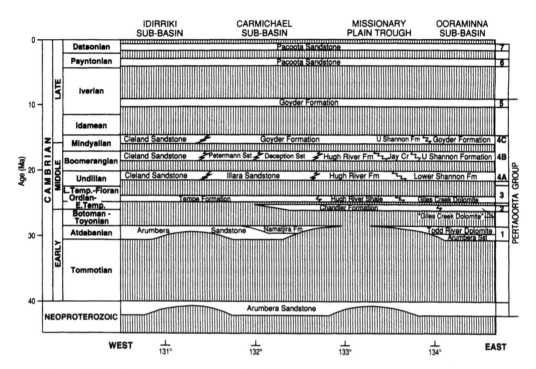

Figure 6.5 Cambrian sequence stratigraphy of the Amadeus Basin (modified after Kennard and Lindsay 1991). *E. Temp.* = Early Templetonian; *L. Temp.* = Late Templetonian.

and 6.5). Most of the Chandler Formation halite (225–470 m thick) might, however, be appreciably older than the thin (10 m) fossiliferous carbonate beds that occur in upper levels. Alternatively, correlation of the salt with the Toyonian regression is untenable, and a basal Middle Cambrian epoch of desiccation may be invoked.

The archaeocyath fossil record in the Georgina Basin is sparse. Kruse (in Kruse and West 1980) described four archaeocyaths and a radiocyath from the Errarra Formation in the Dulcie Syncline. Correlation with the *tardus* zone is favored, but the reported co-occurrence of *Dailyatia ajax* and *Yochelcionella* in drillhole Tobermory 12 (Laurie 1986) suggests a younger, mid-Botoman age at that locality. *Dailyatia ajax* is now known to be long-ranging in the Stansbury Basin. Further studies of brachiopods, mollusks, and small shelly fossils are warranted in the Georgina and Amadeus Basins.

MIDDLE CAMBRIAN TRILOBITE SEQUENCE BIOSTRATIGRAPHY

Background

There remains a fundamental dilemma in Australia as to exactly what is to be regarded as Early Cambrian and what Middle Cambrian. The correlation of the South Australian basins with those of central and northern Australia (Amadeus and Georgina in particular) is fraught with interpretative difficulty due principally to a dearth of South Australian trilobites. In this paper, we base our definition of the Early-Middle Cambrian boundary on the suggestion of Jell (1983), who regarded the first occurrence of the eodiscid genus *Pagetia* to define Middle Cambrian time, following the last occurrence of *Pagetides*. Although *Pagetia* occurs widely in central and northern Australian basins, it has been recorded only recently in South Australia, in the Stansbury Basin. There, Ushatinskaya et al. (1995) have recovered 10 specimens of *Pagetia* sp. from the shallow marine Coobowie Limestone in Port Julia 1A corehole on Yorke Peninsula. This suggests that the epoch boundary should be sought in the Stansbury Basin between sequences €2.3 and €3.1. The Moonan Formation, which underlies the Coobowie Limestone, consists of transgressive black shale followed by highstand siltstone and sandstone, necessitating the addition of a new sequence, €2.3, where previously only one was considered (Gravestock 1995). Stratigraphically beneath are the Stansbury Limestone, Corrodgery Formation and Ramsay Limestone (Daily 1990) (see figure 6.2).

By correlation, the Wirrealpa Limestone is Early Cambrian from the occurrence of *Redlichia guizhouensis* Zhou (Lu et al. 1974) in association with archaeocyaths and a radiocyath (Kruse 1991), and brachiopods, mollusks, and small shelly fossils from both the Ramsay and Wirrealpa limestones (Brock and Cooper 1993). Shales and red beds of the succeeding Moodlatana Formation, containing *Onaraspis rubra* at lower levels, may be Early or Middle Cambrian. The base of the Middle Cambrian in central and northern Australia has been taken traditionally at the beginning of the Ordian stage (sensu Öpik 1967b), in which species of *Onaraspis*, *Redlichia*, *Xystridura*, and *Pagetia* all occur, but archaeocyaths are absent. Because only two Middle Cambrian trilobite taxa are currently known in South Australia, the ensuing text is restricted to central and northern Australian basins, specifically the Amadeus and Georgina basins.

Amadeus Basin

The Amadeus Basin is graced with spectacular outcrops of Cambrian rocks, and its sequences (see figure 6.5) should be pivotal in resolving correlation problems between the South Australian basins and the Georgina Basin, since the archaeocyath-bearing Lower Cambrian rocks allow correlation with the former, and the trilobite-bearing Middle Cambrian permits correlation with the latter. Unfortunately, Lower and Middle Cambrian biostratigraphy of the Amadeus Basin is basically undeveloped, and sequence stratigraphy results largely from seismic stratigraphy, down-hole geophysics, and sedimentation patterns.

Of its four sedimentary compartments, only the easternmost Ooraminna Sub-basin can be correlated with reasonable confidence into the Georgina Basin. This depocenter contains the four diagnostic trilobite genera that define the Ordian stage (sensu stricto) in the lower "Giles Creek Dolomite" (see figure 6.5), which we regard as the carbonate lateral equivalent of the evaporitic Chandler Formation. This last formation,

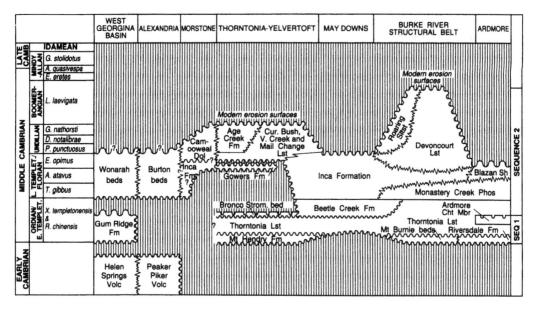

Figure 6.6 Sequence stratigraphic relationship of formations in the Georgina Basin
(after Southgate and Shergold 1991).

comprising 95 percent halite, represents the time of maximum desiccation in central
Australia. Minor evaporites recorded in the Moodlatana Formation of the Arrowie Basin, and those of the Gum Ridge Formation and Thorntonia Limestone of the Georgina
Basin may be correlated through this Ordian event. The Amadeus Basin regrettably
offers no assistance in the resolution of the Early-Middle Cambrian epoch boundary.

Georgina Basin

Our present discussions relate mainly to the eastern (Burke River Structural Belt) and
northern (Thorntonia to May Downs) portions of the Georgina Basin. There are insufficient published data to include the Dulcie and Toko synclines in the southwestern part of the basin. However, this area contains the first recorded Cambrian sequence of the Georgina Basin, which includes archaeocyath-bearing carbonates of late
Atdabanian-Botoman age. There is considerable hiatus between this Early Cambrian
sequence and the first of the two Middle Cambrian sequences, which is of Ordian–
early Templetonian age.

The two Middle Cambrian sequences are essentially those defined by Southgate
and Shergold (1991), who have listed the formations these sequences contain and
their interpreted depositional environments. In this chapter, however, we leave Middle
Cambrian sequence 1 undivided (figure 6.6) and regard it equivalent to subsequence
1 of Southgate and Shergold (1991). The overlying subsequence 1a of these authors
was separated on the basis of the occurrence of the Bronco Stromatolith Bed, ferrugi-

nous surfaces, phosphatic crusts, and widespread coquinite, which they considered to have been deposited in an algal marsh lowstand environment and to represent evidence for subaerial erosion. Southgate (pers. comm., 1995) now considers these features to represent submarine erosion and assigns them, together with basinal and outer shelf correlatives (Beetle Creek Formation and probably contemporaneous Burton beds), to the condensed section of the second Middle Cambrian sequence (sequence 2 of Southgate and Shergold 1991). This interpretation adds a different dimension to the sequence stratigraphic analysis and biostratigraphy of the Middle Cambrian of the Georgina Basin.

Some evidence remains for subaerial erosion at the top of sequence 1 in the northern part of the Thorntonia region, at the top of the Thorntonia Limestone. At the Ardmore Inlier in the Burke River Structural Belt, halite hoppers (Henderson and Southgate 1980; Southgate 1982) occur in the Ardmore Chert Member at the top of the Thorntonia Limestone, concluding sequence 1 sedimentation there.

Middle Cambrian sedimentation is governed by relative subsidence of underlying basement structures. As indicated by Southgate and Shergold (1991), the Middle Cambrian sequences of the eastern and northeastern Georgina Basin formed in response to continuously increasing relative sea level as the Mount Isa Block (and other basement blocks beneath the basin) began to subside. Middle Cambrian sequence 1 represents the initial transgressive event that onlapped from the north or northeast. It is extremely widespread across northern Australia, occurring in the Bonaparte, Ord, Arafura, Daly, and Wiso basins, as well as the Georgina (see figure 6.1). Its stratigraphic equivalents in the Amadeus and Ngalia basins are identified with certainty only in the easternmost Amadeus (lower "Giles Creek Dolomite" in the Gaylad Syncline and at Deep Well).

Middle Cambrian sequence 2 represents renewed transgression. In the Burke River Structural Belt, successive retrogradational parasequence sets onlapped the eastern edge of the Mount Isa Block (Southgate and Shergold 1991), while in the Thorntonia region basement paleotopography continued to influence the distribution of individual lithofacies packets, e.g., phosphatic sediments of the Gowers Formation (Shergold and Southgate 1986; Southgate 1986). Sedimentation along the western margin of the Mount Isa Block was also largely transgressive.

Thorntonia Region

Sequence 1, represented by the Mount Hendry Formation, is characterized initially by sandstone, conglomerate, arkose, siltstone, and shale, representative of coastal plain, fluvial, and valley-fill environments in the lowstand systems tract. These sediments are extremely sparsely fossiliferous, with only occasional ichnofossils occurring. Phosphorite, phosphatic limestone, and limestone of the Border Waterhole Formation and Thorntonia Limestone formed in the transgressive systems tract, and platform car-

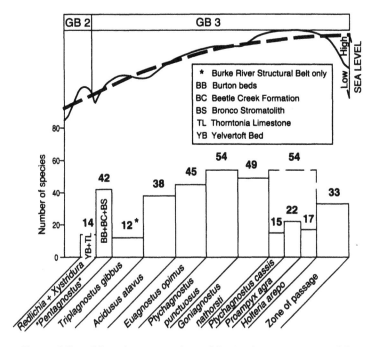

Figure 6.7 Trilobite diversity in the Middle Cambrian sequences of the eastern Georgina Basin. Numbers represent estimated taxa present.

bonate and peritidal carbonate of the upper Thorntonia Limestone represent the highstand systems tract. An oncolitic unit marks the top of sequence 1. Biostratigraphically, sequence 1 is distinguished by the co-occurrence of the trilobite genera *Redlichia* and *Xystridura* and, in certain other basins, *Onaraspis*. *Redlichia* predominates in the Border Waterhole Formation, the Thorntonia Limestone, and the Yelvertoft bed, with 10 species described from the Yelvertoft bed and associated lithofacies by Öpik (1970b). In earlier publications, this fauna has been used to diagnose the Ordian stage (Öpik 1967b) and the *Redlichia chinensis* Zone (biofacies).

The condensed section of sequence 2, containing xystridurid trilobite coquinite, stromatolite, encrinite, and phosphatic hardgrounds of the transgressive Bronco Stromatolith Bed, is dominated by an undetermined species of the *Xystridura* (*Xystridura*) *templetonensis* Zone (biofacies).

On the Barkly Tableland, to the west of the Thorntonia region, the probably contemporaneous Burton beds, containing limestone, shale and chert overlying coquinite, contribute a further 15 species to the *"Pentagnostus"* Zone in figure 6.7.

Upper levels of the transgressive and highstand systems tracts in Middle Cambrian sequence 2 contain the following: Inca Formation, comprising organic-rich black shale belonging to the late Templetonian–Floran Stage, *Triplagnostus gibbus–Acidusus atavus* zones; Gowers Formation, composed of peritidal phosphorites and phosphatic limestones of late Floran, *Euagnostus opimus* age; ramp, platform margin, and platform carbonates of the Currant Bush, Age Creek, Mail Change, and V Creek limestones belonging to the Undillan, *Ptychagnostus punctuosus* and *Goniagnostus nathorsti* zones. Estimated numbers of trilobite taxa recorded from these zones are plotted on figure 6.7.

Burke River Structural Belt

A more complete, different, Middle Cambrian stratigraphy is preserved along the western margin of the Burke River Structural Belt and can be tied into the sequences of the Thorntonia–Mount Isa region through the Ardmore Inlier. At the latter, only Middle Cambrian sequence 1 and the condensed section and part of the initial overlying transgressive systems tract of sequence 2 are preserved. Basal lowstand terrigenous clastics of the Ardmore Inlier are referred to the Riversdale Formation but are thought to correlate with the Mount Hendry Formation of the Thorntonia region and the Mount Birnie beds of the Burke River Structural Belt immediately to the north. The Riversdale Formation passes gradually into the Thorntonia Limestone, whose uppermost unit is the evaporitic Ardmore Chert Member, which contains *Redlichia chinensis*, and terminates sequence 1 (Southgate and Shergold 1991). The base of the overlying condensed section comprises organic-rich black shale, siltstone, and laminated coquinite composed almost entirely of *Xystridura (Xystridura) carteri* (see Öpik 1975:14). This unit is in turn overlain in the transgressive systems tract by phosphorite deposits known unofficially as the Simpson Creek Phosphorite.

In the Burke River Structural Belt per se, an almost complete Middle Cambrian succession crops out between The Monument in the south and Roaring Bore to the northeast of Duchess, a distance of approximately 80 km. Sequence 1 is well exposed at Rogers Ridge, adjacent to The Monument (Shergold and Southgate 1986; Nordlund and Southgate 1988; Southgate and Shergold 1991). Unfossiliferous Mount Birnie beds forming the lowstand systems tract pass gradually into dolostone with chert nodules referred to Unit 1 of the Thorntonia Limestone, and these pass in turn into a stromatolitic unit (unit 2) containing species of *Redlichia*, which marks the highstand of sequence 1. Laminoid fenestral carbonate at the top of unit 2 is suggested to mark the succeeding lowstand systems tract of sequence 2.

The uppermost unit of the Thorntonia Limestone at Rogers Ridge (Unit 3) consists of cyclic phosphatic grainstone, mudstone, skeletal packstone, and dolomitic limestone with hardgrounds and is interpreted as the first of three transgressive parasequence sets of sequence 2. Its fauna (Southgate and Shergold 1991: appendix 3) contains agnostid, oryctocephalid, and xystridurid trilobites, including a probable species of *Pentagnostus*. It may immediately predate the *Triplagnostus gibbus* zone or represent that zone, without the index species. It is succeeded by organic-rich black shale, formerly correlated with the Beetle Creek Formation but now demonstrably younger, from which Öpik (1975) at nearby Galah Creek identified seven trilobite taxa, including the species *Xystridura (Xystridura) carteri, X. (X.) dunstani,* and *Galahetes fulcrosus* occurring with agnostids, importantly *Triplagnostus gibbus.* Shergold (1969) also described an undetermined species of the oryctocephalid *Sandoveria* from this locality. Southgate and Shergold (1991) referred this black shale to the Inca Formation, in the second transgressive parasequence set.

Parasequence set 2 also contains the Monastery Creek Phosphorite. This terminates at a concretionary limestone layer, interpreted as a condensed section, in the basal transgressive systems tract of parasequence set 3 of the overlying Inca Formation. Both the phosphorite and concretions are the same age, which is at the overlap of the late Templetonian and early Floran stages, *Triplagnostus gibbus/Acidusus atavus* zones. A very large phosphatic and phosphatized fauna has been obtained from the Monastery Creek Phosphorite, which includes 11 trilobites (Shergold, in prep.). Only three trilobite taxa occur in the condensed section.

Sedimentation continued in the transgressive systems tract throughout the remainder of Inca Formation, with the deposition of organic-rich black shale containing a predominantly pelagic fauna of agnostid trilobites of the *Acidusus atavus* and *Euagnostus opimus* zones, associated with nepeid and dolichometopid trilobites. This formation passes laterally into and is gradually overlain by Devoncourt Limestone, represented by dark bioclastic highstand ramp carbonate similar to the Currant Bush Limestone of the Thorntonia region, deposited during the Undillan zones of *Ptychagnostus punctuosus* and *Goniagnostus nathorsti.* These zones are again dominated by the occurrence of agnostids, together with nepeid and dolichometopid trilobites (Öpik 1970a, 1982), and also by the appearance of new ptychopariids, such as *Papyriaspis, Asthenopsis,* and mapaniids.

In the northern part of the Burke River Structural Belt, the Roaring Siltstone reflects continuing deposition in the transgressive systems tract. It contains black shales with pelagic trilobite assemblages of the Boomerangian *Lejopyge laevigata* zone, divided by Öpik (1961) into three subzones: *Ptychagnostus cassis, Proampyx agra,* and *Holteria arepo.* Like the earlier Inca Formation, the Roaring Siltstone passes laterally and vertically into Devoncourt Limestone with, at Roaring Bore northeast of Duchess, a layer of calcareous concretions intervening, which suggests the onset of the highstand systems tract. The faunas of the Roaring Siltstone and Devoncourt Limestone, while containing cosmopolitan agnostid trilobites and species of *Centropleura,* become more varied in composition, including ptychopariid, nepeid, dolichometopid, and mapaniid genera.

The Devoncourt Limestone passes gradually into the Selwyn Range Limestone, composed of fine-grained, aphanitic, sparsely fossiliferous limestone totally unlike earlier carbonates in the Burke River region. These appear to terminate the highstand of the Devoncourt Limestone at the Middle-Late Cambrian passage. Elsewhere in the Georgina Basin, the Middle-Late Cambrian transition is characterized by endemism amongst the trilobite faunas, but rich diversity and rapid turnover during the Mindyallan Stage (Öpik 1967a, 1970a).

IMPLICATIONS

Species Diversity and Relative Sea Level Change

The earliest abundant and diverse trace fossil assemblages occur in shallow subtidal, upper highstand clastic sediments of the Uratanna sequence, and the lowstand tract of sequence €1.1. Alexander and Gravestock (1990) have also recorded abundant traces in the lower transgressive tract of sequence €1.1. Mount and McDonald (1992) concluded that such distribution reflects "habitat preference of the earliest Cambrian trace-generating organisms," a habitat that appears best developed during the lowstand half-cycle of relative sea level, when coastal shelves were broadest and siliciclastic sediments in greatest supply from adjacent hinterlands.

Most Australian archaeocyaths ranged from the middle Atdabanian to late Botoman, an epoch of increase and then decline in global archaeocyath generic diversity (A3–B3) (Debrenne 1992). In figure 6.4 the numbers of archaeocyath species are plotted against the sequence stratigraphy and a notional relative sea level curve, to allow comparison of species diversity between the Arrowie and Stansbury basins. Higher diversity of archaeocyaths in the Arrowie Basin results from the persistence of carbonate shelf environments through successive cycles of relative sea level change. Shelves and associated intrashelf depressions supported a variety of archaeocyath-demosponge-"coralline sponge" and archaeocyath-calcimicrobe bioherms (James and Gravestock 1990). In contrast, conditions suited to archaeocyath growth were intermittent in the shelf/ramp setting of the Stansbury Basin. There, stromatolite mudflats and ooid shoals prevailed in *wilkawillinensis* time, and subsequent successive marine transgressions effectively drowned the shelf. Bioherms of the *tardus* Zone on Fleurieu Peninsula were oligotypic (Debrenne and Gravestock 1990), and only during middle to late Botoman time did a major bioherm complex develop, as noted above.

A link between temporal species diversity and relative sea level is evident in the Arrowie Basin. The lowstand tract of sequence €1.1 lacks archaeocyaths but, with initial marine transgression, 14 species of the *wilkawillinensis* zone became established. As transgression continued (while high carbonate production maintained shelf areas), diversity increased to 43 species in the *tenuis* zone. More than half (23 species) occur in a deep-water bioherm with other sponges ("Tor Herm"; James and Gravestock 1990). None of the *wilkawillinensis* zone species survived, suggesting depth-related community replacement similar to that documented for Mongolian Zuune Arts buildups by Wood et al. (1993). Nine *tenuis* zone species ranged into the *tardus* zone, which totals 22 species in the highstand tract.

The transgressive phases of sequences €1.2 and €1.3 lack detailed study but have the following general attributes. Shelf areas that persisted during the €1.2 transgression were covered by small bioherms and oolite-oncolite-calcarenite banks and shoals. Buildups at Wirrealpa Mine contain a diverse archaeocyath assemblage that differs from that in the *tardus* zone. The Moorowie Mine reef and Ten Mile Creek bioherms of sequence €1.3 (Lafuste et al. 1991) also contain species largely distinct from those stratigraphically below and include some giant individuals up to 200 mm in diameter. Late highstand archaeocyath communities are depauperate (see figure 6.4), and lowstand exposure of shelves resulted in species extinction in the central Arrowie Basin. Successive marine transgressions thus appear to have brought new waves of immigrant species.

Compared with Archaeocyatha, the Early Cambrian trilobites of the Arrowie and Stansbury basins are few and of low diversity. Jell (in Bengtson et al. 1990) recorded a mere 18 genera and 30 species contained in four zones: in order of appearance, the *Abadiella huoi, Pararaia tatei, Pararaia bunyerooensis,* and *Pararaia janeae* zones.

In the Stansbury Basin, *A. huoi* occurs near the base of the Parara Limestone in the Curramulka Quarry (sample NMVPL78; Jell in Bengtson et al. 1990). At this locality (south of depositional hinge), *A. huoi* is within sequence €1.1 and probably in the upper transgressive or lower highstand tract.

Elicicola calva is the only species described from sequence €1.1 beneath the Flinders Unconformity in the Arrowie Basin. The boundary between the *huoi* zone (with 5 species) and the *tatei* zone (with 10 species) appears to be in the transgressive tract of sequence €1.2 in both basins. The *bunyerooensis* zone (2 species) occurs in the highstand tract of sequence €1.2 in the Arrowie Basin, and *Atops* (=*Ivshiniellus*) *briandailyi* (Jenkins and Hasenohr 1989) in the Stansbury Basin may be coeval. *Pararaia janeae* zone contains 10 taxa recorded from the upper Mernmerna Formation and Oraparinna Shale and 5 from a bioherm in the upper Wilkawillina Limestone. These trilobites occur in sequence €1.3 in the Arrowie Basin, to which *Atops rupertensis* (Jell et al. 1992) may be added. In the Stansbury Basin, *Redlichia takooensis* from the Emu Bay Shale of Kangaroo Island is also probably in sequence €1.3 or €2.1, the latter containing *Balcoracania flindersi* from the Billy Creek Formation in the Arrowie Basin. Single trilobite taxa, postdating the *janeae* zone, occur in €2.1, Wirrealpa Limestone, and in €2.3, Moodlatana Formation.

This Early Cambrian impoverishment in South Australia is in marked contrast to the Middle Cambrian of the eastern Georgina Basin, where the numbers shown against each zone in figure 6.7 indicate very rapidly increasing diversity associated with successive transgressive systems tracts that developed in response to continuous subsidence of the Mount Isa Block.

Sequence Correlation

Where fossil scarcity hinders biostratigraphic correlation, sequences can be matched between neighboring basins, especially if they share a regime of relatively low tectonic activity and steady subsidence. The eustatic component of relative sea level can be interpreted with greater confidence under these circumstances, and this has enabled correlation of the "Uratanna" and €1.1 sequences in the Stansbury, Arrowie, and

Amadeus basins. The upper boundary of sequence €1.1 (Flinders Unconformity) appears to be the youngest correlative surface.

Thereafter, deep marine deposition, the onset of rifting and foundering of the Kanmantoo Trough outboard of the Stansbury Basin shelf, hampers high-resolution regional correlation. Between the Chandler Formation and underlying Todd River Dolomite in the Amadeus Basin, South Australian sequences €1.2 through €2.3 appear to be missing during the period of the Kangarooian Movements. It is this tectonic activity which is responsible for the difficulty in reconciling correlation of the Ordian Stage.

High rates of siliciclastic sediment supply and overall regression characterize the latest Early and Middle Cambrian of the Moodlatana and Balcoracana formations of the Lake Frome Group in the Arrowie Basin. This condition is interpreted as a consequence of deformation of the neighboring Wonominta Block in western New South Wales (Wang et al. 1989). The Antarctic margin of Gondwana was also tectonically active, and the lower Middle Cambrian stratigraphic record is punctuated by disconformity (Rowell et al. 1992; Evans et al. 1995). As a result, it is difficult to distinguish Australian tectonoeustatic events such as the Kangarooian and Mootwingee movements from Antarctic counterparts, and it would be premature to suggest their correlation with the Canadian Hawke Bay Event on any other than a broad scale.

Geochronology and Speciation Rates

Geochronologic dates based on Cambrian zircons have been most recently summarized by Vidal et al. (1995) and Shergold (1995). These permit us to determine the duration of the Lower and Middle Cambrian sequences. Following Bowring et al. (1993) and Isachsen et al. (1994), we have taken the base of the Cambrian at 545 Ma. SHRIMP (Sensitive High Mass Resolution Ion MicroProbe) dates from tuffs in the upper Heatherdale Shale (Stansbury Basin) and lower Billy Creek Formation (Arrowie Basin) by Cooper et al. (1992) and Compston et al. (1992) of 526 ± 4 and 522.8 ± 1.8 Ma, respectively, effectively constrain the late Botoman stage. Considering the Toyonian stage to be of latest Early Cambrian age in South Australia, the duration of the Early Cambrian is in excess of 25 m.y. and probably closer to 35 m.y. Here we assume a Lower-Middle Cambrian boundary at 510 Ma. Zircons from the Comstock Tuff in northwestern Tasmania, biostratigraphically constrained by the trilobites of the mid-Boomerangian *Lejopyge laevigata* II trilobite assemblage zone, have been dated (Perkins and Walshe 1993) at 494.4 ± 3.8 Ma. Other dates from the Mount Read Volcanics have SHRIMP zircon ages near 503 Ma suggesting that the top of the Middle Cambrian may be as young as 500 Ma. This being so, the duration of the Middle Cambrian is 10–15 m.y. However, all of the dates quoted here from southern Australia have been queried by Jago and Haines (1998) because of doubts about the reliability of the standard (SL13) used to calculate them. Use of the alternative standard QGNG

(Black et al. 1997) produces relatively older ages. At this stage, we are inclined to use the generally accepted original SL13 dates but may need to revise some of our calculated diversity patterns should the QGNG standard be adopted.

These dates have some significance for estimating rates of sedimentation and biotic evolution. Seven third-order depositional sequences are recognized in the Lower Cambrian of South Australia with an estimated average duration of 4–5 m.y. In figure 6.4, in the Arrowie Basin, some 91 archaeocyath species are estimated to occur in the 4–5 m.y. sequence €1.1, and about 113 occur in the 8–10 m.y. sequences €1.2 and €1.3 combined.

Two Middle Cambrian supersequences, containing four parasequences and two condensed sections, are presently recognized in the Georgina Basin, spanning 10–15 m.y. Here, the rate of trilobite speciation is estimated against the sequence stratigraphy and notional relative sea level curve. Georgina Basin supersequence 3 (Middle Cambrian sequence 2) contains three retrogradational parasequence sets that embrace eight trilobite assemblage zones. The time available suggests very rapid rates of evolution in successive transgressive systems tracts, determined by a rapid rate of subsidence of the basement blocks, particularly the Mount Isa Block.

Acknowledgments. An earlier draft of this paper benefited from reviews by Drs. T. Loutit, J. Laurie, and C. B. Foster (Australian Geological Survey Organisation). Drs. Pierre Kruse (Northern Territory Geological Survey) and John Lindsay (AGSO) kindly reviewed the manuscript.

REFERENCES

Alexander, E. M. and D. I. Gravestock. 1990. Sedimentary facies in the Sellick Hill Formation, Fleurieu Peninsula, South Australia. *Geological Society of Australia, Special Publication* 16:269–289.

Bengtson, S., S. Conway Morris, B. J. Cooper, P. A. Jell, and B. N. Runnegar. 1990. Early Cambrian fossils from South Australia. *Association of Australasian Palaeontologists, Memoir* 9:1–364.

Black, L. P., D. B. Seymour, K. D. Corbett, S. E. Cox, J. E. Streit, R. S. Bottrill, C. R. Calver, J. L. Everard, G. R. Green, M. P. McClenaghan, J. Pemberton, J. Taheri, and N. J. Turner. 1997. Dating Tasmania's oldest geological events. *Australian Geological Survey Organisation, Record* 1997/15: 1–57.

Bowring, S. A., J. P. Grotzinger, C. E. Isachsen, A. H. Knoll, S. M. Pelechaty, and P. Kolosov. 1993. Calibrating rates of Early Cambrian evolution. *Science* 261:1293–1298.

Bradshaw, J. 1991. Description and depositional model of the Chandler Formation: A Lower Cambrian evaporite and carbonate sequence, Amadeus Basin, central Australia. In R. J. Korsch and J. M. Kennard, eds., *Geological and Geophysical Studies in the Amadeus Basin, Central Australia*, pp. 227–244. *Bureau of Mineral Resources, Australia, Bulletin* 236.

Brock, G. A. and B. J. Cooper. 1993. Shelly fossils from the Early Cambrian (Toyonian) Wirrealpa, Aroona Creek, and Ramsay Limestones of South Australia. *Journal of Paleontology* 67:758–787.

Brown, L. F., and W. L. Fisher. 1977. Seismic-stratigraphic interpretation of depositional systems: Examples from Brazilian rift and pull-apart basins. *American Association of Petroleum Geologists, Memoir* 26:213–248.

Compston, W. C., I. S. Williams, J. L. Kirschvink, Z. C. Zhang, and G. Ma. 1992. Zircon U-Pb ages for the Early Cambrian timescale. *Journal of the Geological Society of London* 149:171–184.

Cook, P. J. 1988. *Cambrian.* Vol. 1 of *Palaeogeographic Atlas of Australia.* Canberra: Australian Government Publishing Service.

Cooper, J. A., R. J. F. Jenkins, W. Compston, and I. S. Williams. 1992. Ion-probe zircon dating of a mid-Early Cambrian tuff in South Australia. *Journal of the Geological Society of London* 149:185–192.

Daily, B. 1972. The base of the Cambrian and the first Cambrian faunas. *University of Adelaide, Centre for Precambrian Research, Special Paper* 1:13–41.

Daily, B. 1973. Discovery and significance of basal Cambrian Uratanna Formation, Mt. Scott Range, Flinders Ranges, South Australia. *Search* 4:202–205.

Daily, B. 1976a. Novye dannye ob osnovanii kembriya v Yuzhnoy Avstralii [New data on the base of the Cambrian in South Australia]. *Izvestiya Akademii nauk SSSR, Seriya geologicheskaya* 1976 (3):45–52.

Daily, B. 1976b. The Cambrian of the Flinders Ranges. In B. P. Thomson, B. Daily, R. P. Coats, and B. G. Forbes, eds., *Late Precambrian and Cambrian Geology of the Adelaide "Geosyncline" and Stuart Shelf, South Australia,* pp. 15–19. 25th International Geological Congress, Sydney, Excursion Guide 33A.

Daily, B. 1990. Cambrian stratigraphy of Yorke Peninsula. *Geological Society of Australia, Special Publication* 16:215–229.

Daily, B. and B. G. Forbes. 1969. Notes on the Proterozoic and Cambrian, southern and central Flinders Ranges, South Australia. In B. Daily, ed., *Geological Excursions Handbook, ANZAAS,* sec. 3, pp. 23–30. Adelaide: Australia and New Zealand Association for the Advancement of Science.

Dalgarno, C. R. and J. E. Johnson. 1962. Cambrian sequence of the western Flinders Ranges. *Geological Survey of South Australia, Quarterly Geological Notes* 4:1–2.

Debrenne, F. 1992. Diversification of Archaeocyatha. In J. H. Lipps and P. W. Signor, eds., *Origin and Early Evolution of the Metazoa,* pp. 425–443. New York: Plenum Press.

Debrenne, F. and D. I. Gravestock. 1990. Archaeocyatha from the Sellick Hill Formation and Fork Tree Limestone on Fleurieu Peninsula, South Australia. *Geological Society of Australia, Special Publication* 16:290–309.

Debrenne, F. and P. D. Kruse. 1986. Shackleton Limestone archaeocyaths. *Alcheringa* 10:235–78.

Debrenne, F. and P. D. Kruse. 1989. Cambrian Antarctic archaeocyaths. In J. A. Crame, ed., *Origins and Evolution of the Antarctic Biota,* pp. 15–28. Geological Society Special Publication 47.

Debrenne, F. and A. Yu. Zhuravlev. 1992. *Irregular Archaeocyaths.* Paris: Cahiers de Paléontologie, Éditions du Centre National de la Recherche Scientifique.

Debrenne, F., A. Yu. Rozanov, and A. Yu. Zhuravlev. 1990. *Regular Archaeocyaths.* Paris: Cahiers de Paléontologie, Éditions du Centre National de la Recherche Scientifique.

Dyson, I. A., C. G. Gatehouse, and J. B. Jago. 1996. Sequence stratigraphy of the Talisker Calc-siltstone and lateral equivalents in the Cambrian Kanmantoo Group. *Geological Survey of South Australia, Quarterly Geological Notes* 129:27–41.

Evans, K. R., A. J. Rowell, and M. N. Rees. 1995. Sea-level changes and stratigraphy of the Nelson Limestone (Middle Cambrian), Neptune Range, Antarctica. *Journal of Sedimentary Research* B65:32–43.

Fuller, M. K. and R. J. F. Jenkins. 1994. *Moorowipora chamberensis,* a coral from the Early Cambrian Moorowie Formation, Flinders Ranges, South Australia. *Royal Society of South Australia, Transactions* 118:227–235.

Geyer, G. and A. Uchman. 1995. Ichnofossil assemblages from the Nama Group (Neoproterozoic–Lower Cambrian) in Namibia and the Proterozoic-Cambrian boundary problem revisited. In G. Geyer and E. Landing, eds., Morocco '95: The Lower-Middle Cambrian standard of western Gondwana. *Beringeria, Special Issue* 2:175–202.

Gravestock, D. I. 1984. Archaeocyatha from lower parts of the Lower Cambrian carbonate sequence in South Australia. *Association of Australasian Palaeontologists, Memoir* 2:1–139.

Gravestock, D. I. 1995. Early to middle Palaeozoic. In J. F. Drexel and W. V. Preiss, eds., *The Phanerozoic,* vol. 2 of *The Geology of South Australia,* pp. 3–61. South Australian Geological Survey, Bulletin 54.

Gravestock, D. I. and J. E. Hibburt. 1991. Sequence stratigraphy of the eastern Officer and Arrowie basins: A framework for Cambrian oil search. *APEA Journal* 31:177–190.

Gravestock, D. I., C. G. Gatehouse, and J. B. Jago. 1990. Preliminary sequence stratigraphy of the Kanmantoo Group: Implications for South Australian Cambrian correlations. *Third International Symposium on the Cambrian System, Novosibirsk, USSR, Abstracts,* pp. 107–109. Novosibirsk: Institute of Geology and Geophysics, Siberian Branch, USSR Academy of Sciences.

Haines, P. W. 1991. Palaeontological note: Early Cambrian shelly fossils from the Arumbera Sandstone, Amadeus Basin, Northern Territory. *Alcheringa* 15:150.

Haslett, P. G. 1975. The Woodendinna Dolomite and Wirrapowie Limestone—two new Lower Cambrian formations, Flinders Ranges, South Australia. *Royal Society of South Australia, Transactions* 99:211–220.

Henderson, R. A. and P. N. Southgate. 1980. Cambrian evaporitic sequences from the Georgina Basin. *Search* 11:247–249.

Hill, D. 1965. Archaeocyatha from Antarctica and a review of the phylum. *Geology* 3:1–151. Transantarctic Expedition 1955–1958, Scientific Report 10.

Isachsen, C. E., S. A. Bowring, E. Landing, and S. D. Samson. 1994. New constraint on the division of Cambrian time. *Geology* 22:496–498.

Jago, J. B. and P. W. Haines. 1998. Recent radiometric dating of some Cambrian rocks in southern Australia: Relevance to the Cambrian time scale. *Revista Española de Paleontología,* no. extraordinario, Homenaje al Prof. Gonzalo Vidal, pp. 115–122.

Jago, J. B., B. Daily, C. C. von der Borch, A. Cernovskis, and N. Saunders. 1984. First reported trilobites from the Lower Cambrian Normanville Group, Fleurieu Peninsula, South Australia. *Royal Society of South Australia, Transactions* 108:207–211.

Jago, J. B., I. A. Dyson, and C. G. Gatehouse. 1994. The nature of the sequence boundary between the Normanville and Kanmantoo Groups on Fleurieu Peninsula, South Australia. *Australian Journal of Earth Sciences* 41:445–453.

James, N. P. and D. I. Gravestock. 1990. Lower Cambrian shelf and shelf margin buildups, Flinders Ranges, South Australia. *Sedimentology* 37:455–480.

Jell, P. A. 1975. Australian Middle Cambrian

eodiscoids with a review of the superfamily. *Palaeontographica A* 150:1–97.

Jell, P. A. 1983. The Early to Middle Cambrian boundary in Australia. *Lithosphere Dynamics and Evolution of the Continental Crust, Sixth Australian Geological Convention, Canberra, Geological Society of Australia, Abstracts* 9:236.

Jell, P. A., J. B. Jago, and J. G. Gehling. 1992. A new conocoryphid trilobite from the Lower Cambrian of the Flinders Ranges, South Australia. *Alcheringa* 16:189–200.

Jenkins, R. J. F. and P. Hasenohr. 1989. Trilobites and their trails in a black shale: Early Cambrian of the Fleurieu Peninsula, South Australia. *Royal Society of South Australia, Transactions* 113:195–203.

Kennard, J. M. 1991. Lower Cambrian archaeocyathan buildups, Todd River Dolomite, northeast Amadeus Basin, central Australia: Sedimentology and diagenesis. *Bureau of Mineral Resources, Australia, Bulletin* 236:195–225.

Kennard, J. M. and J. F. Lindsay. 1991. Sequence stratigraphy of the latest Proterozoic-Cambrian Pertaoorta Group, northern Amadeus Basin, central Australia. *Bureau of Mineral Resources, Australia, Bulletin* 236:171–194.

Kruse, P. D. 1982. Archaeocyathan biostratigraphy of the Gnalta Group at Mt. Wright, New South Wales. *Palaeontographica A* 177:129–212.

Kruse, P. D. 1991. Cyanobacterial-archaeocyathan-radiocyathan bioherms in the Wirrealpa Limestone of South Australia. *Canadian Journal of Earth Sciences* 28:601–615.

Kruse, P. D. and P. W. West. 1980. Archaeocyatha of the Amadeus and Georgina basins. *BMR Journal of Australian Geology and Geophysics* 5:165–181.

Lafuste, J., F. Debrenne, A. Gandin, and D. I. Gravestock. 1991. The oldest tabulate corals and the associated Archaeocyatha, Lower Cambrian, Flinders Ranges, South Australia. *Géobios* 24:697–718.

Laurie, J. R. 1986. Phosphatic fauna of the Early Cambrian Todd River Dolomite, Amadeus Basin, central Australia. *Alcheringa* 10:431–545.

Laurie, J. R. and J. H. Shergold. 1985. Phosphatic organisms and the correlation of Early Cambrian carbonate formations in central Australia. *BMR Journal of Australian Geology and Geophysics* 9:83–89.

Lindsay, J. F. 1987. Sequence stratigraphy and depositional controls in late Proterozoic–Early Cambrian sediments of the Amadeus Basin, central Australia. *American Association of Petroleum Geologists, Bulletin* 71:1387–1403.

Lindsay, J. F., J. M. Kennard, and P. N. Southgate. 1993. Application of sequence stratigraphy in an intracratonic setting. *International Association of Sedimentologists, Special Publication* 18:605–631.

Lu, Y. H., C. L. Chu, Y. Y. Chien, H. L. Lin, Z. Y. Zhou, and K. X. Yuan. 1974. Bioenvironmental control hypothesis and its application to the Cambrian biostratigraphy and palaeozoogeography. *Memoirs of the Nanking Institute of Geology and Palaeontology* 5:27–110.

McDonald, C. 1992. Origin and sequence stratigraphic significance of the Uratanna channels, basal Cambrian Uratanna Formation, Northern Flinders Ranges, South Australia. Master's thesis, University of California, Davis.

Mount, J. F. 1993. Uratanna Formation and the base of the Cambrian System, Angepena Syncline. IGCP Project 320, Excursion Guide, pp. 86–90. Unpublished.

Mount, J. F. and D. Kidder. 1993. Combined flow origin of edgewise intraclast conglomerates: Sellick Hill Formation (Lower

Cambrian), South Australia. *Sedimentology* 40:315–329.

Mount, J. F. and C. McDonald. 1992. Influence of changes in climate, sea level, and depositional systems on the fossil record of the Neoproterozoic–Early Cambrian metazoan radiation, Australia. *Geology* 20: 1031–1034.

Myrow, P. M. and R. N. Hiscott. 1993. Depositional history and sequence stratigraphy of the Precambrian-Cambrian boundary stratotype section, Chapel Island Formation, southeast Newfoundland. *Palaeogeography, Palaeoclimatology, Palaeoecology* 104: 13–35.

Nedin, C. 1995. The Emu Bay Shale, a Lower Cambrian fossil Lagerstätten, Kangaroo Island, South Australia. *Association of Australasian Palaeontologists, Memoir* 18:31–40.

Nordlund, U. and P. N. Southgate. 1988. Depositional environment and importance of the early Middle Cambrian sequence at Rogers Ridge, W. Queensland. *Acta Universitatis Upsaliensis, Comprehensive Summaries of Uppsala Dissertations from the Faculty of Science* 170.

Oaks, R. Q., J. A. Deckelman, K. T. Conrad, L. T. Hamp, J. O. Phillips, and A. J. Stewart. 1991. Sedimentation and tectonics in the northeastern and central Amadeus Basin, central Australia. *Bureau of Mineral Resources, Australia, Bulletin* 236:73–90.

Öpik, A. A. 1961. The geology of palaeontology of the headwaters of the Burke River, Queensland. *Bureau of Mineral Resources, Australia, Bulletin* 53.

Öpik, A. A. 1967a. The Mindyallan fauna of north-western Queensland. *Bureau of Mineral Resources, Australia, Bulletin* 74, vols. 1 and 2.

Öpik, A. A. 1967b. The Ordian Stage of the Cambrian and its Australian Metadoxididae. *Bureau of Mineral Resources, Australia, Bulletin* 92:113–170.

Öpik, A. A. 1970a. Nepeiid trilobites of the Middle Cambrian of northern Australia. *Bureau of Mineral Resources, Australia, Bulletin* 113.

Öpik, A. A. 1970b. *Redlichia* of the Ordian (Cambrian) of northern Australia. *Bureau of Mineral Resources, Australia, Bulletin* 114.

Öpik, A. A. 1975. Templetonian and Ordian xystridurid trilobites of Australia. *Bureau of Mineral Resources, Australia, Bulletin* 121.

Öpik, A. A. 1982. Dolichometopid trilobites of Queensland, Northern Territory and New South Wales. *Bureau of Mineral Resources, Australia, Bulletin* 175.

Perkins, C. and J. L. Walshe. 1993. Geochronology of the Mount Read Volcanics, Tasmania, Australia. *Economic Geology* 88: 1176–1197.

Rowell, A. J., M. N. Rees, and K. R. Evans. 1992. Evidence of major Middle Cambrian deformation in the Ross Orogen, Antarctica. *Geology* 20:31–34.

Rowland, S. M. and R. A. Gangloff. 1988. Structure and paleoecology of Lower Cambrian reefs. *Palaios* 3:111–135.

Savarese, M., J. F. Mount, and J. E. Sorauf. 1993. Paleobiologic and paleoenvironmental context of coral-bearing Early Cambrian reefs: Implications for Phanerozoic reef development. *Geology* 21:917–920.

Shergold, J. H. 1969. Oryctocephalidae (Trilobita: Middle Cambrian) of Australia. *Bureau of Mineral Resources, Australia, Bulletin* 104.

Shergold, J. H. 1995. *Timescales. 1: Cambrian.* Australian Phanerozoic Timescales, Biostratigraphic Charts, and Explanatory Notes, 2d ser. Australian Geological Survey Organisation Record 1995/30.

Shergold, J. H. and P. N. Southgate, eds. 1986. *Middle Cambrian Phosphatic and Calcareous Lithofacies along the Eastern Margin of the Georgina Basin, Western Queensland.* Australasian Sedimentologists Group Field Excursion Series Guide 2. Sydney: Geological Society of Australia.

Sorauf, J. E. and M. Savarese. 1995. A Lower Cambrian coral from South Australia. *Palaeontology* 38:757–770.

Southgate, P. N. 1982. Cambrian skeletal halite crystals and experimental analogues. *Sedimentology* 29:391–407.

Southgate, P. N. 1986. The Gowers Formation and Bronco Stromatolith Bed, two new stratigraphic units in the Undilla portion of the Georgina Basin. *Queensland Government Mining Journal* (October): 407–411.

Southgate, P. N. and J. H. Shergold. 1991. Application of sequence stratigraphic concepts to Middle Cambrian phosphogenesis, Georgina Basin, Australia. *BMR Journal of Australian Geology and Geophysics* 12: 119–144.

Ushatinskaya, G., E. Zhegallo, N. Esakova, W. L. Zang, and D. Gravestock. 1995. Preliminary report on well Port Julia 1A. Stansbury Basin Project, Progress Report 1. Unpublished.

Vail, P. R., R. M. Mitchum, R. G. Todd, J. M. Widmier, S. Thompson, J. B. Sangree, J. N. Bubb, and W. G. Hatlelid. 1977. Seismic stratigraphy and global changes of sea level. *American Association of Petroleum Geologists, Memoir* 26:49–212.

van Wagoner, J. C., H. W. Posamentier, R. M. Mitchum, P. R. Vail, J. F. Sarg, T. S. Loutit, and J. Hardenbol. 1988. An overview of the fundamentals of sequence stratigraphy and key definitions. In C. K. Wilgus, B. S. Hastings, C. G. St. C. Kendall, H. W. Posamentier, C. A. Ross, and J. C. van Wagoner, eds., Sea-level changes: An integrated approach, *Society of Economic Paleontologists and Mineralogists, Special Publication* 42:39–45.

Vidal, G., M. Moczydłowska, M. and V. R. Rudavskaya. 1995. Constraints on the Early Cambrian radiation and correlation of the Tommotian and Nemakit-Daldynian regional stages of eastern Siberia. *Journal of the Geological Society of London* 152:499–510.

Wallace, M. W., R. R. Keays, and V. A. Gostin. 1991. Stromatolitic iron oxides: Evidence that sea-level changes can cause sedimentary iridium anomalies. *Geology* 19:551–554.

Walter, M. R., R. Elphinstone, and G. R. Heys. 1989. Proterozoic and Early Cambrian trace fossils from the Amadeus and Georgina basins, central Australia. *Alcheringa* 13:209–256.

Wang, Q. Z., K. J. Mills, B. D. Webby, and J. H. Shergold. 1989. Upper Cambrian (Mindyallan) trilobites and stratigraphy of the Kayrunnera Group, western New South Wales. *BMR Journal of Australian Geology and Geophysics* 11:107–118.

Wood, R. A., A. Yu. Zhuravlev, and F. Debrenne. 1992. Functional biology and ecology of Archaeocyatha. *Palaios* 7:131–156.

Wood, R., A. Yu. Zhuravlev, and A. Chimed Tseren. 1993. The ecology of Lower Cambrian buildups from Zuune Arts, Mongolia: Implications for early metazoan reef evolution. *Sedimentology* 40:829–858.

Zhuravlev, A. Yu. 1986. Evolution of archaeocyaths and palaeobiogeography of the Early Cambrian. *Geological Magazine* 123:377–385.

Zhuravlev, A. Yu. and D. I. Gravestock. 1994. Archaeocyaths from Yorke Peninsula, South Australia, and archaeocyathan Early Cambrian zonation. *Alcheringa* 18:1–54.

CHAPTER SEVEN

Mary L. Droser and Xing Li

The Cambrian Radiation and the Diversification of Sedimentary Fabrics

The Cambrian represents a pivotal point in the history of marine sedimentary rocks. Cambrian biofabrics that are directly a product of metazoans include ichnofabrics, shell beds, and constructional frameworks. The development and distribution of biofabrics is strongly controlled by sedimentary facies. In particular, terrigenous clastics and carbonates reveal very different early records of biofabrics. This is particularly obvious with ichnofabrics but equally important with shell beds. Ichnofabrics in high-energy sandstones (e.g., Skolithos *piperock) and fine-grained terrigenous clastic sediments can be well bioturbated at the base of the Cambrian, whereas other settings show less well developed bioturbation in the earliest Cambrian. Nearly all settings demonstrate an increase in extent of bioturbation and tiering depth and complexity through the Cambrian. Shell beds appear with the earliest skeletonized metazoans. Data from the Basin and Range Province of the western United States demonstrate that shell beds increase in thickness, abundance, and complexity through the Cambrian. The study of biofabrics is an exciting venue for future research. This is particularly true of the latest Precambrian and Cambrian, where biofabrics have been relatively underutilized in our exploration to find the relationships between physical, chemical, and biological processes and the Cambrian explosion. Biofabrics provide a natural link between these processes.*

WITH THE CAMBRIAN RADIATION of marine invertebrates, sedimentary rocks on this planet changed forever. The advent of skeletonized metazoans introduced shells and skeletons as sedimentary particles, and the tremendous increase in burrowing metazoans resulted in the partial or complete mixing of sediment and/or in the production of new sedimentary structures. Whereas constructional frameworks formed by stromatolites were common in the Precambrian (e.g., Awramik 1991; Grotzinger and Knoll 1995), metazoan reef builders first appeared near the Precambrian-Cambrian boundary, initiating complex reef fabrics in Early Cambrian time (Riding and Zhuravlev 1995). Diverse and well-defined calcified cyanobacteria and calcified algae appearing in the Cambrian (Riding 1991b; Riding, this volume), along with increased fecal material, represent additional important biological contributors to sedimentary fabrics. Thus, at the Precambrian-Cambrian boundary and continuing into the Cambrian, there was a major shift in sediments and substrates and a dramatic increase in diversity of those sedimentary rocks that gain their final sedimentary fabric from biological sources, either through in situ (autochthonous) processes or through the allochthonous processes of transport and concentration of biogenic sedimentary particles (see also Copper 1997). This shift has important and clear implications for the ecology of the diversifying fauna as well as for sedimentology and stratigraphy.

There is a wide range of sedimentary macrofabrics that result from a biological source or process. Such fabrics can be broadly attributed to three fabric-producing processes: (1) construction by organisms of structures that are then preserved in situ in the rock record—such as reefs, stromatolites, and thrombolites; (2) concentration of individual sedimentary particles that are biological in origin (e.g., skeletal material and oncoids), through primarily depositional but also erosional (winnowing) processes, producing shell beds, oncolite beds, oozes, etc. (additionally, biofabrics produced through baffling appear to be particularly important in the late Precambrian); and (3) bioturbation (and bioerosion), which is due to postdepositional processes. These processes serve as only a starting point for examination of biologically generated fabrics, and at different scales they are not exclusive of one another. For example, oncoids themselves are a constructional microfabric. However, they are then transported and concentrated to produce a depositional macrofabric. Fecal pellets, likewise, are a constructional microfabric but are commonly concentrated (along with abiotic sources) to form peloidal limestones.

Study of Neoproterozoic and Cambrian sedimentary fabrics is further complicated by the presence of nonactualistic sedimentary structures (e.g., Seilacher and Pflüger 1994; Pflüger and Gresse 1996) and by the effects of changing biogeochemical cycles, which are reflected by isotope data as well as the distribution of specific facies types such as black shales, phosphorites, and carbonate precipitates (e.g., Brasier 1992; Grotzinger and Knoll 1995; Logan et al. 1995; Brasier et al. 1996). While the events of the Neoproterozoic and Early Cambrian are becoming better understood, it remains difficult to tease apart the different components—in particular, cause and effect. In this chapter we focus on one aspect of the sedimentological record, that is, those macrofabrics that directly result from the radiation of marine invertebrates. These types of sedimentary fabrics have received remarkably little attention, given their impact on the stratigraphic record, and this chapter represents only a starting point.

Although there is no encompassing terminology that covers all of these types of fabrics, different terminologies have been independently developed for description and interpretation of sedimentary fabrics resulting from a strong biological input.

Efficient and easily applied descriptive terminologies for various aspects of shell beds (fossil concentrations) have been developed (Kidwell 1986, 1991; Kidwell et al. 1986; Kidwell and Holland 1991; Fürsich and Oschmann 1993; Goldring 1995). The ichnofabric concept and associated terminology are well entrenched for dealing with the record of bioturbation (e.g., Ekdale and Bromley 1983; Bromley and Ekdale 1986; Droser and Bottjer 1993; Taylor and Goldring 1993; Bromley 1996). Classifications for coping with reef fabrics, microbial fabrics, and other types of constructional fabrics have also received extensive discussion (e.g., Riding 1991a; Grotzinger and Knoll 1995; Wood 1995; see also Pratt et al. and Riding, this volume).

In this chapter, we examine various aspects of biologically influenced sedimentary rock fabrics and then specifically discuss Cambrian ichnofabrics and fossil concentrations. Precambrian biofabrics resulting from early metazoans are briefly discussed. We are not including constructional frameworks, which are discussed elsewhere in this volume. In order to facilitate communication, when we refer to all biologically effected fabrics as a group, we use the term *biofabrics*. While this term has been used with various definitions in the literature and therefore has a relatively vague meaning, it does serve a purpose here as an inclusive term that does not imply any specific type of process but rather implies a final product that is largely the result of either allochthonous and/or autochthonous processes involving a substantial biological input. In no way does this term serve as a substitute for the terminology for each of these fabric types.

ECOLOGIC SIGNIFICANCE

The production and preservation of biologically influenced sedimentary fabrics are functions of local and large-scale physical, biological, and chemical processes (e.g., Droser 1991; Kidwell 1991; Goldring 1995). Biological controls include life habits and behavior of the infauna and epifauna, mineralogy, fecundity, nature of clonality, growth rates, size of organisms, molting frequency, and rates at which organisms colonize substrates. Local physical controls include frequency and character of episodic sedimentation, overall rate and steadiness of flow and sedimentation, bedding thickness, sediment size and sorting, and rates and nature of erosion. Large-scale processes include sea level changes, climate, tectonics, subsidence, ocean geochemistry, biogeography, and, of course, evolution. These processes acting on various scales dictate the final nature of the sedimentary rocks.

Autochthonous biofabrics represent the response of animals to changing or static environmental conditions or are the result of local physical processes such as winnowing. Allochthonous biofabrics result directly from physical processes. Thus, biofabrics have important implications for sedimentological and stratigraphic interpretations of the rock record. The effects of processes governing the character and distribution of Phanerozoic shell beds and ichnofabrics have been extensively reviewed recently elsewhere (Fürsich and Oschmann 1993; Goldring 1995; Kidwell and Flessa 1995; Savrda 1995) and thus will not be further discussed here.

Biofabrics have an interesting and unique ecologic role. First, the processes that lead to the production of biofabrics result in a change of the original substrate or local environmental and ecologic conditions. Thus, the depositional fabric itself is part of a "taphonomic feedback" (Kidwell and Jablonski 1983). The advent of a new (biofabric-producing) community may result in the development of new or expanded ecologies or may exclude other animals. For example, the process of bioturbation results in the extensive alteration of the physical and chemical properties of the substrate and thus alters the habitat (Aller 1982; Ziebis et al. 1996). As such, the bioturbating community will also be modified. For example, a bioturbating organism may introduce oxygen into the substrate or provide an open burrow system in which others can live symbiotically (Bromley 1996). In contrast, burrowing organisms may create conditions that exclude other animals and, thus, change the community in that way.

Kidwell and Jablonski (1983) recognized two types of taphonomic feedback associated with shell beds: (1) abundant hard parts—shell beds—may restrict infaunal habitat space and/or alter sediment textures; and (2) dead hard parts provide a substrate for firm-sediment dwellers. The importance of this for the development of Ordovician hardground communities has been discussed by Wilson et al. (1992) and might be equally important for the Cambrian. For example, many stromatolite-thrombolite buildups in the Cambrian of the western United States, particularly Upper Cambrian carbonate platform facies, are underlain and/or overlain by trilobite-echinoderm–dominated composite/condensed shell beds. The association of the stromatolite-thrombolite buildups with shell-rich beds suggests that shell beds provide a firm or hard substratum for the stromatolite-building microorganisms to colonize. Thus, many well-developed Cambrian shell beds provided an additional hard substrate that did not exist in the Precambrian for the development of microbial buildups. The spatial distribution of the stromatolite-thrombolite buildups may partly be controlled by the distribution of shell beds.

Cambrian habitat and substrate changes resulting from bioturbation and the production of shell beds are a fruitful area for future research. The effects of the initiation of vertical bioturbation and the development of the infaunal habitat, in particular, have already been cited for destroying nonactualistic Precambrian sedimentary structures, microbial mat surfaces, and possibly the preservation window of the Ediacaran faunas (e.g., Gehling 1991; Seilacher and Pflüger 1994; Pflüger and Gresse 1996; Jensen et al. 1998; Gehling 1999). Increased levels of bioturbation have also been credited with increasing nutrient levels in the water column (Brasier 1991).

The second way in which biofabrics are significant ecologically is that they are uniquely poised for ecologic interpretation from the stratigraphic record. Autochthonous biofabrics, including ichnofabrics, reef fabrics, stromatolites, thrombolites, and other types of microbial fabrics, as well as autochthonous shell beds, essentially preserve in situ ecologic relationships; that is, they record a particular ecology or ecologic event. These types of fabrics are ecologically most significant. However, some of these fabrics may preserve time-averaged assemblages or communities, albeit in situ, as discussed below in the section "Stratigraphic Range and Uniformitarianism." So care must be taken when making ecologic interpretations from biofabrics (e.g., Goldring 1995; Kidwell and Flessa 1995). Nonetheless, these types of fabrics offer an opportunity to examine ecologic relationships that are not otherwise widely available to the paleontologist. (Hardgrounds provide another such example.) Many shell beds are of course allochthonous, and so the viability for ecologic studies must be evaluated only after taphonomic and stratigraphic analysis (e.g., Kidwell and Flessa 1995). Traditionally, studies of reef fabrics have made use of in situ ecologic relationships. However, Cambrian shell beds and ichnofabrics have been underutilized for ecologic studies (but see Droser et al. 1994).

At a temporally larger scale, the stratigraphic distribution of a particular sedimentary fabric can yield insight into the abundance or significance of a particular group of organisms, as discussed below. In these types of studies, the problems of transport may be less important.

STRATIGRAPHIC RANGE AND UNIFORMITARIANISM

Uniformitarianism is an essential part of the geologist's approach to the rock record. However, superimposed on the relative predictability of physical processes are evolution and the ever-changing biota on this planet. Indeed, in a physical world where sedimentological successions reflecting similar types of local physical energies appear differently in various climatic or tectonic regimes, changing biotas through time add even more complications. Biologically generated sedimentary fabrics have distributions that are tied directly to the stratigraphic distribution of the organism. However, commonly, the range of the biofabric will be less than that of the actual organism. For example, articulate brachiopods are present for nearly the entire Phanerozoic, but articulate brachiopod shell beds are a common stratigraphic component from only the Ordovician through the Jurassic (Kidwell 1990; Kidwell and Brenchley 1994; Li and Droser 1995). The trace fossil *Skolithos* is present throughout the Phanerozoic, but *Skolithos* piperock is most common in the Cambrian and declines thereafter (Droser 1991). Thus, the distribution or abundance of a particular biofabric can give insight into the relative importance or abundance of that animal or of a particular depositional setting at any given time. Because biofabrics will be sensitive to biological, physical, and even chemical variations, they provide a unique insight into environmental conditions. In seemingly similar depositional settings, biofabrics may be quite different, depending on several factors; potentially, we can use studies of biofabrics for better understanding of these various parameters. For example, biofabrics may be quite instructive in the recognition of unusual biological or physical conditions. Schubert and Bottjer (1992) suggested that Triassic stromatolites were formed under normal marine conditions and that their abundance at that time is indicative of the removal of other metazoan-imposed barriers to the nearshore normal-marine environments at the end Permian extinction. Zhuravlev (1996) recently discussed other mechanisms that regulate the distribution of stromatolites. Grotzinger and Knoll (1995) have examined Permian reef microfabrics and found them to be more similar to Precambrian ones rather than to those of modern reefs or even other types of Phanerozoic reefs. They suggest, in this situation, that the Precambrian, rather than the recent, provides the key to understanding the dynamics that produced these widespread but poorly understood reef fabrics.

In the past decade, numerous workers have documented paleoenvironmental trends in the origin and diversification of marine benthic invertebrates (e.g., Sepkoski and Miller 1985; Bottjer and Jablonski 1986). If an animal changes its environments through time, then a biofabric produced by that animal may similarly shift, and thus, tight sedimentological and stratigraphic controls are necessary for use of these fabrics for environmental analyses.

Uniformitarian models are commonly applied to the interpretation of sedimentary structures and strata. However, recent work on Precambrian and Cambrian sedimentary structures indicates that a uniformitarian approach may be inappropriate because of the effects of possible widespread microbial mat surfaces as well as the lack of bioturbation in the Neoproterozoic and Early Cambrian (e.g., Gehling 1991, 1999; Sepkoski et al. 1991; Seilacher and Pflüger 1994; Goldring and Jensen 1996; Hagadorn and Bottjer 1996; Pflüger and Gresse 1996; Droser et al. 1999a,b). Continued investigation of these unique Precambrian and Cambrian nonactualistic structures will yield insight into the interactive physical and biological processes operating during this time. Bottjer et al. (1995) have noted that paleoecologic models are most effective when freed from the strict constraints of uniformitarianism. So, too, analyses of biologically generated fabrics will be most useful when similarly viewed.

ICHNOFABRIC: THE POSTDEPOSITIONAL BIOFABRIC

The ichnological record of the Neoproterozoic and Cambrian has received considerable attention (e.g., see review in Crimes 1994). In particular, trace fossils provide important biostratigraphic markers, such as designating the base of the Cambrian (Narbonne et al. 1987), as well as demonstrating increases in the complexity of behavior, types of locomotion, and environmental patterns in diversity and distribution across this boundary. However, another important aspect of the ichnological record is ichnofabric—sedimentary rock fabric that results from all aspects of bioturbation (Ekdale and Bromley 1983). It includes discrete identifiable trace fossils, along with mottled bedding (figures 7.1 and 7.2). Although discrete identifiable trace fossils provide important information, a great deal of data is lost by recording only this aspect of the ichnological record. Studies of ichnofabrics have concentrated on the record of bioturbation as viewed in vertical cross section. Thus, the contribution to ichnofabric of burrows that have a vertical component has been emphasized because they are most important to the final sedimentary rock fabric.

Ichnofabric studies have proven to be instrumental in determining the nature of the infaunal habitat at a given time and in a given environment. However, there have been only a few extensive systematic studies examining Cambrian ichnofabrics (e.g., Droser 1987, 1991; Droser and Bottjer 1988; McIlroy 1996; Droser et al. 1999a; McIlroy and Logan 1999). Trace fossils are relatively common in the late Neoproterozoic, but ichnofabric studies of these strata are lacking. In studying the Cambrian radiation, it is instructive to examine the types of ichnofabrics that characterize the Cambrian as well as how these ichnofabrics compare with those of later times. Although our understanding of Cambrian ichnofabrics is still in its infancy, some generalizations can be made.

Tiering, Extent, and Depth of Bioturbation, and Disruption of Original Physical Sedimentary Structures

A critical factor determining the nature of ichnofabric is tiering, or the vertical distribution of organisms above and below the sediment-water interface (Ausich and Bottjer 1982). In the infaunal realm, trace fossils can provide data on depth of bioturbation and vertical distribution of animals and their activity in the sediment. Infaunal tiering results in the juxtapositioning of several trace fossils as animals burrow to different depths. This produces an ichnofabric composed of crosscutting burrows.

Because infauna are strongly tiered, the upward migration of the sediment column creates what has been termed a "composite ichnofabric" (Bromley and Ekdale 1986) where burrows of organisms in the lower tiers crosscut burrows in the shallower tiers with steady-state accretion. In some sedimentary settings, under certain conditions, the original tiering pattern is preserved. This is termed a "frozen tier profile" (Savrda and Bottjer 1986). Such profiles provide a "snapshot" view of the tiering structure of the infaunal community. Frozen tiered profiles result when (1) organisms do not move vertically upward following sedimentation, and (2) sediments are not subsequently reburrowed. Thus, the documentation of original tiering relationships from composite ichnofabric, through analyses of crosscutting relationships, provides information otherwise not available about the ecology of the infaunal habitat.

Tiering complexity, as well as depth of bioturbation, varies across environments. In nearshore and shallow marine Cambrian sandstones, *Skolithos*, *Diplocraterion*, and *Monocraterion* are common and have depths of up to 1 m (Droser 1991) (figures 7.1

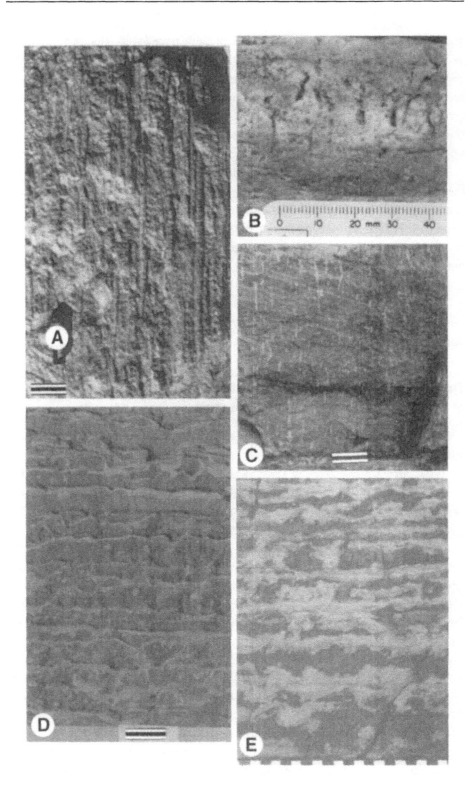

Figure 7.1 Examples of Cambrian ichnofabric. *A, Skolithos* piperock from Lower Cambrian Zabriski Quartzite (Emigrant Pass, Nopah Range, southeastern California, USA) with an ichnofabric index of 4 (ii4); scale bar 4 cm. *B,* Small *Skolithos* burrows in the Lower Member of the Eriboll Sandstone (Skaig Burn, Ordinance Survey #15, Loch Assynt, Scotland); scale bar is in millimeters. *C,* Cross-sectional view of *Skolithos* ichnofabric in the Eriboll Sandstone (Skaig Bridge, Loch Assynt, Scotland); scale bar 15 cm. *D,* Ichnofabric of the Upper Cambrian Dunderberg Shale (Nopah Range, California, USA); ichnofabric index 3 is recorded from this thin-bedded limestone and mudstone unit; scale bar 5 cm. *E,* Ichnofabric of Lower Cambrian Poleta Formation (White-Inyo Mountains, California, USA); differential dolomitization enhances burrows in this limestone; scale at base of photo in centimeters.

and 7.2). This may or may not reflect original depth of bioturbation (because animals adjust to sediment deposition and erosion). Nonetheless, these burrows clearly represent the deepest tiers of the Cambrian. Additionally, *Teichichnus* occurs as a relatively deep tier burrow in the earliest Cambrian and remains important throughout the Cambrian. Other than these burrows, Cambrian infaunal tiering in general was relatively shallow; recorded depth of bioturbation is most commonly under 6 cm.

The extent to which original sedimentary structures will be disrupted and destroyed by bioturbation is a function of sedimentation rate and rate of bioturbation. If sedimentation rate is slow enough, then shallow or even horizontal bioturbation will result in the complete destruction of physical sedimentary structures. A totally bioturbated rock simply shows that the rate of biogenic reworking exceeded that of sedimentation. Thus, thorough bioturbation is possible in virtually any setting. Environmental control is very important, and we see that ichnofabrics vary accordingly. It is critical to examine similar facies when comparing changes in amount or depth of bioturbation through time (Droser and Bottjer 1988). By way of characterizing the Cambrian, complete to nearly complete disruption of physical sedimentary structures is common in only a few settings: (1) in high-energy sandy settings where vertical burrows were common, and (2) in finer-grained sediments when rate of sedimentation was slow enough for shallow-tiered animals to keep up with sedimentation.

Cambrian infaunas produce ichnofabrics that are comparatively simple when contrasted with those of later times but are far more complex than those of the Precambrian. *Skolithos, Diplocraterion, Teichichnus,* and *Monocraterion* all commonly produce a monospecific ichnofabric with a record ichnofabric index (ii) of up to 4 or 5 (see figures 7.1 and 7.2). Shallow-tiered burrows may have been present but are not commonly preserved in these ichnofabrics. Ichnofabrics produced by these burrows are present in lowermost Cambrian strata, and although there may be wide variability— even within the Cambrian—these monotypic ichnofabrics remain essentially unchanged throughout their stratigraphic ranges.

Outside the realm of *Skolithos, Teichichnus,* and *Diplocraterion,* ichnofabrics are in general less well developed than environmentally comparative ones of later times. In pure carbonates, for example, until the advent of boxwork *Thalassinoides* in the Late

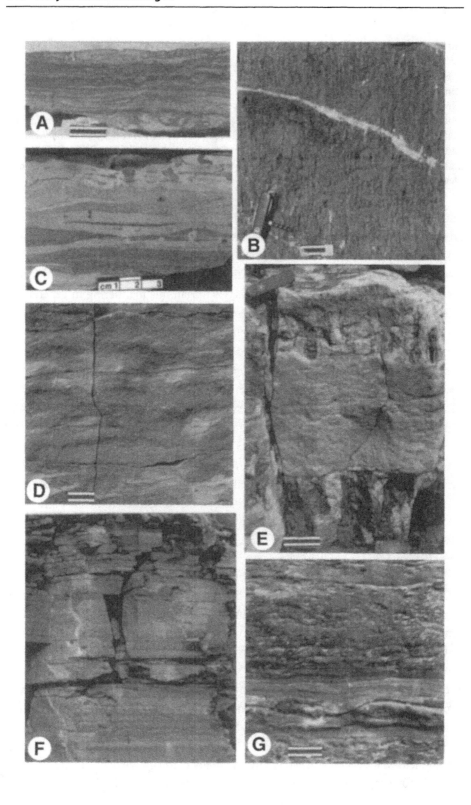

Figure 7.2 Examples of Cambrian ichnofabric. *A, Treptichnus pedum* ichnofabric from the Uratanna Formation from the Castle Rock locality, Flinders Ranges, South Australia; scale bar in centimeters. *B,* Densely packed *Diplocraterion,* producing an index of ii5 in the Lower Cambrian Parachilna Formation (Parachilna Gorge, Flinders Range, Australia); scale bar 6 cm. *C,* Glauconite-rich sandstone from Upper Cambrian St. Lawrence Formation (Upper Mississippi Valley, Wisconsin, USA), showing sediment-starved ripple lamination and small horizontal bioturbation. *Source:* Photograph courtesy of Nigel Hughes. *D,* Tommotian Petrosvet Formation (middle Lena River, Siberian Platform, Russia) with a *Teichichnus* ichnofabric; preserved ripple lamination also occurs; scale bar 3 cm. *E, Diplocraterion* ichnofabric from the Lower Cambrian Hardeberga Formation (Scania, Sweden); scale bar 6 cm. *F,* Outcrop view of Tommotian Petrosvet Formation (middle Lena River, Siberian Platform, Russia); note that overall bedding is preserved but within beds, primary stratification is commonly completely destroyed by *Teichichnus;* field of view approximately 50 cm across. *G,* Laminated sandstones interbedded with bioturbated finer-grained sediments from the Upper Cambrian St. Lawrence Formation (Upper Mississippi Valley, Wisconsin, USA); burrows are nearly all horizontal, but individual fine-grained beds are destroyed, although overall bedding is preserved; scale bar 5 cm. *Source:* Photograph courtesy of Stephen Hesselbo.

Ordovician, tiering was relatively simple, and although complete disruption of original sedimentary fabric occurred (Droser and Bottjer 1988), centimeter-scale bedding is generally still discernible. In shallow marine subtidal terrigenous clastics, tiering was similarly shallow, and although mudstones may be thoroughly bioturbated, sedimentary packages representing storm deposition are commonly preserved.

Cambrian trace fossils are well known, and Cambrian trace fossil assemblages have been extensively documented (e.g., Jensen 1997). These assemblages likely produce distinct ichnofabrics. For example, a type of Cambrian ichnofabric is produced by the *Plagiogmus-Psammichnites-Didymaulichnus* group. Although these burrows are shallow, they are relatively large and generate a great deal of sediment destruction (S. Jensen, pers. comm., 1997). These burrows are widespread, but the resulting ichnofabric has not been described. Trace fossil assemblage data are useful; however, ichnofabric studies of these assemblage-bearing strata will provide even more insight into interacting physical and biological processes and the ecology of Cambrian infaunal metazoans.

Precambrian-Cambrian Transition Ichnofabrics

Trace fossils are common in certain facies in the Precambrian, in particular in shallow marine subtidal terrigenous clastics. However, preliminary study of Precambrian strata in Australia and the western United States indicates that these trace fossils do not result in the production of ichnofabrics (Droser et al. 1999a,b). The earliest ichnofabrics in these sections occur with the first appearance of *Treptichnus pedum* (figure 7.2A). Thus, *T. pedum,* which defines the base of the Cambrian, also marks the initial development of preservable infaunal activity. Preserved depth of bioturbation is on the order of 1 cm, with a maximum of 2 cm; only one tier is present. Because of the three-dimensional nature of *T. pedum,* ichnofabric index 3 (ii3) can be very locally recorded (Droser et al. 1999a). The trace fossils *Gyrolithes* and *Planolites* may also contribute to this ichnofabric. With the recognition of treptichnid trace fossils in the terminal Proterozoic (Jensen et al. 2000), it is also possible that a similar ichnofabric may be present in Precambrian strata.

Characteristic Cambrian Ichnofabrics

Piperock

Perhaps the best-known Cambrian ichnofabric is *Skolithos* piperock, which is a ubiquitous ichnofabric of Cambrian sandstones representing deposition in high-energy shallow marine settings (Droser 1991). The term *piperock* was first used in reference to dense assemblages of *Skolithos* in the Lower Cambrian Eriboll Sandstone in Scotland (figure 7.1C) (Peach and Horne 1884) and popularized by Hallam and Swett (1966). Piperock is a classic Cambrian biofabric. Indeed, in the literature, workers commonly describe post-Cambrian occurrences as "typical Cambrian piperock."

Piperock first appears in the Early Cambrian and represents the advent of deep bioturbation by marine metazoans (figures 7.1A,C). An analysis of the temporal distribution of piperock confirms previous observations that piperock is "typical" of the Cambrian but also demonstrates that piperock occurs throughout the Paleozoic, decreasing in abundance after the Cambrian (Droser 1991).

The term *piperock* is commonly associated with *Skolithos* or *Monocraterion,* but several other vertical trace fossils also form piperock. *Diplocraterion,* in particular, commonly forms piperock in Cambrian sandstones. For example, the base of the Parachilna Formation in Australia has a laterally continuous bed of densely packed (ii5) *Diplocraterion* (figure 7.2B). In the Hardeberga cropping out in Sweden and Denmark, *Diplocraterion* occurs in amalgamated sandstones with a wide range of ichnofabric indices represented (figure 7.2E).

Teichichnus Ichnofabric

A common and well-developed Cambrian ichnofabric is produced by *Teichichnus* (figures 7.2D,F), a burrow that has been recorded from Lower Cambrian strata around the world (see discussion by Bland and Goldring 1995). When it occurs, *Teichichnus* commonly dominates the ichnofabric; ii4 and ii5 are locally common (Bland and Goldring 1995: figure 3). The trace fossil occurs from shallow marine to outer shelf settings. For example, in the Tommotian Petrosvet Formation that crops out along the Lena River in Siberia, *Teichichnus* occurs in an argillaceous limestone with common ripple lamination (figures 7.2D,F). Depth of bioturbation of up to 6 cm is common. Burrows may be reburrowed by *Chondrites.* Ripple marks are commonly preserved on bedding tops, along with other discrete trace fossils that do not contribute to the ichnofabric as recorded on vertical section.

"Mottled" Shallow Marine Limestones

Lower Paleozoic shallow marine carbonates are typically "mottled." Terms such as *rubbley bedding, burrow mottled,* and *mottled limestone* have been used to describe this sedimentary fabric. In most cases, this mottling is due to bioturbation but is often enhanced by diagenesis (figures 7.1D,E).

Trace fossils that significantly contribute to the ichnofabric of pure carbonates include *Thalassinoides, Planolites,* and *Bergaureria.* The *Ophiomorpha*-like trace fossil *Aulophycus* has also been reported from shallow marine Cambrian carbonates of the Siberian Platform (Astashkin 1983, 1985). For the most part, the result of Cambrian bioturbation in this setting was not the complete destruction of original physical sedimentary structures. In subtrilobite Lower Cambrian strata of the Basin and Range, ichnofabric indices 1 and 2 are most commonly recorded; bedding is preserved. For the rest of the Cambrian, generally, although rocks may be completely bioturbated or, in contrast, relatively unbioturbated, on average, ichnofabric index 3 is recorded (figures 7.1D,E). In studies of Cambrian carbonate strata from parts of the Appalachians as well as Kazakhstan, typical carbonate shallow marine strata have mottled bedding where ichnofabric indices from 1 to 5 are recorded but average at about ii3. In Kazakhstan, for example, strata nearly identical to those in the Basin and Range occur.

Thus, until we have the advent of extensive boxwork *Thalassinoides* in the pure carbonates, we have simple tiering and shallow bioturbation. In this setting, complete disruption of original sedimentary fabric occurs (Droser and Bottjer 1988), but on average, bedding is still discernible. Tiering is relatively shallow; mazelike *Thalassinoides* and *Bergaueria* are the most common components. *Chondrites* may be locally common.

Ichnofabrics of Shallow Marine Terrigenous Clastics

Shallow marine terrigenous clastic settings are commonly represented by event beds. In the high-energy end of this setting, amalgamated nearshore sandstones are common with *Skolithos* and *Diplocraterion* piperock. Shallow marine terrigenous clastic strata representing deposition below normal wave base are characterized by storm beds with fining upward successions.

In the lowermost Cambrian, *Treptichnus pedum* ichnofabric characterizes this setting (Droser et al. 1999a). Younger Lower Cambrian rocks show more-complex ichnofabrics. In the Lower Cambrian *Mickwitzia* Sandstone of Sweden, thin-bedded, interbedded sandstones and mudstones that are centimeters in thickness are common. The sandstones have abundant and diverse trace fossils, and the mudstones can be completely bioturbated, but the centimeter-scale bedding is commonly preserved (Jensen 1997).

Goldring and Jensen (1996) examined a Neoproterozoic-Cambrian succession in Mongolia. They describe four types of bed preservation from the Cambrian-aged strata, two of which are similar to Phanerozoic beds deposited under equivalent conditions. These include millimeter-to-centimeter-thick sand event beds with sharp soles and bioturbated upper parts and thin units of heterolithic alternations of sand and mud with *Planolites* and *Palaeophycus* (Goldring and Jensen 1996). The two that are unmatched in younger Phanerozoic deposits include features such as intraformational conglomerates and the absence of gutters and tooled lower surfaces to "event" beds. They suggest that organic binders (Seilacher and Pflüger 1994; Pflüger and Gresse 1996) are the control of these and other unusual sedimentary features.

McIlroy (1996) examined ichnofabric in a Lower Cambrian offshore shelf succession in Wales and documented sediments that were completely homogenized through much of the succession. Data from the Lower Cambrian of the Digermul Peninsula additionally show that, on average, the size of bioturbating organisms and the depth of infaunal tiering both increase through time (McIlroy 1996). Droser (1987) similarly documented an increase in extent of bioturbation in shallow marine terrigenous clastics through the Cambrian of the Basin and Range (western United States).

A heterolithic dolomicrite, siltstone, and sandstone facies representing deposition below fair-weather wave base in the Upper Cambrian, the St. Lawrence Formation of Wisconsin, USA, is dominated by horizontal burrows, including a number of unusual forms such as *Raaschichnus,* a trace made by aglaspidid arthropods (Hughes and Hesselbo 1997). Extensive bioturbation occurs in the finer-grained sediments, and lamination is commonly preserved in the sandstones (figures 7.2C,G). Complete homogenization occurs in some beds, but generally ichnofabric indices 1 to 4 are recorded. Body fossils are found in beds that have not been extensively bioturbated (Hughes and Hesselbo 1997).

In nearly all of these units, depth of bioturbation is relatively shallow; in fact, burrows are generally horizontal and tiering is relatively simple. Thus, although bioturbation may be complete within an event bed, particularly in finer-grained facies, overall bedding is commonly preserved. In contrast, centimeter-thick event beds in the Ordovician and Silurian are not commonly preserved (Sepkoski et al. 1991). Interestingly, while the early record of bioturbation and trace fossils is best preserved in this shallow subtidal terrigenous clastic facies, so too are the sedimentary structures (nonactualistic) indicative of unique Precambrian and Cambrian conditions, such as flat pebble conglomerates, wrinkle marks, and sand chips (e.g., Sepkoski et al. 1991; Seilacher and Pflüger 1994; Goldring and Jensen 1996; Hagadorn and Bottjer 1996; Pflüger and Gresse 1996). And, indeed, these structures remain common throughout the Cambrian (e.g., Hughes and Hesselbo 1997).

A particularly well-developed ichnofabric occurs in a succession of thick sand-

stones, some with interbedded mudstones, in the Cambro-Ordovician Bynguano Formation examined by Droser et al. (1994), cropping out in the Mootwingee area of
western New South Wales, Australia. This deposit represents a higher-energy setting
than those described above. In these strata, *Arenicolites, Skolithos, Trichichnus, Monocraterion,* and *Thalassinoides* are most common. *Thalassinoides* have burrow diameters
of 1–2 mm, which are much smaller than those typical of this ichnogenus. Depth of
bioturbation for the *Thalassinoides* can be estimated to be at least 20–30 cm. In the
Bynguano Formation some trace fossils are preserved in a "frozen tiered profile" that
can be generalized as follows. Three tiers are recognized: (1) the deepest tier is formed
by *Thalassinoides*; (2) an intermediate tier is characterized by *Skolithos* and *Arenicolites*
type A; and (3) a shallow tier is represented by *Trichichnus, Arenicolites* types B and C,
and bedding plane trace fossils. Ichnofabric indices (ii) (Droser and Bottjer 1986) in
these beds range from ii3 to ii5. Thus, by the Cambro-Ordovician, in this setting,
well-developed ichnofabrics occur that exhibit complex tiering patterns as well as
preserve extensive bioturbation.

Deep-Water Facies

Ichnofabrics of outer shelf and deep basin deposits have not received much attention.
However, analysis of outer shelf Cambrian carbonates of the Basin and Range of the
western United States suggests that ichnofabrics were not well developed and that
trace fossils are usually confined to bedding surfaces (Droser 1987). In general, extent
of bioturbation in these strata increased through the Cambrian (Droser 1987). The
Botoman lower Kutorgina Formation at Labaya on the Siberian Platform is likewise
relatively unbioturbated. This is consistent with the suggestions that extensive colonization of the deep sea did not occur until the Early Ordovician (Crimes 1994; Crimes
and Fedonkin 1994). Deeper-water mudstones remain a fruitful area for research.

Ichnofabrics of Carbonates versus Terrigenous Clastics

Ichnofabrics record a differential paleoenvironmental history in the development of
the infaunal biological benthic boundary layer. The most significant environmental
trend is the difference between the record of shallow marine terrigenous clastics and
carbonates. This may be largely a taphonomic artifact. Neoproterozoic and lowermost
Cambrian trace fossils and ichnofabrics are best developed in terrigenous clastics.
Indeed, in successions where terrigenous clastics are interbedded with carbonates,
the terrigenous clastics show a record of bioturbation whereas the carbonates do not
(Droser 1987; Goldring and Jensen 1996). Droser (1987) noted a stepwise increase
in bioturbation in Lower Cambrian carbonates between subtrilobite and trilobite-
bearing strata but a gradual increase in the shallow marine terrigenous clastic setting.
McIlroy (1996) similarly noted a gradual increase in terrigenous clastic shelfal de-

posits. Goldring and Jensen (1996), examining an interbedded siliciclastic-carbonate Precambrian-Cambrian succession in Mongolia, recorded ichnofabrics and trace fossils in terrigenous clastics but noted that there was virtually no record in the carbonates. This discrepancy may be due to diagenetic effects and the nature of bed-junction preservation in pure carbonates versus terrigenous clastics. The best records in terrigenous clastics come from heterolithic beds or event beds. The most consistently well bioturbated strata are shallow-water fine-grained sediments. There are not equivalent-type beds in shallow marine carbonate strata. In carbonates, it appears that, until there is an infauna with a vertical dimension, there is little record. In the Basin and Range, this is represented by the appearance of *Thalassinoides* in the Atdabanian (Droser and Bottjer 1988).

However, it is not entirely a preservational artifact, in that the deep-tier burrows that are common in terrigenous clastics such as *Skolithos, Teichichnus, Diplocraterion, Monocraterion,* and tiny *Thalassinoides* are simply not present in carbonate strata. The significance of facies control on all aspects of the Neoproterozoic-Cambrian record has been discussed by Lindsay et al. (1996).

FOSSIL CONCENTRATIONS

Fossil concentrations represent another type of biofabric that is directly a result of the radiation of marine animals. These fossil-rich accumulations not only are important sources of paleontological and paleoenvironmental data but provide a natural link between biological and environmental processes (Brett and Baird 1986; Kidwell 1986, 1991; Parsons et al. 1988; Kidwell and Bosence 1991).

Precambrian Fossil Concentrations

Ediacaran fossils are known throughout the world. They are common and, in places, abundant. Bedding planes can be covered with *Pteridinium* as figured by Seilacher (1995; see also Crimes, this volume: figure 13.3A). These Precambrian deposits may be analogous to some types of fossil concentrations that have been described from the Phanerozoic. However, many of these biofabrics may be produced by baffling organisms. In the terminal Proterozoic, thick shelfal siliciclastic buildups were enhanced by microbial binding of sand, including baffling benthic organisms such as *Ernietta, Beltanelliformis, Aspidella,* and *Pteridinium* (Droser et al. 1999b; Gehling 1999). The preservation of dense masses of these cup-shaped and winged forms, along with many anactualistic sedimentary structures, is a monument to the absence of benthic predators, scavengers, and penetrative burrowing below the Precambrian-Cambrian boundary (e.g., Seilacher 1995, 1999; Droser et al. 1999a; Gehling 1999). Likewise, weakly calcified benthic metazoans, probably suspension feeders, such as *Cloudina* and goblet-shaped forms, formed closely packed in situ monospecific communities with limited topographic relief in the Nama Group, Namibia (e.g., Germes 1983; Grotzinger et al. 1995; Droser et al. 1999b). These have been considered "reefs" (Germes 1983). Such forms were probably able to bind sediment, but the complementary role of early cements and microbial precipitates is not clear. Anabaritids also formed similar mounded aggregations in the Nemakit-Daldynian (e.g., Droser et al. 1999b).

Cambrian Shell Concentrations

Introduction

Shell concentrations are relatively dense accumulations of biomineralized animal remains (nonreefal skeletal deposits) with various amounts of sedimentary matrix and cement, irrespective of taxonomic compositions and degree of postmortem modification (Kidwell et al. 1986). Shell-rich accumulations have been part of the sedimentary record since the beginning of Early Cambrian (Li and Droser 1997). However, our current understanding of the development and distribution of shell concentrations is primarily from shell accumulations in modern shallow-water environments and from post-Paleozoic shell deposits (e.g., see review by Kidwell and Flessa 1995). Questions related to the formation and distribution of Cambrian shell beds have only recently been addressed (Li and Droser 1997), but occurrences of Cambrian shell beds are reported in the literature. The development of shell beds is related to the evolutionary changes in behavior, diversity, and environmental distribution of organisms (Kidwell 1990, 1991). In that the Cambrian radiation is a critical event in the development of metazoan history, with the advent of skeletonization and the establishment of the Cambrian Evolutionary Fauna, it is an equally important time for the development of fossil concentrations.

In this chapter, we use data primarily collected from the Basin and Range of western United States and west-central Wisconsin to discuss (1) the characteristics of Cambrian shell beds, (2) the characteristics of shell beds from different depositional regimes, and (3) the distribution of shell beds throughout the Cambrian. These represent only two areas but serve as a basis for future comparison.

Cambrian Shell Bed Types

Cambrian shell concentrations consist of skeletal grains and of nonskeletal allochems such as intraclasts, peloids, ooids, oncoids, and sedimentary matrix. The sedimentary matrix of Cambrian shell concentrations consists primarily of carbonate and siliciclastic muds, silts, and sands. Carbonate intraclasts, including flat pebble clasts, are a common component of the shell beds in various facies. In carbonate facies, ooids and/or oncoids are commonly mixed with shell fragments to form thick composite/condensed shell beds (Li and Droser 1997).

Table 7.1 **Characteristic Features of Cambrian Shell Concentrations**

COMPOSITION	GEOMETRY	THICKNESS/TRACEABILITY
Trilobite exclusively	Pavement, lens, stringer, pod, or bed	Varies from mm to 10s of cm, usually not laterally traceable
Trilobite dominated	Stringer, pod, lens, bed, or bedset	Varies from cm up to m, many beds are laterally persistent
Brachiopod exclusively	Pavement, lens, or stringer	Varies from mm up to cm, locally traceable
Brachiopod dominated	Lens, pod, or bed	cm, usually not laterally traceable
Gastropod dominated	Lens, stringer, or pod	cm, usually not laterally traceable
Echinoderm dominated	Usually lens, bed, or bedset	Varies from cm up to m, many beds are laterally persistent
Trilobite/echinoderm mixed	Usually lens, bed, bedset	Varies from cm to m, many beds are laterally persistent
Small tubular shells	Lens and beds	cm to 10s of cm

As expected, the skeletal grains represent typical elements of the Cambrian Fauna (Sepkoski 1981a,b), that is, trilobites, lingulate brachiopod valves, hyoliths, and "small shelly fossils," as well as echinoderm debris and gastropod shells. Shell accumulations composed of reef-building organisms deposited around the reefal buildups are not typically included in the shell bed studies, because they commonly form a component of complex reef fabrics (Kidwell 1990). However, archaeocyath debris is locally common in the Cambrian and can form composite beds that are centimeters to tens of centimeters in thickness. Archaeocyath debris beds in the Lower Cambrian of California are well defined and densely packed beds intercalated in wackestones and grainstones.

Cambrian shell beds are diverse, and each taxonomic type has a distinct stratigraphic and taphonomic signature (table 7.1). The most common Cambrian shell concentrations are trilobite-dominated shell beds (e.g., Li and Droser 1997). They are found throughout many depositional facies and are usually lenticular to planar-bedded deposits. Trilobites occur in different states of preservation that range from highly fragmented sclerites to intact cranidia and pygidia, and they are generally mixed with intraclasts (Westrop 1986; Kopaska-Merkel 1988; Li and Droser 1997). Shell con-

INTERNAL CLOSE-PACKING	COMPLEXITY	TAPHONOMIC FEATURES
Loosely-densely packed	Simple to complex fabric, usually simple	Most trilobites are disarticulated, but intact free cheeks, cranidia, and pygidia are common
Dispersed to densely packed, usually loosely-densely	Simple to complex, complex common	Intact trilobites are rare, many fragmented parts, but intact cranidia and pygidia are common
Dispersed to loosely packed	Simple fabric	Disarticulated shells are common, most shells are intact and concordant to bedding
Dispersed to loosely packed	Simple fabric	Disarticulate shells are common, low abrasion and fragmentation
Dispersed to loosely packed	Simple fabric	Many intact shells, low fragmentation, internal molds are common
Loosely-densely, usually densely packed	Simple to complex, complex is common	High fragmentation and disarticulation, usually recrystallized
Loosely-densely, usually densely packed	Simple to complex, usually complex	High disarticulation and fragmentation, are rare, poor sorting, intact shell parts usually recrystallized
Loosely to densely packed	Usually simple	Highly fragmented shells, usually recrystallized

centrations composed exclusively of trilobites are also common in various lithofacies, particularly in Lower and Middle Cambrian shale and mudstone (figure 7.3C). These trilobite-only beds usually occur as pavements and thin lenticular beds with relatively good preservation of trilobites. Trilobite sclerites in beds from thin-bedded interbedded carbonate and shale successions are commonly fragmented; however, intact free cheeks, genal spines, cranidia, and pygidia are common, and with their original cuticles preserved (Li and Droser 1997).

Lingulate brachiopod shell accumulations are another common type of Cambrian shell bed, particularly in shale, siltstone, or fine-grained sandstone (McGee 1978; Hiller 1993; Li and Droser 1997). They usually occur as loosely to densely packed pavements and lenses, and even beds (Ushatinskaya 1988; Popov et al. 1989; Li and Droser 1997). Although most beds are thin, they are relatively laterally persistent. Fragmentation is usually low, although most valves are disarticulated. Original shells composed of calcium phosphate are well preserved, with fine growth lines on their surfaces. A Botoman example comes from the Bystraya Formation of eastern Trans-baikalia (Ushatinskaya 1988). An impressive example of Late Cambrian (not Early Ordovician, as previously claimed) shell beds consisting of lingulate brachiopods are

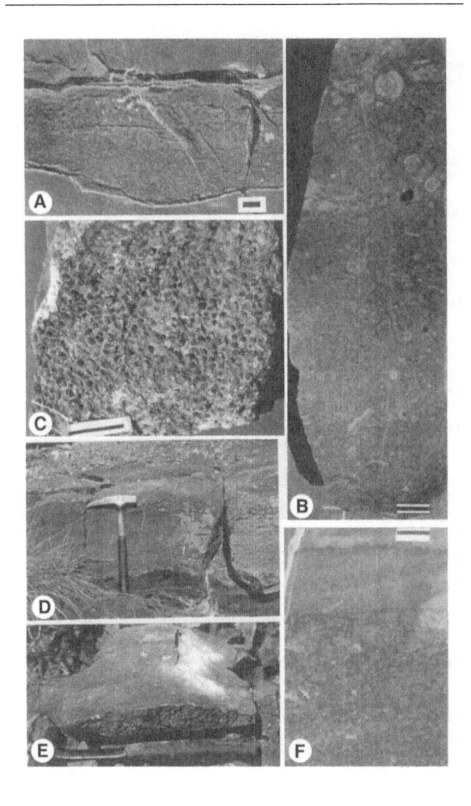

Figure 7.3 Examples of Cambrian fossil concentrations. *A,* A small-shell fossil accumulation from the middle part of the pretrilobite Lower Cambrian Deep Spring Formation (Mount Dunfee, White-Inyo region, California, USA); it occurs as a lenticular bed composed of densely packed, whole and broken, small tubular shells; note the homogenous internal fabric and the sharp contacts with surrounding strata; scale bar 1 cm. *B,* Cross-section of a trilobite-gastropod shell bed from the Upper Cambrian Whipple Cave Formation (central Egan Range, Nevada, USA); note the taxonomic variation within the bed; lower half is composed primarily of trilobite fragments, while the number of gastropod shells increases in the upper half of the bed; scale bar 1 cm. *C,* Bedding plane view of agnostid concentrations from Member A of the Emigrant Springs Limestone (Patterson Pass, southern Schell Creek Range, Nevada, USA); shell bed is densely packed with disarticulated agnostid cephala and pygidia oriented either convex-down or convex-up; this bed was deposited in an outer shelf setting; scale bar 2.5 cm. *D,* Well-developed condensed-composite shell bed from Upper Cambrian Big Horse Limestone of the Orr Formation, central House Range, Utah; this bed is densely packed with trilobite remains, rests on top of massive stromatolite-thrombolite buildups, and is essentially a trilobite grainstone; it is cross-stratified, and sand waves are preserved on the upper surface. Rock hammer for scale. *E,* Isolated lens of a trilobite concentration from the Middle Cambrian Whirlwind Formation, Marjum Canyon, central House Range, Utah; this lenticular deposit is intercalated with green shales and is densely packed with the trilobite *Ehmaniella;* these lenses result from starved ripple migration in shallow subtidal to intertidal settings; field of view is approximately 32 cm across; rock hammer for scale. *F,* Cross-section of a trilobite-dominated shell bed from the Upper Cambrian Orr Formation (central House Range, Utah); the densely packed shell bed is truncated by a hash bed composed of small (<2mm) shell fragments overlain by a mudstone; scale bar 2.5 cm.

the *Obolus* Beds, which are best developed in eastern Europe but also extend into Sweden. They consist largely of *Ungula ingrica.* This occurrence is notable in that the concentration of brachiopods in places is so high that it forms economically exploitable seams of phosphorite (Popov et al. 1989; Hiller 1993; Puura and Holmer 1993; S. Jensen, pers. comm., 1996).

Echinoderm-dominated and trilobite-echinoderm mixed shell beds are very common, particularly in Upper Cambrian shallow marine carbonates (Li and Droser 1997). They usually form amalgamated composite to condensed beds with highly disarticulated, fragmented, recrystallized trilobite sclerites and echinoderm debris. These beds are commonly greater than 10 cm in thickness and have an extensive lateral stratigraphic distribution (figures 7.3D,E).

Small shelly fossil concentrations (figure 7.3A) have been reported from lowermost Cambrian strata throughout the world (Brasier and Hewitt 1979; McMenamin 1985; Brasier 1986; Gevirtzman and Mount 1986; Landing 1988, 1989, 1991; Rozanov and Zhegallo 1989; Brasier et al. 1996; Khomentovsky and Gibsher 1996; Li and Droser 1997). However, only a few of these beds have been described in detail (McMenamin 1985; Gevirtzman and Mount 1986; Li and Droser 1997). These shell accumulations are primarily lenticular to tabular deposits intercalated in shallow marine sandstone and limestone facies. The small shelly fossil concentrations found in the Deep Spring Formation of the White-Inyo Mountains, Nevada, are composed of coleolids, hyoliths, and *Sinotubulites* (most are millimeter-scale tubular to conical skeletons). (*Sinotubu-*

lites has been synonymized tentatively with *Cloudina* [Grant 1990].) These beds have been interpreted as storm-generated lag deposits (Gevirtzman and Mount 1986; Li and Droser 1997).

Internal Fabric of Cambrian Shell Beds

Internal fabric includes the orientation, arrangement, packing, and sorting of the skeletal elements in shell beds. The major components of Cambrian shell beds are small and thin trilobite sclerites and fragmented echinoderm debris (Li and Droser 1997; figures 7.3B,D–F). Hence in cross-sectional view, they lack the typical "interlocking" fabric displayed in Mesozoic and Cenozoic shell beds that are composed of relatively large, thick, bivalved or univalved shells (Norris 1986; Kidwell 1990). Thus, in the field, Cambrian shell beds appear less conspicuous than the brachiopod-dominated shell beds in the post-Cambrian Paleozoic (Li and Droser 1995) and mollusk-dominated shell beds in the Mesozoic and Cenozoic. Although trilobite and trilobite-echinoderm beds are abundant in different Cambrian lithologies (Li and Droser 1997), they are easily overlooked because of the atypical internal fabric of the beds.

In general, Cambrian shell concentrations are primarily loosely packed and loosely to densely packed; the packing of the shell fragments varies vertically and laterally within the composite shell beds (figure 7.3B) because of physical and probably also biological reworking. Discrete shell concentrations commonly display either homogeneous fabric or fine upward. Amalgamated shell beds generally exhibit complex internal structure with lateral and vertical variations in close packing, shell orientation, and sorting (figure 7.3B). In carbonate strata, composite/condensed shell beds usually display irregular bedding between accreted beds and stylolite seams.

Shell Beds from Different Depositional Regimes

Shell concentrations are found in almost all Cambrian shelf lithologies (Li and Droser 1997). The characteristics of shell beds vary across different lithofacies in terms of taphonomic, paleontological, sedimentological, and stratigraphic features (Li and Droser 1997). An instructive comparison can be made between data from the Basin and Range and data from the Upper Mississippi Valley of west-central Wisconsin. Both were passive margins during the Cambrian. However, the two regions were situated in different depositional regimes. The Basin and Range was dominated by carbonate facies, whereas west-central Wisconsin was dominated by siliciclastic facies.

The Big Horse Limestone of Orr Formation (Horse Range, Utah) and the Eau Claire Formation (west-central Wisconsin) both range from the Late Cambrian *Cedaria* trilobite zone into the *Crepicephalus* Zone, and both are interpreted as shallow marine intertidal to subtidal deposits. The Big Horse Limestone consists primarily of thin-bedded to thick-bedded wackestones, packstones, and grainstones; ooids, oncoids, and intraclasts are also common, whereas the Eau Claire Formation consists mainly of thin-bedded to medium-bedded laminated and cross-bedded siltstones and sandstones.

The primary types of shell beds in the Eau Claire Formation are lingulate brachiopod-dominated and trilobite-dominated shell beds. These are mainly event and composite shell beds that formed under storm processes. The trilobite and hyolith remains in the beds are preserved as molds, whereas the original calcareous-phosphatic brachiopod shells are excellently preserved. Most of the concentrations are pavements (millimeters to centimeters in thickness) and lenticular beds (usually less than 15 cm in thickness) with simple internal variations.

Shell beds in the Big Horse Limestone are mainly trilobite-dominated-event, composite, and condensed beds. Many are well-developed deposits (ranging from centimeters to tens of centimeters in thickness) with complex internal fabric, usually showing the features of amalgamation and accretion.

Shell beds in both units show similar taphonomic features such as relatively low fragmentation, poor to moderate sorting, and high disarticulation. Moreover, many shell beds in the Big Horse Limestone were formed at either the bases or tops of sedimentary cycles. In the Eau Claire Formation, shell beds are commonly formed at the bases of thick sandstone beds, and shell fragments are concentrated at the bases of the shell beds.

Shell beds in both units are relatively common. In general, shell beds in the Big Horse Limestone are well developed and thicker than those in the Eau Claire Formation. However, because the shell beds in carbonate facies are not easily distinguished from the nonbioclastic packstone and grainstone beds because of the weathering pattern, they are not visually impressive in outcrop. Moreover, shell beds, particularly trilobite-dominated and echinoderm-trilobite beds, do not split evenly along bedding planes. In siliciclastic facies, although shell beds are relatively thin, they are more easily recognized in outcrop, because shell beds in shale and siltstones usually stand out from the surrounding strata and split along bedding planes.

Temporal Distribution: Evidence from the Great Basin Changes in Types of Shell Beds. Systematically collected Cambrian shell bed data are available only for the Basin and Range (Li and Droser 1997). Although the data are from a single basin, they provide a first look at the trends in the development of Cambrian shell beds.

The oldest shell concentrations in the Basin and Range are "small shelly fossil" accumulations (figure 7.3A) in the subtrilobite Lower Cambrian strata. They are not common, although they are widely known throughout the world from other subtrilobite Cambrian deposits. These pretrilobite shell beds are loosely to densely packed lenticular deposits that range from 4 to 13 cm in thickness (Li and Droser 1997). Similar types of shell beds have been reported from La Ciénega Formation, Caborca region, Sonora, Mexico, by McMenamin (1985). He noted that the small shell fossils

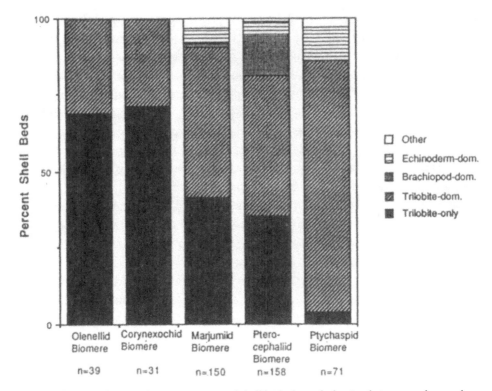

Figure 7.4 Distribution of taxonomic type of shell beds through the Cambrian. n = the number of shell beds described from each biostratigraphic interval; total is 449.

usually occur as shell beds in cross-bedded, sandy dolomitic limestones. They are discontinuous tabular and lenticular deposits that range in thickness from a few centimeters to more than a meter. Thus, they are much thicker than those in the White-Inyo region.

Trilobite shell concentrations first appear within the earliest trilobites (Olenellid biomere) in several stratigraphic units composed of pure carbonate and interbedded carbonates and terrigenous clastics in southwestern Nevada and southeastern California. However, most Lower Cambrian trilobite-rich beds are very thin (<5 cm), and some are just a single layer of trilobite remains on a bedding surface. Thereafter, shell concentrations are a fairly common stratigraphic element in the Cambrian rocks of the Basin and Range.

In the Olenellid and Corynexochid biomeres (upper Lower to Middle Cambrian), about 70 percent of shell beds are composed exclusively of trilobites (figure 7.4), and the others are trilobite-dominated and trilobite-echinoderm. In the rocks of the Marjumiid and Pterocephaliid biomeres (Middle to Upper Cambrian), inarticulate brachiopod- and echinoderm-dominated shell beds occur, but more than 80 percent of the shell beds counted are trilobite accumulations. Trilobite-dominated (as opposed to trilobite-only) shell beds are the dominant type of shell beds in Upper Cambrian Ptychaspid biomere strata (figure 7.4). Echinoderm-dominated and echinoderm-trilobite mixed shell beds are very common in the Ptychaspid biomere, along with rare gastropod- and hyolith-dominated shell beds. Thus, although echinoderms, gastropods, hyoliths, and brachiopods were present in the Early Cambrian, they did not become important components of shell beds in the Basin and Range Cambrian until the late Middle and Late Cambrian, whereas trilobite beds are the most important beds throughout the entire postsubtrilobite Cambrian.

Changes in Thickness and Abundance. In order to document the temporal patterns in the distribution of Cambrian shell beds, thickness and abundance of the shell beds were collected from comparable lithofacies and from comparable thicknesses of stratigraphic intervals from the Early to Late Cambrian of the Basin and Range (Li and Droser 1997).

Data used for examining temporal changes in thickness and abundance of shell beds were collected from thinly bedded argillaceous wackestones and packstones interbedded with shale and siltstone. These rocks represent deposition in shelf environments below fair-weather wave base but above the maximum storm wave base. However, this facies is rare in the upper Upper Cambrian Ptychaspid Biomere strata. Thus, the data were collected from the carbonate strata in this interval that represent similar depositional energies.

Most Cambrian shell concentrations are less than 10 cm thick. Single beds thicker than 20 cm are rare and occur primarily in the Upper Cambrian. The thickness of trilobite-dominated beds increases from the Early to Late Cambrian with a slight decrease in the Late Cambrian Pterocephaliid biomere (figure 7.5A). Shell beds collected from Lower and lower Middle Cambrian strata are predominantly thin pavements, lenses, and discontinuous thin beds. In contrast, many upper Middle and Upper Cambrian shell concentrations are well-developed planar beds or bed sets, and some are laterally persistent; they can be distinguished readily in the field. Overall, the physical dimension of trilobite-dominated shell concentrations increases from the Lower Cambrian to Upper Cambrian, with the major shift in the upper Middle Cambrian. Moreover, the abundance data (figure 7.5B) show that the frequency of occurrence of shell beds also increases from the Early to Late Cambrian with a decrease in the latest Late Cambrian (Ptychaspid biomere).

Variations in the physical dimensions of shell beds through the Phanerozoic have been discussed by Kidwell (1990) and Kidwell and Brenchly (1994), based on post-Cambrian data. Their data show that thickness and abundance of shell beds increase through the Phanerozoic; many shell beds from the post-Paleozoic are meters in thickness. A similar temporal trend in development of Cambrian shell beds has been shown above. However, data used in our study were collected from discrete beds and thus do not reflect a combined thickness of composite beds. Particularly in Upper Cambrian carbonates, composite shell beds (formed by multiple events) form accumula-

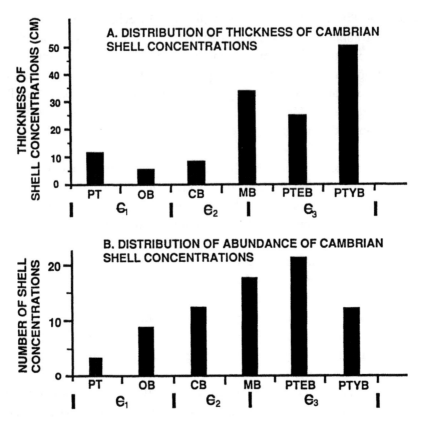

Figure 7.5 Temporal distribution of Cambrian shell concentrations. *PT* = pretrilobite Early Cambrian, *OB* = Olenellid biomere, *CB* = Corynexochid biomere, *MB* = Marjumiid biomere, *PTEB* = Pterocephaliid biomere, and *PTYB* = Ptychaspid biomere. The horizontal axis is arbitrarily divided into six equal intervals. Most data were collected from thinly bedded argillaceous wackestone/packstone interbedded with siltstone and shale. For the Upper Cambrian Ptychaspid biomere, data were collected from pure carbonates. *A,* Distribution of maximum thickness of Cambrian shell concentrations from each biomere; thickness was measured from discrete individual shell beds. *B,* Distribution of abundance of Cambrian shell concentrations; relative abundance data were collected by counting the number of shell beds in a 10 m interval of stratigraphic section considered to contain the most-abundant shell beds from each interval.

tions meters in thickness (Li and Droser 1997). These thick shell beds usually appear as ledges in the outcrop in the Basin and Range. Because they are primarily composed of trilobite and echinoderm fragments and lack typical interlocking fabric, they are easily unnoticed as shell beds.

Shell bed data from the Basin and Range demonstrate an increase in abundance and physical dimension of shell beds, as well as compositional shifts through Cambrian time (figures 7.4 and 7.5). By the Late Cambrian, shell beds are more abundant and thicker, and thus, at least in this particular shallow marine facies, they form a more significant part of the stratigraphic record. There is also an increase in the taxonomic diversity of shell beds though the Cambrian (figure 7.4). However, more than 80 percent of the shell beds are dominated by trilobites. Through the Cambrian there is an increase in diversity of trilobites. This diversification may have been accompanied by an increase in abundance of trilobites that contribute to shell beds.

Controls on the Temporal Distribution of Cambrian Shell Beds. The distribution and stratigraphic nature of shell beds are primarily a function of physical and biological controls (Kidwell 1986, 1991). Diagenesis and other chemical controls may also be important. However, workers have found that changing chemistries are not an overriding process for post-Cambrian shell beds (Kidwell and Brenchley 1994).

Analysis of data from the Basin and Range suggests that variations in thickness and abundance of Late Cambrian shell beds may, in large part, result from the lithological changes associated with a decrease in accommodation space in the Late Cambrian (Li and Droser 1997). The trend of increasing thickness and abundance throughout the rest of the Cambrian is most conservatively interpreted as a result of biological factors, in particular, an increase in skeletonized material (see discussion in Li and Droser 1997). In addition, workers have noted that Lower Cambrian cuticles appear to be thinner and "weaker" than Upper Cambrian forms (Lochman 1947). Thus an increase in cuticle thickness may also contribute to an increased thickness in Cambrian shell beds.

DISCUSSION

The Cambrian represents a critical point in the development of nearly all biofabrics. Near the base of the Cambrian, there was a significant shift in the nature of sedimentary fabrics. From then on, the nature and distribution of biofabrics continued to change through the Phanerozoic but not necessarily in concert. Perhaps, then, the Precambrian-Cambrian boundary represents the only time when there was a wholesale change in sedimentary fabrics.

Regardless of their origin, biofabrics are nevertheless a result of the interaction of physical and biological processes. The development and distribution of biofabrics is strongly controlled by sedimentary facies. This is particularly obvious with ichnofabrics but equally important with shell beds. Understanding these biases is critical to being able to use biofabrics fully as a resource to tease apart ecological and environmental relationships. However, initial studies indicate that Cambrian biofabrics have significant paleoecological and stratigraphic utility (e.g., Li and Droser 1997). Only a few studies (e.g., Droser and Bottjer 1988; Kidwell 1990; Kidwell and Brenchley 1994; Li and Droser 1997) have examined the long-term trends in biofabrics. However, with the increasing interest in, and appreciation of, the significance of biofabrics, this field remains an exciting venue for future research. This is particularly true of the latest Precambrian and Cambrian, where biofabrics have been relatively under-utilized in our exploration of the relationships between physical, chemical, and biological processes and the Cambrian explosion. Biofabrics provide a natural link between these processes.

Acknowledgments. This paper benefited from discussions with colleagues as well as from fieldwork and trips at various Cambrian locales facilitated by numerous people. In particular, we thank Mikhail Apollonov, David Bottjer, Richard Bromley, Jim Gehling, Roland Goldring, Jerry Gunderson, Nigel Hughes, Sören Jensen, Susan Kidwell, Isabel Montañez, Jeff Mount, Robert Riding, Mike Savarese, and Rachel Wood. Roland Goldring, Sören Jensen, and Andrey Zhuravlev made very helpful comments on an earlier draft of this paper. Many of the ideas in this paper stemmed from work supported by the National Science Foundation (EAR 9219731) and the National Geographic Society. Acknowledgment is also made to the White Mountain Research Station. This paper is a contribution to IGCP Project 366.

REFERENCES

Aller, R. C. 1982. The effects of macrobenthos on chemical properties of marine sediment and overlying water. In P. L. McCall and M. J. S. Tevesz, eds., *Animal-Sediment Relationships*, pp. 52–102. New York: Plenum Press.

Astashkin, V. A. 1983. Aulofikus [*Aulophycus*]. In B. S. Sokolov and I. T. Zhuravleva, eds., *Yarusnoe raschlenenie nizhnego kembrita Sibiri: Atlas okamenelostey* [Early Cambrian stage subdivision of Siberia: Atlas of fossils]. *Trudy, Institut geologii i geofiziki, Sibirskoe otdelenie, Akademiya nauk SSSR* 558: 176–177.

Astashkin, V. A. 1985. Problematichnye organizmy—porodoobrazovateli v nizhnem kembrii Sibirskoy platformy [Rock-forming problematic organisms in the Lower Cambrian of the Siberian Platform]. *Trudy, Institut geologii i geofiziki, Sibirskoe otdelenie, Akademiya nauk SSSR* 632: 144–149.

Ausich, W. I. and D. J. Bottjer. 1982. Tiering in suspension-feeding communities on soft substrata throughout the Phanerozoic. *Science* 216: 173–174.

Awramik, S. M. 1991. Archean and Proterozoic stromatolites. In R. Riding, ed., *Calcareous Algae and Stromatolites*, pp. 289–304. Berlin: Springer Verlag.

Bland, B. H. and R. Goldring. 1995. *Teichichnus* Seilacher 1955 and other trace fossils (Cambrian?) from the Charnian of Central England. *Neues Jahrbuch Geologie und Paläontologie Abhandlungen* 195: 5–23.

Bottjer, D. J. and D. Jablonski. 1986. Paleoenvironmental patterns in the history of post-Paleozoic benthic marine invertebrates. *Palaios* 3: 540–560.

Bottjer, D. J., K. A. Campbell, J. K. Schubert, and M. L. Droser. 1995. Palaeoecological models, non-uniformitarianism, and tracking the changing ecology of the past. In D. W. J. Bosence and P. A. Allison, eds., *Marine Palaeoenvironmental Analysis from Fossils*, pp. 7–26. Geological Society Special Publication, London 83.

Brasier, M. D. 1986. The sequence of small shelly fossils (especially conoidal microfossils) from English Precambrian-Cambrian boundary beds. *Geological Magazine* 123:237–256.

Brasier, M. D. 1991. Nutrient flux and the evolutionary explosion across the Precambrian-Cambrian boundary interval. *Historical Biology* 5:85–93.

Brasier, M. D. 1992. Global ocean-atmosphere change across the Precambrian-Cambrian transition. *Geological Magazine* 129:161–168.

Brasier, M. D. and R. A. Hewitt. 1979. Environmental setting of fossiliferous rocks from the uppermost Proterozoic–Lower Cambrian of central England. *Palaeogeography, Palaeoclimatology, Palaeoecology* 27:35–57.

Brasier, M. D., D. Dorjnamjaa, and J. F. Lindsay. 1996. The Neoproterozoic to Early Cambrian in southwest Mongolia: An introduction. *Geological Magazine* 133:365–369.

Brett, C. E. and G. C. Baird. 1986. Comparative taphonomy: A key to paleoenvironmental interpretation based on fossil preservation. *Palaios* 1:207–227.

Bromley, R. G. 1996. *Trace Fossils: Biology, Taphonomy, and Applications.* London: Unwin Hyman.

Bromley, R. G. and A. A. Ekdale. 1986. Composite ichnofabrics and tiering of burrows. *Geological Magazine* 123:59–65.

Copper, P., 1997. Articulate brachiopod shell beds: Silurian examples from Anticosti, Eastern Canada. *Géobios* 20:133–148.

Crimes, T. P. 1994. The period of early evolutionary failure and the dawn of evolutionary success: The record of biotic changes across the Precambrian-Cambrian boundary. In S. K. Donovan, ed., *The Paleobiology of Trace Fossils,* pp. 105–133. London: John Wiley and Sons.

Crimes, T. P. and M. A. Fedonkin. 1994. Evolution and dispersal of deepsea traces. *Palaios* 9:74–83.

Droser, M. L. 1987. *Trends in Extent and Depth of Bioturbation in Great Basin Precambrian-Ordovician Strata, California, Nevada, and Utah.* Dissertation, University of Southern California, Los Angeles.

Droser, M. L. 1991. Ichnofabric of the Paleozoic *Skolithos* ichnofacies and the nature and distribution of piperock. *Palaios* 6:316–325.

Droser, M. L. and D. J. Bottjer. 1986. A semiquantitative field classification of ichnofabric. *Journal of Sedimentary Petrology* 56:558–559.

Droser, M. L. and D. J. Bottjer. 1988. Trends in depth and extent of bioturbation in Cambrian carbonate marine environments, western United States. *Geology* 16:233–236.

Droser, M. L. and D. J. Bottjer. 1993. Trends and patterns of Phanerozoic ichnofabrics. *Annual Review of Earth and Planetary Sciences* 21:205–225.

Droser, M. L., N. C. Hughes, and P. A. Jell. 1994. Infaunal communities and tiering in early Palaeozoic nearshore environments: Trace fossil evidence from the Cambro-Ordovician of New South Wales. *Lethaia* 27:273–283.

Droser, M. L., J. G. Gehling, and S. Jensen. 1999a. When the worm turned: Concordance of Early Cambrian ichnofabric and trace-fossil record in siliciclastic rocks of South Australia. *Geology* 27:625–628.

Droser, M. L., S. Jensen, R. Wood, J. Gehling, and B. Runnegar. 1999b. The Precambrian-Cambrian radiation and the diversification of sedimentary fabrics. *Geological Society of America Abstracts with Programs* 31:A-334.

Ekdale, A. A. and R. G. Bromley. 1983. Trace fossils and ichnofabric in the Kjolby Gaard

Marl, Upper Cretaceous, Denmark. *Bulletin of the Geological Society of Denmark* 31: 107–119.

Fürsich, F. T. and W. Oschmann. 1993. Shell beds as tools in basin analysis: The Jurassic of Kachchh, western India. *Journal of the Geological Society of London* 150: 169–185.

Gehling, J. G. 1991. The case for Ediacaran fossil roots to the metazoan tree. *Journal of Geological Society of India, Memoir* 20: 181–224.

Gehling, J. G. 1999. Microbial mats in terminal Proterozoic siliciclastics; Ediacaran death masks. *Palaios* 14: 40–57,

Germes, G. J. B. 1983. Implications of a sedimentary facies and depositional environmental analysis of the Nama Group in South West Africa / Namibia. *Geological Society of South Africa, Special Publications* 11: 89–114.

Gevirtzman, D. A. and J. F. Mount. 1986. Paleoenvironments of an earliest Cambrian (Tommotian) shelly fauna in the southwestern Great Basin. *Journal of Sedimentary Petrology* 56: 412–421.

Goldring, R. 1995. Organisms and the substrate: Response and effect. In D. W. J. Bosence and P. A. Allison, eds., *Marine Palaeoenvironmental Analysis from Fossils*, pp. 151–180. Geological Society Special Publication 83.

Goldring, R. and S. Jensen. 1996. Trace fossils and biofabrics at the Precambrian-Cambrian boundary interval in western Mongolia. *Geological Magazine* 133: 403–415.

Grant, S. W. F. 1990. Shell structure and distribution of *Cloudina*, a potential index fossil for the terminal Proterozoic. *American Journal of Earth Sciences* 290A: 261–294.

Grotzinger, J. P. and A. H. Knoll. 1995. Anomalous carbonate precipitates: Is the Precambrian the key to the Permian? *Palaios* 10: 578–596.

Grotzinger, J. P., S. A. Bowring, B. Z. Saylor, and A. J. Kaufman. 1995. Biostratigraphic and geochronologic constraints on early animal evolution. *Science* 270: 598–604.

Hagadorn, J. W. and D. J. Bottjer. 1996. Wrinkle-marks: Evidence for extensive development of microbial mats in subtidal siliciclastic settings at the Proterozoic-Phanerozoic transition. *Geological Society of America Abstracts with Programs* 28: A-406.

Hallam, A. and K. Swett. 1966. Trace fossils from the Lower Cambrian Pipe Rock of the north-west Highlands. *Scottish Journal of Geology* 2: 101–106.

Hiller, N. 1993. A modern analogue for the Lower Ordovician *Obolus* conglomerate of Estonia. *Geological Magazine* 130: 265–267.

Hughes, N. C. and S. P. Hesselbo. 1997. Stratigraphy and sedimentology of the St. Lawrence Formation, Upper Cambrian of the northern Mississippi Valley. *Milwaukee Public Museum Contributions in Biology and Geology* 91: 1–50.

Jensen, S. 1993. *Trace Fossils, Body Fossils, and Problematica from the Lower Cambrian Mickwitzia Sandstone, South-Central Sweden.* Dissertation, Uppsala Universitet.

Jensen, S. 1997. Trace fossils from the Lower Cambrian *Mickwitzia* sandstone, south-central Sweden. *Fossils and Strata* 42: 3–10.

Jensen, S., J. Gehling, and M. Droser. 1998. Ediacara-type fossils in Cambrian sediments. *Nature* 393: 567–569.

Jensen, S., B. Z. Saylor, J. G. Gehling, and G. J. B. Germes. 2000. Complex trace fossils from the terminal Proterozoic of Namibia. *Geology* 28: 143–146.

Khomentovsky, V. V. and A. S. Gibsher. 1996. The Neoproterozoic–Lower Cambrian in northern Govi-Altay, western Mongolia: Regional setting, lithostratigraphy, and biostratigraphy. *Geological Magazine* 133: 371–390.

Kidwell, S. M. 1986. Models for fossil concentrations: Paleobiologic implications. *Paleobiology* 12 : 6–24.

Kidwell, S. M. 1990. Phanerozoic evolution of macroinvertebrate shell accumulations: Preliminary data from the Jurassic of Britain. In W. Miller III, ed., *Paleocommunity Temporal Dynamics: The Long-Term Development of Multispecies Assemblages*, pp. 305–327. *Paleontological Society Special Publication 5*.

Kidwell, S. M. 1991. The stratigraphy of shell concentrations. In P. A. Allison and D. E. G. Briggs, eds., *Taphonomy: Releasing the Data Locked in the Fossil Record*, pp. 211–290. New York: Plenum Press.

Kidwell, S. M. and D. W. J. Bosence. 1991. Taphonomy and time-averaging of marine shelly faunas. In P. A. Allison and D. E. G. Briggs, eds., *Taphonomy: Releasing the Data Locked in the Fossil Record*, pp. 115–209. New York: Plenum Press.

Kidwell, S. M. and P. J. Brenchley. 1994. Patterns in bioclastic accumulation through the Phanerozoic: Changes in input or in destruction? *Geology* 22 : 1139–1143.

Kidwell, S. M. and K. W. Flessa. 1995. The quality of the fossil record: Populations, species, and communities. *Annual Review of Ecology and Systematics* 26 : 269–300.

Kidwell, S. M. and S. M. Holland. 1991. Field description of coarse bioclastic fabrics. *Palaios* 6 : 426–434.

Kidwell, S. M. and D. Jablonski. 1983. Taphonomic feedback: Ecological consequences of shell accumulation. In M. J. S. Tevesz and P. L. McCall, eds., *Biotic Interactions in Recent and Fossil Benthic Communities*, pp. 195–248. New York: Plenum Press.

Kidwell, S. M., F. T. Fürsich, and T. Aigner. 1986. Conceptual framework for the analysis and classification of fossil concentrations. *Palaios* 1 : 228–238.

Kopaska-Merkel, D. C. 1988. Depositional environments and stratigraphy of a Cambrian mixed carbonate/terrigenous platform deposit, west-central Utah, USA. *Carbonates and Evaporites* 2 : 133–147.

Landing, E. 1988. Lower Cambrian of eastern Massachusetts: Stratigraphy and small shelly fossils. *Journal of Paleontology* 62 : 661–695.

Landing, E. 1989. Paleoecology and distribution of the Early Cambrian rostroconch *Watsonella crosbyi* Grabau. *Journal of Paleontology* 63 : 566–573.

Landing, E. 1991. Upper Precambrian through Lower Cambrian of Chapel Island: Faunas, paleoenvironments, and stratigraphic revision. *Journal of Paleontology* 65 : 570–595.

Li, X. and M. L. Droser. 1995. Stop 5: Fossil Mountain, Section K. In J. D. Cooper, ed., *Ordovician of the Great Basin*, pp. 88–93. Fullerton, Calif.: Pacific Section Society of Economic Paleontologists and Mineralogists.

Li, X. and M. L. Droser. 1997. Nature and distribution of Cambrian shell concentrations: Evidence from the Basin and Range Province of the western United States (California, Nevada, and Utah). *Palaios* 12 : 111–126.

Lindsay, J. F., M. D. Brasier, D. Dorjnamjaa, R. Goldring, P. D. Kruse, and R. A. Wood. 1996. Facies and sequence controls on the appearance of the Cambrian biota in southwestern Mongolia: Implications for the Precambrian-Cambrian boundary. *Geological Magazine* 133 : 417–428.

Lochman, C. 1947. Analysis and revision of eleven Lower Cambrian trilobites. *Journal of Paleontology* 21 : 59–71.

Logan, G. A., J. M. Hayes, G. B. Hieshima, and R. E. Summons. 1995. Terminal Proterozoic reorganization of biogeochemical cycles. *Nature* 376 : 53–56.

McGee, J. W. 1978. *Depositional Environments*

and Inarticulate Brachiopods of the Lower Wheeler Formation, East-Central Great Basin, Western United States. Master's thesis, University of Kansas, Lawrence.

McIlroy, D. 1996. Paleobiology and Stratigraphy of the Pre-Trilobitic Lower Cambrian of Avalonia and Baltica: Contributions from Ichnology, Macropalaeontology, and Provenance Studies. Dissertation, University of Oxford.

McIlroy, D. and G. A. Logan. 1999. The impact of bioturbation on infaunal ecology and evolution during the Proterozoic-Cambrian transition. Palaios 14:58–72.

McMenamin, M. A. S. 1985. Small shelly fossils from the basal Cambrian La Ciénega Formation, northwestern Sonora, Mexico. Journal of Paleontology 59:1414–1425.

Narbonne, G., P. M. Myrow, E. Landing, and M. M. Anderson. 1987. A candidate stratotype for the Precambrian-Cambrian boundary, Fortune Head, Burin Peninsula, southeast Newfoundland. Canadian Journal of Earth Sciences 24:1277–1293.

Norris, R. D. 1986. Taphonomic gradients in shelf fossil assemblages: Pliocene Purisima Formation, California. Palaios 1:256–270.

Parsons, K. M., C. E. Brett, and K. B. Miller. 1988. Taphonomy and depositional dynamics of Devonian shell-rich mudstones. Palaeogeography, Palaeoclimatology, Palaeoecology 63:109–139.

Peach, B. N. and J. Horne, 1884. Report on the geology of the north-west of Sutherland. Nature 31:31–34.

Pflüger, F. and P. G. Gresse. 1996. Microbial sand chips—a non-actualistic sedimentary structure. Sedimentary Geology 102: 263–274.

Popov, L. Ye., K. K. Khazanovich, N. G. Borovko, S. P. Sergeeva, and R. F. Sobolevskaya. 1989. Opornye razrezy i stratigrfiya kembro-ordovikskoy fosforitonosnoy obolovoy tolshchi na severo-zapade Russkoy platformy [Reference sections and stratigraphy of the Cambrian-Ordovician phosphorite-bearing Obolus Unit on the northwest of the Russian Platform]. Leningrad: Nauka.

Puura, I. and L. E. Holmer. 1993. Lingulate brachiopods from the Cambrian-Ordovician boundary beds in Sweden. Geologiska Föreningens i Stockholm Förhandlingar 115:215–237.

Riding, R. 1991a. Classification of microbial carbonates. In R. Riding, ed., Calcareous Algae and Stromatolites, pp. 21–51. Berlin: Springer Verlag.

Riding, R. 1991b. Cambrian calcareous cyanobacteria and algae. In R. Riding, ed., Calcareous Algae and Stromatolites, pp. 305–334. Berlin: Springer Verlag.

Riding, R. and A. Yu. Zhuravlev. 1995. Structure and diversity of oldest sponge-microbe reefs: Lower Cambrian, Aldan River, Siberia. Geology 23:649–652.

Rozanov, A. Yu. and E. A. Zhegallo. 1989. K probleme genezisa drevnikh fosforitov Azii [To the problem of the genesis of ancient phosphorites in Asia]. Litologiya i poleznye iskopaemye 1989 (3):67–82.

Savrda, C. E. 1995. Ichnologic applications in paleoceanographic, paleoclimatic, and sea-level studies. Palaios 10:565–578.

Savrda, C. E. and D. J. Bottjer. 1986. Trace-fossil model for reconstruction of paleooxygenation in bottom waters. Geology 14: 3–6.

Schubert, J. K. and D. J. Bottjer. 1992. Early Triassic stromatolites as post-mass extinction disaster forms. Geology 20:883–886.

Seilacher, A. 1995. Fossile Kunst: Aklbumblätter der Erdgeschichte. Goldschneck-Verlag.

Seilacher, A. 1999. Biomat-related lifestyles in the Precambrian. Palaios 14:86–93.

Seilacher, A. and F. Pflüger. 1994. From

biomats to benthic agriculture: A biohistoric revolution. In W. E. Krumbein, D. M. Paterson, and L. J. Stal, eds., *Biostabilization of Sediments,* pp. 97–105. Bibliotheks und Informationsstem der Universitat Oldenburg.

Sepkoski, J. J., Jr. 1981a. A factor analytic description of the Phanerozoic marine fossil record. *Paleobiology* 7:36–53.

Sepkoski, J. J., Jr. 1981b. The uniqueness of the Cambrian Fauna. In M. E. Taylor, ed., *Short Papers for the Second International Symposium on the Cambrian System,* pp. 203–207. U.S. Geological Survey Open-File Report 81-743.

Sepkoski, J. J., Jr., and A. I. Miller. 1985. Evolutionary faunas and the distribution of Paleozoic benthic communities. In J. W. Valentine, ed., *Phanerozoic Diversity Patterns,* pp. 181–190. Princeton, N.J.: Princeton University Press.

Sepkoski, J. J., Jr., R. Bambach, and M. L. Droser. 1991. Secular changes in Phanerozoic event bedding and the biological overprint. In G. Einsele, W. Ricken, and A. Seilacher, eds., *Cyclic and Event Stratification,* pp. 298–312. Berlin: Springer Verlag.

Taylor, A. M. and R. Goldring. 1993. Description and analysis of bioturbation and ichnofabric. *Journal of the Geological Society of London* 150:141–148.

Ushatinskaya, G. T. 1988. Obolellidy (brakhiopody) s zamkovym sochleneniem stvorok iz nizhnego kembriya Zabaykal'ya [Obolellids (brachiopods) with articulated valves from the Lower Cambrian of Transbaikalia]. *Paleontologicheskiy zhurnal* 1988: 34–39.

Westrop, S. R. 1986. Taphonomic versus ecologic controls on taxonomic relative abundance patterns in tempestites. *Lethaia* 19: 123–132.

Wilson, M. A., T. J. Palmer, T. E. Guensburg, C. D. Finton, and L. E. Kaufman. 1992. The development of an Early Ordovician hardground community in response to rapid sea-floor calcite precipitation. *Lethaia* 25:19–34.

Wood, R. A. 1995. The changing biology of reef-building. *Palaios* 10:517–529.

Zhuravlev, A. Yu. 1996. Reef ecosystem recovery after the Early Cambrian extinction. In M. B. Hart, ed., *Biotic Recovery from Mass Extinction Events,* pp. 79–96. Geological Society Special Publication, London 102.

Ziebis, W., S. Forster, M. Huettel, and B. B. Jørgensen. 1996. Complex burrows of the mud shrimp *Callianassa truncata* and their geochemical impact in the sea bed. *Nature* 382:619–622.

PART II

Community Patterns
and Dynamics

CHAPTER EIGHT

Andrey Yu. Zhuravlev

Biotic Diversity and Structure During the Neoproterozoic-Ordovician Transition

Diversity of 4,122 metazoan genera, 31 calcimicrobial genera, and 470 acritarch species are plotted for the Nemakit-Daldynian–early Tremadoc interval at zonal level. Generally congruent plots of diversity of metazoan genera, acritarch species, calcified cyanobacteria, and ichnofossils reflect Nemakit-Daldynian–early Botoman diversification, middle Botoman crisis leading to further late Botoman–Toyonian diversity decrease, and Middle-Late Cambrian low-diversity stabilization. All three sources of overall diversity (alpha, beta, and gamma diversity) contributed to the development of generic diversity at the beginning of the Cambrian. The apparent niche partitioning and several levels of tiering, observed in reefal and level-bottom communities, indicate that the biotic structure of these was already complex in the late Tommotian. A wide spectrum of communities was established in the Atdabanian. Ecologic, lithologic, and isotopic features are indicative of a nutrient-rich state of the oceans at the beginning of the Cambrian. The radiation of benthic and planktic filter and suspension feeders considerably refined the ocean waters and led to less nutrient-rich conditions for later, more diverse, evolutionary faunas. The inherent structure of the biota, expressed in relative number of specialists and degree of competition, was responsible for its stability. Extrinsic factors could amplify crises but could hardly initiate them.

AT THE END of the Neoproterozoic and beginning of the Phanerozoic, there was a rapid succession of distinct faunas and a diversity increase that involved the brief flourishing of the enigmatic Ediacaran fauna, subsequent expansion of the Tommotian small shelly taxa, and finally replacement by the more standard Cambrian and Ordovician groups. Discussions of Vendian to Cambrian diversification by Sepkoski (1979, 1981) treated the fauna of this interval as homogeneous. Most of the important Cambrian classes, including archaeocyaths, trilobites, inarticulate brachiopods (mainly lingulates in the present sense), hyoliths, monoplacophorans (now, principally, helcionelloids), stenothecoids, cribricyaths, volborthellids, eocrinoids and some other echinoderm classes, sabelliditids, soft-bodied and lightly skeletonized animals, and various Problematica, were assembled into the "Cambrian Evolutionary Fauna." This fauna dominated the early phase of metazoan diversification. It attained maximum diversity in the Cambrian and then began a long decline. Very few members of the Cambrian fauna participated in the Ordovician radiation or persist today. The Paleozoic Evolutionary Fauna began to radiate during the latest Cambrian and virtually exploded in the Ordovician. The Modern Evolutionary Fauna originated during the Cambrian Period but radiated in the Mesozoic.

The three great evolutionary faunas were identified through Q-mode factor analysis of familial diversity through the Phanerozoic (Sepkoski 1981). The factors of familial data differed significantly from expectation for stochastic phylogenies and therefore reflected some underlying organization in the evolution of Phanerozoic marine diversity (Sepkoski 1991a). Smith (1988), noted that several important classes in the Cambrian Fauna—namely, Inarticulata, Monoplacophora, and Eocrinoidea—are paraphyletic, and he therefore suggested that the distinction between the Cambrian and Paleozoic faunas, and the apparently separate radiations of the Early Cambrian and the Ordovician, might be an artifact of taxonomy coupled with a poor fossil record in the Late Cambrian. He ably demonstrated that eocrinoids represent a poorly defined stem group for later pelmatozoans and cystoids (but see Guensburg and Sprinkle, this volume). In contrast, monoplacophorans and inarticulates are split into several holophyletic clades (class Helcionelloida, class Lingulata) (Gorjansky and Popov 1986; Peel 1991), the bulk of which further increase the distinction mentioned above. Thus, although taxonomic practice may contribute scatter to the pattern, the histories of Cambrian classes continue to remain distinct from members of the Paleozoic and Modern faunas. In addition, the Monte Carlo simulations did not reveal a significant bias produced by paraphyletic taxa (Sepkoski and Kendrick 1993). A distinct pattern is observed in the stratigraphic distribution of fossils treated as earliest pelecypods, rostroconchs, and gastropods: their first representatives disappeared during the middle Botoman extinction event, but the classes apparently diversified at the very end of the Cambrian and Ordovician. Such a pattern emphasizes the distinction between elements that contributed to the Cambrian and Ordovician radiations.

Further investigations by Q-mode factor analysis, performed on generic diversity data, recognized at least three evolutionary faunas at the start of metazoan diversification—the Ediacaran, Tommotian, and Cambrian sensu stricto faunas—and archaeocyaths received their own factor (Sepkoski 1992). The Tommotian Evolutionary Fauna factor received maximum loadings from the Nemakit-Daldynian, Tommotian, and early Atdabanian, and the fauna included orthothecimorph hyoliths, helcionelloids, paragastropods, sabelliditids, and a variety of short-ranging Problematica that originated during this time interval. Finally, the restricted Cambrian Evolutionary Fauna factor received maximum loadings from the late Atdabanian through Sunwaptan; it consisted of trilobites, bradoriids, and some other arthropods, lingulates, and echinoderm classes. This latter assemblage actually represents a mixture of members of the Cambrian sensu stricto, Paleozoic, and Modern faunas.

Metazoans of all taxonomic levels from genus to class exhibit, in general, congruent diversity patterns through the Cambrian-Ordovician (Sepkoski 1992). The major Cambrian radiation of large metazoans with mineralized skeletons was accompanied by a continued radiation of soft-bodied burrowing infauna in both nearshore siliciclastic and carbonate shelf settings expressed in increased diversity of trace fossils and intensity of bioturbation from the Vendian through Early Cambrian; thereafter there was little change in the Early Paleozoic (Crimes 1992a,b, 1994; Droser and Bottjer 1988a,b).

The same pattern is repeated broadly by calcified cyanobacteria (Sepkoski 1992; Zhuravlev 1996) and acritarchs (Rozanov 1992; Knoll 1994; Vidal and Moczydłowska-Vidal 1997). Preliminary data on calcified cyanobacteria and algae allowed Chuvashov and Riding (1984) to establish three major marine Paleozoic floras—the Cambrian, Ordovician, and Carboniferous floras. Quantitative and taxonomic analyses of these entities are needed. However, the diversity pattern of their Cambrian Flora is congruent with that of the Early Cambrian Biota, as has been shown by quantitative data (Zhuravlev 1996). This flora was dominated by calcified probable bacteria (e.g., *Girvanella, Obruchevella, Epiphyton, Renalcis, Acanthina, Bija, Proaulopora*), to which a few problematic calcified algae were added during the Middle to Late Cambrian (see Riding, this volume). Some elements of this flora have a discontinuous record to the Cretaceous. In contrast, the Ordovician Flora, which diversified in the Middle Ordovician, contained a large variety of calcified green and red algae and new groups of calcified cyanobacteria.

Thus, all patterns are remarkably similar as indicated in figure 8.1A–D.

DIVERSITY ANALYSIS

New and revised biostratigraphic data for the Cambrian permits quantitative analysis of changes in biotic diversity, which I accept here as simple taxonomic diversity. Global generic diversity data are calculated on the basis of my literature compilation of stratigraphic ranges and paleogeographic distributions of genera from the Nemakit-Daldynian to Tremadoc for all groups (4,122 genera), with the exception of spicular sponges (figure 8.1A). These data are calibrated by Russian (Siberian) stage and zonal scales for the Early and early Middle Cambrian, North American (Laurentian) stage and zonal scales for the late Middle and Late Cambrian, and Australian Datsonian as the terminal Cambrian interval (from the base of the *proavus* Zone to the base of the *lindstromi* Zone). The global correlation of these stratigraphic units is given by Zhuravlev (1995) for the Early Cambrian and by Shergold (1995) for the Middle and Late Cambrian (Zhuravlev and Riding, this volume: tables 1.1 and 1.2).

As is already well known, the Neoproterozoic–Early Cambrian metazoan explosion was relatively rapid, spanning a period of about 20 m.y. from the Nemakit-Daldynian

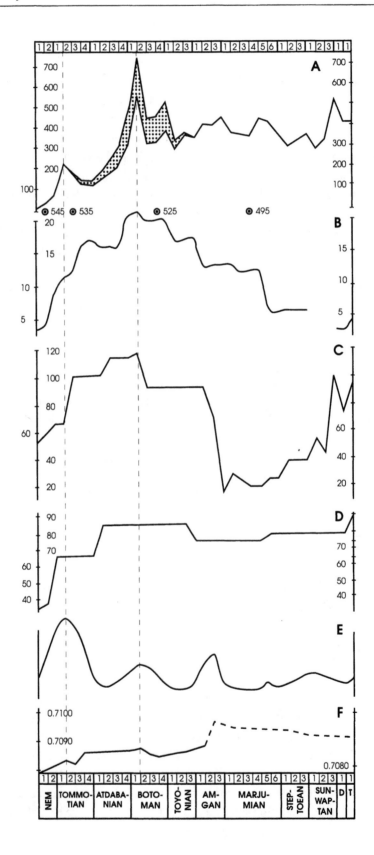

Figure 8.1 Pattern of diversity through the Cambrian-Tremadoc. *A,* Diversity curve for metazoan genera (stippled area shows archaeocyath diversity). *B,* Diversity curve for calcimicrobe genera. *C,* Diversity curve for acritarch species. *D,* Plot of total trace fossil diversity (modified after Crimes 1992a, 1994). *E,* Phosphorite abundance curve (modified after Cook 1992). *F,* $^{87}Sr/^{86}Sr$ plot (compiled from Donnelly et al. 1990; Derry et al. 1994; Saltzman et al. 1995; Montañez et al. 1996; Nicholas 1996). *NEM* = Nemakit-Daldynian; *D* = Datsonian; *T* = Tremadoc.

to the early Botoman (Bowring et al. 1993; Shergold 1995; Landing and Westrop 1997). This is short relative to subsequent Phanerozoic radiations, and the per taxon rate of diversification was much higher (Sepkoski 1992).

The general intensity of extinction in the oceans has declined through the Phanerozoic (Sepkoski 1994). Cambrian intensities are quite high. Detailed field biostratigraphy resolves some of this into three extinction events during the Early Cambrian and four extinction events during the Middle-Late Cambrian, including that at the former Cambrian-Ordovician boundary (*Saukia-Missisquoia* boundary) (Palmer 1965, 1979; Stitt 1971, 1975; Brasier 1991, 1995a; Zhuravlev and Wood 1996). The latter were recognized first by Palmer (1965, 1979), who called them biomere extinctions.

Quantitative analysis of global generic diversity reveals striking changes through the Cambrian. If extinction rates are plotted separately, they exhibit no additional characteristics (Zhuravlev and Wood 1996: figure 1). First, diversity decline occurs in the mid-Tommotian (Brasier 1991). However, the scale of this extinction is likely, in part, to reflect taxonomic oversplitting of scleritome taxa. More striking are two further extinction events noted in the mid-Early Cambrian: in the middle and late Botoman. The later of these events was predicted by selected data (Bognibova and Shcheglov 1970; Newell 1972; Burrett and Richardson 1978; Sepkoski 1992; Signor 1992a; Brasier 1995a) and is related to the well-known Hawke Bay Regression (Palmer and James 1979) or to the "Olenellid biomere event" that affected trilobites at about that time (A. Palmer 1982). Ecologic Evolutionary Unit I of Boucot (1983) was terminated by this extinction (Sheehan 1991). A more pronounced extinction occurred in the middle Botoman (approximately at the *micmacciformis/Erbiella–gurarii* zone boundary) and has been named the Sinsk event (Zhuravlev and Wood 1996). It was responsible for a major disturbance of the Early Cambrian Biota, after which many groups composing the Tommotian Fauna either disappeared or became insignificant. Metazoans attained their highest generic diversity of the Cambrian during the early Botoman, and contrary to Sepkoski's (1992) calculation, this was not exceeded until the Arenig. Archaeocyaths were not the principal group contributing to this pattern (figure 8.1A). At the generic level, they compose only 24 percent rather than about 50 percent (contra Sepkoski 1992) of total early Botoman generic diversity and 18 percent of extinct genera. These differences in data may be explained by the coarser stratigraphic scale and the smaller database that were used by Sepkoski (1992). In comparison, trilobite genera contribute 27 percent and 16 percent, respectively. This decline is well expressed at the species level on all major continents and terranes of the Cambrian world (Zhuravlev and Wood 1996: figure 2). Calcified cyanobacteria (31 genera) and acritarchs (470 species) show a similar decline in diversity (figures 8.1B,C). A slight fall in trace fossil diversity is observed during the Middle and Late Cambrian (figure 8.1D), followed by a steady rise through the Ordovician, resulting from an increase in deep-water trace fossil diversity (Crimes 1992a); the levels of Early Cambrian diversity were not reached again until the Early Ordovician (Crimes 1994). In general outline, this pattern resembles the diversification of body fossils across the same interval.

Four extinction events during the Middle-Late Cambrian are confirmed by global data but are most pronounced among trilobites (figure 8.1A). However, the latest of them affected cephalopods and rostroconchs too. Both rostroconchs and cephalopods produced their first diversification peak in the Datsonian (Pojeta 1979; Chen and Teichert 1983).

The dynamics of three additional indices is quantified for the Nemakit-Daldynian–early Tremadoc interval. These are (1) average monotypic taxa index (MTI), (2) average geographic distribution index (AGI), and (3) average longevity index (ALI). These are calculated for genera in each zone (Zhuravlev and Riding, this volume: tables 1.1 and 1.2, Arabic numerals; and figures 8.2A–C herein). Initially, average indices were determined for each taxonomic group separately. Then average indices were counted for each of the following biotas: Tommotian Biota (anabaritids, sabelliditids, coeloscleritophorans, helcionelloids, orthothecimorph hyoliths, and minor problematic sclerital groups); Early Cambrian Biota (archaeocyath sponges, radiocyaths, cribricyaths, coralomorphs, paragastropods, hyolithomorph hyoliths, bradoriids, anomalocaridids, tommotiids, hyolithelminths, cambroclaves, mobergellans, coleolids, paracarinachitiids, salterellids, and stenothecoids); Middle-Late Cambrian Biota (trilobites, lingulates, calciates, echinoderms, and lightly skeletonized arthropods); and combined Paleozoic-Modern Biota (rostroconchs, cephalopods, gastropods, tergomyans, polyplacophorans, pterobranchs, graptolites, paraconodonts, and euconodonts). These biotas display broadly congruent fluctuations of the indices for most of the Cambrian. Principal deviations from this common pattern will be emphasized below.

The last two indices usually display a similar coherent pattern because the wider the spectrum of conditions under which a genus is able to survive, the wider is its area and the longer it exists (Markov and Naimark 1995; Markov and Solov'ev 1995). AGI is calculated as follows. An appearance of a genus on a single craton is accepted arbitrarily as 1 unit; an appearance of genus in several regions of the same province is scored as 5 units; a global distribution is scored as 10 units. (As has been shown by Markov and Naimark [1995], the change of unit value does not influence the general pattern of the geographic distribution.) The Early Cambrian provinces are confined to Avalonia, Baltica, Laurentia (including Occidentalia), East Gondwana (Australia-Antarctica, China, Mongolia-Tuva, Kazakhstan), West Gondwana (southern and cen-

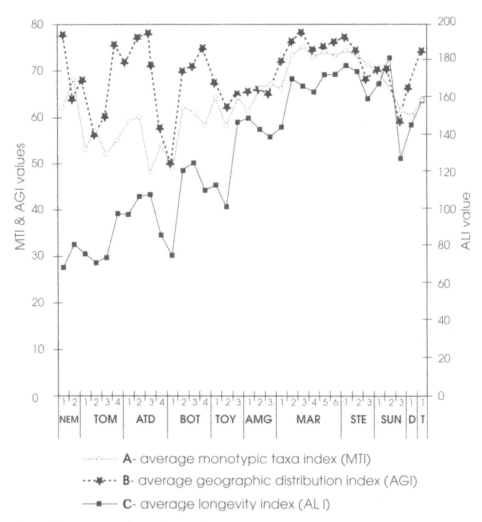

········○······· **A**- average monotypic taxa index (MTI)

··✦·· **B**- average geographic distribution index (AGI)

──■── **C**- average longevity index (AL I)

Figure 8.2 Dynamics of cumulative indices for the Cambrian biotas. *A*, Average monotypic taxa index; the ordinate in this graph represents the percentage of monotypic families that contain a single genus per time unit indicated on the abscissa. *B*, Average geographic distribution index. *C*, Average longevity index. Early Cambrian: *NEM* = Nemakit-Daldynian, *TOM* = Tommotian, *ATD* = Atdabanian, *BOT* = Botoman, *TOY* = Toyonian. Middle Cambrian: *AMG* = Amgan, *MAR* = Marjuman; Late Cambrian: *STE* = Steptoean, *SUN* = Sunwaptan, *D* = Datsonian; *T* = Tremadoc.

tral Europe, Morocco, and the Middle East), and Siberia (Siberian Platform, Altay Sayan Foldbelt). The Middle and Late Cambrian paleobiogeographic subdivisions adopted here are after Jell (1974) and Shergold (1988), respectively. AGI is low during the Tommotian, early Botoman, and Toyonian (figure 8.2B). Thus, our data are broadly similar to the generalization by Signor (1992b), who counted endemic genera on major cratons for early Cambrian stages: more than 50 percent for the Tommotian, about 45 percent for the Atdabanian, almost 60 percent for the Botoman, and 60 percent for the Toyonian.

PATTERN OF BIOTA DEVELOPMENT

Early Cambrian Radiation versus Middle Ordovician Radiation

Many comprehensive reviews discuss different aspects of the origin of the Cambrian biotas (Axelrod 1958; Glaessner 1984; Conway Morris 1987; Valentine et al. 1991; Signor and Lipps 1992; Erwin 1994; Kempe and Kazmierczak 1994; Vermeij 1995; Marin et al. 1996). On the whole, biotic rather than abiotic explanations of this event are preferred here. Among them, ideas about increased predator pressure first offered by Evans (1912) and Hutchinson (1961) and cropper pressure introduced by Stanley (1973) look more attractive in the light of recent observations (Müller and Walossek 1985; Vermeij 1990; Sepkoski 1992; Burzin 1994; Butterfield 1994, 1997; Chen et al. 1994; Conway Morris and Bengtson 1994; Zhuravlev 1996; see also chapters by Butterfield and Burzin et al., this volume). In addition to a direct influence, predator pressure can promote local elimination of a stronger competitor and, respectively, increase community diversity (Vermeij 1987). As the major Cambrian radiation of skeletal metazoans was accompanied by a continued radiation of soft-bodied burrowing organisms, skeletal mineralization was hardly a key innovation: the implied geochemical triggers were not necessary for the radiation (Droser and Bottjer 1988a). Penetration into substrate has several advantages, including the escape from predator pressure. In such a case, the substrate itself plays the role of a hard shield.

The basic sigmoidal patterns of metazoan, phytoplanktic, calcimicrobial, and ichnogeneric taxonomic diversity (see figures 8.1A–D) are consistent with the equilibrium model of taxonomic diversification developed by Sepkoski (1992). This model predicts that early phases of radiations into ecologically vacant environments should be exponential and should be followed by declining diversification resulting from decreased origination and increased extinction as the environment fills with species. The high AGI at the beginning of the Cambrian explosion (see figure 8.2B) is consistent with the suggestion that empty adaptive space allowed extensive divergence and low probability of extinction. This index shows that the diversification is related to extensive divergence (appearance of new genera during occupation of new areas in relatively empty adaptive zones) rather than to a high degree of geographic isolation.

The Ordovician evolutionary radiation represents another major pivotal point in the history of life, when the nature of marine faunas was almost completely changed and both global and local taxonomic diversity increased two- to threefold (Sepkoski and Sheehan 1983; Sepkoski 1995); ecological generalists were suggested to be replaced by specialists even within the same lineages (Fortey and Owens 1990; Leigh 1990; Sepkoski 1992). In addition, the appearance of new groups of predators (euconodonts, cephalopods) and grazers (polyplacophorans, gastropods) and their rapid diversification at the very end of the Cambrian might be among major factors that predetermined the great Ordovician explosion. In contrast to the Cambrian, the Ordovician radiation resembles that of the Mesozoic. With the exception of the Bryozoa, no phyla first appear as part of the Ordovician radiation. This could be because eco-space was sufficiently filled at the beginning of each subsequent radiation to preclude survival of new body plans (cf. Erwin et al. 1987). During the Ordovician radiation the Paleozoic Fauna proliferated while the Cambrian Fauna waned. We see a transition both ecological and taxonomic between the two faunas in the Early Ordovician. Actually, the Ordovician radiation started soon after the Early Cambrian extinction, from the Middle Cambrian onward, and was associated with changes to a new evolutionary fauna that largely involved groups that appeared as unimportant classes during the Cambrian. Nonetheless, euconodonts, graptolites, and new molluscan (rostroconchs, cephalopods, gastropods, polyplacophorans), brachiopod, and trilobite groups entered Cambrian communities and became their most ubiquitous elements by the end of the Cambrian period and even produced their first diversity peak in the late Sunwaptan. A similar contrast pattern of temporal diversity trends is observed among trilobites of the Ibex and Whiterock faunas (Adrain et al. 1998).

The sources of overall diversity are the richness of taxa in a single community (alpha diversity), the taxonomic differentiation of fauna between communities (beta diversity), and the geographic taxonomic differentiation (gamma diversity) (see Sepkoski 1988 and references therein). All three contributed to the growth of generic diversity at the beginning of the Cambrian.

The apparent niche partitioning and several levels of tiering observed in reefal and level-bottom communities (McBride 1976; Conway Morris 1986; Kruse et al. 1995; Zhuravlev and Debrenne 1996; see also Burzin et al. and Debrenne and Reitner, this volume), indicate that the biotic structure of these communities was already complex by the late Tommotian. These complexities provided a basis for an increase in alpha diversity. The Early Cambrian reefal communities contained 50–80 species, whereas their Middle and Late Cambrian counterparts have yielded only about 10 species (Zhuravlev and Debrenne 1996; Pratt et al., this volume). Indeed, without archaeocyaths (stippled on figure 8.1A), cribricyaths, coralomorphs, and other reef dwellers, the entire plot of the Cambrian generic diversity would be a plateau, fluctuating slightly around the level of about 400 genera per zone, since late Atdabanian time.

A wide spectrum of communities providing the basis for beta diversity increase was established in the Atdabanian (see Burzin et al. and Pratt et al., this volume).

Faunal provinciality is estimated as very high since the Early Cambrian (Signor 1992b). Mean values of the Jaccard coefficient of similarity measured for generic sets of major Early Cambrian provinces listed above vary from 0.08 to 0.11 for different stage slices (Debrenne et al. 1999). This supports the suggestion of high endemicity and thus reveals high gamma diversity for the Early Cambrian Biota.

Comparison with the Ordovician radiation indicates that the low magnitude of the Cambrian radiation has to be attributed to comparatively low alpha and beta diversity (Sepkoski 1988). Gamma diversity was hardly important in the Ordovician radiation, because the mutual position of continents did not change much from Middle

Cambrian (low overall diversity) to Middle Ordovician (high diversity) (see Seslavinsky and Maidanskaya, this volume: figures 3.3 and 3.6), and the provinciality of Cambrian faunas was already high (Jell 1974; Shergold 1988; Signor 1992b; Debrenne et al. 1999; Hughes, this volume). On the contrary, the appearance of hardground communities, bryozoan thickets, crinoid gardens, and, probably, offshore deep-water communities, as well as the recovery of metazoan reefal communities (Fortey 1983; Sepkoski and Sheehan 1983; Bambach 1986; Fortey and Owens 1987; Sepkoski 1988, 1991a; Crimes and Fedonkin 1994; see also Crimes, this volume), reveals that the Ordovician radiation was brought about by alpha and beta diversity rise (Sepkoski 1988). Hardground communities already appeared in the late Middle Cambrian (Zhuravlev et al. 1996) but were not diverse until the Middle Ordovician (T. Palmer 1982; see also Rozhnov, this volume).

However, what factors limited the alpha and beta sources of overall diversity? Theoretically, the Early Cambrian radiation might have been explosive because the number of "empty" niches was almost unlimited (Erwin et al. 1987), the morphological plasticity of organisms was significant (Conway Morris and Fritz 1984; Hughes 1991), and the radiation involved considerable morphological innovation (Erwin 1992). Nonetheless, the diversity peak actually achieved by the Cambrian biota was much lower than those for the Paleozoic and Modern biotas, despite the fact that these later biotas were not developed in empty ecospace and thus were much more restricted (Bambach 1983; Bottjer et al. 1996).

The beta diversity of a marine biota is to a certain extent related to cratonic flooding (e.g., Burrett and Richardson 1978; see also Gravestock and Shergold, this volume). However, if a drop in sea level could reduce the shelf area flooded by the oceans and cause a standing crop reduction, then sea level fluctuations would hardly be responsible for the significant increase in Ordovician diversity, because areas flooded during the largest Cambrian and Ordovician transgressions did not differ much in size (see Seslavinsky and Maidanskaya, this volume: figures 3.2 and 3.6).

Brasier (1991) and Vermeij (1995) used increase in nutrient supply to explain both the Cambrian and Ordovician radiations. The Early–early Middle Cambrian and the Early Ordovician (Tremadoc) may be ascribed, indeed, to intervals of a great phosphate availability (see figure 8.1E), but soon afterward, in the Middle Ordovician when major radiation actually commenced, phosphorite abundance drastically decreased (Cook 1992). In addition, field observations reveal a reverse pattern: in Iran a poor Nemakit-Daldynian *Anabarites-Cambrotubulus* assemblage is present in phosphorite-rich sediments above a diverse Tommotian-Meshucunian fauna, and its stratigraphic appearance may instead reflect persistence of conditions unfavorable for the development of a richer shelly fossil assemblage (Zhuravlev et al. 1996). Thus, contrary to a current view linking nutrient flux and evolutionary explosion, the Iranian sedimentary record indicates a drastic diversity decrease during episodes of enhanced nutrient supply.

However, judging from overall phosphorite abundance, and high continental erosion rates indicated by ^{87}Sr/^{86}Sr ratios (Cook 1992; Derry et al. 1994; Nicholas 1996; see also Brasier and Lindsay, this volume), general mesotrophic-eutrophic conditions could have existed during the Early Cambrian. The same can be inferred from the fact that at present all the oceans' waters are filtered by marine biota in only a half year, and the upper 200 m of the water column is filtered in just a few weeks (Bogorov 1974; Karataev and Burlakova 1995). At the beginning of the Cambrian, in the absence of such active filter and suspension feeders as pelecypods, bryozoans, and stromatoporoid sponges, the ocean was hardly likely to resemble the mostly oligotrophic modern ocean.

Although Signor and Vermeij (1994) suggested that the proportion of filter and suspension feeders in Cambrian communities was small, this has been challenged by many observations (Wood et al. 1993; Burzin 1994; Butterfield 1994, 1997; Kruse et al. 1995; Logan et al. 1995; Savarese 1995; Debrenne and Zhuravlev 1997; see also Butterfield, this volume). For example, Butterfield (1994) identified an elaborate and essentially modern crustacean filter apparatus among Early Cambrian arthropods exploiting planktic habitats. The analysis of the contribution of trophic guilds to the Early Cambrian radiation shows that the trophic nucleus of Early Cambrian communities was sessile passive filter and suspension feeders (archaeocyaths and other sponges, radiocyaths, chancelloriids, hyoliths, stenothecoids, brachiopods, many tube-dwelling taxa, early mollusks and echinoderms, *Skolithos*- and *Aulophycus*-producers, and many others) well-adapted to such conditions (Smith 1990; Droser 1991; Debrenne and Zhuravlev 1997; see also Burzin et al., this volume). The proportion of suspension feeders increased from Nemakit-Daldynian to Botoman (Crimes 1992a; Lipps et al. 1992: figure 8.4.3). When observing such a feeding strategy orientation of the Early Cambrian Biota, we should be not surprised that during the Early Cambrian the diversity curve of metazoan genera shows some similarity to acritarch diversity and phosphorite abundance as well as to ^{87}Sr/^{86}Sr excursions plotted by Derry et al. (1994) and Nicholas (1996) (see figure 8.1F). The latter curve reveals major positive shifts in ^{87}Sr/^{86}Sr, signifying high erosion rates (and, indirectly, enhanced nutrient supply) during the early Tommotian and early Botoman, when the Early Cambrian Biota, which consists of the groups listed above, achieved two diversity peaks. Indeed, passive feeding requires an unlimited food supply. Reduced water clarity would shift primary production toward phytoplankton, whereas secondary production would be shifted to filter and suspension feeders at the expense of benthic algae and deposit feeders and grazers (Brasier 1995b). This is exactly the pattern observed among Early Cambrian communities. The bloom-prone spiny Early Cambrian phytoplankton contributed disproportionately to the direct export of cells to benthic habitats through the rapid sinking of aggregates formed by simple adhesion and collision (Butterfield 1997). Such aggregates are plentiful in Early and early Middle Cambrian sediments (Butterfield and Nicholas 1996; Zhegallo et al. 1996; Zhuravlev and Wood 1996).

If the proliferation of the Early Cambrian Biota may be explained to a certain extent in terms of its adaptation to mesotrophic-eutrophic conditions, the same is hardly applicable to the Paleozoic Biota that radiated in the Ordovician. The principal difference between Cambrian filter and suspension feeders and those of the Ordovician, which are represented by crinoids, stromatoporoid and chaetetid sponges, pelecypods, and bryozoans, is that the latter are active filtrators. Passive suspension feeders rely mainly on ambient currents to bring food particles to sites of entrapment, whereas active ones produce their own currents (LaBarbera 1984), allowing them to utilize more dispersed resources. This may be attributed to decreased rather than increased nutrient availability. The contemporary increase in tiering of epifaunal communities (Ausich and Bottjer 1982) and a shift of the former benthic filtrators and microcarnivores (graptolites, some trilobites, radiolarians) to the pelagic realm (Fortey 1985; Underwood 1993; Rigby and Milsom 1996) might also indicate increasing competition due to decreasing nutrient supply. The major increase in the amount of bioturbation that occurred between the Middle and Late Ordovician coincided with the Ordovician radiation, when the average ichnofabric index jumped from 3.1 to 4.5 (Droser and Bottjer 1988b). This also reflects higher infaunal tiering achieved in communities during this time interval. Increased utilization or finer subdivision of ecospace should be manifested in increased alpha diversity (the richness of species in local communities) that measures packing within a community and thus reflects how finely species are dividing ecological resources. Indeed, Bambach's (1977) data are consistent with this.

Another problem created by non-nutrient-limited conditions is limited water transparency. The Early Cambrian reefal fauna was, probably, not light limited (Wood et al. 1992, 1993; Surge et al. 1997). Equally, the principal Early Cambrian primary producers were calcified cyanobacteria adapted to dim conditions (Rowland and Gangloff 1988; Zhuravlev and Wood 1995) and planktic acritarchs, which are relatives of mesotrophic dinoflagellates (Moldowan et al. 1996). The acritarch species diversity plot (see figure 8.1C) fluctuates in some coordination with the relative phosphorite abundance curve (see figure 8.1E), which may reflect relative nutrient supply. The Nemakit-Daldynian–Tommotian highest phosphorite peak corresponds to the beginning of acritarch speciation, and the early Botoman moderate phosphorite peak correlates with the highest acritarch species diversity. Both these curves show low values during the Marjuman–early Sunwaptan interval. It is noteworthy that the triaromatic dinosteroid record, which can be attributed either to dinoflagellates or to acritarchs themselves, shows a hiatus during the same interval (see Moldowan et al., this volume: figure 21.3). On the contrary, better lighting would have been required by the Middle and Late Ordovician reefal communities that consisted of true calcified algae, and photosymbiont-bearing stromatoporoid–chaetetid sponges and tabulate corals (Chuvashov and Riding 1984; Wood 1995 and references therein). However, these are the non-light-limited conditions that allow longer trophic webs and, thus, a higher species richness (Hallock 1987; Wood 1993), and the Modern Biota achieves its highest diversity in well-illuminated oligotrophic environments.

Thus, the major factor in alpha diversity growth in the Ordovician could have been progressive oligotrophication of the world ocean, while Tommotian and Early Cambrian biotas had to be adapted to nutrient-rich conditions.

Ecological Properties of the Early Cambrian Biota

General adaptation of Tommotian and Early Cambrian biotas to nutrient-rich conditions may explain the major features that distinguish them from later biotas. These include (1) relatively low within-habitat species richness, (2) low trophic guild diversity, (3) low ecological differentiation of communities, and (4) relatively low diversity-disparity ratio.

1. A low within-habitat species richness in marine level-bottom communities for the Cambrian, in comparison with that for the Paleozoic, was noted by Bambach (1977; see also Sepkoski 1988). The same is evident in reefal communities. Early Cambrian reefal communities contain 30–80 species, Paleozoic ones average 60–400 species, and Modern ones may exceed 1,200 species (Fagerstrom 1987; Zhuravlev and Debrenne 1996). In these examples, diversity might be controlled in part by tiering, which is estimated as low. A relatively simple tiering of the Cambrian epifaunal and infaunal suspension-feeding communities on soft substrata (Ausich and Bottjer 1982, 1991; Bottjer and Ausich 1986; see also Sepkoski 1982 for lithological evidence) might thus be indicative of the absence of a motivation for food competition (unlimited resources) (cf. Valentine 1973).

On the other hand, Early Cambrian reefal communities were formed by relatively small, non-phytosymbiont-bearing, solitary or low modular forms that anchored in soft substrates (Wood et al. 1993; Kruse et al. 1995). Such soft-substrate reefal communities were probably prone to the "bulldozing" effect (sensu Thayer 1983) because they were not large, did not occur in dense populations, and were relatively short-lived, ephemeral settlements in which often a single species, characterized by rapid dispersal (e.g., by larval spats), dominated and produced almost homogeneous thickets, which unevenly occupied the sea floor. As a result, such communities closely resemble the pioneer communities of later epochs (sensu Ramenskiy 1971; Copper 1989).

2. Among the three great evolutionary faunas, differences in diversity can be related qualitatively to differences in basic ecological strategy. Bambach (1983, 1986) made an extensive study of life modes among the commonly fossilized constituents of the three faunas and argued that the amount of utilized ecospace increased with each. He used a simple classificatory system with three dimensions: trophic guild, life zone, and mode of mobility or attachment. He found that members of the Cambrian faunas occupied fewer than half of the categories in this system (mostly epifaunal guilds) (see also Burzin et al., this volume: tables 10.1 and 10.2). Members of the Paleozoic fauna occupied all previously utilized categories plus about 50 percent more,

and each succeeding evolutionary fauna was characterized by exploitation of more ecospace than was typical of the preceding fauna.

3. Change from a few ecologically widely distributed communities to a large number of communities with narrower ecological ranges occurred after the Cambrian (Sheehan 1991), indicating low community packing or low ecological differentiation of communities (see also Sepkoski 1988). A steady increase in number of communities from the Early to Late Cambrian is observed (Zhuravlev and Debrenne 1996). The Cambrian, especially Early Cambrian communities, occupied a restricted spectrum of conditions (see Burzin et al., this volume: figures 10.3 and 10.4).

4. Because disparity is a measure of the range of morphology in a given sample of organisms—as opposed to diversity, which expresses the number of taxa (Wills et al. 1994)—the diversity/disparity ratio may reflect the taxonomic diversity/ecological diversity ratio. A relatively low diversity/disparity ratio is established in the Cambrian for trilobites, other arthropods, priapulids, and echinoderms (Runnegar 1987; Foote 1992, 1993; Foote and Gould 1992; Wills et al. 1994; Wills 1998). The very fact that average morphological disparity remained constant for some groups since Cambrian time, or grew more slowly than diversity, suggests an increase in the density of species "packing" in the morphospace. The same probably follows from a uniquely high ratio of phyla, classes, and orders to families during this time interval (which are qualitative impressions of disparity as determined historically by taxonomists), which reflects a wide array of invertebrate body plans and subplans—a range of invertebrate types significantly broader than exists at present, despite the relative paucity of species then (Valentine et al. 1991).

Cambrian Extinctions

Reduction of the Tommotian Biota (orthothecimorph hyoliths, helcionelloids, anabaritids, siphogonuchitids, paracarinachitiids, and some minor problematic groups) is already observed at the beginning of Tommotian (see figure 8.1A). This has been attributed to nutrient depletion based on contrasting stratigraphic distributions of phosphatic skeletons and archaeocyaths, which are assumed to have been adapted to oligotrophic conditions (Brasier 1991). The first relative cessation of nutrient supply is reflected by a negative $^{87}Sr/^{86}Sr$ isotope shift and a decline in phosphorite abundance, which occurred during that time (figures 8.1E,F). (Nonetheless, the overall distribution of phosphatic skeletal fossils, which included lingulates, is basically congruent with that of archaeocyaths, with a maximum during the latest Atdabanian–earliest Botoman [Zhuravlev and Wood 1996: figure 2], and archaeocyath sponges can hardly be considered adapted to oligotrophic conditions by their ecological responses [Wood et al. 1992, 1993].) The increased rates of bioturbation in the late Tommotian (Droser and Bottjer 1988a,b) indicate that deposit feeders had to expand their field in a search for additional food sources.

The substantial decline in genera and families during the late Early Cambrian represents an extinction event, following the early Botoman maximum (see figure 8.1A). A similar sigmoidal pattern with the Botoman highstand and Toyonian lowstand is seen in the relative sea level curve plotted by Gravestock and Shergold (this volume: figure 6.2). Indirectly, this coincidence may confirm the conclusion of Zhuravlev and Wood (1996) that two mid–Early Cambrian extinction events can be related to global transgression-driven anoxia and subsequent regression. The first event occurred during the early Botoman (Sinsk event) and is marked by a significant reduction in diversity of almost all groups. The global distribution of varved black shales containing abundant monospecific acritarchs at low latitudes and pyritiferous green shales in temperate regions, as well as features confined to these facies communities, suggest that the Sinsk extinction event may have been caused by anoxia related to phytoplankton bloom and hypertrophy (Brasier 1995a; Zhuravlev and Wood 1996). The well-established early Toyonian Hawke Bay event could have been due to a regression and a restriction of shelf area (Palmer and James 1979; Zhuravlev 1986), as more evidence for a global regression is known from South Europe, Morocco, Laurentia, Baltica, Siberia, South China, and Australia (Seslavinsky and Maidanskaya, this volume). A causal link might exist between anoxia and later regression as C_{carb} production might have decreased during eutrophic times, whereas C_{org} burial rates might have simultaneously increased, as expressed by the middle Botoman positive $\delta^{13}C$ shift (Brasier et al. 1994: figure 1). Such a temporal increase in the export and burial rates of $C_{org} + C_{carb}$ for the biosphere and climate might have contributed to the promotion of a more effective removal of atmospheric CO_2, thereby exerting a negative feedback on climate warming (Föllmi et al. 1994).

The repeating pattern of biomere extinctions is superimposed on a broad Middle-Late Cambrian diversity plateau (see figure 8.1A). Various scenarios have been proposed to explain the mass extinctions of biomere type (see Hughes, this volume). The ultimate cause, however, might have been a global event, given that extinctions of this type occurred in Australia and China as well (Henderson 1976; Rowell and Brady 1976; A. Palmer 1982; Loch et al. 1993) and are pronounced on the overall generic diversity plot (see figure 8.1A). In addition, increased cladogenesis near the origin of new major groups might elevate rates of taxonomic pseudoextinctions by virtue of preponderance of paraphyletic taxa and produce the apparent pattern of a biomere extinction close to the beginning of the Ordovician radiation on the Sunwaptan-Datsonian boundary (Fortey 1989; Edgecombe 1992).

There are some general inconsistencies in the pure extrinsic explanations, including poorly elaborated physical-chemical models. Often similar environmental factors have been used to explain different biotic patterns. For instance, the Early Cambrian Biota was highly vulnerable to a common anoxia-regression couplet, a factor that has become nearly insignificant for the diverse Modern Evolutionary Fauna. The power of a killing mechanism sufficient to destroy the biota (mass extinction) depends, probably, on the stability (resilience and resistance) of the biota itself rather than on any external event that may only enhance or weaken an extinction. The strength of an external "kick" (extrinsic factor) needed for the destruction of a system is indicative of the stability of the system (Robertson 1993). Even if fluctuations of abiotic conditions do not exceed the limits of vulnerability for a community, the community might be disrupted as a result of the evolution of its own elements (Zherikhin 1987).

In order to understand the inherent dynamics of the biota, we have to use the indices selected above. The AGI (average geographic distribution index) and ALI (average longevity index) may be used as approximations of the degree of specialization, because specialists commonly are short-lived endemics (low AGI and ALI values) and generalists usually are widespread eurybionts (high AGI and ALI values) (Markov and Naimark 1995; Markov and Solov'ev 1995). The percentage of monotypic families per time unit (MTI) correlates inversely with fluctuations of AGI and ALI (see figures 8.2A–C). If the fluctuations of MTI values merely reflected subjective taxonomy, they would hardly display (1) a similar temporal pattern for different animal groups and (2) inverse correlation with AGI and ALI plots. Thus, fluctuations of MTI values may approximate the degree of competition in a biota, because the closer the phylogenetic relatives, the higher the probability of niche overlap that leads to competitive relationships (Naimark and Zhuravlev 1995). Together these fluctuations may serve as an approximation to the structure of the entire biota, which is expressed in the specialization and degree of competitive interaction. The values of these indices may indicate relative stability of the biota. We may expect that, with high specialization load (low AGI and ALI values and low MTI value), a biota would be unstable and prone to mass extinction, and vice versa.

Each interval preceding an extinction event (Botoman 1, Toyonian 1, Steptoean 3, Sunwaptan 3) was characterized by a similar fluctuation in indices: decreasing MTI, AGI, and ALI values (see figure 8.2). The severest (Botoman 2) extinction was preceded by the lowest MTI, AGI and ALI values for the entire Cambrian. Thus, accumulation of a certain nonadaptive load (increased degree of both specialization and competition) expressed by low MTI, AGI, and ALI values precedes the extinction event. In accordance, Stitt (1975) noted very short stratigraphic ranges for trilobites that composed preextinction biomere communities. Thus, extinctions, including biomere extinctions, were natural phenomena passed on to the existing biotic system, which was slightly destabilized. Such a system could be overturned with relative ease by an extrinsic trigger, such as an anoxia/transgression couplet followed by regression. Specialists (e.g., ajacicyathids) were affected, but generalists (e.g., archaeocyathids, lingulates, echinoderms) went through the crisis almost unchanged (Wood et al. 1992; Zhuravlev and Wood 1996). At the same time, the entire Middle–Late Cambrian Biota possessed a higher reserve of stability because it was more resilient and resistant than the Early Cambrian Biota: the biomere extinctions were not so severe as Early Cambrian extinctions, and communities restored quickly. Resilience was probably maintained by the replacement of the former community dominants by their close phylogenetic and trophic relatives, because each pioneer biomere community consisted of trilobites of similar appearance (Stitt 1975; Westrop 1989).

Progressive increase in biotic stability is reflected in reduction of turnover rates. Cambrian invertebrates appear to have higher average turnover rates at family (Sepkoski 1984) and generic (Raup and Boyajian 1988) levels than do later Phanerozoic invertebrates. Furthermore, Valentine et al. (1991) recognized a general trend among invertebrate families from fast-turnover taxa as dominants during the Cambrian to intermediate-turnover taxa that dominate during the post-Cambrian. At the generic level, this is expressed in an increase in average longevity: median generic longevity for Cambrian trilobites is 2.1 m.y., 6.3 m.y. for those of the Ordovician, and 10.6 m.y. for Paleozoic invertebrates as a whole (Foote 1988). The same pattern is displayed by more-detailed trilobite data, revealing an average genus duration increase from 0.9 to 3.0 m.y. for the Marjuman-Sunwaptan interval to 16.4 to 22.1 m.y. for the Tremadoc-Ashgill interval (Sloan 1991), and by steady ALI value increase through the Nemakit-Daldynian–Tremadoc (see figure 8.2C).

At any time when there was a local extinction and a more specialized member of the Paleozoic fauna happened to invade and repopulate, it might have been difficult for Cambrian species to regain their preempted share of resources (Sepkoski 1991b). For instance, rostroconch communities, which proliferated during the Datsonian, were never restored following the rise of burrowing pelecypods, which could better establish and maintain their position in the sediment and could burrow into a wider variety of substrates than could rostroconchs (Pojeta 1979; Runnegar 1979).

DISCUSSION: EARLY CAMBRIAN WORLD AND ANTHROPOGENIC LANDSCAPES

High nutrient supply due to both biotic factors (presence of highly cohesive plankters and absence of active filter feeders) and abiotic factors (enhanced erosion rates) might actually modify the character of ecospace occupation, much as occurs in anthropogenic landscapes. In anthropogenic landscapes, initial communities consist of generalists with high niche overlap and are unsaturated in species, leading to a weakening of biotic barriers with predators and competitors (Vakhrushev 1988). This destabilization is accompanied by a sharp increase in interspecific variability, similar to that observed in Cambrian organisms. Pronounced morphological plasticity, which may express high interspecific variability, is observed among pelecypods (Runnegar and Bentley 1983), tommotiids (Conway Morris and Fritz 1984), trilobites (McNamara 1986; Foote 1990; Hughes 1991), and lingulates (Ushatinskaya 1995). It has been interpreted either as arising from diffuse genetic control in the absence of an adequate regulatory mechanism, or as a result of reduced levels of competition. The array of findings on molecular bases of development, however, suggests that genome hypotheses are unlikely to explain the restriction of evolutionary novelties (Valentine 1995). On the contrary, the initial populations would be in the unusual position of occupying a competitive free ecospace that would allow not only a population explosion but also the survival of highly abnormal individuals.

The relative stability of modern anthropogenic communities is supported by an unlimited nutrient supply (provided by humanity) and by the weakness of biotic barriers. The Cambrian communities survived in conditions of a nutrient-rich ocean.

Thus, cessation of nutrient input, coupled with the ecological properties of the Early Cambrian Biota outlined above, would lead to destabilization, which would intensify severe competition because of high niche overlap.

CONCLUSION

Certain problems arise in attempts to understand Cambrian biotic events in terms of purely extrinsic forces. The major weakness of such hypotheses is in the explanation of unique events by nonunique extrinsic factors. There are two possibilities for solving this paradox: either none of the environmental factors is strong enough to affect a global biota, or the effect of an environmental factor influencing a global biota depends on the inherent features of the biota itself. In the opinion of Sepkoski (1994), if a hypothesized perturbation caused a specific extinction event, the perturbation ought to have produced other extinction events every time it occurred through geologic history. Thus, the cause of a biotic elimination has to be looked for in the ecological properties of the biota, in its structure, and in the cumulative pattern of evolutionary cycles of development of groups composing the biota (cf. Sepkoski 1989). From study of Cambrian biotic events, it is possible to conclude that the structure of the biota played the principal role in the apparent pattern and magnitude of radiations and extinctions. Although extrinsic, mainly environmental conditions were significant for rates of biotic development or elimination, the biota itself was responsible for creation of new environmental qualities. Bioturbation of sediment resulted in better aeration and also insertion of organic matter into sediment, which, in turn, allowed progressive subsurface colonization of sediment by a wider variety of organisms. Biomineralization allowed larger hard substrate surface areas, which are a limiting factor for many organisms. Filtration of ocean waters by filter feeders and suspension feeders, together with pelletization, radically changed the properties of the sediment and water habitats, and the rise of these groups in the Early Cambrian should have made ocean waters clearer and the photic zone deeper, providing additional opportunities for photosynthetic organisms to occupy lower levels of the water column, and more opportunities for further extension of adaptive space.

In accordance with the "principle of the essential diversity" of Ashby (1956), only a diversity of selection possibilities may minimize the diversity of outcomes. In other words, the progressive growth of biotic diversity increases biotic integration and, thereby, biotic stability. From this principle, it is not difficult to draw a conclusion concerning the basic features of biotic stability. Diversity is connected with the stability of a system through the duplication of intrasystem connections. "Narrow specialists," commonly being close relatives, represent such duplications (e.g., "biomere speciation"). A replacement of a single species has an insignificant effect on the entire community, because its former function continues to be provided by remaining duplicate species. As a result of such duplication, the total effectiveness of coenotic structure noticeably increases (Zherikhin 1987). The more duplications that exist, the less probable is breakage of the whole system. Because these duplications make it possible to pack the community more optimally (O'Neil et al. 1986), they are responsible for unequal probability of the two strategies and, hence, for the numerical predominance of small genera over larger ones that provides higher biotic stability.

Acknowledgments. This work was supported by a PalSIRP award and Russian Foundation for Basic Research, Project 00-04-484099. Kirill Es'kov, John J. Sepkoski, Jr., and Alan Smith are thanked for constructive comments, and Robert Riding for editing. This paper is a contribution to IGCP Project 366.

REFERENCES

Adrain, J. M., R. A. Fortey, and S. R. Westrop. 1998. Post-Cambrian trilobite diversity and evolutionary faunas. *Science* 280: 1922–1925.

Ashby, W. R. 1956. *An Introduction to Cybernetics.* London: Chapman and Hall.

Ausich, W. I. and D. J. Bottjer. 1982. Tiering in suspension-feeding communities on soft substrata throughout the Phanerozoic. *Science* 216:173–174.

Ausich, W. I. and D. J. Bottjer. 1991. History of tiering among suspension-feeders in the benthic marine ecosystem. *Journal of Geological Education* 39:313–319.

Axelrod, D. I. 1958. Early Cambrian marine fauna. *Science* 128:7–9.

Bambach, R. K. 1977. Species richness in marine benthic habitats through Phanerozoic. *Paleobiology* 3:152–167.

Bambach, R. K. 1983. Ecospace utilization and guilds in marine communities through the Phanerozoic. In M. J. S. Tevesz and P. L. McCall, eds., *Biotic Interactions in Recent and Fossil Benthic Communities,* pp. 719–746. New York: Plenum Press.

Bambach, R. K. 1986. Phanerozoic marine communities. In D. M. Raup and D. Jablonski, eds., *Patterns and Processes in the History of Life,* pp. 407–428. Berlin: Springer Verlag.

Bognibova, R. T. and A. P. Shcheglov. 1970. Osobennosti trilobitovykh soobshchestv na rubezhe rannego i srednego kembriya v Altae-Sayanskoy skladchatoy oblasti [Features of trilobite communities on the Early and Middle Cambrian boundary in the Altay-Sayan Foldbelt]. *Trudy, Sibirskiy nauchno-issledovatel'skiy institut geologii, geofiziki, i mineral'nogo syr'ya* 110:82–87.

Bogorov, V. G. 1974. *Plankton mirovogo okeana* [Plankton of the World Ocean]. Moscow: Nauka.

Bottjer, D. J. and W. I. Ausich. 1986. Phanerozoic development of tiering in soft substrate suspension feeding communities. *Paleobiology* 12:400–420.

Bottjer, D. J., J. K. Schubert, and M. L. Droser. 1996. Comparative evolutionary palaeoecology: Assessing the changing ecology of the past. In M. B. Hart, ed., *Biotic Recovery from Mass Extinction Events*, pp. 1–13. Geological Society Special Publication 102.

Boucot, A. J. 1983. Does evolution take place in an ecological vacuum? 2. *Journal of Paleontology* 56:1–30.

Bowring, S. A., J. P. Grotzinger, C. E. Isachsen, A. H. Knoll, S. M. Pelechaty, and P. Kolosov. 1993. Calibrating rates of early Cambrian evolution. *Science* 261:1293–1298.

Brasier, M. D. 1991. Nutrient flux and the evolutionary explosion across the Precambrian-Cambrian boundary interval. *Historical Biology* 5:85–93.

Brasier, M. D. 1995a. The basal Cambrian transition and Cambrian bio-events (from terminal Proterozoic extinctions to Cambrian biomeres). In O. H. Walliser, ed., *Global Events and Event Stratigraphy in the Phanerozoic*, pp. 113–118. Berlin: Springer Verlag.

Brasier, M. D. 1995b. Fossil indicators of nutrient levels. 1: Eutrophication and climate change. In D. W. Bosence and P. A. Allison, eds., *Marine Palaeoenvironmental Analysis from Fossils*, pp. 113–132. Geological Society Special Publication 83.

Brasier, M. D., R. M. Corfield, L. A. Derry, A. Yu. Rozanov, and A. Yu. Zhuravlev. 1994. Multiple $\delta^{13}C$ excursions spanning the Cambrian to the Botomian crisis in Siberia. *Geology* 22:455–458.

Burrett, C. F. and R. G. Richardson. 1978. Cambrian trilobite diversity related to cratonic flooding. *Nature* 272:717–719.

Burzin, M. B. 1994. Osnovnye tendentsii v istoricheskom razvitii fitoplanktona v pozdnem dokembrii i rannem kembrii [Principal trends in the historical development of the phytoplankton in the Late Precambrian and Early Cambrian]. In A. Yu. Rozanov and M. A. Semikhatov, eds., *Ekosistemnye perestroyki i evolyutsiya biosfery* [Ecosystem restructures and the evolution of biosphere], pp. 51–62. Moscow: Nedra.

Butterfield, N. J. 1994. Burgess Shale–type fossils from a Lower Cambrian shallow-shelf sequence in northwestern Canada. *Nature* 369:477–479.

Butterfield, N. J. 1997. Plankton ecology and the Proterozoic-Phanerozoic transition. *Paleobiology* 23:247–262.

Butterfield, N. J. and C. J. Nicholas. 1996. Burgess Shale–type preservation of both non-mineralizing and "shelly" Cambrian organisms from the Mackenzie Mountains, northwestern Canada. *Journal of Paleontology* 70:893–899.

Chen J.-Y. and C. Teichert. 1983. Cambrian Cephalopoda of China. *Palaeontographica A* 181:1–102.

Chen J.-Y., L. Ramsköld, and G.-Q. Zhou. 1994. Evidence for monophyly and arthropod affinity of Cambrian giant predators. *Science* 264:1304–1308.

Chuvashov, B. I. and R. Riding. 1984. Principal floras of Palaeozoic marine calcareous algae. *Palaeontology* 27:487–500.

Conway Morris, S. 1986. The community structure of the Middle Cambrian Phyllopod Bed (Burgess Shale). *Palaeontology* 29:423–467.

Conway Morris, S. 1987. The search for the Precambrian-Cambrian boundary. *Scientific American* 75:156–167.

Conway Morris, S. and S. Bengtson. 1994. Cambrian predators: Possible evidence from boreholes. *Journal of Paleontology* 68:1–23.

Conway Morris, S. and W. H. Fritz. 1984. *Lapworthella filigrana* n.sp. (incertae sedis) from the Lower Cambrian of the Cassiar Mountains, northern British Columbia, Canada, with comments on possible levels of competition in the early Cambrian. *Paläontologische Zeitschrift* 58:197–209.

Cook, P. J. 1992. Phosphogenesis around the Proterozoic-Phanerozoic transition. *Jour-*

nal of the Geological Society, London 149: 615–620.

Copper, P. 1989. Enigmas in Phanerozoic reefs development. Association of Australasian Palaeontologists, Memoir 8:371–385.

Crimes, T. P. 1992a. Changes in the trace fossil biota across the Proterozoic-Phanerozoic boundary. Journal of the Geological Society, London 149:637–646.

Crimes, T. P. 1992b. The record of trace fossils across the Proterozoic-Cambrian boundary. In J. H. Lipps and P. W. Signor, eds., Origins and Early Evolution of the Metazoa, pp. 177–202. New York: Plenum Press.

Crimes, T. P. 1994. The period of early evolutionary failure and the dawn of evolutionary success: The record of biotic changes across the Precambrian-Cambrian boundary. In S. K. Donovan, ed., The Palaeobiology of Trace Fossils, pp. 105–133. London: John Wiley and Sons.

Crimes, T. P. and M. A. Fedonkin. 1994. Evolution and dispersal of deepsea traces. Palaios 9:74–83.

Debrenne, F. and A. Yu. Zhuravlev. 1997. Cambrian food web: A brief review. Géobios, Mémoir spécial 20:181–188.

Debrenne, F., I. D. Maidanskaya, and A. Yu Zhuravlev. 1999. Faunal migrations of archaeocyaths and Early Cambrian plate dynamics. Bulletin de la Société géologique de France 170:189–194.

Derry, L. A., M. D. Brasier, R. M. Corfield, A. Yu. Rozanov, and A. Yu. Zhuravlev. 1994. Sr and C isotopes in Lower Cambrian carbonates from the Siberian craton: A paleoenvironmental record during the "Cambrian explosion." Earth and Planetary Science Letters 128:671–681.

Donnelly, T. H., J. H. Shergold, P. N. Southgate, and C. J. Barnes. 1990. Events leading to global phosphogenesis around the Proterozoic/Cambrian boundary. In A. J. S. Notholt and I. Jarvis, eds., Phosphorite Research and Development, pp. 273–287. Geological Society Special Publication 52.

Droser, M. L. 1991. Ichnofossils of the Paleozoic Skolithos ichnofacies and the nature and distribution of Skolithos piperock. Palaios 6:316–325.

Droser, M. L. and D. J. Bottjer. 1988a. Trends in depth and extent of bioturbation in Cambrian carbonate marine environments, western United States. Geology 16:233–236.

Droser, M. L. and D. J. Bottjer. 1988b. Trends in extent and depth of early Paleozoic bioturbation in Great Basin (California, Nevada, and Utah). In D. L. Weide and M. L. Faber, eds., This Extended Land: Geological Journeys in the Southern Basin and Range, Field Trip Guidebook, pp. 123–135. Geological Society of America, Cordilleran Section.

Edgecombe, G. D. 1992. Trilobite phylogeny of the Cambrian-Ordovician "event": Cladistic reappraisal. In M. J. Novacek and Q. D. Wheeler, eds., Extinction and Phylogeny, pp. 144–177. New York: Columbia University Press.

Erwin, D. H. 1992. A preliminary classification of evolutionary radiations. Historical Biology 6:133–147.

Erwin, D. H. 1994. Early introduction of major morphological innovations. Acta Palaeontologica Polonica 38:281–294.

Erwin, D. H., Valentine, J. W. and J. J. Sepkoski, Jr. 1987. A comparative study of diversification events: The early Paleozoic versus the Mesozoic. Evolution 41:1177–1186.

Evans, J. S. 1912. The sudden appearance of the Cambrian fauna. 11th International Geological Congress, Stockholm 1912, Compte Rendu 1:543–546.

Fagerstrom, J. A. 1987. *The Evolution of Reef Communities*. New York: John Wiley and Sons.

Föllmi, K. B., H. Weissert, M. Bisping, and H. Funk. 1994. Phosphogenesis, carbon-isotope stratigraphy, and carbonate-platform evolution along the Lower Cretaceous northern Tethyan margin. *Geological Society of America Bulletin* 106:729–746.

Foote, M. 1988. Survivorship analysis of Cambrian and Ordovician trilobites. *Paleobiology* 14:258–271.

Foote, M. 1990. Nearest-neighbor analysis of trilobite morphospace. *Systematic Zoology* 39:371–382.

Foote, M. 1992. Paleozoic record of morphological diversity in blastozoan echinoderms. *Proceedings of the National Academy of Sciences, USA* 89:7325–9.

Foote, M. 1993. Contributions of individual taxa to overall morphological disparity. *Paleobiology* 19:403–419.

Foote, M. and S. J. Gould. 1992. Cambrian and Recent morphological disparity. *Science* 258:1816.

Fortey, R. A. 1983. Cambrian-Ordovician trilobites from the boundary beds in western Newfoundland and their phylogenetic significance. *Special Papers in Palaeontology* 30:179–211.

Fortey, R. A. 1985. Pelagic trilobites as an example of deducing the life habits of extinct arthropods. *Transactions of the Royal Society of Edinburgh (Earth Sciences)* 76:219–230.

Fortey, R. A. 1989. There are extinctions and extinctions: Examples from the Lower Palaeozoic. *Philosophical Transactions of the Royal Society of London B* 325:327–355.

Fortey, R. A. and R. M. Owens. 1987. The Arenig Series in South Wales. *British Museum (National History) Bulletin* 41:69–307.

Fortey, R. A. and R. M. Owens. 1990. Evolutionary radiations in the Trilobita. In P. D. Taylor and G. P. Larwood, eds., *Major Evolutionary Radiations*, pp. 139–164. Systematics Association Special Volume 42. Oxford: Clarendon Press.

Glaessner, M. F. 1984. *The Dawn of Animal Life: A Biohistorical Study*. Cambridge: Cambridge University Press.

Gorjansky, V. Yu. and L. E. Popov. 1986. On the origin and systematic position of the calcareous shelled inarticulate brachiopods. *Lethaia* 19:233–240.

Hallock, P. 1987. Fluctuations in the trophic resource continuum: A factor in global diversity cycles? *Paleoceanography* 2:457–471.

Henderson, R. A. 1976. Upper Cambrian (Idamean) trilobites from western Queensland, Australia. *Palaeontology* 19:325–364.

Hughes, N. C. 1991. Morphological plasticity and genetic flexibility in a Cambrian trilobite. *Geology* 19:913–916.

Hutchinson, G. E. 1961. The biologist poses some problems. In M. Sears, ed., *Oceanography*, pp. 85–94. American Association for the Advancement of Science Publications 67.

Jell, P. A. 1974. Faunal provinces and possible planetary reconstruction of the Middle Cambrian. *Journal of Geology* 82:319–352.

Kalandadze, N. N. and A. S. Rautian. 1993. Simptomatika ekologicheskikh krizisov [Symptomatics of ecological crises]. *Stratigrafiya, Geologicheskaya korrelyatsiya* 1:3–8.

Karataev, A. Yu. and L. E. Burlakova. 1995. Rol' dreysseny v ozernykh ekosistemakh [Role of *Dreissena* in lacustrine ecosystems]. *Ekologiya* 3:232–236.

Kempe, S. and J. Kazmierczak. 1994. The role of alkalinity in the evolution of ocean chemistry, organization of living systems, and biocalcification processes. *Bulletin de l'Institut océanographique, Monaco, no. spécial* 13:61–117.

Knoll, A. H. 1994. Proterozoic and Early Cambrian protists: Evidence for accelerating evolutionary tempo. *Proceedings of the National Academy of Sciences, USA* 91: 6743–6750.

Kruse P. D., A. Yu. Zhuravlev, and N. P. James. 1995. Primordial metazoan-calcimicrobial reefs: Tommotian (Early Cambrian) of the Siberian Platform. *Palaios* 10:291–321.

LaBarbera, N. 1984. Feeding currents and particle capture mechanisms in suspension feeding animals. *American Zoologist* 24: 71–84.

Landing, E. and S. R. Westrop. 1997. Cambrian faunal sequence and depositional history of Avalonian Newfoundland and New Brunswick: Field workshop. *New York State Museum Bulletin* 492:5–75.

Leigh, E. G., Jr. 1990. Community diversity and environmental stability: A reexamination. *Trends in Ecology and Evolution* 5: 340–344.

Lipps, J. H., S. Bengtson, and M. A. Fedonkin. 1992. Ecology and biogeography. In J. W. Schopf and C. Klein, eds., *The Proterozoic Biosphere: A Multidisciplinary Study*, pp. 437–441. Cambridge: Cambridge University Press.

Loch, J. D., J. H. Stitt, and J. R. Derby. 1993. Cambrian-Ordovician boundary interval extinctions: Implications of revised trilobite and brachiopod data from Mount Wilson, Alberta, Canada. *Journal of Paleontology* 67:497–517.

Logan, G. A., J. M. Hayes, G. B. Hieshima, and R. E. Summons. 1995. Terminal Proterozoic reorganization of biogeochemical cycles. *Nature* 376:53–56.

Marin, F., M. Smith, Y. Isa, G. Muyzer, and P. Westbroek. 1996. Skeletal matrices, muci, and the origin of invertebrate calcification. *Proceedings of the National Academy of Sciences, USA* 93:1554–1559.

Markov, A. V. and E. B. Naimark. 1995. Vzaimosvyaz' urovnya raznoobraziya starshikh taksonov so stepen'yu spetsializirovannosti vidov i rodov (na primere nekotorykh grupp paleozoyskikh bespozvonochnykh) [Connections of the level of diversity of higher taxa with the degree of specialization of genera and species (illustrated by certain groups of Paleozoic invertebrates)]. *Zhurnal Obshchey Biologii* 56:97–107.

Markov, A. V. and A. N. Solov'ev. 1995. Kolichestvennyy analiz makroevolyutsionnykh protsessov u morskikh ezhey nadotryada Spatangacea (Quantitative analysis of the macroevolutionary processes in sea urchins of the superorder Spatangacea). In A. Yu. Rozanov and M. A. Semikhatov, eds., *Ekosistemnye perestroyki i evolyutsiaya biosfery*, vypusk 2 [Ecosystem restructures and the evolution of biosphere, issue 2], pp. 88–94. Moscow: PIN RAN.

McBride, D. J. 1976. Outer shelf communities and trophic groups in the Upper Cambrian of Great Basin. *Brigham Young University Geological Studies* 23:139–152.

McNamara, K. J. 1986. The role of heterochrony in the evolution of Cambrian trilobites. *Biological Reviews* 61:121–156.

Moldowan, J. M., J. Dahl, S. R. Jacobson, B. J. Huizinga, F. J. Fago, R. Shetty, D. S. Watt, and K. E. Peters. 1996. Chemostratigraphic reconstruction of biofacies: Molecular evidence linking cyst-forming dinoflagellates with pre-Triassic ancestors. *Geology* 24: 159–162.

Montañez, I. P., J. L. Banner, A. David, and J. K. Bosserman. 1996. Cambrian carbonates of the southern Great Basin: A record of physical isotopic paleo-oceanographic secular variation. In *Carbonates and Global*

Change: An Interdisciplinary Approach, pp. 106–107. SEPM/IAS Research Conference, 22–27 June 1996, Wildhaus, Switzerland, Abstract Book.

Müller, K. J. and D. Walossek. 1985. A remarkable arthropod fauna from the Upper Cambrian "orsten" of Sweden. *Transactions of the Royal Society of Edinburgh (Earth Sciences)* 76:161–172.

Naimark, E. B. and A. Yu. Zhuravlev. 1995. Level of complexity of the Early Cambrian Biota: Constraints from the monotypic taxa index. In *Sardinia 95, Sixth Paleobenthos Symposium, 25–31 October 1995, Abstracts Book*, pp. 42–43.

Newell, N. D. 1972. The evolution of reefs. *Scientific American* 226:54–65.

Nicholas, C. J. 1996. The Sr isotopic evolution of the oceans during the "Cambrian Explosion." *Journal of the Geological Society, London* 153:243–254.

O'Neil, R. V., D. L. DeAngelis, J. B. Waide, and T. F. H. Allen. 1986. *A Hierarchical Concept of Ecosystems*. Princeton: Princeton University Press.

Palmer, A. R. 1965. Biomere—a new kind of biostratigraphic unit. *Journal of Paleontology* 39:149–153.

Palmer, A. R. 1979. Biomere boundaries reexamined. *Alcheringa* 3:33–41.

Palmer, A. R. 1982. Biomere boundaries: A possible test for extraterrestrial perturbation of the biosphere. *Geological Society of America, Special Paper* 190:469–476.

Palmer, A. and N. P. James. 1979. The Hawke Bay event: A circum-Iapetus regression near the lower Middle Cambrian boundary. In D. R. Wones, ed., *The Caledonides in the USA*, pp. 15–18. IGCP Project 27, Caledonide Orogen. Blacksburg, Va.: Department of Geological Sciences, Virginia Polytechnic Institute and State University.

Palmer, T. J. 1982. Cambrian to Cretaceous changes in hardground communities. *Lethaia* 15:309–323.

Peel, J. S. 1991. Functional morphology, evolution, and systematics of early Palaeozoic univalved molluscs. *Grønlands Geologiske Undersøgese, Bulletin* 161:1–116.

Pojeta, J., Jr. 1979. Geographic distribution of Cambrian and Ordovician rostroconch molluscs. In J. Gray and A. J. Boucot, eds., *Historical Biogeography, Plate Tectonics, and the Changing Environment*, pp. 27–36. Eugene, Oreg.: Oregon State University Press.

Ramenskiy, L. G. 1971. *Izbrannye raboty: Problemy i metody izucheniya rastitel'nogo pokrova* [Selected papers: Problems and methods in the study of vegetation]. Leningrad: Nauka.

Raup, D. M. and G. E. Boyajian. 1988. Patterns of generic extinction in the fossil record. *Paleobiology* 14:109–125.

Rigby, S. and C. Milsom. 1996. Benthic origins of zooplankton: An environmentally determined microevolutionary effect. *Geology* 24:52–54.

Robertson, D. S. 1993. Mass extinction and the simulated annealing algorithm. *Evolutionary Theory* 10:203–207.

Rowell, A. J. and M. J. Brady. 1976. Brachiopods and biomeres. *Brigham Young University Geological Studies* 23:165–180.

Rowland, S. M. and R. A. Gangloff. 1988. Structure and paleoecology of Lower Cambrian reefs. *Palaios* 3:111–135.

Rozanov, A. Yu. 1992. Some problems concerning the Precambrian-Cambrian boundary and the Cambrian fauna radiation. *Journal of the Geological Society, London* 149:593–598.

Runnegar, B. 1979. Origin and evolution of the class Rostroconchia. *Philosophical Transactions of the Royal Society of London B* 284:319–333.

Runnegar, B. 1987. Rates and modes of evolution in the Mollusca. In K. S. W. Campbell and M. F. Day, eds., *Rates of Evolution*, pp. 39–60. London: Allen and Unwin.

Runnegar, B. and C. Bentley. 1983. Anatomy, ecology, and affinities of the Australian early Cambrian bivalve *Pojetaia runnegari* Jell. *Journal of Paleontology* 57:73–92.

Saltzman, M. R., J. P. Davidson, P. Holden, B. Runnegar, and K. C. Lohmann. 1995. Sea-level–driven changes in ocean chemistry at an Upper Cambrian extinction horizon. *Geology* 23:893–896.

Savarese, M. 1995. Functional significance of regular archaeocyathan central cavity diameter: A biomechanical and paleoecological test. *Paleobiology* 21:356–378.

Sepkoski, J. J., Jr. 1979. A kinetic model of Phanerozoic taxonomic diversity. 2: Early Paleozoic families and multiple equilibria. *Paleobiology* 5:222–252.

Sepkoski, J. J., Jr. 1981. The uniqueness of the Cambrian fauna. In M. E. Taylor, ed., *Short Papers for the Second International Symposium on the Cambrian System, 1981*, pp. 203–207. U.S. Department of the Interior Geological Survey Open-File Report 81-743.

Sepkoski, J. J., Jr. 1982. Flat-pebble conglomerates, storm deposits, and the Cambrian bottom fauna. In G. Einsele and A. Seilacher, eds., *Cyclic and Event Stratification*, pp. 371–385. Berlin: Springer Verlag.

Sepkoski, J. J., Jr. 1984. A kinetic model of Phanerozoic taxonomic diversity. 3: Post-Paleozoic families and mass extinctions. *Paleobiology* 10:246–267.

Sepkoski, J. J., Jr. 1988. Alpha, beta, and gamma: Where does all the diversity go? *Paleobiology* 14:221–234.

Sepkoski, J. J., Jr. 1989. Periodicity in extinction and the problem catastrophism in the history of life. *Journal of the Geological Society, London* 146:7–19.

Sepkoski, J. J., Jr. 1991a. Diversity in the Phanerozoic oceans: A partisan review. In E. C. Dudley, ed., *The Unity of Evolutionary Biology*, pp. 210–236. Proceedings of ICSEB IV, 1. Portland, Oreg.: Dioscorides Press.

Sepkoski, J. J., Jr. 1991b. A model of onshore-offshore change in faunal diversity. *Paleobiology* 17:58–77.

Sepkoski, J. J., Jr. 1992. Proterozoic–Early Cambrian diversification of metazoans and metaphytes. In J. W. Schopf and C. Klein, eds., *The Proterozoic Biosphere: A Multidisciplinary Study*, pp. 553–561. Cambridge: Cambridge University Press.

Sepkoski, J. J., Jr. 1994. Extinction and the fossil record. *Geotimes* (March): 15–17.

Sepkoski, J. J., Jr. 1995. The Ordovician radiations: Diversification and extinction shown by global genus-level taxonomic data. In J. D. Cooper, M. L. Droser, and S. C. Finney, eds., *Ordovician Odyssey: Short Papers for the Seventh International Symposium on the Ordovician System (Las Vegas, Nevada, USA, June 1995)*, pp. 393–396. Fullerton, Calif.: Pacific Section Society for Sedimentary Geology.

Sepkoski, J. J., Jr., and D. C. Kendrick. 1993. Numerical experiments with model monophyletic and paraphyletic taxa. *Paleobiology* 19:168–184.

Sepkoski, J. J., Jr., and P. M. Sheehan. 1983. Diversification, faunal change, and community replacement during the Ordovician radiation. In M. J. S. Tevesz and P. L. McCall, eds., *Biotic Interactions in Recent and Fossil Benthic Communities*, pp. 673–717. New York: Plenum Press.

Sheehan, P. M. 1991. Patterns of synecology during the Phanerozoic. In E. C. Dudley, ed., *The Unity of Evolutionary Biology*, pp. 103–118. Proceedings of ICSEB IV, 1. Portland, Oreg.: Dioscorides Press.

Shergold, J. H. 1988. Review of trilobite biofacies distributions at the Cambrian-Ordovician boundary. *Geological Magazine* 125:363–380.

Shergold, J. H. 1995. *Timescales. 1: Cambrian.* Australian Geological Survey Organisation Record 1995/30.

Signor, P. W. 1992a. Taxonomic diversity and faunal turnover in the Early Cambrian: Did the most severe mass extinction of the Phanerozoic occur in the Botomian stage? In *Fifth North American Paleontological Convention, Abstracts with Programs,* p. 272.

Signor, P. W. 1992b. Evolutionary and tectonic implications of Early Cambrian faunal endemism. In C. A. Hall, Jr., V. Doyle-Jones, and B. Widawski, eds., *The History of Water: Eastern Sierra Nevada, Owens Valley, White-Inyo Mountains,* pp. 1–13. White Mountain Research Station Symposium, vol. 4. Los Angeles: Regents of the University of California.

Signor, P. W. and J. H. Lipps. 1992. Origin and early radiation of the Metazoa. In J. H. Lipps and P. W. Signor, eds., *Origins and Early Evolution of the Metazoa,* pp. 3–23. New York: Plenum Press.

Signor, P. W. and G. J. Vermeij. 1994. The plankton and the benthos: Origins and early history of an evolving relationship. *Paleobiology* 20:297–319.

Sloan, R. E. 1991. A chronology of North American Ordovician trilobite genera. *Geological Survey of Canada Paper* 90-9:165–177.

Smith, A. B. 1988. Patterns of diversification and extinction in early Palaeozoic echinoderms. *Palaeontology* 32:799–828.

Smith, A. B. 1990. Evolutionary diversification of echinoderms during the early Palaeozoic. In P. D. Taylor and G. P. Larwood, eds., *Major Evolutionary Radiations,* pp. 265–286. Oxford: Clarendon Press.

Stanley, S. M. 1973. An ecological theory for the sudden origin of multicellular life in the Late Precambrian. *Proceedings of the National Academy of Sciences, USA* 70:1486–1489.

Stitt, J. H. 1971. Repeating evolutionary pattern in Late Cambrian trilobite biomere. *Journal of Paleontology* 45:178–181.

Stitt, J. H. 1975. Adaptive radiation, trilobite paleoecology, and extinction, Ptychaspidid Biomere, Late Cambrian of Oklahoma. *Fossils and Strata* 4:381–390.

Surge, D. M., M. Savarese, J. D. Dodd, and K. C. Lohmann. 1997. Carbon isotopic evidence for photosynthesis in Early Cambrian oceans. *Geology* 25:503–506.

Thayer, C. W. 1983. Sediment-mediated biological disturbance and the evolution of marine benthos. In M. J. S. Tevesz and P. L. McCall, eds., *Biotic Interactions in Recent and Fossil Benthic Communities,* pp. 479–625. New York: Plenum Press.

Underwood, C. J. 1993. The position of graptolites within Lower Palaeozoic planktic ecosystems. *Lethaia* 26:189–202.

Ushatinskaya, G. T. 1995. Drevneyshie lingulyaty [Oldest lingulates]. *Trudy, Paleontologicheskiy institut, Akademiya nauk SSSR* 262:1–91.

Vakhrushev, A. A. 1988. Nachal'nye etapy formirovaniya soobshchestv na primere sinantropizatsii ptits [Initial stages of the community formation illustrated by synanthropic birds]. In V. A. Krassilov, ed., *Evolyutsionnye issledovaniya: Vavilovskie temy* [Evolutionary studies: Vavilov's themes], pp. 34–46. Vladivostok: Institute of Biology and Pedology, Far East Branch, USSR Academy of Sciences.

Valentine, J. W. 1973. *Evolutionary Paleoecology of the Marine Biosphere.* Englewood Cliffs, N.J.: Prentice-Hall.

Valentine, J. W. 1995. Why no new phyla after the Cambrian? Genome and ecospace hypotheses revisited. *Palaios* 10:190–194.

Valentine, J. W., S. M. Awramik, P. W. Signor, and P. M. Sadler. 1991. The biological explosion at the Precambrian-Cambrian boundary. *Evolutionary Biology* 25:279–356.

Vermeij, G. J. 1987. *Evolution and Escalation: An Ecologic History of Life.* Princeton, N.J.: Princeton University Press.

Vermeij, G. J. 1990. The origin of skeletons. *Palaios* 4:585–589.

Vermeij, G. J. 1995. Economics, volcanoes, and Phanerozoic revolutions. *Paleobiology* 21:125–152.

Vidal, G. and M. Moczydłowska-Vidal. 1997. Biodiversity, speciation, and extinction trends of Proterozoic and Cambrian phytoplankton. *Paleobiology* 23:230–246.

Westrop, S. R. 1989. Trilobite diversity patterns in an Upper Cambrian stage. *Paleobiology* 14:401–409.

Wills, M. A. 1998. Cambrian and Recent disparity: The picture from priapulids. *Paleobiology* 24:177–199.

Wills, M. A., D. E. G. Briggs, and R. A. Fortey. 1994. Disparity as an evolutionary index: A comparison of Cambrian and Recent arthropods. *Paleobiology* 20:93–130.

Wood, R. 1993. Nutrients, predation, and the history of reefs. *Palaios* 8:526–543.

Wood, R. 1995. The changing biology of reef-building. *Palaios* 10:517–529.

Wood, R., A. Yu. Zhuravlev, and F. Debrenne. 1992. Functional biology and ecology of Archaeocyatha. *Palaios* 7:131–156.

Wood, R., A. Yu. Zhuravlev, and A. Chimed Tseren. 1993. The ecology of Lower Cambrian buildups from Zuune Arts, Mongolia: Implications for early metazoan reef evolution. *Sedimentology* 40:829–858.

Zhegallo, E. A., A. G. Zamiraylova, Yu. N. Zanin, and G. M. Pisareva. 1996. Akritarkhi v nizhnekembriyskikh goryuchikh slantsakh kuonamskoy svity Sibirskoy platformy (r. Molodo) [Acritarchs in the Lower Cambrian Kuonamka Formation combustible shales of the Siberian Platform (Molodo River)]. *Doklady Akademii nauk* 347:69–71.

Zherikhin, V. V. 1987. Biotsenoticheskaya regulyatsiya evolyutsii [Biocoenotic regulation of evolution]. *Paleontologicheskiy zhurnal* 1987 (1):3–11.

Zhuravlev, A. Yu. 1986. Evolution of archaeocyaths and palaeobiogeography of the Early Cambrian. *Geological Magazine* 123:377–385.

Zhuravlev, A. Yu. 1995. Preliminary suggestions on the global Early Cambrian zonation. *Beringeria Special Issue* 2:147–160.

Zhuravlev, A. Yu. 1996. Reef ecosystem recovery after the Early Cambrian extinction. In M. B. Hart, ed., *Biotic Recovery from Mass Extinction Events*, pp. 79–96. Geological Society Special Publication 102.

Zhuravlev, A. Yu. and F. Debrenne. 1996. Pattern of evolution of Cambrian benthic communities: Environments of carbonate sedimentation. *Rivista Italiana di Paleontologia e Stratigrafia* 102:333–340.

Zhuravlev, A. Yu. and R. Wood. 1995. Lower Cambrian reefal cryptic communities. *Palaeontology* 38:443–470.

Zhuravlev, A. Yu. and R. A. Wood. 1996. Anoxia as the cause of the mid-Early Cambrian (Botomian) extinction event. *Geology* 24:311–314.

Zhuravlev, A. Yu., B. Hamdi, and P. D. Kruse. 1996. Ecological aspects of Cambrian radiation—Field meeting. *Episodes* 19:136–137. IGCP Project 366.

CHAPTER NINE

Nicholas J. Butterfield

Ecology and Evolution of Cambrian Plankton

Probable eukaryotic phytoplankton first appear in the fossil record in the Paleopro-terozoic but undergo almost no morphologic change until the Early Cambrian. The radiation of diverse acanthomorphic phytoplankton in exact parallel with the Cambrian explosion of large animals points to an ecologic linkage, probably effected by the introduction of small herbivorous metazoans into the plankton. By establishing the second tier of the Eltonian pyramid in the marine plankton, such mesozooplankton might be considered a proximal and ecologic cause of the Cambrian explosion.

THE PLANKTON COMPRISES the majority of all modern marine biomass and metabolism, is the ultimate source of most exported carbon, and plays an essential role at the base of most marine ecosystems (Nienhuis 1981; Berger et al. 1989). Thus, it is hardly surprising to find it figuring in broad-scale considerations of Early Cambrian ecology (e.g., Burzin 1994; Signor and Vermeij 1994; Butterfield 1997), biogeochemical cycling (e.g., Logan et al. 1995), and evolutionary tempo and mode (e.g., Knoll 1994; Rigby and Milsom 1996). The Cambrian is of course of particular interest in that it constitutes one side of the infamous Precambrian-Cambrian boundary, the preeminent shift in ecosystem structure of the last 4 billion years. The question is, what role, if any (cf. Signor and Vermeij 1994), did the plankton play in the Cambrian explosion of large animals? The answer entails a critical analysis of the fossil record, combined with a consideration of indirect lines of evidence and a general examination of plankton ecology and how it relates to large-animal metabolism. There is in fact a good case to be made that developments in the plankton gave rise to both the evolutionary and the biogeochemical perturbations that characterize the Proterozoic-Phanerozoic transition.

THE FOSSIL RECORD AND AN ECOLOGIC HYPOTHESIS

The fossil record of Proterozoic-Cambrian protists has been most recently reviewed in detail by Knoll (1992, 1994). Simple, small to moderately sized spheromorphic acri-

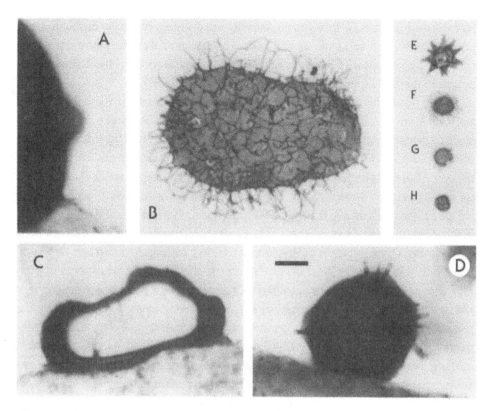

Figure 9.1 Neoproterozoic and Lower Cambrian acritarchs; all except *A* are figured at the same scale. Neoproterozoic examples include silicified *Trachyhystrichosphaera* (*A, D*) and *Cymatiosphaeroides* (*C*) from the ca. 750 Ma Svanbergfjellet Formation, Spitsbergen; an unnamed form from the ca. 850 Ma Wynniatt Formation, Victoria Island, Canada (*B*); and a leiosphaerid from the ca. 1250 Ma Agu Bay Formation, Baffin Island, Canada (*H*). Lower Cambrian forms include an unidentified acanthomorph from the Mural Formation, Alberta, Canada (*E*), and species of *Skiagia* from the Tokammane Formation, Spitsbergen (*F,G*). *A–D* are inferred to have had a benthic habit, because of their large size and/or obvious attachment to the sediment; note the thin sheath connecting the vesicle and substrate in *A*. *E–H* are inferred to have been planktic. Scale bar in *D* equals 13 μm for *A* and 50 μm for *B–H*.

tarchs (leiosphaerids) first appeared in the Paleoproterozoic around 1800 Ma and remained the predominant constituent of shale-hosted microfossil assemblages for the rest of the Proterozoic (figure 9.1H). Acritarch diversity began to rise in the late Mesoproterozoic and accelerated substantially through the Neoproterozoic with the introduction of various ornamented and acanthomorphic acritarchs (figures 9.1A–D), vaseshaped microfossils, and "scale" microfossils reminiscent of certain chrysophyte or prymnesiophyte algae (Allison and Hilgert 1986; Kaufman et al. 1992). This same interval also witnessed a marked increase in the size and diversity of spheromorphic acritarchs (Mendelson and Schopf 1992: figure 5.5.12), and the first appearance of identifiable seaweeds (Hermann 1981; Butterfield et al. 1990, 1994). Following a major extinction/disappearance during the Varanger ice age, acanthomorphic acritarchs recovered to reach their Proterozoic diversity maximum, only to be decimated in a terminal Neoproterozoic extinction. Against a background of extinction-resistant leiosphaerids, a new class of small, rapidly diversifying acanthomorphic acritarchs appeared in the Early Cambrian (figures 9.1E–G) (Knoll 1994).

At first glance there appears to be considerable evolutionary activity in the Proterozoic plankton. It is important to realize, however, that the acritarchs are an entirely artificial group united only by their organic constitution and indeterminate taxonomic affiliation. Although there is a good case for identifying most Paleozoic acritarchs as the cysts of unicellular phytoplankton, such broad-brush categorization does not hold for the Proterozoic. Notably, most of the increases in Proterozoic acritarch diversity collated by Mendelson and Schopf (1992) and Knoll (1994) are contributed by forms that are exceptionally large relative to their Paleozoic counterparts (several hundreds or thousands of micrometers versus several tens of micrometers diameter; Knoll and Butterfield 1989) (figure 9.1). Given the inverse exponential relationships of both buoyancy and nutrient absorption with cell size, such forms are unlikely to have been planktic (Kiørboe 1993; Butterfield 1997). Such a conclusion is supported by the general restriction of these large acritarchs to conspicuously shallow-water environments (Butterfield and Chandler 1992) and/or a commonly clustered arrangement on bedding planes (e.g., *Chuaria-Tawuia* assemblages; Butterfield 1997). A benthic interpretation is unambiguous in instances where there is direct evidence of attachment to sediment surfaces; e.g., the common Late Riphean taxa *Trachyhystrichosphaera* (figures 9.1A,D) and *Cymatiosphaeroides* (figure 9.1C) (Butterfield et al. 1994).

The record of Proterozoic-Cambrian plankton thus differs markedly from that of acritarchs or protists as a whole: leiosphaerid plankters first appear in the Paleoproterozoic and persist more or less unchanged for 1300+ million years. Then, near the base of the Tommotian, and in remarkable parallel with the Cambrian explosion of large organisms, a whole range of complex new forms are introduced, and the rate of evolutionary turnover increases by perhaps two orders of magnitude (cf. Knoll 1994; Zhuravlev, this volume: figures 8.1A,C). Certainly there was an earlier "big bang of eukaryotic evolution" in the Neoproterozoic (Knoll 1992), but the exceptionally large acritarchs, seaweeds, tawuiids, and Ediacara-type metazoans that defined it were predominantly, if not entirely, benthic. The plankton appears to have remained profoundly monotonous until the Early Cambrian.

The coincidence of the first important shift in plankton evolution with the Cambrian explosion of large animals points compellingly to a causal connection. Most "large" animals, however, do not operate at a microscopic or unicellular level. In modern aquatic ecosystems, the primary productivity of unicellular phytoplankton is generally transmitted to large animals via small grazing planktic animals, the mesozooplankton (e.g., small crustaceans such as copepods and cladocerans). The size of organisms increases incrementally along this food chain simply because most planktic heterotrophs are whole-organism ingesters and typically larger than their prey.

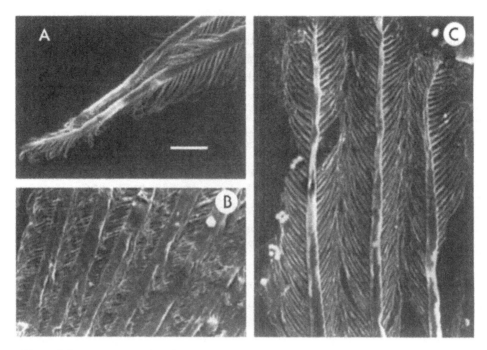

Figure 9.2 SEM micrographs of disarticulated filter-feeding mesozooplankton (cladoceran-type branchiopods) from the Lower Cambrian (ca. Botoman) Mount Cap Formation, western Northwest Territory, Canada. Scale bar in *A* equals 14 μm for *A*, 10 μm for *B*, and 8 μm for *C*.

Given that the transfer efficiency between trophic levels is only about 10% (Pauly and Christensen 1995), it is clear that the pathway between phytoplanktic primary production and larger metazoans must be short and direct (in this context it is important to recognize that optimum predator:prey size ratio is low for *micro*zooplankton [1:1 to 3:1 for flagellates and 8:1 for ciliates] but high for *meso*zooplankton [18:1 for rotifers and copepods and about 50:1 for cladocerans and meroplanktic larvae [Hansen et al. 1994]). The ability to convert microscopic particles to macroscopic ones rapidly (i.e., in one step) places the mesozooplankton in a key position with respect to large-animal marine ecology.

No mesozooplankton have been recognized among Proterozoic fossils, and in the absence of obvious macrozooplankton or nekton at this time, this is perhaps not unexpected. In the Cambrian, however, there are two occurrences of millimeter-sized branchiopod crustaceans, one in the Upper Cambrian orsten deposits of Sweden (Walossek 1993), and the other in the Lower Cambrian (ca. Botoman) Mount Cap Formation of northwestern Canada (Butterfield 1994) (figure 9.2). Both exhibit unambiguous specializations for small-particle filter feeding, and both are reasonably interpreted as planktic, although Walossek (1993) prefers a demersal or epiplanktic habit for the orsten assemblage. Here then is the direct evidence of an early Cambrian mesozooplankton and a potentially causal link between the coincident radiation of unicellular phytoplankton and large animals. The sudden shift from a long, monotonous record of leiosphaerid phytoplankton through the Proterozoic to the diverse, rapidly evolving acanthomorphic phytoplankton of the Cambrian can be readily interpreted as an evolutionary response to the introduction of mesozooplanktic grazing (Burzin 1994; Butterfield 1997). By establishing the second tier of the Eltonian pyramid in the pelagic realm, the Early Cambrian introduction of mesozooplankton would have set off a cascade of ecological and evolutionary events, now recognized as the Cambrian explosion (Butterfield 1997).

Previous hypotheses for the Cambrian explosion have also focused on the cascading ecological and evolutionary effects of herbivory (Stanley 1973, 1976) and/or predation (McMenamin 1986; Vermeij 1989; Bengtson and Zhao 1992). The "zooplankton" hypothesis presented here falls broadly into this same category but differs in recognizing the distinct evolutionary histories of the early plankton and benthos. In his "cropping" hypothesis, Stanley (1973, 1976) characterized the whole of the Proterozoic biosphere as profoundly monotonous, with the benthos limited to cyanobacterial mats and the plankton choked with simple unicellular eukaryotes. The rich diversity of Neoproterozoic fossils discovered over the past 20 years clearly belies such a premise; certainly it is not the case that multicellular seaweeds appeared in concert with the Cambrian radiation of metazoans (see review by Knoll 1992). Nevertheless, a "cropping hypothesis" may still stand for the plankton, which did indeed remain undistinguished until the Early Cambrian; to reiterate, Neoproterozoic diversity appears to have been centered overwhelmingly in the benthos.

THE PRACTICE OF EVOLUTIONARY PALEOECOLOGY

Evolutionary paleoecology presents the unique challenge of reconstructing ecosystems occupied largely or entirely by extinct organisms. In the first instance, such analysis will entail the interpretation of organism autecology from fossil form and phylogeny (Fryer 1985; Bryant and Russell 1992); e.g., the filter-feeding and planktic habit of the Mount Cap branchiopods (Butterfield 1994). Synecological assessment, however, is a much more complex issue. Accurate reconstruction here is confounded not only by a limited understanding of comparable modern ecosystems but also by the fundamental loss of resolution through taphonomic processes. The problem of time averaging, in particular, has attracted considerable recent attention (e.g., Kidwell and Flessa 1995; Bambach and Bennington 1996; Jablonski and Sepkoski 1996); however, it is the taphonomic loss of "soft-bodied" constituents that stands as the overarching bias of the fossil record. These typically unfossilized forms comprise a majority of taxa and individuals in almost all communities and occupy a host of key ecological positions (e.g., Stanton and Nelson 1980; McCall and Tevesz 1983; Conway Morris 1986; Butterfield 1990).

Given this preservational filter, the reconstruction of any ancient community will necessarily involve a range of more or less uniformitarian assumptions concerning its unpreserved attributes (e.g., Stanton and Nelson 1980). These assumptions are unproblematic when dealing with the relatively recent past, and it is undoubtedly the case that the early Tertiary oceans operated in a manner broadly comparable to those of today. Such uniformitarian reasoning, however, becomes progressively less certain with age, and it is not at all clear that a pre-Mesozoic marine biosphere can be modeled on the same basis. New production in the modern oceans, for instance, is dominated by diatoms, dinoflagellates, and haptophytes, and secondary production by calanoid copepods; each of these groups contributes uniquely to the overall ecology and eventual fate of the modern plankton (Verity and Smetacek 1996), but none has a significant body-fossil record prior to the Mesozoic.

Signor and Vermeij (1994) have further emphasized the sparse fossil record of Cambrian plankton and suspension feeders, inferring profound differences between the pre- and post-Late Cambrian biospheres, possibly to the non-uniformitarian extent of a decoupled plankton and benthos in the early Paleozoic. Certainly there is some important information in this analysis, but it is not clear that the paleoecological resolution of the data is sufficient to support their conclusions, at least at the scale they propose. Notably, the conclusions are based largely on negative evidence—a dearth of Cambrian plankton and suspension feeders—as recorded in the *conventional* fossil record.

Taphonomic filters are not distributed evenly across communities or ecosystems. The plankton, for example, can be seriously underrepresented because of the vertical transport required before burial. Indeed, the most abundant constituents of the modern marine plankton—the 0.2–2.0 μm diameter picoplankton that dominate oligotrophic water masses (Azam et al. 1983)—are not registered in the fossil record, simply because they are too small to sink; their nonappearance does not imply an absence of picoplankton in the Cambrian or even the Archean. By contrast, dinoflagellates have a good Triassic to Recent fossil record, represented by relatively large (typically several tens of micrometers in diameter) degradation-resistant cysts. Most dinoflagellates, however, do not form cysts, and their tendency to do so appears to have shifted over time; hence the approximately 70-million-year "disappearance" of *Ceratium* between the Cretaceous and Recent. Indeed, recent analyses of pre-Triassic acritarchs and biomarker molecules point to a dinoflagellate record extending well back into the Proterozoic (Summons et al. 1992; Moldowan et al. 1996, this volume; Butterfield and Rainbird 1998; Moldowan and Talyzina 1998).

The record of fossil zooplankton is even patchier. The preservation potential of nonloricate ciliates and amoebae (microzooplankton), for example, is vanishingly small because of the insubstantial nature of their integument (but see Reid 1987 and Poinar et al. 1993). And metazoan mesozooplankton and macrozooplankton fare little better: copepods, for example, dominate modern marine animal biomass (Nienhuis 1981; Verity and Smetacek 1996), but as fossils they are limited to localized occurrences in Holocene marine sediments (van Waveren and Visscher 1994), a non-marine assemblage in the Miocene (Palmer 1960), and parasitic forms on the gills of two lower Cretaceous fish (Cressey and Boxshall 1989). Euphausiids (krill) and salps are likewise of fundamental importance in the modern ocean but lack any fossil record, and the record of cnidarian medusae is extremely sparse.

To some degree, this taphonomic screen can be lifted by recognizing the contribution of fossil Lagerstätten, fossil mother lodes whose paleobiological importance vastly outweighs their rare occurrence. With their exceptional preservation of non-mineralizing organisms, occurrences such as the Chengjiang biota or the Burgess Shale paint a picture of Cambrian diversity and paleoecology fundamentally different from that of the conventional fossil record (Conway Morris 1986). Burgess Shale–type assemblages, for example, reveal an Early-Middle Cambrian abundance of carnivores (priapulids, anomalocarids), relatively high-level suspension feeders (sponges, chancelloriids, pennatulaceans), filter-feeding mesozooplankton (branchiopods), macro-zooplankton (ctenophores, eldoniids), and probable nekton (chordates, chaetognaths, various arthropods) (Briggs and Whittington 1985; Conway Morris 1986; Rigby 1986; Briggs et al. 1994; Butterfield 1994). These "adaptive strategies" are left largely unrecorded by the conventional fossil record; hence the conventional view of Cambrian ecology's being dominated by detritivores and low-level suspension feeders (e.g., Bambach 1983; Signor and Vermeij 1994). Although not modern in detail, Burgess Shale–type assemblages show the Early-Middle Cambrian biosphere to have been at least qualitatively so (Briggs and Whittington 1985; Conway Morris 1986); in the terminology of Droser et al. (1997), it included all marine ecosystems of the "first level," and a considerably greater range of second-level "adaptive strategies" than conventionally appreciated.

Even so, there is good reason to doubt that the Burgess Shale, the Chengjiang, or indeed any fossil Lagerstätte accurately documents a complete and functional paleocommunity. Although there is little likelihood of significant time-averaging in the case of nonmineralizing macroorganisms, differential preservation is still very much in effect. Under Burgess Shale–type conditions, for example, the fossilization of organisms lacking some sort of extracellular cuticle remains highly improbable; if *Amiskwia* is correctly interpreted as a chaetognath (Butterfield 1990), it is probably the only true soft-bodied organism in the Burgess Shale, and one of the rare nekton. By the same token, body fossils of unshelled mollusks or lophophorates are not expected in the Burgess Shale, nor are nemerteans, flatworms, mesozoans, or nonloricate ciliates and amoebae. Other groups, such as rotifers, gastrotrichs, kinorhynchs, nematodes, nematomorphs, gnathostomulids, entoprocts, loriciferans, sipunculans, echiurans, and tardigrades, are known to produce organically preservable structures but, for whatever reason, are not recognized in Burgess Shale–type biotas. Given the presence of most larger-bodied phyla, the (admittedly uniformitarian) suspicion is that this absence is more likely a product of taphonomy than evolution.

In other words, Lagerstätten are not a panacea. Apart from their obvious restriction to certain environments (Conway Morris 1986), key ecologic constituents are inevitably left unrepresented or undiscovered, thus preventing a uniformitarian-free assessment of ancient community structure. Lagerstätten are also rare, leaving little confidence as to the first appearance of key ecologic groups (e.g., Marshall 1990). Moreover, these instances of exceptional preservation are not distributed evenly, or even randomly, through time (Allison and Briggs 1993; Butterfield 1995). In the last 700 million years, for example, Burgess Shale–type preservation appears to have been limited to a critical interval in the Lower and Middle Cambrian. Nonoccurrence of this preservational mode in the Vendian would seem to preclude any definitive statements about the rise of Burgess Shale–type organisms (and modern metazoan ecosystems) other than that it occurred sometime between 750 and 550 million years ago (Butterfield 1995). There is of course a trace fossil record documenting the introduction of a large energetic infauna beginning in the terminal Proterozoic, but this does not rule out the possibility of sophisticated ecosystems comprised of small, nonmineralizing and/or pelagic metazoans (Fortey et al. 1996). Such a possibility is of some concern, given molecular clock arguments for a deep Proterozoic divergence of metazoan phyla (Wray et al. 1996; Wang et al. 1999; but see Ayala et al. 1998).

Fortunately, paleoecologic inference is not limited solely to a capricious fossil record. Large-scale structures, at least, are potentially detected by proxy. There is, for example, clear biogeochemical evidence for a long-term large-scale continuity in marine phytoplankton: the organic carbon content of an "average" sedimentary rock, which today derives almost exclusively from planktic primary productivity, has remained more or less constant from at least the Paleoproterozoic (Strauss et al. 1992). In the absence of alternate sources and in view of the long-term record of leiosphaerid acritarchs, it is clear that phytoplankton have been occupying the photic zone and accumulating in bottom sediments for at least the past two billion years.

At another level, Logan et al. (1995) have argued for a sudden introduction of herbivorous zooplankton in the Early Cambrian based on a shift in hydrocarbon signatures across the Precambrian-Cambrian boundary: an improved preservation of algal-lipid chemistry beginning in Cambrian is explained as a consequence of increased vertical transport, brought on by the introduction of fecal pellet production. Although there are some difficulties with these data and the proposed mechanism (Butterfield 1997), the conclusion is consistent with the zooplankton hypothesis outlined here. More speculative are suggestions that secular shifts in ^{13}C through the latest Proterozoic and Early Cambrian reflect major ecologic innovations (Margaritz et al. 1991; Brasier et al. 1994), including the possibility that the evolution of herbivorous meso-zooplankton was responsible for the marked fall in ^{13}C at the base of the Tommotian (Butterfield 1997).

Proxy evidence of underlying ecologic structures can also be drawn from the available body fossil record. Thus, the recognition of a broadly modern aspect to Early and

Middle Cambrian marine ecosystems (Briggs and Whittington 1985; Conway Morris 1986) in and of itself argues for a modern-style Cambrian mesozooplankton. Such "addition by inference" (Scott 1978) is justified simply on the basis of metabolic requirements: a diverse and energetic metazoan ecology of modern aspect must have had a direct link to the principal source of primary productivity, i.e., phytoplankton. The general introduction of large-animal ecosystems in the Cambrian thus implies an underlying superabundance of small animals, especially herbivorous zooplankton capable of efficiently exploiting and repackaging unicellular phytoplankton.

The coincidence of a fundamental increase in phytoplankton diversity and evolutionary turnover with the Cambrian explosion of large animals offers further indirect evidence for an involvement of mesozooplankton. The Early Cambrian radiation of planktic acanthomorphic acritarchs is readily interpreted as a response to small herbivores, with the acquisition of spines and processes increasing effective cell size (an effective strategy against whole-organism predation) without decreasing buoyancy or capacity for nutrient absorption (Burzin 1994; Butterfield 1997). At the same time, it is difficult to come up with an alternative mechanism for this burst of morphological diversification in planktic primary producers: a long and successful Proterozoic history of leiosphaerid phytoplankton belies the suggestion that ornamentation was necessary for or contributed significantly to flotation, and it is hard to see how it might have been induced by enhanced nutrient availability as implied by the "nutrient stimulus scenario" of Brasier (1992).

Neither metazoan herbivory nor predation is likely to have been limited to the Phanerozoic, but any earlier occurrences may well have been limited to the benthos. All Ediacaran body and trace fossils, for example, now appear to represent benthos, and the declining diversity of stromatolites through the Vendian is reasonably interpreted as a consequence of increased benthic grazing (Grotzinger and Knoll 1999). More speculatively, the early Neoproterozoic radiation of large acritarchs, "scale" microfossils, seaweeds, and tawuiids, all of which appear to be benthic, may be proxy evidence for earlier metazoan activities in the benthos, possibly coincident with early metazoan cladogenesis (cf. Wray et al. 1996; Wang et al. 1999).

ECOLOGIC MODELS AND SCALING

As with most hypotheses, the present one is inevitably simplistic, both in the ecologic scenario presented and in the tacit assumption that such responses can be scaled up to yield large-scale evolutionary effects. There is, however, a case to be made for both. The long and monotonous history of Proterozoic plankton, for example, points clearly to a highly simplified pelagic ecology, apparently devoid of metazoan herbivory or predation. These activities, moreover, would presumably have been added in increments at the onset of the Phanerozoic, such that the early stages of the modern marine biosphere would have followed a relatively simple, potentially reconstructible path.

At its lowest level, plankton ecology is controlled by basic physics. Size, for example, is of fundamental importance to buoyancy and nutrient uptake, with both of these decreasing exponentially with increasing size (Kiørboe 1993). The next level of complexity, although not so obvious from first principles, can also be appreciated actualistically. The most simple planktic ecosystems today occur in lakes, apparently because of their limited phylogenetic diversity (Neill 1994) (perhaps not unlike an Early Cambrian plankton). At the appropriate scale, many of the properties of limnetic communities can be successfully modeled on the basis of their relatively simple size-class structure: a similarity of morphology, physiology, life history, and environmental sensitivity within three or four basic size classes places strong constraints on the community organization of lakes. A comparable situation is reasonably invoked for the primitive planktic ecosystems of the Early Cambrian.

Two basic models have been promoted for explaining the structure and control of limnetic ecology: "bottom-up" models argue that biomass and/or productivity at a particular trophic level are controlled by primary production: increased nutrients boost primary production, which in turn boosts secondary consumers, and so on up the food chain. "Top-down" models, by contrast, argue that the principal control comes from consumers at the top of the food chain, a view that has given rise to the concept of a "trophic cascade." Here the addition of a new level of predation to the top of the Eltonian pyramid translates to reduced productivity and biomass in the underlying tier, which increases productivity and biomass in the next lower tier, and so on, eventually cascading down to affect the quantity and quality of primary productivity (McQueen et al. 1986; Carpenter and Kitchell 1993; Ramcharan et al. 1996; Brett and Goldman 1997). Top-down and bottom-up effects of course both contribute importantly to plankton ecology, the contribution of each depending largely on local circumstances; for example, trophic cascades are not developed under extremely oligotrophic or extremely eutrophic conditions and may be disrupted by secondary effects such as increased water clarity resulting from enhanced grazing (McQueen et al. 1986; Verity and Smetacek 1996). Trophic cascades are not well developed in modern marine ecosystems, apparently because of the greater phylogenetic complexity and generally more oligotrophic conditions in the sea (Neill 1994; Verity and Smetacek 1996).

How might any of this apply to the Proterozoic-Phanerozoic transition? Brasier (1992) notes the widespread occurrence of phosphorites, black shales, and carbon isotope shifts associated with this interval and suggests a bottom-up increase of nutrients as the impetus for the Cambrian explosion. If, however, the terminal Proterozoic lacked a grazing mesozooplankton, as argued here, then it is difficult to see how increased nutrients would do anything except induce eutrophication; in the plankton, there would have been nothing to take advantage of the increased productivity. Thus it appears that any increase in trophic complexity would have had to come from novel additions to the top of the food chain. Both the direct and indirect evidence of fossil record point to an early Cambrian introduction of herbivorous mesozooplankton.

McQueen et al. (1986) and Brett and Goldman (1997) have shown that the transmissibility of both top-down and bottom-up effects in a pelagic food chain is affected by the number of steps (trophic levels) through which it must pass, with each additional step substantially attenuating the signal. Thus the potential for top-down effects to impinge on primary productivity and biogeochemical cycling in any pronounced way is limited to situations in which the trophic structure is both simple and short. This, I would argue, was the case during that unique interval in the earliest Cambrian when the modern pelagic ecosystem was under construction. The direct, top-down effect of a newly introduced mesozooplankton on primary productivity is powerfully expressed both in the marked shift in phytoplankton evolution and in the fluctuating biogeochemistry of the Proterozoic-Phanerozoic transition. Subsequent addition of higher-level tiers to the Eltonian pyramid may have induced subsequent top-down cascades, but these would have dissipated before impinging significantly on primary producers. From this angle, then, the transition between the Proterozoic and Phanerozoic was uniquely susceptible to ecologic and biogeochemical perturbation, the accompanying sedimentary expressions (e.g., phosphorites, black shales, carbon isotope shifts) are more likely to represent consequences than causes of the Cambrian explosion.

All this is interesting, but do effects that register at the ecologic level translate into evolutionary, particularly macroevolutionary, change? Gould (1985), for example, has allowed that although ecology may be the principal evolutionary motor at one level (the first tier), these effects are largely overprinted by higher-order selection at the second tier (i.e., species selection), which is in turn subordinate to a third tier of mass extinction. Be that as it may, a reasonable case has been made for biotic interactions playing an important macroevolutionary role, albeit in a diffuse, protracted, and not always obvious manner (Aronson 1992; Jablonski and Sepkoski 1996). The best examples are perhaps those relating to the hypothesis of escalation presented by Vermeij (1987), i.e., that increases in predation intensity through geologic time have induced evolutionary counterresponses among prey.

In the present context, the introduction of herbivorous metazoans into a plankton previously devoid of such organisms would have had a profound effect on contemporary plankton ecology, and the burst of phytoplankton diversification in the Early Cambrian is readily interpreted as its evolutionary effect. There was nothing inherently special about metazoans entering the plankton, which was probably first achieved by a small, possibly neotenic constituent of the benthos in the process of evading (benthic) predation (Butterfield 1997). Nevertheless, this particular innovation was a *key* innovation; it contributed to changes not only at the level of "community" (i.e., a fourth-level change in the terminology of Droser et al. 1997) but also at the levels of "community-type," "adaptive strategy," and ecosystem (third, second, and first levels, respectively). By establishing a new ecosystem—pelagic metazoans—

the simple ecologic derivation of the mesozooplankton scales up to a macroevolutionary level, where it has survived the length of the Phanerozoic, including its series of "third tier" mass extinctions.

DISCUSSION

Numerous hypotheses have been offered to explain the Cambrian explosion of large animals, ranging from major intrinsic innovations in developmental programs (e.g., Erwin 1993) to extrinsic causes such as increased levels of oxygen (Knoll 1992) or nutrients (Brasier 1992). By contrast, the zooplankton hypothesis presented here invokes a relatively minor ecological shift in animal activity triggering cascades of interconnected effects at a number of scales, most importantly through the introduction of a new ecosystem.

The method and purpose of assessing higher-level paleoecologic categories differ considerably from those directed at reconstructing ancient "communities." Time averaging, for example, is not an issue at this scale, so the presence of an "adaptive strategy" or ecosystem can be readily documented on the basis of a single Lagerstätten occurrence of a key innovation; for example, the unique discovery of filter-feeding mesozooplankton in the Early Cambrian (Butterfield 1994) establishes the significant presence of pelagic metazoans in the earliest Phanerozoic. Because the effects of higher-level categories tend to cascade down through lower levels (Droser et al. 1997), the overall impact of a newly introduced zooplankton would have been profound.

Although a single fossil occurrence may document the minimum age of a particular habit, it suggests little about first appearance (Marshall 1990), particularly given the narrow temporal distribution of Burgess Shale–type preservation (Butterfield 1995; Fortey et al. 1996). Certainly the phytoplankton record provides proxy evidence in support of a first appearance of mesozooplankton in the Tommotian, but it might still be argued that the evidence remains largely negative, i.e., a *lack* of observed diversity among pre-Cambrian phytoplankton. The counterargument is that most acritarchs—certainly those that represent phytoplankton cysts—do not require exceptional conditions for their preservation and extend more or less continually from the Paleoproterozoic into the Paleozoic; unlike almost all other groups (including metazoans), they show no fundamental change in preservation potential across the Precambrian-Cambrian boundary. Combined with geochemical evidence (e.g., Logan et al. 1995), the acritarch record points to a true absence of pre-Cambrian mesozooplankton and the reality of a Cambrian "explosion," albeit as an ecologic rather than a deep-seated phylogenetic phenomenon.

With the case for an Early Cambrian introduction of mesozooplankton relatively strong, it remains to be shown that the Phanerozoic plankton is closely coupled to the benthos, that this was not the case prior to the Cambrian, and that the difference between a coupled and a decoupled plankton-benthos is significant. Certainly it is possible to base a metazoan ecosystem solely on benthic primary productivity and detritivory, but by not directly exploiting the phytoplankton, such ecosystems are liable to be of limited diversity and activity; the terminal Proterozoic Ediacaran fauna and associated simple trace fossils set the obvious example. Actualistic studies show that the modern benthos is indeed closely coupled to the plankton, with benthic metazoan communities responding to phytoplankton blooms in a matter of days (Graf 1989). By extension, Levinton (1996) has argued that even a short-term cessation of phytoplanktic productivity, such as is often invoked as a proximal cause for the Cretaceous-Tertiary mass extinction, should (but notably did not) devastate deposit-feeding benthos. That some vertical transport in the modern oceans is traveling via copepod fecal pellets (e.g., Graf 1989) is consistent with the geochemical argument made by Logan et al. (1995) for a pre-Cambrian absence of pellet-producing zooplankton. Fecal pellets, however, are certainly not the only link (indeed, not even the principal link) between the plankton and benthos in the modern oceans, and it remains to be resolved what particular role they may have played in the Cambrian explosion; McIlroy and Logan (1999) offer some interesting possibilities.

Signor and Vermeij (1994) have stressed the possible decoupling of an early Paleozoic plankton and benthos, but they place the transition at the end of the Cambrian rather than the beginning. Certainly the Cambro-Ordovician transition was of major importance, but in ecologic terms it was simply not on the same scale as the Cambrian explosion. Whereas the Ordovician witnessed the appearance of numerous new "adaptive strategies" and their cascading effects (Droser et al. 1997), it was the Cambrian that first introduced animals to the plankton, thereby establishing the two "first level" ecosystems that arguably define the Phanerozoic: pelagic metazoans and benthic metazoans coupled closely to the plankton.

Acknowledgments. I thank the editors for their helpful advice, the reviewers for useful comments, and Andy Knoll for figures 9.1F and 9.1G. This work was carried out in the Department of Earth Sciences, University of Western Ontario, with the support of the Natural Sciences and Engineering Research Council of Canada.

REFERENCES

Allison, C. W. and J. W. Hilgert. 1986. Scale microfossils from the Early Cambrian of northwest Canada. *Journal of Paleontology* 60:973–1015.

Allison, P. A. and D. E. G. Briggs. 1993. Exceptional fossil record: Distribution of soft-tissue preservation through the Phanerozoic. *Geology* 21:527–530.

Aronson, R. B. 1992. Biology of a scale-independent predator-prey interaction. *Marine Ecology Progress Series* 89:1–13.

Ayala, F. J., A. Rzhetsky, and F. J. Ayala. 1998.

Origin of the metazoan phyla: Molecular clocks confirm paleontological estimates. *Proceedings of the National Academy of Sciences, USA* 95:606–611.

Azam, F., T. Fenchel, J. G. Gray, L. A. Meyer-Reil, and T. Thingstad. 1983. The ecological role of water-column microbes in the sea. *Marine Ecology Progress Series* 10:257–263.

Bambach, R. K. 1983. Ecospace utilization and guilds in marine communities through the Phanerozoic. In M. J. S. Tevesz and P. L. McCall, eds., *Biotic Interactions in Recent and Fossil Benthic Communities,* pp. 719–746. New York: Plenum Press.

Bambach, R. K. and J. B. Bennington. 1996. Do communities evolve: A major question in evolutionary paleoecology. In D. Jablonski, D. H. Erwin, and J. H. Lipps, eds., *Evolutionary Paleobiology,* pp. 123–160. Chicago: Chicago University Press.

Bengtson, S. and Y. Zhao. 1992. Predatorial borings in Late Precambrian mineralized exoskeletons. *Science* 257:367–369.

Berger, W. H., V. S. Smetacek, and G. Wefer, eds. 1989. *Productivity of the Ocean: Present and Past.* Chichester: John Wiley and Sons.

Brasier, M. D. 1992. Nutrient-enriched waters and the early skeletal fossil record. *Journal of the Geological Society of London* 149:621–629.

Brasier, M. D., A. Yu. Rozanov, A. Yu. Zhuravlev, R. M. Corfield, and L. A. Derry. 1994. A carbon isotope reference scale for the Lower Cambrian succession in Siberia: Report of IGCP Project 303. *Geological Magazine* 131:767–783.

Brett, M. T. and C. R. Goldman. 1997. Consumer versus resource control in freshwater pelagic food webs. *Science* 275:384–386.

Briggs, D. E. G. and H. B. Whittington. 1985. Modes of life of arthropods from the Bur-gess Shale, British Columbia. *Transactions of the Royal Society of Edinburgh (Earth Sciences)* 76:149–160.

Briggs, D. E. G., D. H. Erwin, and F. J. Collier. 1994. *The Fossils of the Burgess Shale.* Washington, D.C.: Smithsonian Institution Press.

Bryant, H. N. and A. P. Russell. 1992. The role of phylogenetic analysis in the inference of unpreserved attributes of extinct taxa. *Philosophical Transactions of the Royal Society of London B* 337:405–418.

Burzin, M. B. 1994. Osnovnye tendentsii v istoricheskom razvitii fitoplanktona v pozdnem dokembrii i rannem kembrii [Principal trends in the historical development of phytoplankton in the late Precambrian and early Cambrian]. In A. Yu. Rozanov and M. A. Semikhatov, eds., *Ekosistemnye perestroyki i evolyutsiya biosfery* [Ecosystem restructuring and the evolution of the biosphere], pp. 51–62. Moscow: Nedra.

Butterfield, N. J. 1990. Organic preservation of non-mineralizing organisms and the taphonomy of the Burgess Shale. *Paleobiology* 16:272–286.

Butterfield, N. J. 1994. Burgess Shale–type fossils from a Lower Cambrian shallow shelf sequence in northwestern Canada. *Nature* 369:477–479.

Butterfield, N. J. 1995. Secular distribution of Burgess Shale-type preservation. *Lethaia* 28:1–13.

Butterfield, N. J. 1997. Plankton ecology and the Proterozoic-Phanerozoic transition. *Paleobiology* 23:247–262.

Butterfield, N. J. and F. W. Chandler. 1992. Palaeoenvironmental distribution of Proterozoic microfossils, with an example from the Agu Bay Formation, Baffin Island. *Palaeontology* 35:943–957.

Butterfield, N. J. and R. H. Rainbird. 1998.

Diverse organic-walled fossils, including "possible dinoflagellates," from the early Neoproterozoic of Arctic Canada. *Geology* 26:963–966.

Butterfield, N. J., A. H. Knoll, and K. Swett. 1990. A bangiophyte red alga from the Proterozoic of Arctic Canada. *Science* 250: 104–107.

Butterfield, N. J., Knoll, A. H., and Swett, K. 1994. Paleobiology of the Neoproterozoic Svanbergfjellet Formation, Spitsbergen. *Fossils and Strata* 34:1–84.

Carpenter, S. R. and J. F. Kitchell, eds. 1993. *The Trophic Cascade in Lakes.* Cambridge: Cambridge University Press.

Conway Morris, S. 1986. The community structure of the Middle Cambrian Phyllopod Bed (Burgess Shale). *Palaeontology* 29: 423–467.

Cressey, R. and G. Boxshall. 1989. *Kabatarina pattersoni:* A fossil parasitic copepod (Dichelesthiidae) from a Lower Cretaceous fish. *Micropaleontology* 35:150–167.

Droser, M. L., D. J. Bottjer, and P. M. Sheehan. 1997. Evaluating the ecological architecture of major events in the Phanerozoic history of marine invertebrate life. *Geology* 25:167–170.

Erwin, D. H. 1993. The origin of metazoan development: A palaeobiological perspective. *Biological Journal of the Linnean Society* 50:255–274.

Evitt, W. R. 1985. *Sporopollenin Dinoflagellate Cysts: Their Morphology and Interpretation.* Dallas: American Association of Stratigraphic Palynologists Foundation.

Fortey, R. A., D. E. G. Briggs, and M. A. Wills. 1996. The Cambrian evolutionary "explosion": Decoupling cladogenesis from morphological disparity. *Biological Journal of the Linnean Society* 57:13–33.

Fryer, G. 1985. Structure and habits of living branchiopod crustaceans and their bear-

ing on the interpretation of fossil forms. *Transactions of the Royal Society of Edinburgh (Earth Sciences)* 76:103–113.

Gould, S. J. 1985. The paradox of the first tier: An agenda for paleobiology. *Paleobiology* 11:2–12.

Graf, G. 1989. Benthic-pelagic coupling in a deep-sea benthic community. *Nature* 341: 437–439.

Grotzinger, J. P. and A. H. Knoll. 1999. Stromatolites in Precambrian carbonates: Evolutionary mileposts or environmental dipsticks. *Annual Review of Earth and Planetary Sciences* 27:213–358.

Hansen, B., P. K. Bjørnsen, and P. J. Hansen. 1994. The size ratio between planktonic predators and their prey. *Limnology and Oceanography* 39:395–403.

Hermann, T. N. 1981. Nitchatye mikroorganizmy Lakhandinskoj svity reki Mai [Filamentous microorganisms in the Lakhanda Formation on the Maya River]. *Paleontologicheskiy zhurnal* 1981 (2):126–131. [*Paleontological Journal* 1981 (2):100–107.]

Jablonski, D. and J. J. Sepkoski, Jr. 1996. Paleobiology, community ecology, and scales of ecological pattern. *Ecology* 77:1367–1378.

Kaufman, A. J., A. H. Knoll, and S. M. Awramik. 1992. Biostratigraphic and chemostratigraphic correlation of Neoproterozoic sedimentary successions: Upper Tindir Group, northwestern Canada, as a test case. *Geology* 20:181–185.

Kidwell, S. M. and K. W. Flessa. 1995. The quality of the fossil record: Populations, species, and communities. *Annual Review of Ecology and Systematics* 26:269–299.

Kiørboe, T. 1993. Turbulence, phytoplankton cell size, and the structure of pelagic food webs. *Advances in Marine Biology* 29:1–72.

Knoll, A. H. 1992. The early evolution of the

eukaryotes: A geological perspective. *Science* 256:622–627.

Knoll, A. H. 1994. Proterozoic and Early Cambrian protists: Evidence for accelerating evolutionary tempo. *Proceedings of the National Academy of Sciences, USA* 91:6743–6750.

Knoll, A. H. and N. J. Butterfield. 1989. New window on Proterozoic life. *Nature* 337:602–603.

Levinton, J. S. 1996. Trophic group and the end-Cretaceous extinction: Did deposit feeders have it made in the shade? *Paleobiology* 22:104–112.

Logan, G. A., J. M. Hayes, G. B. Hieshima, and R. E. Summons. 1995. Terminal Proterozoic reorganization of biogeochemical cycles. *Nature* 376:53–56.

Margaritz, M., J. L. Kirschvink, A. J. Latham, A. Yu. Zhuravlev, and A. Yu. Rozanov. 1991. Precambrian/Cambrian boundary problem: Carbon isotope correlations for Vendian and Tommotian time between Siberia and Morocco. *Geology* 19:847–850.

Marshall, C. R. 1990. Confidence intervals on stratigraphic ranges. *Paleobiology* 16:1–10.

McCall, P. L. and J. S. Tevesz. 1983. Soft-bottom succession and the fossil record. In M. J. S. Tevesz and P. L. McCall, eds., *Biotic Interactions in Recent and Fossil Benthic Communities*, pp. 157–194. New York: Plenum Press.

McIlroy, D. and G. A. Logan. 1999. The impact of bioturbation on infaunal ecology and evolution during the Proterozoic-Cambrian transition. *Palaios* 14:58–72.

McMenamin, M. A. S. 1986. The garden of Ediacara. *Palaios* 1:178–182.

McQueen, D. J., J. R. Post, and E. L. Mills. 1986. Trophic relationships in freshwater pelagic ecosystems. *Canadian Journal of Fisheries and Aquatic Science* 43:1571–1581.

Mendelson, C. V. and J. W. Schopf. 1992. Proterozoic and Early Cambrian acritarchs. In J. W. Schopf and C. Klein, eds., *The Proterozoic Biosphere*, pp. 219–232. Cambridge: Cambridge University Press.

Moldowan, J. M. and N. M. Talyzina. 1998. Biogeochemical evidence for dinoflagellate ancestors in the Early Cambrian. *Science* 281:1168–1170.

Moldowan, J. M., J. Dahl, S. R. Jacobson, B. J. Huizinga, F. J. Fago, R. Shetty, D. S. Watt, and K. E. Peters. 1996. Chemostratigraphic reconstruction of biofacies: Molecular evidence linking cyst-forming dinoflagellates with pre-Triassic ancestors. *Geology* 24:159–162.

Neill, W. W. 1994. Spatial and temporal scaling and the organization of limnetic communities. In P. S. Giller, A. G. Hildrew, and D. G. Raffaelli, eds., *Aquatic Ecology—Scale, Pattern, and Process*, pp. 189–231. Oxford: Blackwell Scientific.

Nienhuis, P. H. 1981. Distribution of organic matter in living marine organisms. In E. K. Duursma and R. Dawson, eds., *Marine Organic Chemistry—Evolution, Composition, Interactions, and Chemistry of Organic Matter in Seawater*, pp. 31–69. Amsterdam: Elsevier Scientific.

Palmer, A. R. 1960. Miocene copepods from the Mojave Desert, California. *Journal of Paleontology* 34:447–452.

Pauly, D. and V. Christensen. 1995. Primary production required to sustain global fisheries. *Nature* 374:255–257.

Poinar, G. O., Jr., B. M. Waggoner, and U.-C. Bauer. 1993. Terrestrial soft-bodied protists and other microorganisms in Triassic amber. *Science* 259:222–224.

Ramcharan, C. W., R. L. France, and D. J. McQueen. 1996. Multiple effects of planktivorous fish on algae through a pelagic trophic cascade. *Canadian Journal of Fisheries and Aquatic Science* 53:2819–2828.

Reid, P. C. 1987. Mass encystment of a planktonic oligotrich ciliate. *Marine Biology* 95: 221–230.

Rigby, J. K. 1986. Sponges of the Burgess Shale (Middle Cambrian), British Columbia. *Palaeontographica Canadiana* 2: 1–105.

Rigby, S. and C. Milsom. 1996. Benthic origins of zooplankton: An environmentally determined macroevolutionary effect. *Geology* 24: 52–54.

Scott, R. W. 1978. Approaches to trophic analysis of paleocommunities. *Lethaia* 11: 1–14.

Signor, P. W. and G. J. Vermeij. 1994. The plankton and the benthos: Origins and early history of an evolving relationship. *Paleobiology* 20: 297–319.

Stanley, S. M. 1973. An ecological theory for the sudden origin of multicellular life in the late Precambrian. *Proceedings of the National Academy of Sciences, USA* 70: 1486–1489.

Stanley, S. M. 1976. Ideas on the timing of metazoan diversification. *Paleobiology* 2: 209–219.

Stanton, R. J., Jr., and P. C. Nelson. 1980. Reconstruction of the trophic web in paleontology: Community structure in the Stone City Formation (Middle Eocene, Texas). *Journal of Paleontology* 54: 118–135.

Strauss, H., D. J. Des Marais, J. M. Hayes, and R. E. Summons. 1992. Concentrations of organic carbon and maturities and elemental compositions of kerogens. In J. W. Schopf and C. Klein, eds., *The Proterozoic Biosphere*, pp. 95–99. Cambridge: Cambridge University Press.

Summons, R. E., J. Thomas, J. R. Maxwell, and C. J. Boreham. 1992. Secular and environmental constraints on the occurrence of dinosterane in sediments. *Geochimica et Cosmochimica Acta* 56: 2437–2444.

van Waveren, I. M. and H. Visscher. 1994. Analysis of the composition and selective preservation of organic matter in surficial deep-sea sediments from a high productivity area (Banda Sea, Indonesia). *Palaeogeography, Palaeoclimatology, Palaeoecology* 112: 85–111.

Verity, P. G. and V. Smetacek. 1996. Organism life cycles, predation, and the structure of marine pelagic ecosystems. *Marine Ecology Progress Series* 130: 277–293.

Vermeij, G. J. 1987. *Evolution and Escalation: An Ecological History of Life.* Princeton: Princeton University Press.

Vermeij, G. J. 1989. The origin of skeletons. *Palaios* 4: 585–589.

Walossek, D. 1993. The Upper Cambrian *Rehbachiella* and the phylogeny of Branchiopoda and Crustacea. *Fossils and Strata* 32: 1–202.

Wang, D.Y.-C., S. Kumar, and S. B. Hedges. 1999. Divergence time estimates for the early history of animal phyla and the origin of plants, animals, and fungi. *Proceedings of the Royal Society of London B* 266: 163–171.

Wray, G. A., J. S. Levinton, and L. H. Shapiro. 1996. Molecular evidence for deep Precambrian divergences among metazoan phyla. *Science* 274: 568–573.

CHAPTER TEN

Mikhail B. Burzin, Françoise Debrenne, and Andrey Yu. Zhuravlev

Evolution of Shallow-Water Level-Bottom Communities

Features of Cambrian level-bottom communities that inhabited carbonate and silici-clastic substrates are outlined. A high diversity of level-bottom communities with multiple trophic guilds was established in the Early Cambrian, replacing largely microbial-dominated Vendian ecosystems. Taxonomic richness of Early Cambrian communities contrasts with relative impoverishment of their Middle and Late Cambrian counterparts. Displacement of communities was common, and entire communities might migrate into areas with more favorable conditions if their original habitats suffered a crisis.

CAMBRIAN DEPOSITIONAL SYSTEMS can be divided into clastic and carbonate regimes, because substrate type strongly influences community composition. These aspects of sedimentation were in general controlled by climate and the size of the area available for denudation. With few exceptions, environments of carbonate sedimentation were restricted to low latitudes and siliciclastic-dominated settings occurred mostly in temperate conditions. The Siberian Platform throughout the Cambrian exemplified carbonate-dominated habitats. Baltica, Bohemia, and Avalonia represented regions where siliciclastic sedimentation prevailed. Laurentia and Australia were characterized by a mosaic of facies.

TROPHIC GUILDS

Although the entire set of trophic guilds existed from the beginning of the period, Cambrian guilds were different even from their Ordovician successors and probably had already changed significantly by the end of the Cambrian. Tables 10.1 and 10.2 display the ecospace utilization by Cambrian organisms that are preserved now as body fossils.

Benthic primary producers were represented chiefly by probable calcified cyanobacteria (e.g., *Obruchevella*) and by carbonaceous algae (e.g., *Margaretia*) and possible

Table 10.1 Ecospace Utilization by Animals of Level-Bottom
Communities During the Early Cambrian

	Epifaunal	Infaunal
Mobile	Trilobites, nontrilobite arthropods, halkieriids, low conical helcionelloids, paragastropods, orthothecimorphs, "lobopodians," polychaetes, tommotiids	Laterally compressed helcionelloids, trilobites, priapulids, fordillids, polychaetes, palaeoscolecidans?
Sessile low tier (<10 cm)	Demosponges, calcareans, chancelloriids, lingulates, calciates, anabaritids, coleolids, hyolithelminths, tianzhushanellids, edrioasteroids, pterobranchs, hyolithomorphs, orthothecimorphs, stenothecoids	*Plalysolenites*, lingulates
Sessile high tier (>10 cm)	Hexactinellids, heteractinids, cnidarians, eocrinoids, helicoplacoids	

Table 10.2 Ecospace Utilization by Animals of Level-Bottom
Communities During the Middle and Late Cambrian

	Epifaunal	Infaunal
Mobile	Trilobites, nontrilobite arthropods, tergomyans, gastropods, orthothecimorphs, polychaetes, "lobopodians," cephalopods, homoisteleans, stylophorans	Rostroconchs, trilobites, priapulids?, polychaetes?, palaeoscolecidans?
Sessile low tier (<10 cm)	Demosponges, calcareans, lingulates, calciates, edrioasteroids, pterobranchs, hyolithomorphs, gastropods	Lingulates
Sessile high tier (>10 cm)	Hexactinellids?, eocrinoids, crinoids, graptolites, branching hyolithelminths	

cyanobacteria (e.g., *Morania*). Noncalcified bacteria grew abundantly in the Cambrian stromatolites and thrombolites and undoubtedly on most sediment surfaces, as they do in modern marine environments. It has been suggested that bacteria are the main producers of micritic carbonates (Riding 1991), which often possess a typical clotted texture, and of phosphates (Gerasimenko et al. 1996), in the Cambrian. However, planktic primary producers, including free-living and attached bacteria and phytoplankton (acritarchs and prasinophytes, at least), were the main food source for level-bottom filter and suspension feeders. For instance, acritarchs are abundant in pelleted carbonates (Zhegallo et al. 1994; Zhuravlev and Wood 1996). Acritarchs were probable endocysts of polyphyletic origin; they possessed a sporopollenin-like wall, similar to that produced by photosynthetic eukaryotes (Martin 1993; see Moldowan et al., this volume, for biomarker data).

Feeding strategies are considered to be diverse among consumers (Debrenne and Zhuravlev 1997) (figure 10.1).

1. Filtrators consisted of sponges (hexactinellids, heteractinids, demosponges, and probable calcareans), calciate brachiopods, probably the majority of mollusks—including helcionelloids, pelecypods, and rostroconchs—and piperock producers (in this volume, see chapters by Debrenne and Reitner; Kouchinsky; and Ushatinskaya).

2. Suspension feeders were represented by lingulate brachiopods, echinoderms, chancelloriids, hyolithomorph and some orthothecimorph hyoliths, stenothecoids, and Late Cambrian trilobites (in this volume, see chapters by Guensburg and Sprinkle; Hughes; Kouchinsky; and Ushatinskaya). Many of the tubicolous taxa (coleolides, hyolithelminths, anabaritids), as well as brachiopod-like animals (Tianzhushanellidae), were apparently semi-infaunal suspension feeders sensu lato (Bengtson and Conway Morris 1992; Parkhaev 1998). By analogy with living polychaetes, some of them could be pure filter feeders consuming bacterioplankton (Sorokin 1992), but others, such as phosphatic hyolithelminths, with a metabolism probably similar to that of lingulates, could be true suspension feeders. During the earliest Early Cambrian, *Platysolenites* might have been an agglutinated foraminifer (McIlroy et al. 1994), which belonged to suspension feeders, according to the test morphology (Lipps 1983). Since the Middle Cambrian, dendroid graptolites joined the group of sessile filter and suspension feeders (Sdzuy 1974) for a short time before planktic forms were developed, probably in response to the general shift of phytoplankton grazing from the sea floor to the water column. Flow pattern modeling of sessile conical dendroid graptolites shows that such colonies were well designed to use ambient currents to reduce the energetic cost of suspension feeding (Melchin and Doucet 1996). This modeling also supports the suggestion by Rickards et al. (1990) that different dendroid rhabdosomal morphologies may have been adapted to different currents. The aperture of even large graptolite thecae with simple openings rarely exceeded 2 mm, severely restricting the maximum size for food particles; most graptolites had even smaller apertures, and in

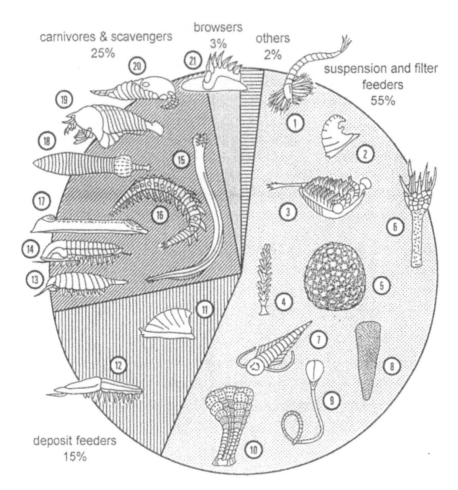

Figure 10.1 Approximate average share of different trophic groups among Cambrian bodied animals and their representatives. Suspension and filter feeders: *1,* crustacean *Skara; 2,* helcionelloid mollusk *Yochelcionella; 3,* arthropod *Sarotrocercus; 4,* graptolite *Archaeolaphoea; 5,* radiocyath *Girphanovella; 6,* eocrinoid echinoderm *Lepidocystis; 7,* hyolithomorph hyolith; *8,* chancelloriid *Chancelloria; 9,* lingulate brachio-pod; *10,* archaeocyath sponge *Coscinocyathus.* Deposit feeders: *11,* helcionelloid mollusk *Helcionella; 12,* arthropod *Naraoia.* Carnivores and scavengers: *13,* arthropod *Sidneyia; 14,* trilobite *Olenoides; 15,* conodont-chordate; *16,* "lobopod" *Xenusion; 17,* halkieriid *Halkieria; 18,* priapulid *Ottoia; 19,* arthropod *Sanctacaris; 20,* anomalocaridid *Laggania.* Browsers: *21,* chitonlike mollusk *Matthevia.*

many species these are reduced by lobes, lappets, or spines, even further restricting the maximum size of particle uptake (Underwood 1993). Pterobranchs were already present in the Early Cambrian.

3. Predator and scavenger guilds consisted of a variety of cnidarians, trilobites, and nontrilobite arthropods, "lobopodians," and giant anomalocaridids, which were large and mostly mobile carnivores (Fortey and Owens 1999; Nedin 1999; in this volume, see chapters by Debrenne and Reitner; Hughes; and Budd). Some polychaetes, priapulids, and their close relatives palaeoscolecidans exploited this feeding strategy

(Conway Morris 1976, 1979; Hou and Bergström 1994). Protoconodonts may have occupied a demersal predator niche by analogy with extant chaetognaths (Szaniawski 1982), as well as later euconodonts (Purnell 1995). Boreholes in shells and scars of healed injuries in trilobite carapaces resulted from the action of unknown predators and scavengers (Jago 1974; Conway Morris and Bengtson 1994; Pratt 1994).

4. Destructors, which attacked hard mineral and cellular substances, were common. Cambrian endolithic borings are known in ooids, echinoderm ossicles, brachiopod shells, archaeocyath cups, various small shelly fossils, and conodonts (Müller and Nogami 1972; Kobluk and Kahle 1978; Li 1997). In some cases, tentative interpretation in favor of cyanobacterial and fungal borings has been provided (Kobluk and Risk 1977). Saprophytes have been recognized in the Cambrian communities, including phycomycetes and actinomycetes (Burzin 1993b).

5. Trace fossil data (Crimes 1992) indicate that the Cambrian biota includes 50% (Nemakit-Daldynian) to 40% (Atdabanian) deposit feeders (feeding traces). Crimes (1992) also suggests that grazing traces account for 10% to 20% of the total trace fossil diversity. But given that these are recorded on soft substrates, in contrast to the feeding strategy of true grazers, they should instead be considered as deposit feeders, the percentage of which had thus increased to 60%. *Chondrites* and many other branching traces exemplify deposit-feeding strategies, some of which were very peculiar and restricted to the Cambrian. For instance, a vermiform *Plagiogmus*-producer burrowed within the substrate but fed on surface detritus by means of a siphon (McIlroy and Heys 1997). Microburrowings may represent detritivorous meiofauna (Wood et al. 1993). Body fossils, however, do not allow us to infer the true producers of these traces. Deposit feeders on silty substrates are recognized among low-spired, widely expanded helcionelloid mollusks, most orthothecimorph hyoliths, some trilobites, and nontrilobite arthropods; small paragastropods were probable mobile epifaunal deposit feeders (in this volume, see chapters by Kouchinsky and by Hughes and Budd).

6. Possible Cambrian algal croppers have been noted by Edhorn (1977) from the Bonavista Group of Avalon. These "croppers" are sessile orthothecimorph hyoliths ("*Ladatheca*" of Landing 1993). However, Kobluk (1985) reported some possible grazer scratches on calcimicrobes from the Upper Shady Dolomite.

7. Among parasites, pentastomes are established in the Cambrian (Walossek et al. 1994). Some borings and skeletal abnormalities may also be interpreted as parasite traces (Conway Morris and Bengtson 1994; in this volume, see chapters by Hughes and by Budd).

CARBONATE-DOMINATED SETTINGS

Evaporite Basins

Evaporite basins, containing carbonates and evaporites, are typified by low clastic input and high evaporation rates. Their coastlines are characterized by chains of islands that shelter hypersaline lagoons with reduced tidal ranges, where microbial mats are formed. They produced extensive stromatolite deposits; the best examples occurred in the Toyonian Angara Formations of the Siberian Platform where stratiform and columnar stromatolites formed low but very wide buildups, up to several kilometers in length, peripherally covered by ooidal grainstones (Korolyuk 1968). This stromatolite community did not change during the Cambrian. However, various mollusks (rostroconchs and chitonlike forms) intruded into barrier complexes formed under generally higher salinities in Australia during the Datsonian (Druce et al. 1982).

On the periphery of evaporite basins, an oligotypic trilobite community occurred locally (e.g., Olekma Formation, Siberian Platform) from the Atdabanian through the remainder of the Cambrian. Rare hyoliths and brachiopods also were present (Repina 1977).

Peritidal Carbonate Environments

Peritidal carbonate environments include oolite shoals, carbonate sand shoals and beaches, and intertidal to subtidal flat settings. Since Atdabanian time, *Ophiomorpha*-like trace producers (*Aulophycus*) occupied shifting lime muds in shoal agitated backreef conditions. *Ophiomorpha*-type burrows represented innovative behavior, in their ability to produce pellet-lined burrows, which prevent collapse in substrates of relatively low cohesive strength (Crimes and Droser 1992). The *Aulophycus* community persisted through the entire Cambrian: Atdabanian Nokhoroy Unit and Kyndyn Formation, Botoman upper *Kutorgina* and Toyonian Keteme formations of the Siberian Platform, and Botoman Poleta Formation and Shady Dolomite of Laurentia (Balsam 1974; Zhuravleva et al. 1982; Astashkin 1985; Droser and Bottjer 1988).

In restricted nutrient-rich lagoons, cyanobacterial communities, chiefly oscillatoriaceans, formed phosphatized mats of helically coiled and prostrate filaments (Rozanov and Zhegallo 1989; Sergeev and Ogurtsova 1989; Soudry and Southgate 1989). Such communities were common during the Nemakit-Daldynian–Tommotian (e.g., Chulaktau Formation, Kazakhstan; Khesen Formation, Mongolia) but became rare later in the Cambrian.

Peritidal limestones were deposited in Avalonia under temperate conditions (Brasier and Hewitt 1979; Landing et al. 1989; Landing 1991, 1993). Here peritidal limestones have stromatolitic, mud-cracked caps and include helcionelloid mollusks (*Igorella, Oelandiella*), phosphatic sclerite-armored animals (*Eccentrotheca, Lapworthella*), phosphatic tube dwellers (*Torellella*), and orthothecimorph hyoliths (*Turcutheca, Laratheca*) that are absent in subtidal shales. In the early Tommotian (Chapel Island Formation upper Member 3 through Member 4), the *Watsonella crosbyi* fauna existed, including "*Ladatheca*" thickets overgrown by stromatolites. Later in the Atdabanian (e.g., Home Farme Member), these thickets were ecologically displaced by *Coleoloides typicalis* thickets of vertically oriented tubes (Brasier and Hewitt 1979). Atdabanian-

Botoman peritidal limestones of the Weymouth Formation contain an especially rich fauna, including coleolids, hyolithelminths, "lobopodians," chancelloriids, halkieriids, tommotiids, hyoliths, helcionelloids, paragastropods, lingulate brachiopods, and eodiscid and olenelloid trilobites. The faunal enrichment of the shallowest environments in Avalonia probably reflects its high latitudinal position and thus a high thermocline.

Shallow Carbonate Seas

Shallow carbonate seas include several carbonate environments, all of which lay at or below fair-weather wave base. A high range of communities inhabited this zone, including level-bottom, reefal, and hardground communities. The latter two are scrutinized elsewhere (Pratt et al. and Rozhnov, both in this volume).

In terms of taxonomic composition and dominant feeding strategies, Early and early Middle Cambrian level-bottom communities were similar to coeval reefal settings but differed by absence of heavily calcified organisms. In both cases, filter and suspension feeders dominated in both number and diversity (Zhuravleva et al. 1982, 1986; Wood et al. 1993; Kruse et al. 1995) (figures 10.2.1 and 10.2.2).

The shallow level-bottom community underwent significant changes during the Cambrian (see tables 10.1 and 10.2). After the demise of the Tommotian Evolutionary Fauna by the end of the Early Cambrian, communities were dominated by trilobites and lingulate brachiopods until the Middle Ordovician in Laurentia and Siberia (Sepkoski and Sheehan 1983; Sukhov and Pegel' 1986; Varlamov and Pak 1993), as well as in Australia, China, and Kazakhstan. During the Steptoean–Early Ordovician, community reorganization proceeded through the addition of new elements, especially gastropods, rostroconchs, and, from the Datsonian, cephalopods (Chen and Teichert 1983). In the Marjuman, trilobites account for two-thirds of the species present, associated with inarticulate brachiopods and hyoliths (Westrop et al. 1995). By the Datsonian–Early Ordovician interval, paleocommunity compositions were split more or less evenly between trilobites and mollusks. Finally, during the Middle Ordovician, trilobites were reduced to about one-third of the species composing communities (Westrop et al. 1995).

The Dysaerobic Community

The dysaerobic community represents an unusual kind of level-bottom community that usually exists in deep waters but, in case of hypertrophy, can also appear in shallow-water conditions.

A typical Early Cambrian example was recognized by Zhuravlev and Wood (1996) from the Botoman Sinsk Formation of the Siberian Platform (figure 10.2.3). The Sinsk biota is represented by the calcified cyanobacterium *Obruchevella* and the abundant

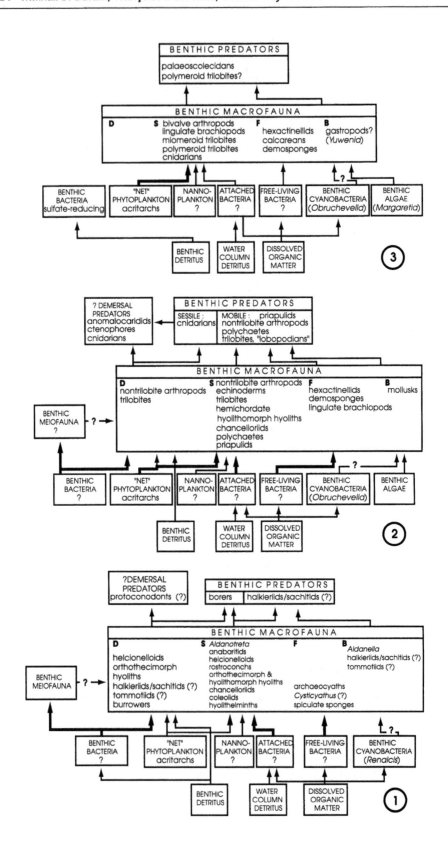

Figure 10.2 Trophic webs in the principal Early Cambrian benthic communities. *1*, Reefal archaeocyath-coralomorph-hyolith community; *2*, level-bottom open marine priapulid-nontrilobite arthropod-spicular sponge community; *3*, level-bottom dysaerobic trilobite-lingulate community (modified after Zhuravlev and Debrenne 1996). *B* = browsers and grazers; *D* = deposit feeders; *F* = filter feeders; *S* = suspension feeders.

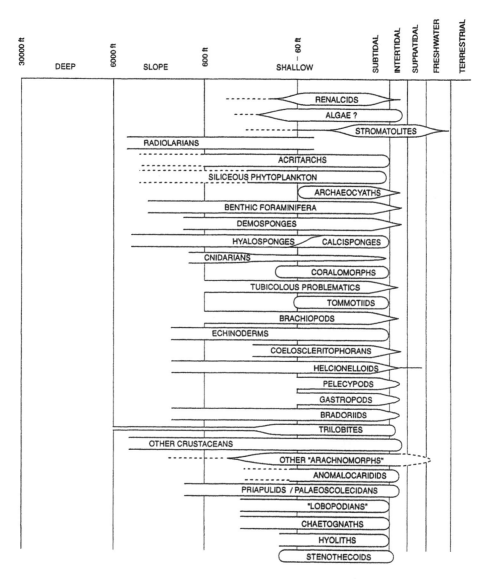

Figure 10.3 Distribution of major marine groups composing the Cambrian biota, relative to water depth. *Source:* Modified after Debrenne and Zhuravlev 1997.

green fleshy alga *Margaretia* as primary producers; by spicular sponges as filter feeders; by hyoliths, lingulate brachiopods, and probable cnidarians as suspension feeders, and rare paragastropods as grazers; and by palaeoscolecidans and, possibly, protolenin trilobites as carnivores. Abundant miomeroid trilobites could feed on minute organic particles, including algae (Fortey and Owens 1999). The absence of burrows reveals extreme reduction of deposit-feeders. Polymeroid trilobites with a wide, thin exoskeleton, a smooth carapace, multiple thoracic segments, and enlarged pleurae were nektobenthic trilobites adapted to low oxygen tension (Repina and Zharkova 1974; Fortey and Wilmot 1991). In turn, two other common groups, lingulates and palaeoscolecidans (closely related to priapulids), could survive dysaerobic conditions because their respiration was maintained by hemerythrin (Runnegar and Curry 1992). Volumetrically, trilobites and lingulates dominated. The latter might have fed on the abundant but monotypic acritarch flora. Despite harsh conditions, a multilevel tiering was developed by hexactinellids and demosponges that ranged in height from 4 to 60 cm (Ivantsov et al. 2000). A similar community occurred on the Siberian Platform during the late Early–early Middle Cambrian (Pel'man 1982). Later, agnostids and olenids replaced eodiscids and protolenins, respectively.

SILICICLASTIC SETTINGS

Deltas

Deltas are major depositional centers that produce thick sedimentary successions. High nutrient input, high turbidity, and decreased salinity are typical of deltaic areas.

In the prograde delta-front sequence of the Chapel Island Formation of the Nemakit-Daldynian of Avalonia, the higher-energy environments show a preponderance of vertical burrows (e.g., *Arenicolites*, *Skolithos*), simple horizontal burrows (*Buthotrephis*, *Planolites*), and few more-complex feeding burrow systems (e.g., *Phycodes*) (Crimes and Anderson 1985; Myrow and Hiscott 1993). Trace fossils from the Middle Cambrian deltaic Oville Sandstones of northern Spain were subdivided into several associations according to their restriction to tidal channel (*Rusophycus*, *Diplocraterion*, *Arenicolites*), sand flat (*Diplocraterion*, *Arenicolites*), mixed flat (*Arenicolites*, *Planolites*, *Rusophycus*, *Skolithos*, *Cruziana*, *Diplocraterion*), bar/beach (*Skolithos*), tidal delta slope (*Planolites*, *Rusophycus*, *Phycodes*), lower delta slope (*Teichichnus*, *Planolites*), or shelf/pro-delta (*Planolites*, *Teichichnus*) facies (Legg 1985). These examples show a diversity of feeding strategies in the deltaic communities, closely correlated with the energy conditions and mud content rather than with water depth.

Due to water column stratification, a dysaerobic bottom layer commonly developed in estuaries. This peculiar environment was deployed by organisms as early as the middle Vendian (Redkinan). In the estuaries of Baltica, the bushy alga *Eoholynia* formed floating mats (Burzin 1996). Their remains accumulated on the pycnocline, where they were further destroyed by sulfate-reducing bacteria before final deposition on the bottom (Burzin 1996). Sabelliditid-like tube-dwelling Saarinidae were the first animals adapted to such conditions in Baltica (Gnilovskaya 1996). During the Nemakit-Daldynian and Tommotian, true sabelliditids occupied similar conditions in this region. Their tubes contain pyrite framboids (Burzin 1993a), which may indicate sabelliditid symbiotic interactions with sulfate-reducing bacteria.

The early Botoman Chengjiang fauna of China has been regarded as an outer shelf to slope-basin assemblage (Hou and Sun 1988; Hou et al. 1991). However, close sedimentological examination has revealed that this impoverished community inhabited estuarine conditions (Lindström 1995). Nonetheless, this community contains diverse fleshy algae, spicular sponges, arthropods, and "lobopodians," as well as some palaeoscolecidans, anomalocaridids, cnidarians, chordates, and brachiopods (Chen and Zhou 1997). Several tiers are observed in this community, at least among sponges (Debrenne and Reitner, this volume). A similar community of trilobites, bivalved arthropods, palaeoscolecidans, anomalocaridids, and brachiopods is recognized in the youngest late Botoman of Australia (Big Gully fauna, Emu Bay Shale) but is interpreted as a lagoonal community (Nedin 1995).

Clastic Shoreline Systems

Clastic shoreline systems encompass a variety of environments—such as tidal flats, low intertidal, beach, shoreface, and barrier islands and very shallow, low-gradient subtidal and sand-shoal environments—that lie within fair-weather and average seasonal storm wave base or are continuously reworked by tidal and oceanic currents.

J-, U- or Y-shaped burrows were especially abundant and might have been produced by filter or suspension feeders that were deploying milieu energy (Vogel and Bretz 1972). Dense assemblages of *Skolithos,* a simple vertical tube, characterized shifting sandy substrates and high-energy conditions during the Cambrian (Hallam and Swett 1966; Droser 1991; Gozalo 1995). *Diplocraterion* probably dominated in warm semiarid climates (Middle Cambrian Riley Formation, Laurentia), where it replaced *Skolithos* under similar conditions (Cornish 1986). High-energy tidal channel sands and associated intertidal sand and mud flats contained only *Arenicolites, Astropolichnus, Bergaueria, Diplocraterion, Skolithos,* and rare *Cruziana, Rusophycus,* and *Planolites* (Crimes et al. 1977). In subtidal to intertidal areas of the Random Formation, with migrating sand bars and intervening channels, a variety of "deep"-water types such as *Helminthoida crassa, Nereites, Paleodictyon, Protopaleodictyon,* and *Squamodictyon* are present in addition to numerous shallow-water traces (Crimes and Anderson 1985). This anomalous distribution of *Nereites* ichnofacies may reflect not only inadequate supplies of food in the deep sea but also the patchy quality of food in shallow waters (Brasier 1995; Brasier and Lindsay, this volume). On the whole, these facies were dominated by soft-bodied suspension and filter feeders, with arthropods (e.g., *Rusophycus* dwellers; Jensen 1990) preying on them.

Shallow Siliciclastic Seas

Shallow siliciclastic seas are below fair-weather wave base but above rare storm wave base (~10–200 m). Major influencing factors are rate and type of sediment supply, type and intensity of the shelf hydraulic regime, sea level fluctuation, climatic and chemical factors, and organism-sediment interactions.

During the Kotlin interval, a wide, brackish restricted lake-like basin was formed in Baltica (Pirrus 1986; Burzin 1993a, 1996). Cyanobacterial, mostly oscillatoriacean, mats covered temporally dry parts of the basin. In more-stable areas, ribbonlike vendotaeniacean algae thrived and oscillatoriacean mats were restricted (Burzin 1993a, 1996). In the absence of animal competition, prokaryotic actinomycetes (*Primoflagella*) and chytrid fungi (*Vendomyces*) were principal consumers of the rich detritus (Burzin 1993b). Fungi and the Vendotaeniaceae still existed in the Nemakit-Daldynian, but the latter were replaced soon by *Tyrasotaenia*.

Since the Early Cambrian, shallow subtidal siliciclastic substrates that are unconsolidated and poorly sorted have been the principal habitats of different trace fossil dwellers. Lower-energy, thinner-bedded sand and mud flats contain abundant ichnofauna, including *Cruziana, Rusophycus, Monocraterion, Diplichnites, Planolites,* and occasional *Plagiogmus, Arenicolites, Diplocraterion, Phycodes, Teichichnus,* and *Skolithos* (Tommotian-Atdabanian Candana Quartzite and Herrería Sandstone, Spain and upper Chapel Island Formation, Avalonia) (Crimes et al. 1977; Crimes and Anderson 1985). With reduction in sediment grain size, a sharp drop in the number of sessile filter feeders (*Skolithos-, Diplocraterion-,* and *Monocraterion*-dwellers) is observed; in contrast, the number of deposit feeders increases (Pacześna 1996). *Cruziana* facies might bear filamentous algae, polymeroid and miomeroid trilobites, and diverse brachiopods (e.g., Middle Cambrian Murero Formation, Spain) (Liñan 1995). By the Late Cambrian, it was predominantly a *Cruziana-Rusophycus* community with additional lingulates (McKerrow 1978). Characteristic organisms therefore included suspension and deposit feeders, mobile carnivores, and scavengers.

During the Nemakit-Daldynian and Tommotian on Baltica and Laurentia, the *Platysolenites* community characterized muddy substrates (Rozanov 1983; Mount and Signor 1985). Later, during the Marjuman-Tremadoc, a shallow epicontinental sea continued to occupy Baltica (Sablinka, Ülgase, Ladoga, and Tosna formations). The dynamic conditions of this sea (shifting sands, turbidity) extremely limited benthic diversity. Only epibenthic obolids (lingulate brachiopods) and *Skolithos*-dwelling animals were adapted to such conditions (Popov et al. 1989). Branching hyolithelminths (*Torellella*) and acrotretid and/or siphonotretid lingulates formed their own community in deeper and less turbid parts of the basin on silty-muddy substrates, which were unfavorable for obolids. *Torellella* occupied a higher tier, and lingulates exploited the lower one. Modern analogies in the formation of economic phosphatic shell concentrations (Hiller 1993) and the Baltica paleoposition both reveal an influence of upwelling, bringing phosphate-rich cold waters onto a shallow shelf.

Amgan to Marjuman communities of shallow subtidal, rough-water, sandy environments of Bohemia (Jince Formation) consisted mostly of lingulates and polymeroid trilobites with few stylophoran echinoderms (*Ceratocystis*). In calmer, muddier, and deeper conditions, the diversity of trilobites increased, and agnostids, bradoriids, calciates, hyoliths, mollusks, and edrioasteroid, eocrinoid, and ctenocystoid echinoderms were associated with them (Mergl and Slehoferová 1990).

Open Shelves

Open shelves, facing the open ocean, offered shelter to level-bottom communities that lived in low-energy conditions. The Middle Cambrian Stephen and Wheeler Formation faunas (Laurentia) as well as Kaili Formation (South China) provide a unique insight into the structure of a level-bottom community, because of exceptional soft-part preservation (Conway Morris 1986; Robison 1991; Zhang et al. 1995) (see figure 10.2.2). A high ecologic complexity is recognized in the Phyllopod Bed (Stephen Formation) community, with a finer niche partitioning: low levels were occupied mostly by brachiopods, rare edrioasteroids, and sponges, together with rarer pelmatozoan echinoderms, which dominated at higher levels. More contrasting is the role of fleshy algae and cyanobacteria as primary producers and deposit-feeding arthropods as consumers. Diverse predators are confirmed by the abundance of effectively soft-bodied carnivores, together with some carnivores with hard parts.

CONCLUSIONS

On the whole, Vendian (pre–Nemakit-Daldynian) benthic communities were dominated by microbial ecosystems. Only in temperate conditions, usually coupled with active hydrodynamics, were more-complex communities developed. Although the Ediacara fauna is known from shallow-water onshore and offshore sandstones, deep-water turbidites, and shallow-water carbonates (Fedonkin 1987), it may be attributed to a single community of sessile filter and/or suspension feeders and to a few deposit feeders independently of the systematic interpretation of its members (Lipps et al. 1992).

Cambrian communities evolved by further partitioning ecologic niches and by replacing older forms by new ones within established trophic guilds. This is especially noticeable in ecospace utilization, as recognized by developed tiers and diverse feeding strategies (tables 10.1 and 10.2; see also figures 10.1 and 10.2). In this respect, the Cambrian biota was very different from the Neoproterozoic one, but also from younger Paleozoic and Meso-Cenozoic biotas (Bambach 1977; Ausich and Bottjer 1982; Bottjer and Ausich 1986). The relatively diverse Early Cambrian benthos was replaced during the Middle Cambrian by a simpler, trilobite-dominated Cambrian sensu stricto Evolutionary Fauna. Taxonomic impoverishment of Cambrian

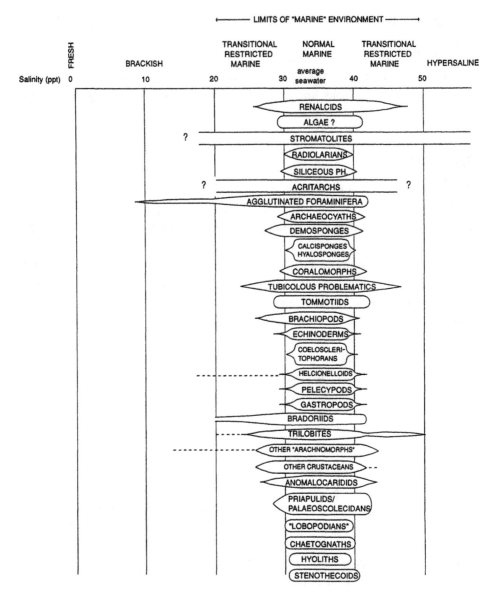

Figure 10.4 Distribution of major marine groups composing the Cambrian biota, relative to salinity. *Source:* Modified after Debrenne and Zhuravlev 1997.

communities is observed in siliciclastic as well as in carbonate level-bottom communities. A similar trend is obvious in reefal communities (Zhuravlev 1996). In terms of taxonomic composition, proximal communities were the most conservative; distal, deep-water communities grew permanently by addition of new elements (Conway Morris 1989); and intermediate shallow shelf subtidal communities were the most changeable.

Displacement of communities was common. The earliest deep-water communities were derived from former shallow-water water elements (Bottjer et al. 1988; Crimes, this volume). On the other hand, a trilobite-lingulate community, which during the Middle Cambrian occupied normal marine conditions, at first appeared in dysaerobic Early Cambrian basins (see figure 10.2.3).

Although Cambrian organisms occupied various depositional regimes, they were largely restricted to normal marine and to shallow subtidal conditions (figures 10.3 and 10.4). Possibly, the further spreading into these environments was one of the sources of higher Ordovician and Mesozoic-Cenozoic radiations.

Acknowledgments. We thank Mary Droser and Peter Crimes for helpful comments and Françoise Pilard and Henri Lavina for the preparation of figures. This paper is a contribution to IGCP Project 366.

REFERENCES

Astashkin, V. A. 1985. Problematichnye organizmy—porodoobrazovateli v nizhnem kembrii Sibirskoy platformy [Problematic rock-forming organisms in the Lower Cambrian of the Siberian Platform]. *Trudy, Institut geologii i geofiziki, Sibirskoe otdelenie, Akademiya nauk SSSR* 632:144–149.

Ausich, W. I. and D. J. Bottjer. 1982. Tiering in suspension-feeding communities on soft substrata throughout the Phanerozoic. *Science* 216:173–174.

Balsam, W. L. 1974. Reinterpretation of an archaeocyathid reef: Shady Formation, South Western Virginia. *Southeastern Geology* 16:121–129.

Bambach, R. K. 1977. Species richness in marine benthic habitats through Phanerozoic. *Paleobiology* 3:152–167.

Bengtson, S. and S. Conway Morris. 1992. Early radiation of biomineralizing phyla. In J. H. Lipps and P. W. Signor, eds., *Origin and Early Evolution of the Metazoa,* pp. 447–481. New York: Plenum Press.

Bottjer, D. J. and W. I. Ausich. 1986. Phanerozoic development of tiering in soft substrata suspension-feeding communities. *Paleobiology* 12:400–420.

Bottjer, D. J., M. L Droser, and D. Jablonski. 1988. Palaeoenvironmental trends in the history of trace fossils. *Nature* 333:252–255.

Brasier, M. D. 1995. Fossil indicators of nutrient levels. 1: Eutrophication and climate change. In D. W. J. Bosence and P. A. Allison, eds., *Marine Palaeoenvironmental Analysis from Fossils,* pp. 113–132. Geological Society Special Publication 83.

Brasier, M. D. and R. A. Hewitt. 1979. Environmental setting of fossiliferous rocks from the uppermost Proterozoic–Lower Cambrian of Central England. *Palaeogeography, Palaeoclimatology, Palaeoecology* 27:35–57.

Burzin, M. B. 1993a. Mikrobnye bentosnye soobshchestva pozdnego venda [Microbial benthic communities of the late Vendian]. In A. Yu. Rozanov, ed., *Problemy doantropogennoy evolyutsii biosfery* [Problems of the preanthropogenic evolution of the biosphere], pp. 282–293. Moscow: Nauka.

Burzin, M. B. 1993b. Drevneyshiy khitridio-mitset (Mycota, Chytridiomycetes incertae sedis) iz verkhnego venda Vostochno-Evropeyskoy platformy [The oldest chytrid (Mycota, Chytridiomycetes incertae sedis) from the Upper Vendian of the East-European Platform]. In B. S. Sokolov, ed., *Fauna i ekosistemy geologicheskogo proshlogo* [Fauna and ecosystems of geological past], pp. 21–33. Moscow: Nauka.

Burzin, M. B. 1996. Late Vendian (Neoproterozoic III) microbial and algal communities of the Russian Platform: Models of facies-dependent distribution, evolution, and reflection of basin development. *Rivista Italiana di Paleontologia e Stratigrafia* 102:307–315.

Chen, J.-Y. and C. Teichert. 1983. Cambrian Cephalopoda of China. *Palaeontographica A* 181:1–102.

Chen, J. and G. Zhou. 1997. Biology of the Chengjiang fauna. *Bulletin of the National Museum of Natural Sciences, Taichung, Taiwan* 10:11–105.

Conway Morris, S. 1976. Fossil priapulid worms. *Special Papers in Palaeontology* 20: 1–101.

Conway Morris, S. 1979. Middle Cambrian polychaetes from the Burgess Shale of British Columbia. *Philosophical Transactions of the Royal Society of London B* 285:227–274.

Conway Morris, S. 1986. The community structure of the Middle Cambrian Phyllopod Bed (Burgess Shale). *Palaeontology* 29: 423–467.

Conway Morris, S. 1989. The persistence of Burgess Shale–type faunas: Implications for the evolution of deeper-water faunas. *Transactions of the Royal Society of Edinburgh (Earth Sciences)* 80:271–283.

Conway Morris, S. and S. Bengtson. 1994. Cambrian predators: Possible evidence from boreholes. *Journal of Paleontology* 68: 1–23.

Cornish, F. G. 1986. The trace-fossil *Diplocraterion:* Evidence of animal-sediment interactions in Cambrian tidal deposits. *Palaios* 1:478–491.

Crimes, T. P. 1992. The record of trace fossils across the Proterozoic-Cambrian boundary. In J. H. Lipps and P. W. Signor, eds., *Origin and Early Evolution of the Metazoa,* pp. 177–202. New York: Plenum Press.

Crimes, T. P. and M. M. Anderson. 1985. Trace fossils from late Precambrian–Early Cambrian strata of southeastern Newfoundland (Canada): Temporal and environmental implications. *Journal of Paleontology* 59:310–343.

Crimes, T. P. and M. L. Droser. 1992. Trace fossils and bioturbation: The other fossil record. *Annual Review on Ecology and Systematics* 23:339–360.

Crimes, T. P., I. Legg, A. Marcos, and M. Arboleya. 1977. Late Precambrian–low Lower Cambrian trace fossils from Spain. *Geological Journal Special Issue* 9:91–138.

Debrenne, F. and A. Yu. Zhuravlev. 1997. Cambrian food web: A brief review. *Géobios,* Mémoir *spécial* 20:181–188.

Droser, M. L. 1991. Ichnofabric of the Paleozoic *Skolithos* ichnofacies and the nature and distribution of *Skolithos* piperock. *Palaios* 6:316–325.

Droser, M. L. and D. J. Bottjer. 1988. Trends in extent and depth of early Paleozoic bioturbation in the Great Basin (California, Nevada, and Utah). In D. L. Weide and M. L. Faber, eds., *This Extended Land: Geological Journeys in the Southern Basin and Range, Field Trip Guidebook,* pp. 123–135. Geological Society of America, Cordilleran Section, Los Angeles.

Druce, E. C., J. H. Shergold, and B. M. Radke. 1982. A reassessment of the Cambrian-Ordovician boundary section at Black Mountain, western Queensland, Australia. In M. G. Bassett and W. T. Dean, eds.,

The Cambrian-Ordovician Boundary: Sections, Fossil Distribution, and Correlations, pp. 193–209. Cardiff: National Museum of Wales, Geological Series 3.

Edhorn, A.-S. 1977. Early Cambrian algae croppers. Canadian Journal of Earth Sciences 14:1014–1020.

Fedonkin, M. A. 1987. Besskeletnaya fauna venda i ee mesto v evolyutsii metazoa [Nonskeletal Vendian Fauna and its place in the evolution of the metazoa]. Trudy, Paleontologicheskiy institut, Akademiya nauk SSSR 226.

Fortey, R. A. and R. M. Owens. 1999. Feeding habits in trilobites. Palaeontology 42:429–465.

Fortey, R. A. and N. V. Wilmot. 1991. Trilobite cuticle thickness in relation to palaeoenvironment. Paläontologische Zeitschrift 65: 141–151.

Gerasimenko, L. M., I. V. Goncharova, E. A. Zhegallo, G. A. Zavarzin, L. V. Zaytseva, V. K. Orleanskiy, A. Yu. Rozanov, and G. T. Ushatinskaya. 1996. Protsess mineralizatsii (fosfatizatsii) nitchatykh tsianobakteriy [Process of the mineralization (phosphatization) of filamentous cyanobacteria]. Litologiya i poleznye iskopaemye 1996 (2): 208–214.

Gnilovskaya, M. B. 1996. Novye saarinidy iz venda Russkoy platformy [New saarinids from the Vendian of the Russian Platform]. Doklady Rossiyskoy Akademii nauk 348:1–5.

Gozalo, R. 1995. El Cámbrico de las Cadenas Ibéricas. In J. A. Gámez Vintaned and E. Liñan, eds., La expansión de la vida en el Cámbrico, pp. 137–167. Zaragoza: Institución "Fernando de Católico."

Hallam, A. and K. Swett. 1966. Trace fossils from the Lower Cambrian Pipe Rock of the north-west Highlands. Scottish Journal of Geology 2:101–106.

Hiller, N. 1993. A modern analogue for the Lower Ordovician Obolus conglomerate of Estonia. Geological Magazine 130:265–267.

Hou, X.-G. and J. Bergström. 1994. Palaeoscolecid worms may be nematomorphs rather than annelids. Lethaia 27:11–17.

Hou, X.-G. and W.-G. Sun. 1988. [Discovery of Chengjiang fauna at Meishucun, Jinning, Yunnan]. Acta Palaeontologica Sinica 27:1–12.

Hou, X.-G., L. Ramsköld, and J. Bergström. 1991. Composition and preservation of the Chengjiang fauna—a Lower Cambrian soft-bodied biota. Zoologica Scripta 20:395–411.

Ivantsov, A. Yu., A. Yu. Zhuravlev, V. A. Krassilov, A. V. Leguta, L. M. Melnikova, L. N. Repina, A. Urbanek, G. T. Ushatinskaya and Ya. E. Malakhovskaya. 2000. Unikal'nye sinskie mestonakhozdeniya rannekembriyskikh organizmov (Sibirskaya platforma) [Extraordinary Sinsk localities of Early Cambrian organisms (Siberian Platform)]. Moscow: Paleontological Institute.

Jago, J. B. 1974. Evidence for scavengers from Middle Cambrian sediments in Tasmania. Neues Jahrbuch für Geologie und Paläontologie Monatshefte 1:13–17.

Jensen, S. 1990. Predation by Early Cambrian trilobites on infaunal worms—evidence from Swedish Mickwitzia Sandstone. Lethaia 23:29–42.

Kobluk, D. R. 1985. Biota preserved within cavities in Cambrian Epiphyton mounds, Upper Shady Dolomite, southwestern Virginia. Journal of Paleontology 59:1158–1172.

Kobluk, D. R. and C. F. Kahle. 1978. Geological significance of boring and cavity-dwelling marine algae. Canadian Petroleum Geology Bulletin 28:362–379.

Kobluk, D. R. and M. J. Risk. 1977. Classification of exposed filaments of endolithic algae, micrite envelope formation, and sediment production. *Journal of Sedimentary Petrology* 47:517–528.

Korolyuk, I. K. 1968. Biogennye obrazovaniya Zapadnogo Pribaikal'ya [Biogenic formations of western Cisbaikalia]. In G. A. Smirnov and M. L. Klyuzhina, eds., *Iskopaemye rify i metody ikh izucheniya* [Fossil reefs and methods of their study], pp. 55–71. Sverdlovsk: Institute of Geology and Geophysics, Uralian Branch, USSR Academy of Sciences.

Kruse, P. D., A. Yu. Zhuravlev, and N. P. James. 1995. Primordial metazoan-calcimicrobial reefs: Tommotian (Early Cambrian) of the Siberian Platform. *Palaios* 10:291–321.

Landing, E. 1991. Upper Precambrian through Lower Cambrian of Cape Breton Island: Faunas, paleoenvironments, and stratigraphic revision. *Journal of Paleontology* 65:570–595.

Landing, E. 1993. In situ earliest Cambrian tube worms and the oldest metazoan-constructed biostrome (Placentian Series, southeastern Newfoundland). *Journal of Paleontology* 67:333–342.

Landing, E., P. A. Myrow, and G. M. Narbonne. 1989. The Placentian Series: Appearance of the oldest skeletonized faunas in southeastern Newfoundland. *Journal of Paleontology* 63:739–769.

Legg, I. C. 1985. Trace fossils from a Middle Cambrian deltaic sequence, north Spain. In H. A. Curran, ed., Biogenic structures: Their use in interpreting depositional environments, *Society of Economic Paleontologists and Mineralogists Special Publication* 35:151–165.

Li, G. 1997. Early Cambrian phosphate-replicated endolithic algae from Emei, Sichuan, SW China. *Bulletin of the National Museum of Natural Sciences, Taichung, Taiwan* 10:193–216.

Liñan, E. 1995. Una aproximación a los ecosistemas marinos cámbricos. In J. A. Gámez Vintaned, and E. Liñan, eds., *La expansión de la vida en el Cámbrico*, pp. 27–48. Zaragoza: Institución "Fernando de Católico."

Lindström, M. 1995. The environment of the Early Cambrian Chengjiang fauna. In *International Cambrian Explosion Symposium (April 1995, Nanjing), Programme and Abstracts*, p. 17.

Lipps, J. H. 1983. Biotic interactions in benthic Foraminifera. In M. J. S. Tevesz and P. L. McCall, eds., *Biotic Interactions in Recent and Fossil Benthic Communities*, pp. 331–376. New York: Plenum Press.

Lipps, J. H., S. Bengtson, and M. A. Fedonkin. 1992. Ecology and biogeography. In J. W. Schopf and C. Klein, eds., *The Proterozoic Biosphere: A Multidisciplinary Study*, pp. 437–441. Cambridge: Cambridge University Press.

Martin, F. 1993. Acritarchs: A review. *Biological Reviews* 68:475–538.

McIlroy, D. and G. R. Heys. 1997. Palaeobiological significance of *Plagiogmus arcuatus* from the Lower Cambrian of central Australia. *Alcheringa* 21:161–178.

McIlroy, D., O. R. Green, and M. D. Brasier. 1994. The world's oldest foraminiferans. *Microscopy and Analysis* (November): 13–15.

McKerrow, W. S., ed. 1978. *The Ecology of Fossils*. Cambridge: MIT Press.

Melchin, M. J. and K. M. Doucet. 1996. Modelling flow patterns in conical dendroid graptolites. *Lethaia* 29:39–46.

Mergl, M. and P. Slehoferová. 1990. Middle Cambrian inarticulate brachiopods from Central Bohemia. *Sbornik geologických Věd, Paleontologie* 31:67–104.

Mount, J. F. and P. W. Signor. 1985. Early Cambrian innovation in shallow subtidal environments: Paleoenvironments of Early Cambrian shelly fossils. *Geology* 13:730–733.

Müller, K. J. and Y. Nogami. 1972. Entöken und Bohrspuren bei den Conodontophorida. *Paläontologische Zeitscrift* 46:68–86.

Myrow, P. M. and R. N. Hiscott. 1993. Depositional history and sequence stratigraphy of the Precambrian-Cambrian boundary stratotype section, Chapel Island Formation, southeastern Newfoundland. *Paleogeography, Palaeoclimatology, Palaeoecology* 104:13–35.

Nedin, C. 1995. The Emu Bay Shale: A Lower Cambrian fossil Lagerstätten, Kangaroo Island, South Australia. *Association of Australasian Palaeontologists, Memoir* 18:31–40.

Nedin, C. 1999. *Anomalocaris* predation on nonmineralized and mineralized trilobites. *Geology* 27:987–990.

Pacześna, J. 1996. The Vendian and Cambrian ichnocoenoses from the Polish part of the East-European Platform. *Prace Państwowego Institutu Geologicznego* 72:1–77.

Parkhaev, P. Yu. 1998. Siponoconcha—A new class of Early Cambrian bivalved organisms. *Paleontological Journal* 32 (1):1–15.

Pel'man, Yu. L. 1982. Ekologicheskie gruppirovki fauny v srednekembriyskikh domanikoidnykh otlozheniyakh kuonamskoy svity (Sibirskaya platforma, r. Muna) [Ecological groups of fauna in the Middle Cambrian domanicoid sediments of the Kuonamka Formation (Siberian Platform, Muna River)]. *Trudy, Institut geologii i geofiziki, Sibirskoe otdelenie, Akademiya nauk SSSR* 628:60–74.

Pirrus, E. 1986. [The groundlines of the lithogenesis of the Cambrian and Vendian deposits in the Baltic syneclise and its neighboring area]. In E. Pirrus, ed., *Fatsii i stratigrafiya venda i kembriya zapada Vostochno-Evropeyskoy platformy* [Facies and stratigraphy of the Vendian and Cambrian of the western East-European Platform], pp. 171–177. Tallin: Akademiya nauk Estonskoy SSR.

Popov, L. Ye., K. K. Khazanovich, N. G. Borovko, S. P. Sergeeva, and R. F. Sobolevskaya. 1989. *Opornye razrezy i stratigrafiya kembro-ordovikskoy fosforitonosnoy obolovoy tolshchi na severo-zapade Russkoy platformy* [Reference sections and stratigraphy of the Cambrian-Ordovician phosphate-bearing *Obolus* Unit on the northwest of the Russian Platform]. Leningrad: Nauka.

Pratt, B. R. 1994. Possible predation of early Late Cambrian trilobites and its disappearance in the Marjuman-Steptoan extinction event. *Terra Nova* 6 (Abstract Supplement 3): 6.

Purnell, M. A. 1995. Microwear on conodont elements and macrophagy in the first vertebrates. *Nature* 374:798–800.

Repina, L. N. 1977. Biofatsii trilobitov tarynskogo urovnya nizhnego kembriyskom Sibirskoy platformy [Trilobite biofacies at the Lower Cambrian Taryn level of the Siberian Platform]. *Trudy, Institut geologii i geofiziki, Sibirskoe otdelenie, Akademiya nauk SSSR* 302:51–74.

Repina, L. N. and T. M. Zharkova. 1974. Ob usloviyakh obitaniya trilobitov v rannekembriyskom basseyne Sibiri [On the environmental conditions of trilobites in the Early Cambrian basin of Siberia]. *Trudy, Institut geologii i geofiziki, Sibirskoe otdelenie, Akademiya nauk SSSR* 84:100–108.

Rickards, R. B., P. W. Baillie, and J. B. Jago. 1990. An Upper Cambrian (Idamean) dendroid assemblage from near Smithton, northwestern Tasmania. *Alcheringa* 14:207–232.

Riding, R. 1991. Classification of microbial carbonates. In R. Riding, ed., *Calcareous Algae and Stromatolites*, pp. 21–51. New York: Springer Verlag.

Robison, R. A. 1991. Middle Cambrian biotic

diversity: Examples from four Utah Lager-stätten. In A. M. Simonetta and S. Conway Morris, eds., *The Early Evolution of Metazoa and the Significance of Problematic Taxa*, pp. 77–98. Cambridge: Cambridge University Press.

Rozanov, A. Yu. 1983. *Platysolenites*. In A. Urbanek and A. Yu. Rozanov, eds., *Upper Precambrian and Cambrian Palaeontology of the East-European Platform*, pp. 94–100. Warsaw: Wydawnictwa Geologiczne.

Rozanov, A. Yu. and E. A. Zhegallo. 1989. K probleme genezisa drevhikh fosforitov Azii [To the problem of the genesis of ancient phosphorites in Asia]. *Litologiya i poleznye iskopaemye* 1989 (3):67–82.

Runnegar, B. and G. B. Curry. 1992. Amino acid sequences of hemerythrin from *Lingula* and a priapulid worm and the evolution of oxygenic transport in the Metazoa. In *29th International Geological Congress, Kyoto, Japan, 24 August–3 September 1992, Abstracts*, vol. 2, p. 346.

Sdzuy, K. 1974. Mittelkambrische Graptolithen aus NW-Spanien. *Paläontologische Zeitschrift* 48:110–139.

Sepkoski, J. J., Jr., and P. M. Sheehan. 1983. Diversification, faunal change, and community replacement during the Ordovician radiation. In M. J. S. Tevesz and P. L. McCall, eds., *Biotic Interactions in Recent and Fossil Benthic Communities*, pp. 673–717. New York: Plenum Press.

Sergeev, V. N. and R. N. Ogurtsova. 1989. Mikrobiota nizhnekembriyskikh fosforitonosnykh otlozheniy Malogo Karatau (Yuzhnyy Kazakhstan) [Lower Cambrian microbiota from phosphorite-bearing sediments of Maly Karatau (southern Kazakhstan)]. *Izvestiya Akademii nauk SSSR, Seriya geologicheskaya* 1989 (3):58–66.

Sorokin, Yu. I. 1992. K kharakteristike planktona vod Bol'shogo Bar'ernogo rifa [To the plankton characteristics from the Great Barrier Reef waters]. *Zhurnal Obshchey Biologii* 53:557–570.

Soudry, D. and P. N. Southgate. 1989. Ultrastructure of a Middle Cambrian primary nonpelletal phosphorite and its early transformation into phosphate vadoids: Georgina Basin, Australia. *Journal of Sedimentary Petrology* 59:53–64.

Sukhov, S. S. and T. V. Pegel'. 1986. Lito- i biofatsial'nyy analiz srednekembriyskikh otlozheniy vostoka Sibirskoy platformy dlya rekonstruktsii usloviy karbonato-nakopleniya [Lithological and biofacies analysis of the Middle Cambrian strata on the east of the Siberian Platform for the carbonate deposition environment reconstruction]. In V. I. Krasnov, ed., *Paleoekologicheskiy i litologo-fatsial'nyy analizy dlya obosnovaniya detal'nosti regional'nykh stratigraficheskikh skhem* [Palaeoecological and lithological facies analyses for the grounds of detailed regional stratigraphic schemes], pp. 35–50. Novosibirsk: Siberian Scientific-Research Institute of Geology, Geophysics, and Mineral Resources.

Szaniawski, H. 1982. Chaetoghnath grasping spines recognized among Cambrian protoconodonts. *Journal of Paleontology* 56:806–810.

Underwood, C. J. 1993. The position of graptolites within Lower Palaeozoic planktic ecosystems. *Lethaia* 26:189–202.

Varlamov, A. I. and K. L. Pak. 1993. Soobshchestva trilobitov i fatsii verkhnekembriyskikh otlozheniy severo-zapada Sibirskoy platformy [Upper Cambrian trilobite communities and facies on the northwest of the Siberian Platform]. *Stratigrafiya, Geologicheskaya korrelyatsiya* 1:104–110.

Vogel, S. and W. L. Bretz. 1972. Interfacial organisms: Passive ventilation in the velocity gradients near surface. *Science* 175:210–211.

Walossek, D., J. Repetski, and K. Müller.

1994. An exceptionally preserved parasitic arthropod *Heymonsicambria taylori* n. sp. (Arthropoda incertae sedis: Pentastomida), from Cambrian-Ordovician boundary beds of Newfoundland, Canada. *Canadian Journal of Earth Sciences* 31:1664–1671.

Westrop, S. R., J. V. Tremblay, and E. Landing. 1995. Declining importance of trilobites in Ordovician nearshore paleocommunities: Dilution or displacement? *Palaios* 10:75–79.

Wood, R., A. Yu. Zhuravlev, and A. Chimed Tseren. 1993. The ecology of Lower Cambrian buildups from Zuune Arts, Mongolia: Implications for early metazoan reef evolution. *Sedimentology* 40:829–858.

Zhang, Z.-G., X.-Y., Gong, Y.-L., Zhao, J.-R. Mao, and J.-W. Shen. 1995. Palaeoenvironment of the Kaili Lagerstätten. *International Cambrian Explosion Symposium (April 1995, Nanjing), Programme and Abstracts,* pp. 44–45.

Zhegallo, E. A., A. G. Zamiraylova, and Yu. N. Zanin. 1994. Mikroorganizmy v sostave porod kuonamskoy svity nizhnego—srednego kembriya Sibirskoy platformy (r. Molodo) [Microorganisms in the Lower–Middle Cambrian Kuonamka Formation rocks on the Siberian Platform (Molodo River)]. *Litologiya i poleznye iskopaemye* 1994 (5):23–27.

Zhuravlev, A. Yu. 1996. Reef ecosystem recovery after the Early Cambrian extinction. In M. B. Hart, ed., *Biotic Recovery from Mass Extinction Events,* pp. 79–96. Geological Society Special Publication, London 102.

Zhuravlev, A. Yu. and F. Debrenne. 1996. Pattern of evolution of Cambrian benthic communities: Environments with carbonate sedimentation. *Rivista Italiana de Paleontologia e Stratigrafia* 102:333–340.

Zhuravlev, A. Yu. and R. A. Wood. 1996. Anoxia as the cause of the mid-Early Cambrian (Botomian) extinction event. *Geology* 24:311–314.

Zhuravleva, I. T., N. P. Meshkova, V. A. Luchinina, and L. N. Kashina. 1982. Biofastsii Anabarskogo morya v pozdnem dokembrii i rannem kembrii [Biofacies of the Anabar Sea in the Late Precambrian and Early Cambrian]. *Trudy, Institut geologii i geofiziki, Sibirskoe otdelenie, Akademiya nauk SSSR* 510:74–103.

Zhuravleva, I. T., V. A. Luchinina, N. P. Meshkova, Yu. L. Pel'man, L. N. Repina, and Z. V. Borodaevskaya. 1983. Ekologiya naseleniya rannekembriyskogo basseyna Sibirskoy platformy na primere Atdabanskogo rifoida [Ecology of the inhabitants of an Early Cambrian basin from the Siberian Platform exemplified by the Atdaban Reefoid]. *Trudy, Paleontologicheskiy institut, Akademiya nauk SSSR* 194:33–43.

Evolution of the Hardground Community

Hardground communities first appeared in the late Middle Cambrian but they were not common before the Ordovician. Two factors had a major influence on the early development of hardgrounds and resulted in abrupt and rapid increase in hardground area as well as in community density and diversity. The first factor was the change from an aragonite to a calcite sea epoch; the second factor was positive feedback between the expansion of hardgrounds and the increase in carbonate production by members of hardground communities. Stemmed echinoderms played a key role in the development of hardgrounds.

HARDGROUNDS, areas of synsedimentarily lithified carbonate sea floor, occurred for the first time in the late Middle Cambrian and were widely distributed in the Ordovician. The time of their occurrence and wide distribution coincided with the Ordovician radiation of marine biota, which resulted in the replacement of the Cambrian Evolutionary Fauna by the Paleozoic Evolutionary Fauna (Sepkoski 1979, 1981, 1984) that was to dominate the remainder of the Paleozoic. A significant increase in biodiversity was connected with this radiation.

The lack of appearance of new taxa of rank higher than class and subphylum, apart from the Bryozoa, was characteristic of the Ordovician radiation. In comparison, the previous major radiation, during the Precambrian-Cambrian interval, led to the formation of new phyla and subphyla. After the Permian extinction, no new taxa above subclass, and generally not higher than ordinal rank, arose. New taxa of marine biota at the Cretaceous-Tertiary boundary did not exceed superfamilial and subordinal rank (Valentine 1992) (figure 11.1).

The Cambrian-Ordovician transition is the most interesting interval for the study of the evolution of higher taxa of marine biota. One of the major radiations at high taxonomic level in the history of the marine fauna took place at this time. Because ancestors of many Ordovician organisms already had skeletons in the Cambrian, it is possible to study the Ordovician radiation. We can thus compare these two consec-

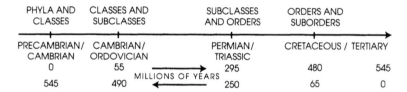

Figure 11.1 Maximum taxonomic rank among marine Metazoa during major evolutionary radiations of the Phanerozoic.

utive faunas effectively and trace the trends in the formation of taxa of higher rank: classes and subclasses.

Valentine (1992), in examining the macroevolution of phyla, suggested that phyla and other higher taxa remain cryptogenetic whether studied from the perspectives of comparative developmental and/or adult morphology, of molecular evolution, or of the fossil record. This suggestion is accepted by many authors, and the explanation for it is usually that many branches of the evolutionary tree "originated relatively abruptly and within a narrow window of geologic time" (Valentine 1992:543).

When explaining high rates of evolution at the moment of occurrence of higher taxa, various authors draw attention to internal aspects of evolution, such as significant fast genome reorganization and various kinds of heterochrony, or to external aspects—characteristics of the environment. Both these kinds of aspects can be seen in the Ordovician radiation (Droser et al. 1996). The pattern of their interaction is discussed in this chapter.

Change in marine substrate structure represented the abiotic factor that directly influenced the Ordovician radiation of benthic fauna: hardgrounds became widely distributed and many soft substrates became enriched by bioclastic debris. The main purpose of this chapter is to demonstrate the connection between radiation of marine biota and change in substrate type, as well as to show the interrelationships of these processes.

TYPES OF HARD SEA FLOOR

The faunas of hard sea floors always differ strongly in composition and number from those of soft sea floors. There are two main types of hard sea floor, differing in their mechanism of formation and in hydraulic energy: rockgrounds and hardgrounds. Consequently, these kinds of substrate differ strongly in their environmental conditions.

Rockgrounds

Rockgrounds are formed during transgressions accompanied by erosion of previously accumulated deposits. They represent high-energy environments, and this determines the adaptations of the associated fauna. The rocky sea floor has existed since the appearance of marine basins, framework cavities within reefs and deep-water rocky areas of the bottom; surfaces of submarine lava flows, pebbles, etc., exemplify such rockgrounds. Inhabited reefal cavities are known from at least as early as the Paleoproterozoic (Hofmann and Grotzinger 1985; Turner et al. 1993). The rocky sea floor has always occupied a relatively small part of marine substrates (Johnson 1988) and therefore has not played an important role, although sometimes it has influenced the formation of the marine biota in a very special way, as has happened, for instance, around volcanic vents.

Hardgrounds

Hardgrounds are "synsedimentarily lithified carbonate seafloor that became hardened in situ by the precipitation of a carbonate cement in the primary pore spaces" (Wilson and Palmer 1992:3). Thus, hardgrounds are not necessarily associated with very high hydrodynamic energy.

Hardgrounds occurred for the first time in geologic history not earlier than late Middle Cambrian. Since the Ordovician, hardgrounds have occupied locally extensive areas on the sea floor and have been characterized by an abundant and diverse benthic fauna. Hardgrounds may pass laterally to various debris-rich soft grounds, resulting in the existence of mixed hardground and softground associations.

Wide distribution of hardgrounds from the beginning of the Ordovician can be largely explained by abiotic factors (Wilson et al. 1992; Myrow 1995), the most important of which was lowering of the Mg^{2+}/Ca^{2+} ratio and rise of CO_2 activity in seawater, which can account for change in mineralogy of marine carbonate precipitates. This resulted in the replacement of shallow-water high-magnesium calcite and aragonite precipitation by low-magnesium calcite: so-called aragonite seas were replaced by calcite seas (Sandberg 1983). The original calcite cement grew syntaxially on calcite substrates such as echinoderm ossicles and other calcite bioclasts; early aragonite cement could not do this. Although hardgrounds occur in aragonite seas as well (e.g., at the present day), they appear to have been more widespread in calcite sea times because calcite precipitates faster and more extensively (Wilson and Palmer 1992).

The structure of the echinoderm skeleton is another factor promoting hardground formation, through a significant increase in calcite debris on the sea floor (Wilson and Palmer 1992). First, it is highly porous and hence achieves considerably greater volume for the same weight in comparison with calcite skeletons of other animals. Second, the skeletons of echinoderms are built from separate small skeletal elements joined together by organic ligament. This construction allows rapid postmortem disarticulation and fragmentation of the skeleton, with the accumulation of large amounts of debris on the sea floor. For example, after death and fragmentation, a crinoid skeleton with height of 1 m and a stem diameter of 0.5 cm could produce enough debris to cover at least 0.5 m^2 of sea floor with a layer 1 mm thick.

A certain balance between sediment deposition and lithification is necessary for hardground formation. When sedimentation was faster than lithification, a particular kind of softground with a hardened underlying layer was formed. This phenomenon is responsible for a wide variety of semihard substrates and for their various combinations with true hardgrounds, which has resulted in a high diversity of benthic fauna inhabiting these substrates, as can be observed, for example, in the Early Ordovician of the Baltic paleobasin (Rozhnov 1994).

CHARACTERISTICS OF THE EARLY PALEOZOIC SEA FLOOR

Marine Substrates in the Cambrian

The Cambrian sea floor was covered mainly with soft silt sediments, whereas deposits enriched with bioclastic debris were rare (see also Droser and Li, this volume). In the Early Cambrian, firm bottoms occupied small areas and were represented almost entirely by rockgrounds. Rockground faunas are poorly known on account of their poor preservation.

Nevertheless an unusual fauna was discovered in calcimicrobial-archaeocyath reefs of western Nevada and Labrador (James et al. 1977; Kobluk and James 1979): calcified cyanobacteria, sponges (including juvenile archaeocyaths), possible foraminifers, some problematic organisms, and *Trypanites* borings. These organisms inhabited reefal cavities that were completely or partially protected from wave action. A similar cryptic fauna has been found in cavities of Early Cambrian reefs in many regions of the world, including the Siberian Platform, southern Urals, Altay Sayan Foldbelt, Mongolia, southern Australia, and Antarctica (Zhuravlev and Wood 1995).

Hardgrounds formed by early diagenetic replacement of cyanobacterial mats by phosphatic minerals are known from the Middle Cambrian of Greenland. Numerous small echinoderm(?) holdfasts are attached to these hardgrounds (Wilson and Palmer 1992).

The earliest typical hardground surfaces, with numerous eocrinoid holdfasts and some orthid brachiopods and spicular demosponges, have been found in the late Middle Cambrian part of the Mila Formation in the Elburz Mountains, northern Iran (Zhuravlev et al. 1996) (figure 11.2). In this example, hardgrounds developed on calciate brachiopod shell beds and lithified bacterial (algal?) crusts. Eocrinoid settlement on carbonate flat pebbles is described from intraformational conglomerates of the Late Cambrian of Nevada, Montana, and Wyoming (Brett et al. 1983; Wilson et al. 1989). Such rigid bottoms can be considered as genuine hardgrounds, though they differed in some aspects from Ordovician hardgrounds (Rozhnov 1994).

Thus, in the Cambrian there were no close similarities between the faunas of rockgrounds and the first hardgrounds. However, *Trypanites* may provide an exception, because the most ancient borings of these animals are found in Early Cambrian reefal

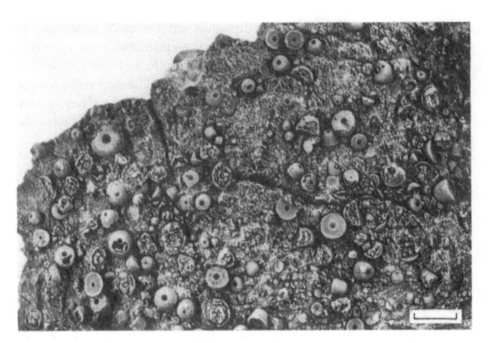

Figure 11.2 Hardground surface with eocrinoid holdfasts, collection of PIN, late Middle Cambrian Mila Formation, Member 3 (Shahmirzad, Elburz Mountains, northern Iran). *Source:* Photograph courtesy of Andrey Zhuravlev. Scale bar equals 1 cm.

cavities of Labrador and western Newfoundland (James et al. 1977; Palmer 1982). These borings are not known from the Middle and Late Cambrian (Wilson and Palmer 1992) but reappear in great numbers in Early Ordovician hardgrounds (Rozhnov 1994), becoming widespread in the Middle and Late Ordovician. However, the real identity of the progenitors of Early Cambrian and Ordovician *Trypanites* raises some doubts, because the Ordovician borings are considered to have been produced by polychaetes, whereas the nature of Cambrian *Trypanites* remains unknown (James et al. 1977; Kobluk et al. 1978). Thus, one can suppose that the majority of the hardground fauna arose independently of the rocky bottom fauna. Attached echinoderms are pioneers and are the most important components of the initial hardground ecosystems.

Cambrian hardgrounds were created presumably by consolidation of cobbles or large shells, on which echinoderms initially settled (Brett et al. 1983; Zhuravlev et al. 1996). The debris, accumulated between pebbles after postmortem destruction of echinoderm skeletons, favored cementation of pebble bottoms. Calcite productivity of echinoderms in the Cambrian was low, and the debris produced by echinoderms was only enough to fill spaces between cobbles. Thus, the community that settled on such hardgrounds could not expand the hardground area beyond the pebbled area. The low abundance of hardgrounds in the Cambrian was determined by these limits and also probably by the reduced distribution of calcite seas at that time.

Marine Substrates in the Ordovician

A considerable part of the Ordovician epicontinental sea floor was also covered with soft silts. Ordovician soft substrates, however, in contrast to the Cambrian ones, commonly contained abundant calcite debris and thus were transformed into hardgrounds that occupied large areas.

Ordovician as well as Cambrian rockgrounds occupied relatively small areas and were colonized only by benthic animals to a limited extent. Abundant and diverse faunas largely developed in framework cavities within various reefs. The framework cavities in bryozoan-algal reefs (Middle Ordovician, Caradoc) from near Vasalemma village in Estonia provide an example; various bryozoans, crinoids, cystoids, edrioasteroids, and brachiopods, often well preserved, are found in these cavities (pers. obs.). Nonetheless, on the whole, this fauna was insignificant for the evolution of the marine benthos, because such ecologic niches were relatively ephemeral, their colonization was rather occasional, and they had no evolutionary future.

Ordovician hardgrounds were very widely distributed. They occupied large areas and were colonized by a characteristic and abundant fauna. This was especially typical of Middle Ordovician hardgrounds (Palmer and Palmer 1977). The faunas of Early Ordovician hardgrounds are considered to be transitional between those of Cambrian and Middle Ordovician hardgrounds, based on detailed analysis of hardgrounds in the Middle Ordovician Kanosh Shale in west-central Utah (Wilson et al. 1992). The formation of hardgrounds in the carbonate part of this sequence can be described by the following succession of steps (Wilson et al. 1992): (1) development of early diagenetic carbonate nodules in fine-grained siliciclastics; (2) storm current winnowing and formation of cobble lags; (3) encrustation of the cobbles by large numbers of stemmed echinoderms (predominantly eocrinoids), trepostome bryozoans, and a few sponges; (4) accumulation of echinoderm debris in lag deposits; (5) and early marine cementation of hardgrounds and the settlement of additional stemmed echinoderms, bryozoans, and sponges.

The community of the third stage of this sequence can be compared with the Late Cambrian community (Rozhnov 1994) found in the Snowy Range Formation of Montana and Wyoming (Brett et al. 1983), as well as with the late Middle Cambrian community of the Mila Formation of Iran. All these communities are similar in the dominance of eocrinoids and the absence of *Trypanites* borings, which are typical of younger hardgrounds.

The presence of bryozoans in Early and Middle Ordovician hardground communities is considered the main ecologic difference from Cambrian hardground communities. In my opinion, however, the most important difference between these hardgrounds is displayed in the mechanism of their formation. Cambrian hardgrounds developed only on pebbles (Snow Range Formation) or large calciate brachiopod shells (Mila Formation), because calcitic debris from echinoderms and other encrusters was sufficient only to fill the space between the pebbles, whereas in the Early

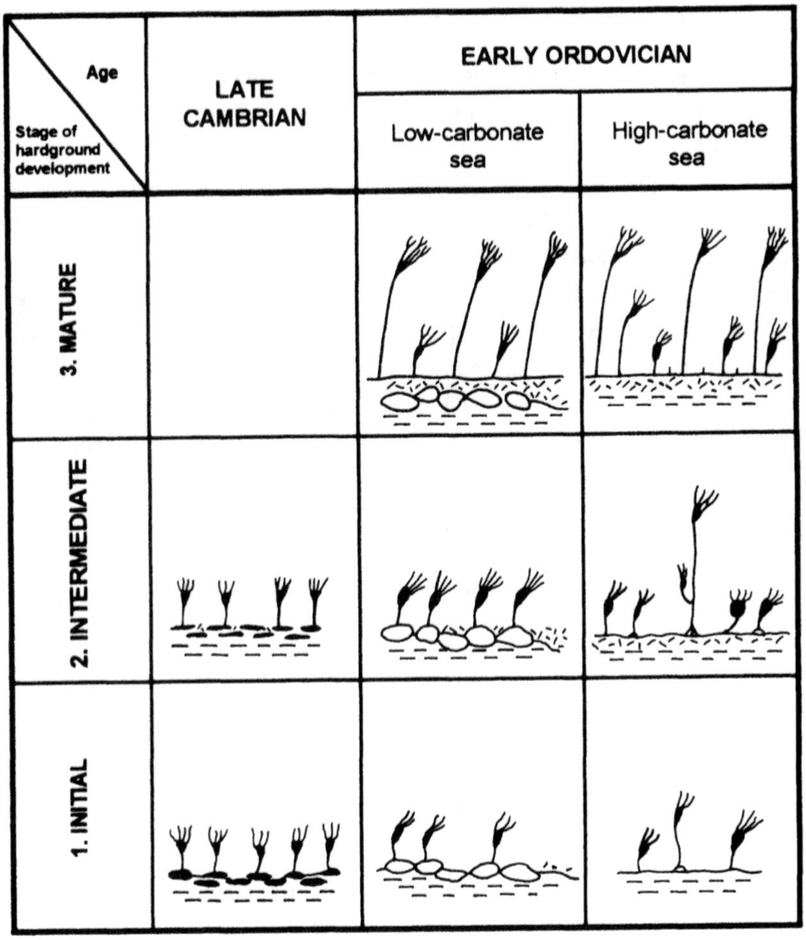

Figure 11.3 Stages of Late Cambrian and Early Ordovician hardground development. *Source:* Modified after Rozhnov 1994.

Ordovician, as demonstrated for the Kanosh Shale, the amount of echinoderm debris was enough for hardground formation even outside the area covered by pebbles (Wilson et al. 1992). Therefore, the analogs of the fourth and fifth stages of development of hardgrounds in the Cambrian described by Wilson et al. (1992) and Zhuravlev et al. (1996) were absent, and these stages can be considered as typically Ordovician phenomena (figure 11.3). The accumulation of abundant debris, initially provided by echinoderms, and fast expansion of these hardgrounds due to the supply of debris coming from new encrusters, are characteristic of these later stages (figure 11.3).

Study of Early Ordovician hardgrounds from the eastern part of the Leningrad Region (Baltic Basin) has revealed further differences from Cambrian hardgrounds and provides an opportunity to establish a pattern of hardground formation based on positive feedback between the development of encrusters (initially echinoderms), and the expansion of hardgrounds themselves (Rozhnov 1994, 1995; Palmer and Rozh-

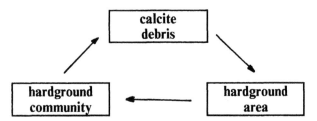

Figure 11.4 Positive feedback between the expansion of
hardgrounds and the increase of calcite debris production
by hardground communities.

nov 1995) (figure 11.4). One of these features is the presence of *Trypanites* borings,
widely distributed in Early Ordovician hardgrounds of the Baltic Basin, as has already
been reported by Hecker (1960) and Vishnyakov and Hecker (1937). The second
important difference is the mass supply of debris, produced mainly by echinoderms
inhabiting hardgrounds, and its accumulation in areas where new hardgrounds or
soft grounds, depending on the sedimentary regime, possessing a hard layer at a given
depth below soft sediments were formed.

Hardgrounds could not develop widely in the Ordovician until the quantity of accumulated calcite debris on the sea floor increased sharply in comparison with that
of the Cambrian. This increase in debris supply in the Ordovician was, first of all,
connected with the change in the structure of benthic communities, especially in the
carbonate-precipitating seas, where echinoderms began to play a dominant, or at least
an important, role. The abrupt increase in the amount of echinoderm debris in post-
Cambrian sediments corroborates this opinion.

Supply of calcite debris produced by other groups of animals, such as ostracodes,
brachiopods, bryozoans, and trilobites, also sharply increased in the Ordovician. This
implies that the production and supply of $CaCO_3$ debris by various organisms in the
Ordovician increased. In any case, the balance of $CaCO_3$ content in marine water
should have been affected because of the redistribution of its production among different groups of organisms (from mostly trilobites in the Cambrian to echinoderms,
brachiopods, bryozoans, and mollusks in the Ordovician) (see also Droser and Li,
this volume).

In the Ordovician, echinoderm calcite productivity increased by at least an order
of magnitude relative to that in the Cambrian. It was connected with an increase in
the general number and variety of echinoderms, as well as with their individual increase in size. In the Cambrian, stemmed echinoderms were represented mainly by
eocrinoids, which almost never reached a height greater than 15 cm above the sea floor
and usually were shorter (Bottjer and Ausich 1986; Ausich and Bottjer 1982; Rozhnov 1993).

In the Ordovician, some eocrinoids reached a height of 25–30 cm (Rozhnov 1989),
and crinoids with long stems could rise 1 m or more above the sea floor. The diverse

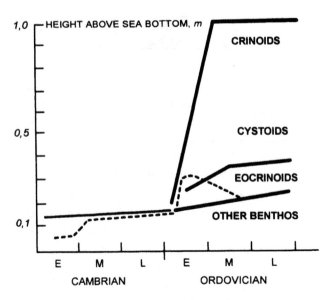

Figure 11.5 Maximum height of food-gathering apparatus above the sea floor among some groups of benthic animals in the Cambrian and Ordovician. *Source:* Modified after Rozhnov 1993.

and numerous cystoids could reach 30–40 cm in height (figure 11.5). This resulted in the deployment of suspension feeding into the basal meter of the water column. It sharply increased the tiering for echinoderms and, as a consequence, caused an increase in the overall number of echinoderms. Simultaneously, the individual sizes of echinoderms sharply increased by almost an order of magnitude. This was connected not only with the replacement of small-sized groups by larger ones but also with a general trend of size increase in all groups of echinoderms. Large crinoids were common in the Ordovician and often formed dense settlements. As a result of these developments, supply of calcite debris to the sea floor increased dramatically.

Therefore, substrates around such settlements mostly consisted of echinoderm debris. Not far from these settlements, echinoderm debris also constituted a substantial proportion of the sediment. For example, as described by Põlma (1982) in the Ordovician of the northern structural-facies province of eastern Baltica, echinoderm fragments compose 25–30 percent of the total amount of debris, increasing in reefal facies up to 95 percent. Such a change in the character of substrates at the Cambrian-Ordovician boundary would likely affect the structure and diversity of the entire benthos. Another feature of echinoderms that influenced sea floor changes in carbonate-precipitating seas at this boundary that should be taken into account is that each skeletal element of an echinoderm is monocrystalline. Calcite cements grew syntaxially on isolated echinoderm ossicles, and thus the cementation rate in sediments enriched by echinoderm debris was very fast. As a result, in suitable conditions abundant echinoderm debris was rapidly cemented on the sea floor to form hardgrounds

(Wilson et al. 1992). When the rate of sedimentation was equal to, or less than, the rate of cementation, substrates became rigid and hardgrounds formed. These new hardgrounds were ideal for the settlement of stemmed echinoderms that needed rigid substrates, and they quickly colonized them. Hardgrounds were also favorable for the settlement of many other benthic groups, such as bryozoans, ostracods, and small brachiopods, as well as for boring organisms, among which *Trypanites* dominated.

FEEDBACK AS A UNIQUE FEATURE OF ORDOVICIAN SUBSTRATES

The formation of the first hardgrounds in geologic history and the origin of hardground communities coincided with the appearance of many new higher taxa and with a sharp increase in diversity and abundance of many marine groups—first, echinoderms (Crinoidea, Diploporita, and Rhombifera), as well as classes of the Bryozoa and numerous new taxa of lower taxonomic rank (Walker and Diehl 1985; Palmer and Wilson 1990; Guensburg and Sprinkle 1992, this volume; Wilson and Palmer 1992; Sprinkle and Guensburg 1993, 1995). This does not seem to have been a random coincidence. The relationships between development of bottom substrates and the evolution of benthic fauna warrant further investigation. Such relationships may be seen in the ability of marine substrates to self-reproduce and expand. The hardground feedback may have been almost unique to the Ordovician or at least appeared during this period for the first time.

MECHANISM OF HARDGROUND FEEDBACK

Ordovician hardgrounds were formed by the accumulation of calcite debris produced by a benthic community inhabiting the very same substrate (Wilson and Palmer 1992). In the Cambrian there was a similar source of debris supply, but the quantity of debris was not sufficient for the expansion of hardgrounds. That is, hardgrounds could appear under suitable conditions, usually when echinoderms settled on cobble lag surfaces, but they could not expand beyond these lags. In contrast, Cretaceous hardgrounds depended on debris of planktic organisms, mainly coccolithophorids. Thus, the phenomenon of early Paleozoic hardground feedback lies in the ability of hardground expansion, which depends on the amount of debris supplied by the benthic community itself. This phenomenon is especially typical of the Ordovician.

Hardgrounds with Low and Medium Hydrodynamic Energy

The feedback mechanism of Ordovician hardgrounds is connected primarily with echinoderms, for which hardgrounds with low and medium hydrodynamic energies represented ideal locations for settlement. The echinoderm larvae were planktic and became attached to some hard surface for further development—for example, to

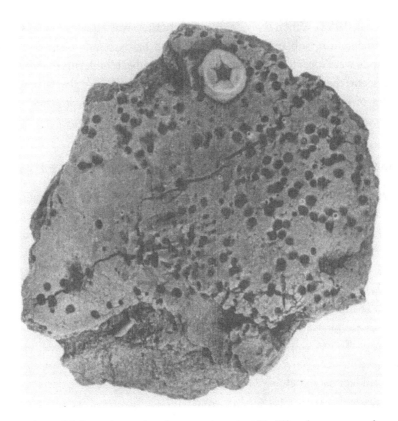

Figure 11.6 Hardground with encrusting crinoid holdfast, bryozoans and *Trypanites* borings, PIN 4565/10, Early Ordovician, Arenig, Volkhov stage (Simonkovo village, Volkhov River right bank, Leningrad region). Natural size.

large bioclasts on softgrounds or to hardgrounds (Guensburg and Sprinkle, this volume). In addition, long-stemmed echinoderms with high crowns, such as crinoids, needed a sufficiently strong support on the sea floor. Stemmed echinoderms easily solved this problem by attaching to hardground surfaces by the simplest, primitive holdfast (figure 11.6). On softer substrates this primitive holdfast was considerably complicated by a ramose root system (figure 11.7).

Echinoderm Debris as a Material for Hardgrounds

Echinoderm debris accumulated around echinoderm communities and was an ideal material for the formation of hardgrounds because of the following features: (1) the single-crystal nature of each echinoderm skeletal element, resulting in fast syntaxial growth of calcite cement precipitating from pore waters in sediment on loose echinoderm debris; (2) the larger volume of echinoderm debris in comparison with calcite debris of the same weight produced by other animals, due to the high porosity of the echinoderm skeleton (stereome structure); and (3) the multiple nature of the echinoderm skeleton, resulting in the postmortem production of numerous calcite ossicles

Figure 11.7 Ramose crinoid holdfast on the surface of a softground (grainstone), PIN 4565/11, Early Ordovician, Arenig, Volkhov stage (Obukhovo village, Volkhov River right bank, Leningrad region). Natural size.

even in quite quiet water. Thus, hardgrounds represented an ideal place for dense settlement of echinoderms, and the postmortem accumulation of their debris favored further hardground development (Guensburg and Sprinkle 1992; Wilson et al. 1992). The more hardgrounds expanded, the greater was the number of echinoderms that settled on their surfaces, and the more calcite debris accumulated around these settlements, which further enhanced hardground expansion. This positive feedback between the expansion of hardground areas and the increase of echinoderm biomass led to very rapid expansion of hardgrounds over large areas and to abrupt and rapid increase in the number of echinoderms and of other sessile organisms (see figure 11.4) because it occurred in shallow-water calcite-precipitating Ordovician seas.

HYPOTHESIS ON THE INFLUENCE OF HARDGROUND FEEDBACK ON THE BENTHIC FAUNA EVOLUTION

The extremely fast growth in the number of echinoderms should have resulted in high rates of evolutionary innovation in this phylum. From the conventional classic point of view, the self-reproducing hardgrounds that appeared for the first time in the

Ordovician represented a new system of ecologic niches facilitating the existence of many benthic groups, particularly and primarily echinoderms. Colonization of these niches should have been accompanied by specialization and ubiquitous morphogeneses of the pioneer fauna. Hardgrounds were likely to be disjunct and patchy and to have appeared suddenly as a result of storm erosion, thereby favoring r-selection strategies, at least among the pioneers. Frequent burial and overturning of cobbles in high-energy seas would certainly select for progenetic lineages. Therefore, r-selection promoted the early sexual maturity of individuals and the subsequent shift of ancestral juvenile features to the mature stages in descendants (paedomorphosis or progenesis). This mode of natural selection might have resulted in the appearance of many new higher taxa, especially among the Echinodermata.

In essence, classes of animals, including those well documented in the paleontologic record, are distinct groups, between which obvious morphologic hiatuses exist, whereas the intermediate forms are absent. Roots of many classes cannot be traced beyond the Ordovician or the Cambrian. It is possible that the classes—for example, echinoderm classes—that are known since the Ordovician had only latent ancestors among Cambrian skeletal echinoderms and did not arise from soft-bodied forms. Therefore, since we do not see their direct ancestors in the paleontologic record of the forms with mineralized skeletons, we may assume that they originated from certain Cambrian forms as a result of changes in ontogeny and that these changes were rapid enough to escape the fossil record. The ontogenetic changes that generated evolutionary transformations are based on heterochronic shifts of the relative rates of different processes in individual development. At present, one kind of heterochrony, paedomorphosis, is recognized as one of the most probable mechanisms for the acceleration of macroevolutionary rate and saltatory speciation, because it provides a large evolutionary potential as the mechanism that permits rapid and profound coordinated changes in morphology, physiology, biochemistry, and behavior through an insignificant initial somatic disturbance (McKinney and McNamara 1991; Smirnov 1991). Thus, paedomorphosis not only could play a role in the main morphogenetic mechanism during the Ordovician radiation of marine biota but also could be the most important mechanism in the origin of higher taxa, especially of echinoderm classes.

CONCLUSIONS

New and very diverse communities—with wider feeding opportunities and a higher degree of niche partitioning, especially among encrusting and boring organisms, in comparison with Cambrian communities—originated at the beginning of the Ordovician in shallow epicontinental seas, occupying significant areas. This phenomenon was due to the development of hardgrounds, which, in turn, was promoted by an interplay of abiotic and biotic factors such as the change from aragonite to calcite sea conditions, and the development of new groups of benthic organisms responsible for high rates of production of calcite debris.

Acknowledgments. I am grateful to T. J. Palmer, P. Taylor, M. L. Droser, A. Yu. Zhuravlev, A. I. Osipova, and A. Yu. Rozanov for discussion of the problems touched on in this paper. I am deeply indebted to Mary Droser and Maria Hecker for help in the translation of the paper. I would also like to thank the anonymous reviewers for critical remarks and helpful suggestions. This research was supported by the International Science Foundation Project MV5000, by the International Science Foundation and Russian Government Project MV5300, and by the Russian Foundation for Basic Research Project 99-04-49468 and 98-05-65065. This paper is a contribution to IGCP Project 366.

REFERENCES

Ausich, W. I. and D. J. Bottjer. 1982. Tiering in suspension-feeding communities on soft substrata throughout the Phanerozoic. *Science* 216:173–174.

Bottjer, D. J. and W. I. Ausich. 1986. Phanerozoic development of tiering in soft substrate suspension-feeding communities. *Paleobiology* 12:400–420.

Brett, C. E., W. D. Liddell, and K. L. Derstler. 1983. Late Cambrian hard substrate communities from Montana/Wyoming: The oldest known hardground encrusters. *Lethaia* 16:281–289.

Droser, M. L., R. A Fortey, and X. Li. 1996. The Ordovician Radiation. *American Scientist* 84:122–131.

Guensburg, T. E. and J. Sprinkle. 1992. Rise of echinoderms in the Paleozoic evolutionary fauna: Significance of paleoenvironmental controls. *Geology* 20:407–410.

Hecker, R. T. 1960. Iskopaemye fatsii gladkogo kamennogo morskogo dna (K voprosu o tipakh kamennogo morskogo dna) [Fossil facies of marine hardground (To the problem of kinds of marine hardground)]. *Trudy, Institut geologii, Akademiya nauk Estonskoy SSR* 5:199–227.

Hofmann, H. J. and J. P. Grotzinger. 1985. Shelf-facies microbiotas from the Odjick and Rocknest formations (Epworth Group: 1.89 Ga), northwestern Canada. *Canadian Journal of Earth Sciences* 22:1781–1792.

James, N. P., D. R. Kobluk, and S. G. Pemberton. 1977. The oldest macroborers: Lower Cambrian of Labrador. *Science* 197: 980–983.

Johnson, M. E. 1988. Why are ancient rocky shores so uncommon? *Journal of Geology* 96:469–480.

Kobluk, D. R. and N. P. James. 1979. Cavity-dwelling organisms in Lower Cambrian patch reefs from southern Labrador. *Lethaia* 12:193–218.

Kobluk, D. R., N. P. James, and S. G. Pemberton. 1978. Initial diversification of macroboring ichnofossils and exploitation of macroboring niche in the lower Paleozoic. *Paleobiology* 4:163–170.

McKinney, M. L. and K. J. McNamara. 1991. *Heterochrony: The Evolution of Ontogeny.* New York: Plenum Press.

Myrow, P. M. 1995. *Thalassinoides* and the enigma of Early Paleozoic open-framework burrow systems. *Palaios* 10:58–74.

Palmer, T. J. 1982. Cambrian to Cretaceous changes in hardground communities. *Lethaia* 15:309–323.

Palmer, T. J. and C. D. Palmer. 1977. Faunal distribution and colonization strategy in a Middle Ordovician hardground community. *Lethaia* 10:179–199.

Palmer, T. J. and S. V. Rozhnov. 1995. The origin of hardground ecosystem and the Ordovician radiation of benthos. In *International Symposium, Ecosystem Evolution, Abstracts, Moscow, Palaeontological Institute, RAS, 1995*, pp. 67–68.

Palmer, T. J. and M. A. Wilson. 1990. Submarine cementation and the origin of intraformational conglomerates in Cambro-Ordovician calcite seas. In *International Sedimentalogical Congress, 13th, Nottingham, England, Abstracts of Papers*, pp. 171–172.

Põlma, L. 1982. *Sravnitel'naya litologiya karbonatnykh porod ordovika severnoy i sredney Pribaltiki* [Comparative lithology of the Ordovician carbonate rocks in the northern and middle Baltica]. Tallin: Valgus.

Rozhnov, S. V. 1989. Novye dannye o ripidotsistidakh (Eocrinoidea) [New data on rhipidocystids (Eocrinoidea)]. In D. L. Kaljo, ed., *Problemy izucheniya iskopaemykh i sovremennykh iglokozhikh* [Problems in the study of fossil and recent echinoderms], pp. 38–57. Tallin: Academy of Sciences of Estonian SSR.

Rozhnov, S. V. 1993. Osvoenie iglokozhimi pridonnogo sloya vody v rannem paleozoe [Deploying by echinoderms of the near-bottom water layer in the early Paleozoic]. *Paleontologicheskiy zhurnal* 1993 (3):125–127.

Rozhnov, S. V. 1994. Izmenenie soobshchestv tverdogo morskogo dna na rubezhe kembriya i ordovika [The change of hardground communities at the Cambrian and Ordovician boundary]. *Paleontologicheskiy zhurnal* 1994 (3):70–75.

Rozhnov, S. V. 1995. Osobennosti stanovleniya vysshikh taksonov iglokozhikh [The peculiarities of origin of echinoderm higher taxa]. In *Programma i tezisy soveshchaniya Faktory taksonomicheskogo i biokhorologicheskogo raznoobraziya* [Program and abstracts for the symposium factors of taxonomic and biohorologic diversity], p. 66. St. Petersburg: Botanic and Zoological institutes, Russian Academy of Sciences.

Sandberg, P. A. 1983. An oscillating trend in Phanerozoic non-skeletal carbonate mineralogy. *Nature* 305:19–22.

Sepkoski, J. J., Jr. 1979. A kinetic model of Phanerozoic taxonomic diversity. 2: Early Phanerozoic families and multiple equilibria. *Paleobiology* 5:222–251.

Sepkoski, J. J., Jr. 1981. A factor analytic description of the Phanerozoic marine fossil record. *Paleobiology* 7:36–53.

Sepkoski, J. J., Jr. 1984. A kinetic model of Phanerozoic taxonomic diversity. 3: Post-Paleozoic families and mass extinctions. *Paleobiology* 10:246–267.

Smirnov, S. V. 1991. Pedomorfoz kak mekhanizm evolyutsionnykh preobrazovaniy organizmov [Paedomorphosis as a mechanism of evolutionary transformations in organisms]. In E. I. Vorob'eva and A. A. Vronskiy, eds., *Sovremennaya evolyutsionnaya morfologiya* [Contemporary evolutionary morphology], pp. 88–103. Kiev: Naukova Dumka.

Sprinkle, J. and T. E. Guensburg. 1993. Between evolutionary faunas: Comparison of Late Cambrian and Early Ordovician echinoderms and their paleoenvironments. *Geological Society of America Abstracts with Programs* 25:149.

Sprinkle, J. and T. E. Guensburg. 1995. Origin of echinoderms in the Paleozoic evolutionary fauna: The role of substrates. *Palaios* 10:437–453.

Turner, E. C., J. M. Narbonne, and N. P. James. 1993. Neoproterozoic reef microstructures from the Little Dal Group, northwestern Canada. *Geology* 21:259–262.

Valentine, J. W. 1992. The macroevolution of phyla. In J. H. Lipps and P. W. Signor, eds., *Origin and Early Evolution of the Metazoa,* pp. 525–553. New York: Plenum Press.

Vishnyakov, S. G. and R. T. Hecker. 1937. Sledy razmyva i vnutriplastovye narusheniya v glaukonitovykh izvestnyakakh nizhnego silura Leningradskoy oblasti [Erosion marks and intrastratal disturbances in Lower Silurian glauconitic limestones of the Leningrad Region]. In G. P. Sinyagin, ed., *K 45-letiyu nauchnoy deyatel'nosti chlena TsNIGRI doktora geologo-mineralogicheskikh nauk F. N. Pogrebova* [To the 45th anniversary of the scientific work of the TsNIGRI member, Doctor of the Geological-Mineralogical Sciences F. N. Pogrebov], pp. 30–45. Leningrad: Chief Editorial Board on Mining-Fuel and Geological-Exploring Literature.

Walker, K. R. and W. W. Diehl. 1985. The role of marine cementation in the preservation of Lower Paleozoic assemblages. *Philosophical Transactions of the Royal Society of London B* 311:143–153.

Wilson, M. A. and T. J. Palmer. 1992. Hardgrounds and hardground faunas. *University of Wales, Aberystwyth, Institute of Earth Studies Publications* 9:1–131.

Wilson, M. A., T. J. Palmer, T. E. Guensburg, and C. D. Finton. 1989. Sea-floor cementation and the development of marine hard substrate communities: New evidence from Cambro-Ordovician hardgrounds in Nevada and Utah. *Geological Society of America Abstracts with Programs* 21:A253.

Wilson, M. A., T. J. Palmer, T. E. Guensburg, C. D. Finton, and L. E. Kaufmann. 1992. The development of an Early Ordovician hardground community in response to rapid sea-floor calcite precipitation. *Lethaia* 25:19–34.

Zhuravlev, A. Yu. and R. Wood. 1995. Lower Cambrian reefal cryptic communities. *Palaeontology* 38:443–470.

Zhuravlev, A. Yu., B. Hamdi, and P. D. Kruse. 1996. IGCP 366: Ecological aspects of the Cambrian radiation—field meeting. *Episodes* 19:136–137.

CHAPTER TWELVE

*Brian R. Pratt, Ben R. Spincer, Rachel A. Wood,
and Andrey Yu. Zhuravlev*

Ecology and Evolution of Cambrian Reefs

*The history of reef building through the Cambrian records the replacement of pre-
dominantly microbial communities by those in which sessile animals participated in
construction, so heralding a new reef ecosystem with elaborate trophic webs, complex
organism interactions, increased niche partitioning, and high taxonomic diversity.
Thus, the domical and branching stromatolites of the Proterozoic (composed mainly
of micron-sized crystals precipitated within or trapped upon laminar biofilms) were
replaced by highly cavernous, sediment-generating structures constructed by the
skeletons of sessile filter feeders, calcified aggregations of coccoid microbes, and mats
of cyanobacterial filaments. A new microbial community (Epiphyton-Renalcis) appeared in the Nemakit-Daldynian and was joined in the Tommotian by archaeocyathan sponges, which colonized both open surfaces and cavities. Other sessile frame
builders, such as corals, as well as abundant obligate cryptobionts, demonstrate increasing complexity of this ecosystem. Metazoan diversity reached its zenith in the
early Botoman. Equally dramatic, however, was the disappearance of this community toward the end of the Early Cambrian. There followed a protracted interval
when reefs were almost entirely microbial. Rigid spiculate demosponges began to
occupy late Middle Cambrian reefs, but a level of complexity comparable to that of
the Early Cambrian was not achieved again until the Middle Ordovician, when an
increase in reef biotic diversity paralleled the radiation of the shelly benthos.*

FOR 3.5 BILLION YEARS tropical sea floors within the photic zone, which were relatively free from the influence of terrestrial runoff—sediment, freshwater, and nutrients—have hosted communities of aggregated sessile organisms. Such communities
—sometimes mainly microbial, at other times mostly skeletal metazoan—acquired
topographic relief and typically were the sites of synsedimentary cementation, thereby
forming reefs.

Here we summarize the composition of Cambrian reefs, outline its community structure, and trace its evolutionary history. It is clear that the Cambrian was a threshold for reef development no less than it was for many other sedimentary environments colonized by early complex life-forms. New microorganisms and skeletal, shelly, and soft-bodied metazoans evolved in reefs. They built an ecological and structural complex that changed the character of the reef ecosystem and its resulting rocks irreversibly from its pre-Cambrian precursors. Reefs record subsequent biotic revolutions, in the form of extinctions and radiations, which took place even in the Cambrian (Zhuravlev and Wood 1996). Although the composition and structure of Cambrian reefs appear relatively simple in comparison with Mesozoic and Cenozoic reefs, it is apparent that the main reef ecological niches were rapidly occupied in the Early Cambrian and that their complexity is not exceeded by younger Paleozoic reefs.

REEF-BUILDING AGENTS

Reefs consist basically of a rigid framework around which collect calcareous sediments generated by the reef itself. For this reason, reefs are not restricted to structures with conspicuous metazoan frame builders; stromatolites and thrombolites compose reefs too (Pratt and James 1982; Pratt 1982, 1995). Since all reefs possess topographic relief, we do not distinguish between reefs and mounds, unlike James and Bourque (1992). In the Cambrian, reefal framework was mostly the product of accreting microbial elements encrusted by animal skeletons, but locally dense settlements of skeletal animals strengthened by microbial crusts and synsedimentary cementation. Reef matrix sediment contains fragments of this framework, bioclasts of animals that lived on and under it, along with peloids and lime mud. Early Cambrian reefs are better studied than reefs of many younger intervals, and detailed descriptions can be found in a number of papers (James and Kobluk 1978; Kobluk and James 1979; Rees et al. 1989; James and Gravestock 1990; Debrenne et al. 1991, 1993; Kruse 1991; Wood et al. 1993; Kruse et al. 1995, 1996; Zhuravlev and Wood 1995).

Stromatoid-Thromboid Associations

Among Cambrian reefs, laminated and unlaminated microbial structures—stromatolites and thrombolites, respectively—are common (Pratt 1995). They are composed of fenestrate micrite and clotted micrite believed to have originated through organically induced precipitation within microbial biofilms or mats. Like modern counterparts, these mats were dominated by photosynthetic taxa but may have contained heterotrophs as part of a complex microbial community of cyanobacteria and other bacteria. $CaCO_3$ precipitation likely occurred during mat decay, and only rarely is some evidence of the mat-dwelling "cells" preserved (Pratt 1995: figure 6D).

Thromboids are generally subordinate elements in Early Cambrian reefs (figure 12.1A), but are usually dominant in those of Middle and Late Cambrian age from shelf

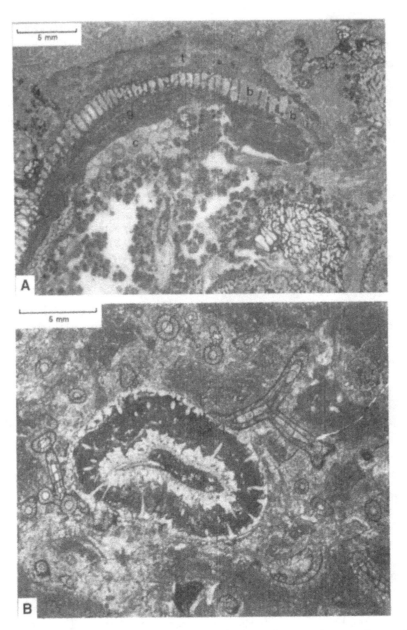

Figure 12.1 Thin-section photomicrographs of Lower Cambrian frameworks. *A*, Upside-down bowl-shaped *Metaldetes profundus* (Billings) encrusted on one side by thromboid (*t*) and on the other side first by *Girvanella* (*g*), second by cryptic *Archaeotrypa* sp. (*c*), and third by *Renalcis*; right side of *Metaldetes* perforated by *Trypanites* macroboring (*b*); stick-shaped *Metaldetes* at right and *Retilamina amourensis* Debrenne and James at lower left and lower right; upper Botoman Forteau Formation (western Newfoundland, Canada). Collection of Pratt. *B*, Radiocyath *Girphanovella georgensis* (Rozanov) with branching archaeocyath *Cambrocyathellus tuberculatus* (Vologdin) and abundant pseudomorphosed aragonite botryoids; Atdabanian Salaany Gol Formation (Zuune Arts Mount, Zavkhan Province, Mongolia); MNHN Collection of Debrenne SM × 27,720.

settings (Astashkin et al. 1984; Kennard and James 1986; Kennard et al. 1989; Pratt 1995). Biomicrite interpreted to have been bound by mats, a common motif in middle Paleozoic deeper-water mud mounds, occurs in some Early Cambrian reefs (James and Gravestock 1990).

Calcified Microbes

A wide variety of "microfossils" of accepted microbial origin is present in Cambrian reefs and indeed is integral to many frameworks. Calcification of filamentous cyano-bacteria in various stages of degradation formed tubules and threads referable to *Girvanella* and similar microfossils, and these are common as tabular crusts or sub-ordinate encrustations (Kruse et al. 1996) (figure 12.1A). Discontinuous, planar to arcuate *Girvanella* crusts composed of tangles and multifilament sheets compose laminar zones up to a meter thick in archaeocyath-*Renalcis* reefs (Rees et al. 1989), and meter- to decameter-scale, conical to hemispheroidal tufa-like masses that are common in Middle and Late Cambrian platform-margin reefs and downslope mud mounds (James 1981; Kobluk 1985; Pratt 1989, 1995, 2000) (figure 12.2B).

Renalcis, Epiphyton, and similar objects are millimeter-sized aggregates of hollow and solid micritic spheroids and clots. In Early Cambrian reefs these formed free-standing masses that grew upward, as well as encrustations on cavity walls—especially the undersides and insides of archaeocyaths and *Girvanella* crusts, and within cavities excavated from matrix bioclastic lime mud (Kobluk and James 1979; James and Gravestock 1990; Wood et al. 1993; Kruse et al. 1995, 1996; Zhuravlev and Wood 1995) (figure 12.1A). In Middle and Late Cambrian reefs in both shelf and platform-margin settings, *Renalcis* is attached to thromboids and *Girvanella* crusts, and dendritic forms like *Epiphyton* are abundant in a pendent habit (James 1981; Pratt 1995: figure 7) (figure 12.2B).

Sponges

Sponges are represented by the extinct calcified class Archaeocyatha and the siliceous Hexactinellida and demosponge groups Anthaspidellidae and Axinellidae. The prob-able calcarean *Gravestockia* has also been reported from the Atdabanian low-energy biomicrite mud mounds of South Australia (Debrenne and Reitner, this volume: fig-ure 14.1E).

Archaeocyaths are the most conspicuous and most abundant metazoan in Early Cambrian reefs, having contributed to patch reefs as old as earliest Tommotian (Rid-ing and Zhuravlev 1995). Modular Archaeocyathida exhibit their greatest generic di-versity in the Botoman (Wood et al. 1992b). Ajacicyathids and branching mono-cyathids formed thickets on soft substrates (figure 12.2A), and archaeocyathids at-tached to firm or hard substrates by means of aporous epitheca (Debrenne and

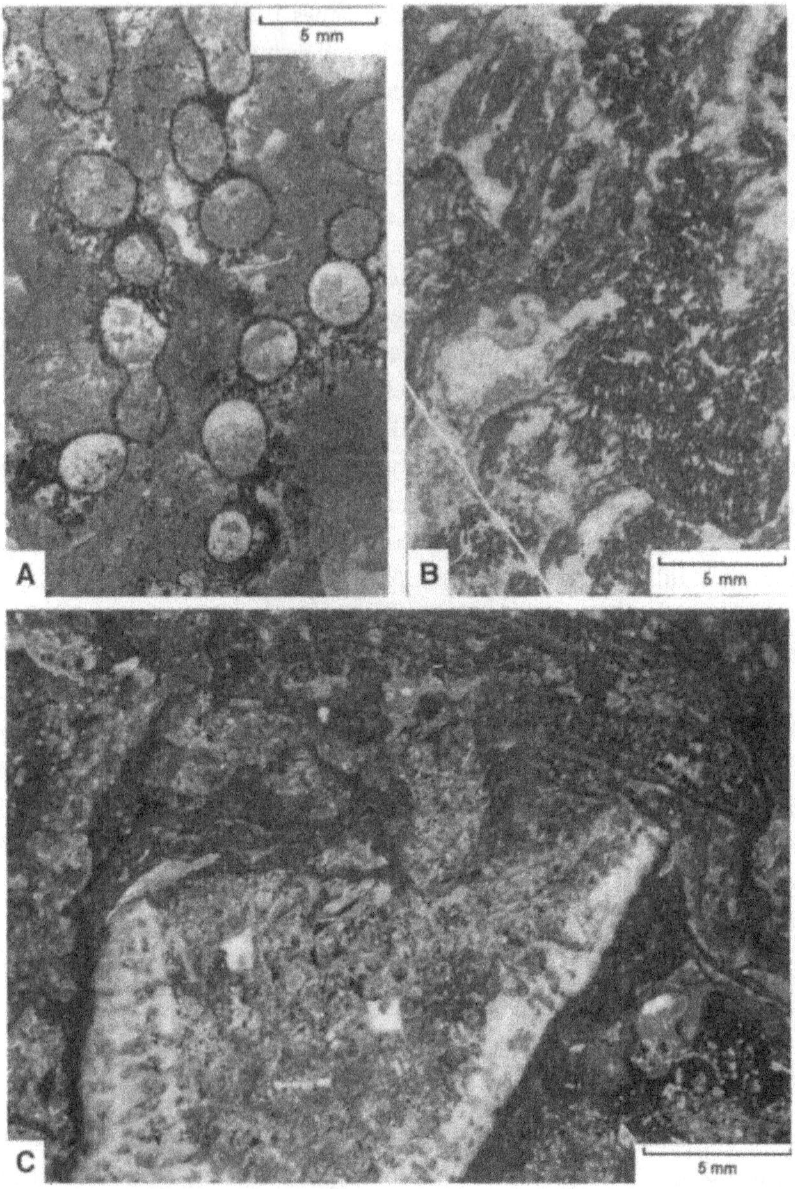

Figure 12.2 Thin-section photomicrographs of Cambrian frameworks. *A*, Branching monocyathid *Archaeolynthus polaris* (Vologdin) intergrowths with subordinate *Renalcis*, MNHN M810034, middle Tommotian Pestrotsvet Formation (Zhurinskiy Mys, middle Lena River, Siberia, Russia). *B*, *Girvanella* crusts with pendent *Epiphyton*; Upper Cambrian, Sunwaptan Cow Head Group (western Newfoundland, Canada). Collection of Pratt. *C*, Anthaspidellid demosponge *Wilbernicyathus donegani* Wilson encrusted from the top by *Girvanella* and eocrinoid holdfasts; Upper Cambrian, Sunwaptan Wilberns Formation, upper Morgan Creek Limestone (Llano Uplift, Texas, USA). Collection of Spincer.

Reitner, this volume: figures 14.3A and 14.4A). Their dominant orientation is sideways or downward growth in cavities (Zhuravlev and Wood 1995). This habit argues against the suggestion that archaeocyaths contained symbiotic photoautotrophs. The skeletons were originally high-Mg calcite (James and Klappa 1983), except for *Dictyocyathus translucidus*, which was aragonitic because it occurs consistently as blocky calcite cement-filled molds (Kruse et al. 1995).

Anthaspidellids with local encrustations of microbial filaments form the framework of late Middle Cambrian (late Marjumian) reefs from Iran (Hamdi et al. 1995; Zhuravlev et al. 1996; Debrenne and Reitner, this volume: figure 14.1B). In the Late Cambrian (Sunwaptan) of Texas, complete sponges are intergrown with thromboids and compose up to 25% of the boundstone (Spincer 1996) (see figure 12.2C). In some Early Cambrian reefs (James and Kobluk 1978; Kobluk and James 1979; Wood et al. 1993) and Late Cambrian (Sunwaptan) intrashelf thrombolite reefs (Pratt 1995: figure 11B), sediment-stabilizing hexactinellids left spicules and rare root tufts in the matrix.

Cnidaria

Modular, tabulate coral-like skeletons have been documented from Early Cambrian reefs. The Botoman of South Australia yields *Flindersipora* and some other tabulate-like corals (Lafuste et al. 1991; Sorauf and Savarese 1995; Debrenne and Reitner, this volume: figure 14.6D). Also, reefs of probable Botoman age from the Canadian Rocky Mountains and Labrador contain centimeter-sized remains of variably polygonal skeletons lacking septa (figure 12.1A). These have been described as *Rosellatana, Archaeotrypa*, and *Labyrinthus* (Kobluk 1979, 1984; Kobluk and James 1979). Their form of increase similar to longitudinal fission, however, does suggest affinities with chaetetids (calcified sponges) rather than mainstream corals (Scrutton 1997).

Corals have not been observed in younger Cambrian reefs; tabulate corals reappear in the lowermost Ordovician reefs of western Newfoundland (Pratt and James 1982, 1989). More material is needed to ascertain whether or not the Late Cambrian (Marjumian) coral-like *Cambrophyllum* Fritz and Howell (1955) is found associated with reefs.

Problematica

Several sessile taxa restricted to Early Cambrian reefs contribute to frameworks but are of uncertain affinity and function. *Cysticyathus* from the middle to late Tommotian is a saclike fossil up to 3 cm in diameter, with porous walls and thin, widely spaced tabulae (Kruse et al. 1995). It has been suggested to be a "coralomorph" (Zhuravlev et al. 1993). The somewhat similar *Tabulaconus* from the Botoman of Cordilleran Laurentia bears many closely spaced arcuate and locally bifurcating tabulae, and it may

too be a coral-like organism (Debrenne et al. 1987). The Hydroconozoa are represented by narrow to broad cones composed of lamellar calcite (Korde 1963), which
have been regarded tentatively as "coralomorphs" (Zhuravlev et al. 1993). They are
common encrusting elements usually found in cavities, in the Atdabanian and Botoman of Siberia and Mongolia (Wood et al. 1993).

Early Cambrian objects with a variably laminated structure have been placed in the
Tannuolaiidae (=Khasaktiidae) and regarded as possible stromatoporoids (Sayutina
1980). Specifically, *Khasaktia* forms arcuate laminar encrustations a few millimeters
thick on framework surfaces in the early Atdabanian of Siberia, and ramose *Rackov-
skia* is found in the late Atdabanian to early Botoman of Mongolia, Siberia, and the
Urals. However, domical encrustations much larger than *Khasaktia* occur in the probable Botoman of the Canadian Rocky Mountains, and these have a remnant microstructure and habit consistent with a high-modular, filter-feeding organism that may
be a stromatoporoid-grade sponge (Pratt 1994).

Radiocyaths are irregularly conical structures up to several centimeters in size,
composed of stout, weakly fused, originally aragonite structures (nesasters) similar to
meromes of Receptaculitida (Zhuravlev 1986). Whole specimens occur worldwide
from the late Tommotian to the Toyonian and are subordinate to archaeocyaths or
form their own thickets (Zhuravlev and Sayutina 1985; Kennard 1991; Kruse 1991;
Wood et al. 1993) (figure 12.1B).

Cribricyaths are 0.5 mm wide, undulating to irregularly coiled tubes that encrust
the sides and undersides of archaeocyaths and *Renalcis*. They occur in the Atdabanian
and Botoman of Siberia and Mongolia (Jankauskas 1972; Wood et al. 1993) and may
have been filter feeders.

Wetheredella consists of small encrustations of coiled, overlapping micrite-walled
tubes about 100 mm wide and has been observed rarely in cavities in the late Botoman of Labrador (Kobluk and James 1979). The subcircular cross section of the tubes
suggests that this object may be a foraminifer rather than a calcified cyanobacterium
(contra Kaźmierczak and Kempe 1992). *Wetheredella* has not been reported from other
Cambrian strata.

Synsedimentary Cementation

All reefs contain some proportion of $CaCO_3$ cement precipitated on the sea floor,
which enhanced rigidity. Cambrian reefs with large pores and growth-framework
cavities exhibit abundant isopachous, acicular to bladed high-Mg calcite cement
(James and Klappa 1983). Micritic calcite cement occurs in small intraskeletal, inter-
particle, and fenestral pores. Early Cambrian reefs differ from younger Cambrian examples in that many contain botryoids of now-calcitized acicular aragonite up to
several centimeters in size (James and Klappa 1983; Wood et al. 1993; figure 12.1B).
Such cements from Mongolia have been interpreted as algae and named *Zaganolomia*

(Drozdova 1980). A temporal control may have been present, in that botryoidal cements have been documented from Nemakit-Daldynian to Botoman reefs but not in Middle and Late Cambrian reefs (Zhuravlev 1993).

REEF-DWELLING ORGANISMS

Trilobites

Trilobites can usually be observed in the matrix of Cambrian reefs. They are essentially absent from *Girvanella* frameworks but present in surrounding facies. In Early Cambrian reefs they are typically scattered as small bioclasts less than a few millimeters in size. Reef-dwelling trilobites are generally believed to have constituted a community that differed from off-reef sites (Mikulic 1981), and this belief is supported by data specifically for the Cambrian (Repina 1983; Sukhov and Pegel' 1986). Thrombolites from the Middle Cambrian (Marjumian) of the Siberian Platform and Late Cambrian (Sunwaptan) of the Canadian Rocky Mountains and Arctic Islands preferentially contain the centimeter-sized convex cephala and pygidia belonging to plethopeltids (Stephen R. Westrop and Pratt, pers. obs.), perhaps a morphology adapted for a libero-sessile burrow-dwelling habit (Stitt 1976). *Stigmacephaloides curvabilis* is another taxon that appears restricted to reefs of Sunwaptan age (Spincer, pers. obs.). Reefal trilobites presumably grubbed on the sediment surface in search of organic particles and perhaps meiofauna.

Bivalved Arthropods

Cambrian bivalved crustaceans display a great deal of convergence, and many taxa are no longer regarded as true ostracodes (Hou et al. 1996). However, some Early Cambrian reefs contain disarticulated valves that possess a finely prismatic shell microstructure identical to that of younger ostracodes, which are also common bioclasts in reefs (Kobluk and James 1979).

Brachiopods

Lingulate brachiopods are rare in Cambrian reefs, and examples probably represent "stray" individuals, although Middle Ordovician mud mounds do host a diverse fauna specific to that setting (Krause and Rowell 1975). Calciate brachiopods were present in reefs since Atdabanian as disarticulated and articulated valves and include kutorginids, obolellids (James and Klappa 1983; Ushatinskaya, this volume: figure 16.5), and orthids (Kruse 1991). In the Middle Cambrian, calciates (billingsellids) formed extensive shell beds, which served as substrate for anthaspidellid reefs and stromatolites (Zhuravlev et al. 1996), but they resumed a common reef-dwelling habit in the Early Ordovician.

Stenothecoids

The stenothecoid group embraces bivalved, brachiopod-like shells that include forms with and without hinge articulation. Their affinity is uncertain, and they are not distinctively molluscan (Yochelson 1969; Rozov 1984). They occur mostly in reefs of Atdabanian to Amgan age and were probably immobile, epifaunal suspension feeders (Spencer 1981; Kouchinsky, this volume).

Hyoliths

Centimeter-sized conical shells with an operculum, originally aragonite composition, and lenticular, triangular, trapezoidal, or subcircular cross sections have been considered mollusks (Marek and Yochelson 1976) or a separate phylum (Runnegar et al. 1975). Hyolithomorph hyoliths are common in matrix biomicrite as well as in non-reefal beds, reaching their maximum diversity in the Botoman (Rozanov and Zhuravlev 1992). They have not been reported to be associated with younger reefs. Their mode of life is unknown, but they may have been semisessile suspension feeders (Kruse et al. 1995; Kouchinsky, this volume). Orthothecimorph hyoliths were common in Nemakit-Daldynian and Tommotian peri-reefal grainstones and sometimes were immured in muddy bioherms as vertically oriented cones (Landing 1993). Thus, a suspension-feeding strategy may be implied for some of them.

Salterellids

Salterellids are another group with centimeter-sized conical shells but are composed largely of lamellar calcite (James and Klappa 1983). Proposed as belonging to the phylum Agmata (Yochelson 1977), they appear to be restricted to the Laurentian Botoman and are also largely peri-reefal fossils.

Mollusks

Millimeter-sized helcionelloids occur in Tommotian to Atdabanian peri-reefal grainstones (e.g., Moreno-Eiris 1987; Kruse et al. 1995), are relatively rare in reefs after the Sinsk event (James and Klappa 1983), and appear to be absent in reefs of the younger Cambrian. The similar-sized gastropod *Sinuella* is abundant in Sunwaptan reefs of Texas, although it appears rarely in correlative rocks elsewhere in Laurentia.

Echinoderms

Echinoderm ossicles are present in reefs from the latest Atdabanian, whereas they may form grainstones in associated beds (Kobluk and James 1979; Rozanov and Zhuravlev 1992). Their precise taxonomic affinity is often uncertain, owing to the rarity of articulated specimens. Eocrinoid ossicles are common in late Middle and Late Cambrian thrombolites, and holdfasts are observed contributing to frameworks (Spincer 1996; Zhuravlev et al. 1996) (figure 12.2C).

Chancelloriids

Disarticulated, hollow, star-shaped sclerites belonging to chancelloriids are abundant in Early Cambrian reefs and peri-reefal sediments (James and Klappa 1983) but have been observed only in nonreefal facies in the Middle Cambrian. Soft-bodied preservation in the Early and Middle Cambrian shows that chancelloriids may have been sessile animals covered in sclerites (Briggs et al. 1994:212–213) and were probably not sponges (Mehl 1996).

Small Shelly Fossils

Various microfossils difficult to assign on the basis of petrographic characteristics occur in Early Cambrian reefs as old as early Tommotian (Riding and Zhuravlev 1995; Kruse et al. 1995) and as young as Atdabanian (Wood et al. 1993). These fossils are more common in nonreefal beds. Such bioclasts are rare after the middle Botoman (after the Sinsk event) and are absent in younger Cambrian reefs.

Boring Organisms

The macroboring ichnogenus *Trypanites,* whose holes are similar to those produced by sipunculid worms, occurs abundantly on the upper surfaces of some Early Cambrian reefs in Labrador (James et al. 1977). However, its rarity on substrates within reefs (James and Gravestock 1990; Kruse 1991) (figure 12.1A) suggests that the borer was only occasionally part of the reef-dwelling community proper. Macroborings have not been described within younger Cambrian reefs.

Endolithic "algal" microborings have been documented from ooids beginning in the Vendian (Green et al. 1988), but these are rare in Cambrian reef-associated bioclasts (Kobluk and Kahle 1978; Conway Morris and Bengtson 1994). Pervasively micritized shell material, characteristic of Cretaceous and Cenozoic endolithic infestation, is not present. Silt-size microspar grains resembling "chips" from clionid-type sponge boring have been identified by Kobluk (1981), and possible sponge borings may occur in Botoman coral-like skeletons from the Canadian Rocky Mountains (Pratt 1994). Scalloped surfaces are locally present and are suggestive of rasping activity (Zhuravlev and Wood 1995).

Burrowing Organisms

Bioturbation is present in all Cambrian reefs, including those of Nemakit-Daldynian age (Zhuravlev and Wood 1995; Kruse et al. 1996), that possess muddy sediment around and within the framework, although repeated phases of total reworking do not seem to be typical. Burrows are discrete, unbranched, meandering, millimeter-wide tubes that commonly have infilling microspar or micrite and, locally, pellets; the concave lamination from deliberate backfilling, as seen in many burrows in siliciclastic facies, is absent. Sub-millimeter-sized burrows are locally branching and typically partly empty. Burrow style suggests a certain firmness of the sediment by the time the observed burrow generations were made.

Miscellaneous Microorganisms

Reef sediments undoubtedly hosted biodegrading bacteria, and the presence of pyrite indicates sulfate-reducing bacteria also. Kobluk and James (1979) reported possible fungi as filaments and as tangled "fecal pellets," which could represent fungal infestation after pellet formation. The fungal interpretation is not yet convincing, however, and similar hematitic filaments in red Late Devonian mud mounds have been interpreted as bacterial (Bourque and Boulvain 1993). Zhuravlev and Wood (1995) illustrated presumably calcified filaments 0.3–1 mm wide that are encrusting the underside of an archaeocyath, and they interpreted them as fungi. The fungal identification of these structures has yet to be confirmed.

COMMUNITIES

Environmental Setting

Cambrian reefs occupied depositional settings in tropical to subtropical, normal marine waters at the shelf-slope break along the margins of carbonate platforms or shelves, in the middle of platforms or near shore in shallow water, as well as in deeper water downslope or in intrashelf basins (James and Kobluk 1978; Astashkin 1981; Kennard et al. 1989; Pratt 1989, 2000; James and Gravestock 1990). The variety of energy levels is mirrored in the kinds of flanking sediments, which may be ooidal and oncoidal, or thin, nodular-bedded argillaceous lime mudstone. The proportion of micrite and biomicrite decreases with increasing energy, but there may be other factors involved in the sediment-generating capacity of the reef.

   Although filter-feeding archaeocyaths indicate the presence of abundant suspended organic material and sufficient nutrients, the dominance of microbial structures in all Cambrian reefs points to the overall clarity of the water. Regional profiles may show lateral variation in framework type (figure 12.3), but the reasons for this are not easy to decipher. Early Cambrian reef-bearing units are typically intercalated with purely siliciclastic units, and reefs in nearshore areas commonly have admixed subangular silt and fine sand. The ecological tolerance of archaeocyaths is difficult to assess, but their filter-feeding capacity and skeletal construction suggest that they were comparatively

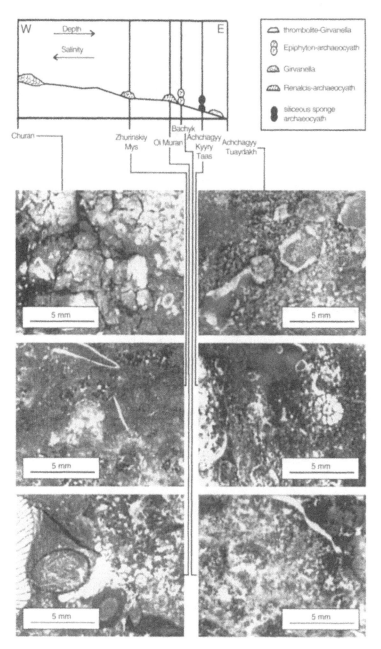

Figure 12.3 Lower Cambrian (lower Atda-banian) profile along the present middle courses of the Lena River (Siberia). Shown are peritidal *Girvanella* reefs (Churan); shallow subtidal *Renalcis* reefs with rare and nondiverse archaeocyaths and hyolithomorphs (Zhurin-skiy Mys); shallow subtidal *Renalcis*-archaeocy-athan reefs with abundant and diverse archaeo-cyaths (Oi Muran); shallow subtidal *Epiphyton* reefs with rare but diverse cryptic archaeocy-aths (Bachyk); subtidal (below fair-weather wave base) siliceous sponge-archaeocyathan mud mound with rare but diverse, mainly cryptic, archaeocyaths (Achchagyy Kyyry Taas); subtidal (below fair-weather wave base) bur-rowed *Girvanella*-bearing thrombolite reefs (Achchagyy Tuoydakh). The distance between Churan and Achchagyy Tuoydach is 120 km.

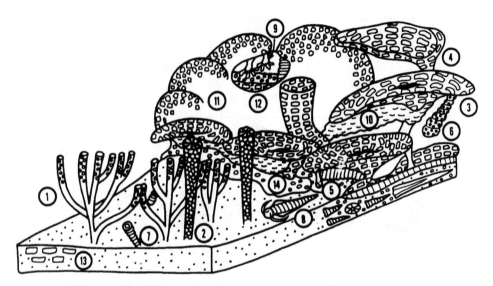

Figure 12.4 Reconstruction of Early Cambrian (late Tommotian) reef of the Siberian Platform. *Foreground: Cambrocyathellus* (Archaeocyathida) thickets (1) on muddy substrate with *Nochoroicyathus* (Ajacicyathida) sticks (2), hyolithomorph (8) and orthothecimorph (7) hyoliths, and *Chondrites* burrows (13). *Middle ground: Okulitchicyathus* (3)–
Sakhacyathus (Archaeocyathida) (4)–*Renalcis* (10) boundstone on lithified substrate with cryptic *Archaeolynthus* (Monocyathida) (5) and *Coscinocyathus* (Capsulocyathida) (6) and microborings (14). *Background: Epiphyton* framework (11) with cavity (9) inhabited by cryptic *Hydroconus* (coralomorph) (9) and microburrows (12).

resilient to clastic input and increased nutrient supply. The ability of *Epiphyton* and related microbial organisms to inhabit dim cavities argues that they may have been able to withstand sporadic turbidity. By contrast, Middle and Late Cambrian thrombolites and stromatolites, passively constructed by benthic biofilms or mats, seem to have flourished where terrigenous influence was far removed or temporarily suppressed.

Ecological Reconstruction

Early Cambrian reefs exhibit evidence for remarkably complex ecological interactions (figure 12.4). The main guilds developed with the diversification of archaeocyaths and accessory frame-building and dwelling elements, although the constructing guild was still dominated by microbes. Most described examples are patch reefs and do not exhibit a distinct lateral zonation. The areal distribution of radiocyaths in the Toyonian of South Australia (Kruse 1991) suggests that some framework elements responded to directed phenomena, presumably turbulence. Because many archaeocyaths were able to secrete extensive stabilizing exotheca, their relative scarcity on the upper surfaces of microbial reefs is possibly due to competition from the microbes that produced *Renalcis*.

Cavities hosted a diverse cryptic community (Zhuravlev and Wood 1995). Competition for space in cavities is demonstrated by chains of individual archaeocyaths or multiple overgrowths, probably because adjacent substrates were occupied by soft-bodied organisms or microbial biofilms that prevented settling. Competition is also seen by distorted archaeocyaths, where the growth of one species has been hampered by another, or even by soft-bodied organisms that have left no other record (Zhuravlev and Wood 1995). Mortality was high among archaeocyaths, judging from the number of juvenile cups seen in most frameworks. Such a mortality may be attributed to competitive interactions because most of the juveniles are sealed by their relatives and other organisms. Bioerosion, although locally present, was insignificant as a destroyer of skeletons and as a producer of fine-grained sediment.

A vertical zonation is not obvious in Early Cambrian reefs; however, different archaeocyathan communities seem to be confined to different stages of an individual reef development (Kruse et al. 1995; Riding and Zhuravlev 1995). Nevertheless, these reefs were undoubtedly capable of rapid accretion, given an intuitive judgment of the growth rates of archaeocyaths and rates of microbe calcification (Pratt 1984). Synoptic relief was generally not great, on the order of a meter or less. Early Cambrian reefs became firmly cemented with an unusually high amount of calcite and at times aragonite, reflecting extensive cavity development.

The complex ecological interactions seen in Early Cambrian reefs vanished by the start of the Middle Cambrian. The Middle and Late Cambrian saw some reefs with anthaspidellid and axinellid demosponges, but otherwise metazoan frame builders were absent, and the dwelling fauna depauperate by comparison.

Trophic Reconstruction

Only a rather generalized trophic analysis for Early Cambrian reefs is possible (Wood et al. 1993; Kruse et al. 1995; Zhuravlev and Wood 1995; Burzin et al., this volume: figure 10.2), as it is limited by the uncertain affinity of many of the faunal elements. The water column hosted primary producers in the form of acritarchs and bacterioplankton. Benthic microbial communities comprised biofilms of photosynthetic cyanobacterial and biodegrading bacteria. *Renalcis* and *Epiphyton* were formed by microbial aggregates, but many of these do not exhibit phototaxis; it is possible that if individual cells were photosynthetic, their light requirements were low enough for them to inhabit near-surface cavities. Archaeocyaths and other sponges were epifaunal filter feeders: most preferred living in cavities or on muddy substrates, a lifestyle perhaps dictated by competition from microbial biofilms. Tabulate-like corals may have been microcarnivores or suspension feeders that captured coarser organic aggregates. Brachiopods and possibly hydroconozoans, chancelloriids, cribricyaths, hyolithomorphs, some orthothecimorphs, and stenothecoids were suspension feeders. Biodegrading and sulfate-reducing bacteria resided in the sediment. Organisms that relied on sedimented detritus and its bacteria include deposit-feeding infaunal worms and epifaunal trilobites and possibly some of the remaining, minor skeletal components, most of which seem to have preferred off-reef areas. There may have been meiofaunal microorganisms, but none has been preserved. Grazers were uncommon.

What is evident from the Early Cambrian reef biota is that almost all the major trophic groups were occupied, as was noted by Kobluk and James (1979). By contrast, Middle and Late Cambrian reefs were ecologically simplified: the benthic community consisted of the microbes, with locally important filter-feeding siliceous sponges and echinoderms, and deposit-feeding worms and trilobites.

EVOLUTION

Precambrian reefs are stromatolites of various shapes and microstructures, exhibiting variably developed lamination and locally clotted micrite in which calcified microbes may be discerned (Turner et al. 1993; Pratt 1995). The framework of Nemakit-Daldynian reefs appears to be purely microbial as well, but exhibiting *Renalcis*-type fabrics (Zadorozhnaya 1974; Luchinina 1985; Kruse et al. 1996). Earliest Tommotian patch reefs from the Siberian Platform exhibit the familiar Early Cambrian motif of archaeocyaths intergrown with *Renalcis,* leading to a cavernous framework with isopachous rinds of submarine calcite cement and reef-derived biomicrite (Riding and Zhuravlev 1995); reefs with upward-oriented *Epiphyton* appeared later in the Tommotian (Luchinina 1985). Bioclasts include hyoliths, chancelloriids, lingulate brachiopods, sponge spicules, other small shelly fossils, and burrows, indicating that the complex reef ecosystem appeared with the evolution of these organisms. This basic community structure persisted until the late Botoman, almost unchanged except for the local contribution to frameworks by calcarean sponges, modular corals, coralomorphs, problematic organisms such as the radiocyaths and cribricyaths, and some added bioclastic components like stenothecoids and calciate brachiopods.

In the early Toyonian, the entire ecosystem collapsed. All but a few species of archaeocyaths vanished (Debrenne et al. 1984; Wood et al. 1992a,b), as did the entire metazoan biota that dwelt within the reefs (Zhuravlev and Wood 1996). The degree of morphological variation exhibited by calcified microbes, probably governed at least in part by the topographic complexity of frameworks, also plummeted (Zhuravlev 1996). The Toyonian Hawke Bay regression marked the disappearance of carbonate generating conditions in Laurentia and Morocco (Palmer and James 1979), and similar facies changes are observed in many other regions (Seslavinsky and Maidanskaya, this volume). The earlier Sinsk event is ascribed to widespread anoxia and may have precipitated the drastic reduction in archaeocyath diversity in the Botoman of the Siberia Platform, South China, and South Australia (Zhuravlev and Wood 1996). The Sinsk event may have brought the reef community to a threshold that left it vulnerable and unable to diversify after further perturbations.

Middle and Late Cambrian reefs are microbial: dominated by *Girvanella* and *Epi-*

phyton at platform margins but thrombolitic in platform interiors. Siliceous sponges, an important contributor in Ordovician through Jurassic deeper-water mud mounds, took hold locally in the late Middle Cambrian and Late Cambrian. Reefs at platform margins tend to be poor in matrix sediment (James 1981; Hamdi et al. 1995; Pratt 1995). Thrombolites have a burrowed micritic matrix, but the bioclastic component is sparse, restricted mainly to trilobite sclerites and eocrinoid ossicles. The Middle Cambrian shift from aragonite-facilitating to -inhibiting conditions may have affected the biomineralization capacities of certain groups, as well as the nature of synsedimentary cementation.

The diversity of the dwelling fauna exhibited by Early Cambrian reefs did not recover until the Middle Ordovician. Colonial corals reappeared—perhaps reevolved— in the earliest Ordovician (Pratt and James 1989), and subsequently encrusting bryozoans, stromatoporoids, and ostracodes occupied the reefal ecospace and displaced microbial components. Why it took so long, relative to the rapidity of the Early Cambrian diversification, is unclear; perhaps the late Early Cambrian black shale and regressive events were too overwhelming. A dearth of nutrients may have characterized the vast epeiric seas that developed during the Middle and Late Cambrian.

Acknowledgments. Our Cambrian reef researches have been funded primarily by the Natural Sciences and Engineering Research Council of Canada (BRP), a Royal Society University Research Fellowship (RAW), the Paleontological Institute of the Russian Academy of Sciences and Russian Foundation for Basic Researches (AYZ), and the National Engineering Research Council of the United Kingdom (BRS). BRP and AYZ also thank the Muséum National d'Histoire Naturelle, Paris, for support during sojourns at the Laboratoire de Paléontologie. M. Debrenne prepared figure 12.3, and F. Pilard (MNHN) prepared figure 12.4.

REFERENCES

Astashkin, V. A. 1981. Paleogeomorfologicheskie usloviya formirovaniya kebriiskogo rifovogo kompleksa Zapadnoi Yakutii [Paleogeomorphological conditions of formation of Cambrian reef complexes of western Yakutia]. *Trudy, Sibirskiy nauchnoissledovatel'skiy institut geologii, geofiziki, i mineral'nogo syr'ya* 292:19–25.

Astashkin, V. A., A. I. Varlamov, N. K. Gubina, A. E. Ekhanin, V. S. Pereladov, V. I. Romenko, S. S. Sukhov, N. V. Umperovich, A. B. Fedorov, B. B. Shishkin, and E. I.

Khobnya. 1984. *Geologiya i perspektivy neftegazonosnosti kembriyskikh rifovykh sistem Sibirskoy platformy* [Geology and oil and gas prospects in Cambrian reef systems of the Siberian Platform]. Moscow: Nedra.

Bourque, P.-A. and F. Boulvain. 1993. A model for the origin and petrogenesis of the red stromatactis limestone of Paleozoic carbonate mounds. *Journal of Sedimentary Petrology* 63:607–619.

Briggs, D. E. G., D. H. Erwin, and F. J. Collier.

1994. *Fossils of the Burgess Shale.* Washington: Smithsonian Institution Press.

Conway Morris, S. and S. Bengtson. 1994. Cambrian predators: possible evidence from boreholes. *Journal of Paleontology* 68: 1–23.

Debrenne, F., A. Yu. Rozanov, and G. F. Webers. 1984. Upper Cambrian Archaeocyatha from Antarctica. *Geological Magazine* 121:291–299.

Debrenne, F., R. A. Gangloff, and J. G. Lafuste. 1987. *Tabulaconus* Handfield: microstructure and its implication in the taxonomy of primitive corals. *Journal of Paleontology* 61: 1–9.

Debrenne, F., A. Gandin, and A. Zhuravlev. 1991. Palaeoecological and sedimentological remarks on some Lower Cambrian sediments of the Yangtze platform (China). *Bulletin de la Societie géologique de France* 162:575–583.

Debrenne, F., A. Gandin, and M. Debrenne. 1993. Calcaires à archéocyathes du Membre de la Vallée de Matoppa (Formation de Nebida), Cambrian inférieur du Sud-Ouest de la Sardaigne (Italie). *Annales de Paléontologie* 79:77–118.

Drozdova, N. A. 1980. *Vodorosli v organogennykh postroykakh nizhnego kembriya Zapadnoy Mongolii* [Algae in Lower Cambrian organogenous buildups of western Mongolia]. *Trudy, Sovmestnaya Sovetsko-Mongol'skaya paleontologicheskaya ekspeditsiya* 10:1–140.

Fritz, M. A. and B. F. Howell. 1955. An Upper Cambrian coral from Montana. *Journal of Paleontology* 29:181–183.

Green, J. W., A. H. Knoll, and K. Swett. 1988. Microfossils from oolites and pisolites of the upper Proterozoic Eleonore Bay Group, central East Greenland. *Journal of Paleontology* 62:835–852.

Hamdi, B., A. Yu. Rozanov, and A. Yu. Zhuravlev. 1995. Latest Middle Cambrian meta-zoan reef from northern Iran. *Geological Magazine* 132:367–373.

Hou, X.-G., D. J. Siveter, M. Williams, D. Walossek, and J. Bergström. 1996. Appendages of the arthropod *Kunmingella* from the early Cambrian of China: its bearing on the systematic position of the Bradoriida and the fossil record of the Ostracoda. *Philosophical Transactions of the Royal Society of London B* 351:1131–1145.

James, N. P. 1981. Megablocks of calcified algae in the Cow Head Breccia, western Newfoundland: vestiges of a Cambro-Ordovician platform margin. *Geological Society of America Bulletin* 92:799–811.

James, N. P. and P.-A. Bourque. 1992. Reefs and mounds. In R. G. Walker and N. P. James, eds., *Facies Models: Response to Sea Level Change*, pp. 323–347. Geological Association of Canada.

James, N. P. and D. I. Gravestock. 1990. Lower Cambrian shelf and shelf margin buildups, Flinders Ranges, South Australia. *Sedimentology* 37:455–480.

James, N. P. and C. F. Klappa. 1983. Petrogenesis of early Cambrian reef limestones, Labrador, Canada. *Journal of Sedimentary Petrology* 53:1051–1096.

James, N. P. and D. R. Kobluk. 1978. Lower Cambrian patch reefs and associated sediments: southern Labrador, Canada. *Sedimentology* 25:1–35.

James, N. P., D. R. Kobluk, and S. G. Pemberton. 1977. The oldest macroborers: Lower Cambrian of Labrador. *Science* 197:980–983.

Jankauskas, T. V. 1972. Kribritsiaty nizhnego kembriya [Cribricyaths of the Lower Cambrian]. In I. T. Zhuravleva, ed., *Problemy biostratigrafii i paleontologii nizhnego kembriya Sibiri* [Problems of Lower Cambrian biostratigraphy and paleontology of Siberia], pp. 161–183. Moscow: Nauka.

Kaźmierczak, J. and S. Kempe. 1992. Recent

cyanobacterial counterparts of Paleozoic *Wetheredella* and related problematic fossils. *Palaios* 7:294–304.

Kennard, J. M. 1991. Lower Cambrian archaeocyathan buildups, Todd River Dolomite, northeast Amadeus Basin, central Australia: sedimentology and diagenesis. In R. J. Korsch and J. M. Kennard, eds., *Geological and Geophysical Studies in the Amadeus Basin, Central Australia*, pp. 195–225. Bureau of Mineral Resources, Australia, Bulletin 236.

Kennard, J. M. and N. P. James. 1986. Thrombolites and stromatolites: two distinct types of microbial structures. *Palaios* 1:492–503.

Kennard, J. M., N. Chow, and N. P. James. 1989. Thrombolite-stromatolite bioherm, Middle Cambrian, Port au Port Peninsula, western Newfoundland. In H. H. J. Geldsetzer, N. P. James, and G. E. Tebbutt, eds., *Reefs, Canada and Adjacent Areas*, pp. 151–155. Canadian Society of Petroleum Geologists, Memoir 13.

Kobluk, D. R. 1979. A new and unusual skeletal organism from the Lower Cambrian of Labrador. *Canadian Journal of Earth Sciences* 16:2040–2045.

Kobluk, D. R. 1981. Lower Cambrian cavity-dwelling endolithic (boring) sponges. *Canadian Journal of Earth Sciences* 18:972–980.

Kobluk, D. R. 1984. A new compound skeletal organism from the Rosella Formation (Lower Cambrian), Atan Group, Cassiar Mountains, British Columbia. *Journal of Paleontology* 58:703–708.

Kobluk, D. R. 1985. Biota preserved within cavities in Cambrian *Epiphyton* mounds, upper Shady Dolomite, southwestern Virginia. *Journal of Paleontology* 59:1158–1172.

Kobluk, D. R. and N. P. James. 1979. Cavity-dwelling organisms in Lower Cambrian patch reefs from southern Labrador. *Lethaia* 12:193–218.

Kobluk, D. R. and C. F. Kahle. 1978. Geological significance of boring and cavity-dwelling marine algae. *Bulletin of Canadian Petroleum Geology* 28:362–379.

Korde, K. B. 1963. Hydroconozoa—novyy klass kishechnopolostnykh zhivotnykh [Hydroconozoa—a new class of coelenterates]. *Paleontologicheskiy zhurnal* 1963 (2):20–25.

Krause, F. F. and A. J. Rowell. 1975. Distribution and systematics of the inarticulate brachiopods of the Ordovician carbonate mud mound of Meiklejohn Peak, Nevada. *University of Kansas Paleontological Contributions* 61:1–74.

Kruse, P. D. 1991. Cyanobacterial-archaeocyathan-radiocyathan bioherms in the Wirrealpa Limestone of South Australia. *Canadian Journal of Earth Sciences* 28:601–615.

Kruse, P. D., A. Yu. Zhuravlev, and N. P. James. 1995. Primordial metazoan-calcimicrobial reefs: Tommotian (Early Cambrian) of the Siberian Platform. *Palaios* 10:291–321.

Kruse, P. D., A. Gandin, F. Debrenne, and R. Wood. 1996. Early Cambrian bioconstructions in the Zavkhan Basin of western Mongolia. *Geological Magazine* 133:429–444.

Lafuste, J., F. Debrenne, A. Gandin, and D. Gravestock. 1991. The oldest tabulate coral and the associated Archaeocyatha, Lower Cambrian, Flinders Ranges, South Australia. *Geobios* 24:697–718.

Landing, E. 1993. In situ earliest Cambrian tube worms and the oldest metazoan-constructed biostrome (Placentian Series, southeastern Newfoundland). *Journal of Paleontology* 67:333–342.

Luchinina, V. A. 1985. Vodoroslevye postroyki rannego paleozoya severa Sibirskoy platformy [Early Paleozoic algal buildups on the north Siberian Platform]. *Trudy, Institut geologii i geofiziki, Sibirskoe otdelenie, Akademiya nauk SSSR* 628:45–50.

Marek, L. and E. L. Yochelson. 1976. Aspects of the biology of Hyolitha (Mollusca). *Lethaia* 9:65–82.

Mehl, D. 1996. Organization and microstructure of the chancelloriid skeleton: implications for the biomineralization of the Chancelloriidae. *Bulletin de l'Institut océanographique, Monaco,* no. spécial 14 (4): 377–385.

Mikulic, D. G. 1981. Trilobites in Paleozoic carbonate buildups. *Lethaia* 14:45–56.

Moreno-Eiris, E. 1987. Los montículos arrecifales de algas y arqueociatos del Cámbrico Inferior de Sierra Morena. *Boletín Geológico y Minero* 98:1–127.

Palmer, A. R. and N. P. James. 1979. The Hawke Bay event: a circum-Iapetus regression near the lower Middle Cambrian boundary. In D. R. Wones, ed., *The Caledonides in the U.S.A., IGCP Project 27: Caledonide Orogen,* pp. 15–18. Blacksburg, Va.: Department of Geological Sciences, Virginia Polytechnic Institute and State University.

Pratt, B. R. 1982. Stromatolite decline—a reconsideration. *Geology* 10:512–515.

Pratt, B. R. 1984. *Epiphyton* and *Renalcis*—diagenetic microfossils from calcification of coccoid blue-green algae. *Journal of Sedimentary Petrology* 54:948–971.

Pratt, B. R. 1989. Deep-water *Girvanella-Epiphyton* reef on a mid-Cambrian continental slope, Rockslide Formation, Mackenzie Mountains, Northwest Territories. In H. H. J. Geldsetzer, N. P. James, and G. E. Tebbutt, eds., *Reefs, Canada and Adjacent Areas,* pp. 161–164. *Canadian Society of Petroleum Geologists, Memoir* 13.

Pratt, B. R. 1994. Lower Cambrian reefs of the Mural Formation, southern Canadian

Rocky Mountains. *Terra Nova* 3 (Abstract Supplement 6):5.

Pratt, B. R. 1995. The origin, biota and evolution of deep-water mud mounds. In C. L. V. Monty, D. W. J. Bosence, P. H. Bridges, and B. R. Pratt, eds., *Carbonate Mud-mounds: their Origin and Evolution,* pp. 49–123. *International Association of Sedimentologists, Special Publication* 23.

Pratt, B. R. 2000. Microbial contribution to reefal mud-mounds in ancient deep-water settings. In R. Riding, ed., *Microbial Sediments,* pp. 282–288. Berlin: Springer Verlag.

Pratt, B. R. and N. P. James. 1982. Cryptalgal-metazoan bioherms in the Early Ordovician St. George Group, western Newfoundland. *Sedimentology* 29:543–569.

Pratt, B. R. and N. P. James. 1989. Coral-*Renalcis*-thrombolite reef complex of Earth Ordovician age, St. George Group, western Newfoundland. In H. H. J. Geldsetzer, N. P. James, and G. E. Tebbutt, eds., *Reefs, Canada and Adjacent Areas,* pp. 224–230. *Canadian Society of Petroleum Geologists, Memoir* 13.

Rees, M. N., B. R. Pratt, and A. J. Rowell. 1989. Early Cambrian reefs, reef complexes, and associated lithofacies of the Shackleton Limestone, Transantarctic Mountains. *Sedimentology* 36:341–361.

Repina, L. N. 1983. Biofatsii trilobitov rannego kembriya Sibirskoy platformy [Early Cambrian trilobite biofacies on the Siberian Platform]. *Trudy, Institut geologii i geofiziki, Sibirskoe otdelenie, Akademiya nauk SSSR* 569:54–76.

Riding, R. and A. Yu. Zhuravlev. 1995. Structure and diversity of oldest sponge-microbe reefs: Lower Cambrian, Aldan River, Siberia. *Geology* 23:649–652.

Rozanov, A. Yu. and A. Yu. Zhuravlev. 1992. The Lower Cambrian fossil record of the

Soviet Union. In J. H. Lipps and P. W. Signor, eds., *Origin and Early Evolution of the Metazoa*, pp. 205–282. New York: Plenum Press.

Rozov, S. N. 1984. Morfologiya, terminologiya, i sistematicheskoe polozhenie stenotecoid [Morphology, terminology, and systematic affinity of stenothecoids]. *Trudy, Institut geologii i geofiziki, Sibirskoe otdelenie, Akademiya nauk SSSR* 597:117–133.

Runnegar, B., J. Pojeta, N. J. Morris, J. D. Taylor, and G. McClung. 1975. Biology of the Hyolitha. *Lethaia* 8:181–191.

Sayutina, T. A. 1980. A new Early Cambrian family Khasaktiidae n. fam.—possible Stromatoporata. *Paleontological Journal* [Paleontologicheskiy zhurnal] 1980 (4): 13–28.

Scrutton, C. T. 1997. The Palaeozoic corals, I: origins and relationships. *Proceedings of the Yorkshire Geological Society* 51:177–208.

Sorauf, J. E. and M. Savarese. 1995. A Lower Cambrian coral from South Australia. *Palaeontology* 38:757–770.

Spencer, L. M. 1981. Palaeoecology of a Lower Cambrian archaeocyathid inter-reef fauna from southern Labrador. In M. E. Taylor, ed., *Short Papers for the Second International Symposium on the Cambrian System*, pp. 215–218. *U.S. Geological Survey Open-File Report* 81-743.

Spincer, B. R. 1996. Paleoecology of some Upper Cambrian microbial-sponge-eocrinoid reef, central Texas. *Paleontological Society Special Publication* 18:367.

Stitt, J. H. 1976. Functional morphology and life habits of the Late Cambrian trilobite *Stenopilus pronus* Raymond. *Journal of Paleontology* 50:561–576.

Sukhov, S. S. and T. V. Pegel'. 1986. Lito- i biofatsial'nyy analiz srednekembriyskikh otlozheniy vostoka Sibirskoy platformy dlya rekonstruktsii usloviy karbonato-nakopleniya [Lithological and biofacies analysis of the Middle Cambrian strata on the east Siberian Platform for carbonate depositional environment reconstruction]. In V. I. Krasnov, ed., *Paleoekologicheskiy i litologo-fatsial'nyy analizy dlya obosnovaniya detal'nosti regional'nykh stratigraficheskikh skhem* [Paleoecological and lithological facies analyses for the grounds of detailed regional stratigraphic schemes], pp. 35–50. Novosibirsk: Sibirskiy nauchno-issledovatel'skiy institut geologii, geofiziki, i mineral'nogo syr'ya.

Turner, E. C., G. M. Narbonne, and N. P. James. 1993. Neoproterozoic reef microstructures from the Little Dal Group, northwestern Canada. *Geology* 21:259–262.

Wood, R., K. R. Evans, and A. Yu. Zhuravlev. 1992a. A new post–early Cambrian archaeocyath from Antarctica. *Geological Magazine* 129:491–495.

Wood, R., A. Yu. Zhuravlev, and F. Debrenne. 1992b. Functional biology and ecology of Archaeocyatha. *Palaios* 7:131–156.

Wood, R., A. Yu. Zhuravlev, and A. Chimed Tseren. 1993. The ecology of Lower Cambrian buildups from Zuune Arts, Mongolia: implications for early metazoan reef evolution. *Sedimentology* 40:829–858.

Yochelson, E. L. 1969. Stenothecoida, a proposed new class of Cambrian Mollusca. *Lethaia* 2:49–62.

Yochelson, E. L. 1977. Agmata, a proposed extinct phylum of Early Cambrian age. *Journal of Paleontology* 51:437–454.

Zadorozhnaya, N. M. 1974. Rannekembriyskie organogennye postroyki vostochnoy chasti Altae-Sayanskoy skladchatoy oblasti [Early Cambrian organogenous buildups of the eastern part of the Altay Sayan Fold-belt]. *Trudy, Institut geologii i geofiziki, Sibirskoe otdelenie, Akademiya nauk SSSR* 84: 159–186.

Zhuravlev, A. Yu. 1986. Radiocyathids. In A. Hoffman and M. H. Nitecki, eds., *Problematic Fossil Taxa*, pp. 35–44. Oxford: Clarendon Press.

Zhuravlev, A. Yu. 1993. Early Cambrian steps of biomineralization: mineralogy. In *Biomineralization 93: Seventh International Symposium on Biomineralization, Monaco, Program and Abstracts*, p. 102.

Zhuravlev, A. Yu. 1996. Reef ecosystem recovery after the Early Cambrian extinction. In M. B. Hart, ed., *Biotic Recovery from Mass Extinction*, pp. 79–96. *Geological Society Special Publication* 102.

Zhuravlev, A. Yu. and T. A. Sayutina. 1985. Radiotsiati Mongolii. K revizii "klassa" Radiocyatha [Radiocyaths of Mongolia. On the revision of the "class" Radiocyatha]. *Trudy, Institut geologii i geofiziki, Sibirskoe otdelenie, Akademiya nauk SSSR* 632:117–133.

Zhuravlev, A. Yu. and R. Wood. 1995. Lower Cambrian reefal cryptic communities. *Palaeontology* 38:443–470.

Zhuravlev, A. Yu. and R. Wood. 1996. Anoxia as the cause of the mid–Early Cambrian (Botomian) extinction event. *Geology* 24:311–314.

Zhuravlev, A. Yu., F. Debrenne, and J. Lafuste. 1993. Early Cambrian microstructural diversification of Cnidaria. *Courier Forschungsinstitut Senckenberg* 164:365–372.

Zhuravlev, A. Yu., B. Hamdi, and P. D. Kruse. 1996. IGCP 366: Ecological aspects of Cambrian radiation—field meeting. *Episodes* 19:136–137.

Evolution of the Deep-Water Benthic Community

Megascopic life evolved in the Archean with the buildup of stromatolitic mounds in shallow-water environments. By the Proterozoic, stromatolites had already extended down to well below fair-weather wave base. During the late Vendian there was an increase in megascopic biota in shallow water, with both soft-bodied fossils and trace fossils becoming relatively abundant. Some of the soft-bodied forms, such as Pteridinium, were large and preserved three-dimensionally, with remarkable detail, in high-energy medium-to-coarse-grained sandstones. This style of preservation resembles that of trace fossils, which were produced within similar sequences during the Phanerozoic, and may suggest that some of these early life-forms grew through already deposited sediment as a unicellular protoplasmic mass. Some Ediacaran body fossils (e.g., Charniodiscus, Ediacaria, Pteridinium) may have survived into the Cambrian by migrating into deeper water, where many of the reported body fossils were exceptionally preserved soft-bodied forms. There was also a slight increase in trace fossil diversity in deep water during the Cambrian, and this too may reflect the activity of a dominantly soft-bodied fauna. There was a major progressive colonization by hard-bodied forms of the outer shelf by the Early Ordovician, and of the slope toward the end of the Middle Ordovician. In contrast, there is a significant increase in trace fossil abundance and diversity in deep-water flysch sequences as early as the Early Ordovician. It appears that soft-bodied animals, including those which produced trace fossils, were involved first in the onshore-offshore migration and were generally well established in deeper-water niches before the arrival of faunas rich in skeletal forms.

INTRODUCTION

The colonization of deep-sea environments appears to have been a slow process (Crimes 1974; Sepkoski and Miller 1985; Bottjer et al. 1988), and a high percentage of Precambrian and Cambrian megascopic body and trace fossils occur in sediments considered to have been deposited in shallow water, mostly above storm wave base.

There are, however, several abiological factors that might emphasize this apparent distribution. First, deep-water sediments, by the nature of their tectonic setting, are more prone to deformation and metamorphism, and these processes will eliminate some forms and make recovery of others difficult. Second, shallow-water shelf seas were dominant late in the Precambrian and early in the Cambrian. Consequently, the exposed area of shallow-water strata representing the period when life was evolving rapidly far exceeds that of deep water, and third, it is easier to find definitive sedimentological evidence for shallow-water environments than for deep-water ones.

Nevertheless, it is generally accepted that many animals evolved in shallow water during the late Precambrian and early Cambrian and then gradually spread into the deep oceans (Crimes 1974; Sepkoski and Miller 1985; Sepkoski 1990). Indeed, it has been claimed that there is something unique about shallow-water environments that promotes the origin of evolutionary novelties or the assembly of novel community types (Sepkoski and Miller 1985). The most distinctive ecological features of shallow-water environments are the frequent disturbances and the high-energy, stressful, ambient conditions, and these factors may be conducive to the evolution of novel taxa and communities (Steele-Petrovic 1979; Jablonski and Bottjer 1983; Sepkoski and Sheehan 1983; Valentine and Jablonski 1983).

The evolution of a deep-water fauna requires adaptation to certain extreme conditions, such as permanent darkness, high pressure, and low temperature (except in the case of hydrotherms). In addition, deep seas show low fertility. In the absence of terrestrial plant debris influencing community structure, early deep benthos would probably suffer from very limited food (Bambach 1977).

The late Precambrian and Cambrian circumstance of high diversity in shallow water and decreasing diversity in progressively deeper water is in marked contrast to that in modern oceans, where unusually high diversity has been found in deep water (Hessler and Sanders 1967). For example, the diversity of polychaetes and bivalves increases with depth below the continental shelf and, at bathyal depths, reaches levels equivalent to those in tropical soft-bottomed communities at subtidal depths (Sanders 1968). Similarly, when considered for a single type of substrate, the diversity of gastropods and several other groups increases from the shelf to bathyal depths (Rex 1973, 1976, 1981).

Trace fossil evidence suggests that significant colonization of the deep sea may have been delayed until the Ordovician (Crimes 1974; Crimes et al. 1992), while analysis of body fossil diversity data implies that a shallow-water "Cambrian fauna" became progressively restricted to deeper-water environments from the Ordovician onward (Sepkoski 1990:38; Sepkoski 1991).

Recent investigations (e.g., Narbonne and Aitken 1990), however, suggest that even during the Precambrian, animals were penetrating at least into intermediate water depths, and by the Cambrian there was a limited colonization of even bathyal depths (e.g., Crimes et al. 1992; Hofmann et al. 1994).

The purpose of this chapter is to review the progressive colonization of the deep sea from the Precambrian to the Ordovician, that is, through the period of Cambrian radiation.

THE ENVIRONMENTAL SETTING OF THE EARLIEST LIFE

It has become fashionable to regard hydrothermal systems as likely sites for organic synthesis and the origin of life (see Chang 1994 and references therein). Indeed, it has been claimed that present-day microorganisms with the oldest lineages based on molecular phylogenies are anaerobic, thermophilic, sulfur-dependent chemolitho-autotrophic archaebacteria (Woese 1987). It has been suggested that deep marine communities had formed around black smokers and white smokers already in the Precambrian (Kuznetsov et al. 1994). Fossil examples of such communities have been reported in Silurian, Devonian, and Carboniferous sulphur-rich, hydrothermal strata in the ophiolitic suites of the Urals and northeastern Russia, where they are accompanied by vestimentiferans (Pogonophora) and calyptogenid pelecypods similar to the inhabitants of present-day smokers (Kuznetsov 1989; Kuznetsov et al. 1994). Recognition of such sites in early Proterozoic sequences is, however, likely to prove difficult, and although it might be argued that they were more common during early Earth history, they must nevertheless have occupied a small percentage of available ecospace. Therefore, unless they were almost uniquely favorable locations, it is statistically unlikely that they would be the "chosen" sites.

The earliest well-documented signs of life come from ~3–3.5 Ga, in early Archean strata in the Swaziland Supergroup of South Africa and the Pilbara Supergroup in western Australia (Schopf 1994). These units contain stromatolites and microfossils, and it is considered that the former, at least, grew in narrow, shallow-water zones along shorelines of volcanic platforms subject to periodic agitation by waves or currents (Groves et al. 1981; Byerly et al. 1986). The similarity between these stromatolites and much more recent ones suggests strongly that they exhibited bacterial or cyanobacterial photosynthesis (Schopf 1994) and were therefore restricted to shallow water.

The first sediments considered to have been deposited on a stable carbonate platform occur in the Middle Archean Nsuze Group, which includes stromatolitic dolomites in a tidally influenced environment (Walter 1983; Grotzinger 1994). By the Late Archean, stromatolite-bearing carbonates were being deposited in cratonic and non-cratonic settings (Grotzinger 1994), but the growth of large cratonic masses of continental lithosphere during the Archean-Proterozoic transition (Veizer and Compston

1976) gave rise to a dramatic increase in carbonate platforms (Grotzinger 1994), and this provided ecospace for a significant increase in abundance and diversity of stromatolites (figure 13.1A), which peaked in the Middle Proterozoic (Awramik 1971; Walter and Heys 1985). This increase was accompanied by the occupation of more-varied niches extending down to well below fair-weather wave base but still presumably within the photic zone (Grotzinger 1990; Walter 1994: figure 4). The colonization of "deeper" water seems to have already commenced.

The decline of stromatolites is commonly ascribed to the advent of soft-bodied Metazoa, as evidenced by the Ediacara fauna and its associated trace fossils (Garrett 1970; Awramik 1971). Some of these forms may have been able to destroy stromatolites by grazing and burrowing, but there has been no significant documentation of stromatolites affected in this way. Competitive exclusion by higher algae may also have contributed to the decline (Hofmann 1985; Butterfield et al. 1988; and see Droser and Li, Pratt et al., Riding, this volume). Many later organisms may have responded to competitive pressures by migrating into deep water (Crimes 1974; Sepkoski 1990), but stromatolites, being limited to the photic zone, had probably occupied much of the available ecospace by the late Proterozoic and, consequently having "nowhere to go and nowhere to hide," might have suffered badly from increased competition with an expanding trophic web.

One of the earliest records of probable metazoan life is *Bergaueria*-like trace fossils (see Crimes 1994:114) from the 800–1100 Ma Little Dal Group of the Mackenzie Mountains, Canada (Hofmann and Aitken 1979). These occur in a carbonate-dominated sequence of varied lithology, considered to be of basinal aspect and deposited in water several tens to 200 m deep (Hofmann and Aitken 1979:153). These fossils may therefore also mark an early colonization of slightly deeper water.

THE COLONIZATION OF DEEPER WATER DURING THE VENDIAN

The Vendian era, extending from ~610–545 Ma (Grotzinger et al. 1995), commences with the Varanger tillites and their equivalents and is the first to yield relatively common and diverse undisputed body fossils and trace fossils.

The oldest Vendian biota, consisting of *Nimbia*, *Vendella?*, and *Irridinitus?*, was found in the intertillite Twitya Formation of the Mackenzie Mountains, Canada (Hofmann et al. 1990). This sequence comprises siliciclastic turbidites associated with major channel-fill conglomerates and is considered to be relatively deep-water (Hofmann et al. 1990).

The majority of post-tillite Vendian biotas have been found in shallow-water sequences, apparently deposited above fair-weather wave base, and in some regions (e.g., Australia, Namibia, Russia, Ukraine), remarkably abundant, diverse, and well-preserved faunas have been found (see reviews in Glaessner 1984; Sokolov and Iwanowski 1985; Fedonkin 1992; Jenkins 1992). Indeed, in some sequences deposited

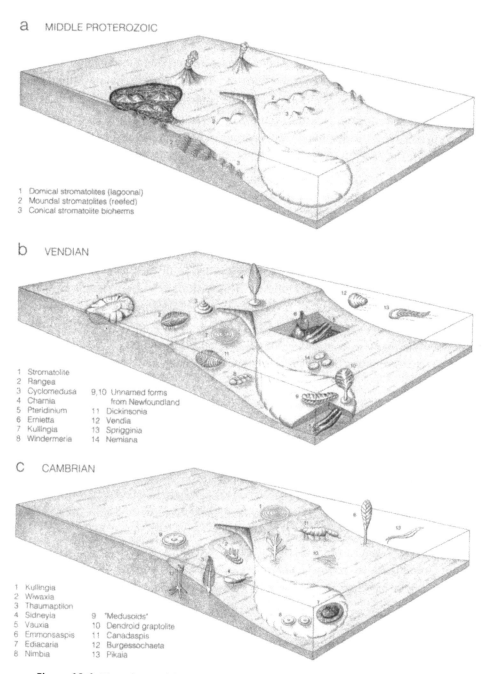

a   MIDDLE PROTEROZOIC

1   Domical stromatolites (lagoonal)
2   Moundal stromatolites (reefed)
3   Conical stromatolite bioherms

b   VENDIAN

1   Stromatolite
2   Rangea
3   Cyclomedusa   9,10   Unnamed forms
4   Charnia   from Newfoundland
5   Pteridinium   11   Dickinsonia
6   Ernietta   12   Vendia
7   Kullingia   13   Sprigginia
8   Windermeria   14   Nemiana

c   CAMBRIAN

1   Kullingia
2   Wiwaxia
3   Thaumaptilon
4   Sidneyia   9   "Medusoids"
5   Vauxia   10   Dendroid graptolite
6   Emmonsaspis   11   Canadaspis
7   Ediacaria   12   Burgessochaeta
8   Nimbia   13   Pikaia

Figure 13.1 "Snapshots" of the ocean floor faunas for Middle Proterozoic, Vendian, and Cambrian, showing the progressive colonization of deeper water based on body fossils.

under varied depths of water, fossils occur only in the shallower-water lithologies. For example, in the Tanafjorden area of Norway, the Vendian Innerelv Member consists of two shallowing-upward sequences, each representing a transition from offshore marine (quiet basin, below wave base) to wave-influenced, shallow, subtidal and intertidal deposition (Banks 1973), but a biota consisting of *Cyclomedusa, Ediacaria?, Beltanella, Hiemalora,* and *Nimbia?* occurs only in sediments interpreted as representing a current-swept, wave-influenced environment (Farmer et al. 1992).

There are, however, a few well-documented examples in which body and/or trace fossils do occur in deeper-water deposits (figures 13.1B and 13.2A). In the case of body fossils, it might be possible to claim that they have been transported from shallow water, but such an argument cannot be applied to trace fossils, which reflect life activity at the precise location where they are now found.

In the Wernecke Mountains, Canada, Narbonne and Hofmann (1987) record a fairly extensive Ediacara fauna, most of which comes from Siltstone Units 1 and 2, deposited under shallow-water conditions. This includes the body fossils *Beltanella, Beltanelliformis, Charniodiscus, Cyclomedusa, Kullingia?, Medusinites, Nadalia, Spriggia,* and *Tirasiana,* as well as the trace fossils *Gordia, Neonereites?,* and *Planolites.* However, *Charniodiscus* was also recorded from the Goz Siltstone, which includes slump and load structures and was deposited on a slope in a deeper-water setting.

A more extensive deeper-water biota has been described by Narbonne and Aitken (1990) from the Sekwi Brook area of northwestern Canada, where the Sheepbed and Blueflower formations include turbidity current–deposited sandstones and common slump deposits and are interpreted as representing a deep-water basin slope setting, below storm wave base. The biota includes the body fossils *Beltanella, Charniodiscus?, Cyclomedusa, Ediacaria, Eoporpita, Inkrylovia, Kullingia, Pteridinium,* and *Sekwia* and the trace fossils *Aulichnites, Helminthoida, Helminthoidichnites, Helminthopsis, Lockeia, Neonereites, Palaeophycus, Planolites,* and *Torrowangea.* More recently, *Hiemalora* and *Windermeria* have been reported from the same sequence (Narbonne 1994).

Pteridinium has also been recorded from the South Carolina Slate Belt in deepwater, thinly bedded to finely laminated pelites and siltstones of the Albermarle Group, which may have been deposited between 586 and 550 Ma (Gibson et al. 1984). This sequence has also yielded the trace fossils *Gordia, Neonereites, Planolites,* and *Syringomorpha* (Gibson 1989).

Surfaces covered with numerous predominantly frondlike and bushlike Ediacaran body fossils, including *Charnia* and *Charniodiscus,* occur within volcaniclastic turbidite sequences interpreted as deep-water submarine fan and slope deposits (Myrow 1995) within the Conception Group on the Avalon Peninsula, Newfoundland, Canada (see Anderson and Misra 1968; Misra 1969; Anderson and Conway Morris 1982; Conway Morris 1989a; Jenkins 1992). Taphonomic and sedimentological data indicate that this is an in situ life assemblage that suffered rapid burial by volcanic ash at

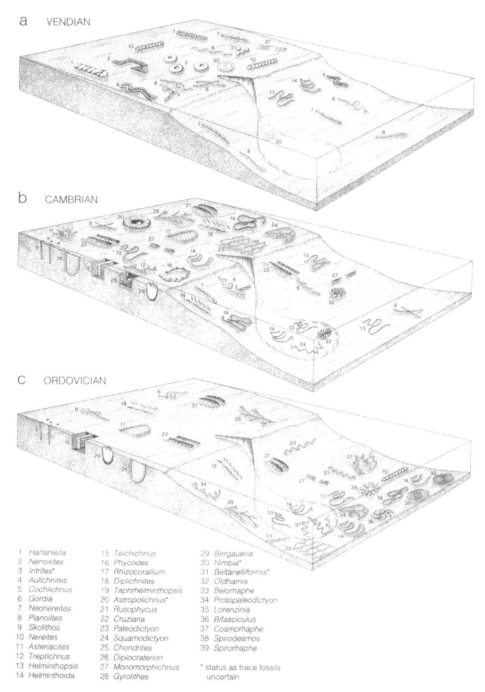

a VENDIAN

b CAMBRIAN

c ORDOVICIAN

1	Harlaniella	15	Teichichnus	29	Bergaueria
2	Nenoxites	16	Phycodes	30	Nimbia*
3	Intrites*	17	Rhizocorallium	31	Beltanelliformis*
4	Aulichnites	18	Diplichnites	32	Oldhamia
5	Cochlichnus	19	Taphrhelminthopsis	33	Belorhaphe
6	Gordia	20	Astropolichnus*	34	Protopaleodictyon
7	Neonereites	21	Rusophycus	35	Lorenzinia
8	Planolites	22	Cruziana	36	Bifasciculus
9	Skolithos	23	Paleodictyon	37	Cosmorhaphe
10	Nereites	24	Squamodictyon	38	Spirodesmos
11	Asteriacites	25	Chondrites	39	Spirorhaphe
12	Treptichnus	26	Diplocraterion		
13	Helminthopsis	27	Monomorphichnus	*	status as trace fossils
14	Helminthoida	28	Gyrolithes		uncertain

Figure 13.2 "Snapshots" of the ocean floor faunas for Vendian, Cambrian, and Ordovician, showing the progressive colonization of deeper water based on ichnofossils.

some horizons (Jenkins 1992; Seilacher 1992; Myrow 1995). The turbidites may not have formed at truly oceanic depths but perhaps on a continental terrace (Benus 1988; Jenkins 1992). A broadly similar setting has been postulated for the occurrence of *Charnia, Charniodiscus,* and *Pseudovendia* within a Vendian sequence at Charnwood Forest, Leicestershire, England, where Jenkins (1992) suggests that the frequency of slumps, together with some current rippling and an absence of oscillation ripples, implies deposition on a slope environment below storm wave base. Boynton and Ford (1995) record three new genera from this sequence (*Ivesia, Shepshedia,* and *Blackbrookia*), but conclude that, despite the presence of graded bedding and absence of shallow-water indicators, water depth may be little more than wave base.

The classic sequence at Ediacara, Australia, which has yielded an abundant and diverse nonskeletal fauna, has been interpreted by Gehling (1991) as deposited in an outer shelf setting below fair-weather wave base, with burial of the organisms by storm surge sands. Seilacher (in Jenkins 1992: 152) considers that the common occurrence of wave oscillation and interference ripples suggests deposition on the shoreface, albeit perhaps by storm events, and a shallow-water tidal environment also seems indicated by the large polygonal desiccation cracks in the highly fossiliferous parts of the section (Jenkins 1992: 153).

Evidence of life at truly bathyal depths is largely absent during the Vendian, although records of the trace fossil *Planolites* within the deep-sea turbidite sequence of the South Stack Formation of the Mona Complex on Anglesey, Wales, by Greenly (1919) have been substantiated during recent fieldwork. The age of these rocks is debatable, but radiometric dates on intrusive granites suggest that it is greater than 600 Ma (Shackleton 1969).

The conclusion appears to be that while most Ediacarian body and trace fossils from the prolific localities in Australia, Namibia, Russia, and Ukraine occurred in shallow-water environments at or above wave base, other localities, including Charnwood Forest, Newfoundland, Sekwi Brook, and Wernecke Mountains, show features suggestive of a slightly deeper-water environment below storm wave base, mostly on the continental slope. There is not, however, any evidence of significant colonization of truly oceanic depths during the Vendian.

Such colonization as took place in intermediate water depths was dominated by sessile body fossils (e.g., *Cyclomedusa, Ediacaria*) and detritus-feeding animals that produced traces either on muddy substrates (e.g., *Helminthoida, Helminthopsis*) or at very shallow depths (e.g., *Paleodictyon*). Significant bioturbation did not occur until the Early Cambrian (Crimes and Droser 1992). In present-day oceans, faunas inhabiting muddy substrates are more abundant and diverse than those of sandy areas (Menzies et al. 1973), whereas in these ancient seas, the absence of algae and the scarcity of large animals increased the survival possibilities of the detritivorous trophic group (Sanders and Hessler 1969; Sokolova 1989).

BIOTIC CHANGES ACROSS THE PRECAMBRIAN-CAMBRIAN BOUNDARY

Diversity curves of metazoan genera show a fall at the Vendian-Tommotian boundary (Sepkoski 1992: figure 11.4.2). This data set includes genera from all depositional environments, and the fall has been interpreted as reflecting a mass extinction. Seilacher (1984) suggested that Vendian biota mark not simply a nonskeletal start to metazoan evolution but a distinct episode to the history of life, terminated by a major extinction. He later suggested that they were quilted constructions that represented an evolutionary experiment that failed with the incoming of macrophagous predators (Seilacher 1989).

There is, however, also a remarkable change in the style of preservation of many of the body fossils in passing across the Precambrian-Cambrian boundary (cf. Seilacher 1984). The Vendian shallow-water sequences are dominated by relatively large forms, commonly exceedingly well preserved in three dimensions and found within fine-to-coarse-grained, well-washed, matrix-poor sandstones (figure 13.3). Such three-dimensional preservation is almost unknown in the Phanerozoic (cf. Seilacher 1984, 1989). By that time, these high-energy sandstones commonly lack body fossils and are dominated by trace fossils, many of which are produced within or between beds. Explanations for the three-dimensional preservation of Vendian body fossils include early mineral precipitation within the matrix (Jenkins 1992), low rates of microbial decomposition (Runnegar 1992), absence of scavengers (Conway Morris 1993), and the supposed existence of mineral crusts formed by cyanobacterial mats (Gehling 1991).

The parallels between the three-dimensional preservation of these body fossils in the Vendian and the trace fossils produced *within* similar sandstones in the Phanerozoic is remarkable but is consistent with the conclusions of Crimes and Fedonkin (1996) that many of these three-dimensionally preserved Vendian body fossils actually formed by growth within the sediment by a process of plasmic permeation. Such animals would then undoubtedly suffer from the incoming of macrophagous predators in the Phanerozoic as envisaged by Seilacher (1989). They seem to have responded by onshore-offshore migration, and a few appear in deeper-water environments during the Cambrian (Conway Morris 1993; Crimes and Fedonkin 1996). Crimes (1994) has argued that Vendian trace fossils also include many unusual and short-ranging forms. The trace fossil diversity data (Crimes 1992: figure 2; Crimes 1994: figure 4.1) do not, however, support a mass extinction, nor indeed is this indicated by the body fossil data set of Sepkoski (1992: figure 11.4.1) when considered in terms of families, orders, or classes. Evidence for such an event is perhaps best shown when the fauna is divided into "Ediacaran, "Tommotian," and "Cambrian sensu stricto" (Sepkoski 1992: figure 11.4.2). There is, however, increasing evidence that some, or perhaps many, elements of the Ediacara fauna continue through the Nemakit-Daldynian (see Brasier 1989) and into later Cambrian strata (Conway Morris 1992;

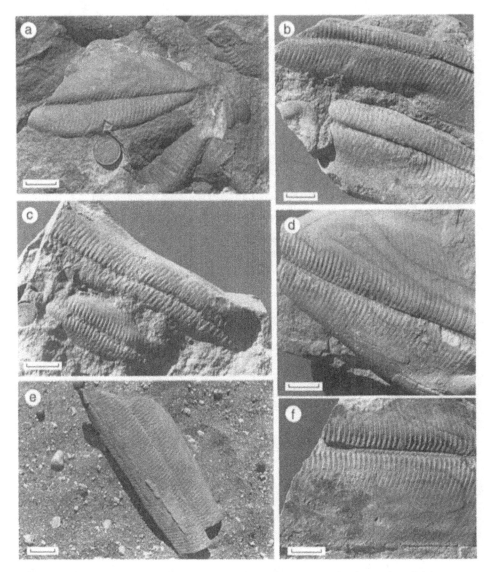

Figure 13.3 Three-dimensional nature of *Pteridinium* from the Kliphoek Member of the Neoproterozoic Nama Group (South Namibia). *A,* Field photograph at Plateau Farm, near Aus; *B–F,* specimens lodged in a small museum at Aar Farm, by permission of Mr. H. Erni. All scale bars 2 cm.

Crimes et al. 1995; Crimes and Fedonkin 1996). Additionally, the data are imprecise because of correlation problems at this level.

Although an overall reduction in diversity cannot be discounted, the picture is far from clear, and, interestingly, Jablonski (1995) places the first of his "Big Five" mass extinctions at the end of the Ordovician. One might also anticipate that any extinction event could have greater consequences in shallow water than in the more constant slope environments considered here. In contrast, the dramatic increase in diversity of both body and trace fossils in the earliest Cambrian strata is obvious (Sepkoski 1992;

Crimes 1994) and has led to the concept of "explosive evolution." There are a significant number of short-ranging forms in the late Precambrian, but the Cambrian is dominated by much longer-ranging forms of Phanerozoic type, and this has prompted Crimes (1994) to suggest that the major change is a biological one in which a period of early evolutionary failure, as represented by a high proportion of short-ranging forms, is replaced by evolutionary success.

CHANGES IN DEEP-WATER BIOTA DURING THE CAMBRIAN

There was a considerable increase in body and trace fossil abundance and diversity from the Late Vendian through the Early Cambrian (see figures 13.1c and 13.2b) (Crimes 1974, 1992, 1994; Sepkoski and Miller 1985; Signor 1990; Sepkoski 1991, 1992; Crimes and Fedonkin 1994). Most of these developments took place in shallow water, but this must have resulted in a dramatic increase in dispersal pressures. The first great evolutionary fauna (Sepkoski and Miller 1985) evolved during the Cambrian in shelf seas, with many of the first appearances probably in subtidal environments, below fair-weather wave base (Mount and Signor 1985). This fauna was dominated by trilobites but with associated hyoliths, eocrinoids, helcionelloid mollusks, lingulate brachiopods, and a variety of lightly sclerotized arthropods. Maximum diversity was achieved in the late Middle to early Late Cambrian, according to Sepkoski (1992; but see Zhuravlev, this volume: figure 8.1a).

By the Tommotian, all the main Phanerozoic trace fossil lineages were well established in shallow water, and they achieved a high degree of behavioral perfection by the end of the Atdabanian (Crimes 1992). These lineages include forms that later were to invade the deep oceans and retreat from shallow-water seas (Crimes 1994). Examples include the network structure *Paleodictyon*, which made an initial appearance in the Vendian (as *Catellichnus* in Bekker and Kishka 1989: plates 1–6) and subsequently appeared with better behavioral programming as *Paleodictyon* and *Squamodictyon* in the Early Cambrian (Crimes and Anderson 1985; Pacześna 1985), and meandering forms such as *Helminthoida, Parahelminthoida,* and *Taphrhelminthoida,* which appear in the Vendian (Narbonne and Aitken 1990; Gehling 1991) and are well developed by the Early Cambrian (Crimes and Anderson 1985; Hofmann and Patel 1989; Goldring and Jensen 1996).

Nevertheless, it has long been recognized that this evolutionary burst, and the dispersal pressures that it must have created, did not immediately lead to dramatic colonization of the deep sea (Crimes 1974, 1994; Sepkoski and Miller 1985). For example, numerous investigations over the last 150 years have revealed few records of body fossils within the deep-water turbidite sequences of the classic Cambrian outcrops in Wales. In northern Gwynedd, a strong cleavage has hampered collecting, but in the Harlech Dome to the south and, more particularly, on St. Tudwal's Peninsula to the west, deformation is much less. The only significant records are small restricted faunas of Lower Cambrian trilobites from the Hell's Mouth Grits of St. Tudwal's Peninsula (Bassett and Walton 1960) and Green Slates of northern Gwynedd (Wood 1969). There may also be doubt as to whether even these meager faunas are in situ. The earliest well-documented subthermocline fauna is the Botoman *Elliptocephala asaphoides* fauna from the Taconics of New York and Vermont, which has some affinities with typical Laurentian faunas (Theokritoff 1985).

Several fossil assemblages have been recognized in alternating flaggy limestones, argillaceous limestones, marlstones, and mudstones, deposited in outer slope and open-marine facies that occurred distally in the Yudoma-Olenek Basin of the Siberian Platform during the late Middle Cambrian (Fedorov et al. 1986). Pterobranchs, sessile graptolites, trilobites, lingulate brachiopods, hyoliths, echinoderms, and possible *Brooksella* have been described here from the Zelenotsvet, Dzhakhtar, and Siligir formations (Lazarenko and Nikiforov 1972; Astashkin et al. 1991; Pel'man et al. 1992; Durham and Sennikov 1993).

Sessile dendroid graptolites were ubiquitous elements of muddy substrates. They have been reported in the Middle Cambrian Amgan Oville Formation in northern Spain and the Late Cambrian Idamean (Steptoean) of Tasmania (Sdzuy 1974; Rickards et al. 1990). The graptolites are accompanied mostly by hexactinellid sponges, lingulates, and trilobites, such as solenopleuropsids in the Middle Cambrian and agnostids, olenids, and ceratopygids in the Late Cambrian.

There was, however, also some colonization of the deep sea by trace fossils (figure 13.2B), presumably representing mainly a soft-bodied fauna.

In southeastern Ireland, the Lower to Middle Cambrian Cahore Group is a deep-sea proximal turbidite sequence that has yielded *Arenicolites, Helminthopsis, Helminthoida, Monocraterion, Oldhamia, Palaeophycus, Planolites,* and *Protopaleodictyon* (Crimes and Crossley 1968; Crimes et al. 1992). In North Wales, proximal turbidites of the late Early Cambrian Hell's Mouth Grits on St. Tudwal's Peninsula contain *Palaeophycus, Phycodes,* and *Planolites* (Crimes 1970; Crimes et al. 1992), whereas the Middle Cambrian Cilan Grits have *Bergaueria, Cruziana, Planolites* and *Protopaleodictyon* (Crimes et al. 1992). Deep-water turbidites yielding *Oldhamia,* and of known or inferred Early to Middle Cambrian age, occur in many localities, including Belgium, the United States, and various parts of Canada (see Dhonau and Holland 1974; Hofmann et al. 1994 and references therein). In Quebec, Canada, Sweet and Narbonne (1993) recorded *Oldhamia* from a deep-water channel-fan environment that is directly overlain by strata containing the Early Cambrian brachiopod *Botsfordia pretiosa.* In the Yukon and Alaska, *Oldhamia* is accompanied by *Bergaueria, Cochlichnus, Helminthoidichnites, Helminthorhaphe, Monomorphichnus, Planolites,* and *Protopaleodictyon* in deep-sea sediments (Hofmann et al. 1994). The most diverse collection of Lower to Middle Cambrian deep-water trace fossils occurs in the Puncoviscana and Suncho formations in northwest Argentina, where Aceñolaza and Durand (1973), Aceñolaza (1978), and Aceñolaza and Toselli (1981) have described *Cochlichnus, Dimorphichnus, Diplichnites, Glockerichnus, Gordia, Helminthopsis, Nereites, Oldhamia, Planolites, Protichnites, Proto-*

virgularia, Tasmanadia, and *Torrowangea.* Deep-water "Cambrian" sediments in North Greenland have also yielded *Helminthopsis, Gordia, Planolites,* and *Protopaleodictyon* (Pickerill et al. 1982). In the Early Cambrian sequence of the Holy Cross Mountains of Central Poland, there is a diverse ichnofauna in the shallow-water units and a restricted one, comprising *Oldhamia, Planolites,* and *Scolicia,* in the Czarna Shale Formation, which was deposited below wave base in the deeper part of the basin (Orłowski 1989). There are also body fossils in this unit, including hyoliths, mollusks, bradoriids, coleolids, sabelliditids, vendotaeniaceans, and *Platysolenites* (Orłowski 1989). Jenkins and Hasenohr (1989) have also described conocoryphid trilobites and their unnamed traces from black shales deposited below wave base in South Australia.

Some of the exceptionally preserved Cambrian faunas, such as the Chengjiang fauna at Meishucun, Yunnan, China, occur in a shallow-water setting (Hou and Sun 1988). However, the Early Cambrian Kinzers Formation of Pennsylvania and Middle Cambrian Burgess Shale of British Columbia comprise fine-grained sediments deposited under anoxic or dysaerobic conditions in deeper-water open-shelf situations and have yielded abundant and diverse faunas of which only 14% of genera and 2% of individuals have hard parts that would be capable of fossilization under normal taphonomic conditions (Conway Morris 1985, 1992). The typical Burgess Shale fauna comprises representatives of the principal major groups: arthropods, polychaetes, priapulids, sponges, brachiopods, mollusks, hyoliths, echinoderms, cnidarians, chordates, hemichordates, cyanobacteria, acritarchs, and probable red and green algae. In terms of genera, arthropods (including trilobites) and sponges are the dominant groups. The sessile epifauna is dominated by sponges, brachiopods, chancelloriids, echinoderms, and some enigmatic forms, all inferred to be suspension feeders. The vagrant epifauna is mostly composed of arthropods, together with mollusks, wiwaxiids, and hyoliths. Diverse predators are indicated by the abundance of carnivores, mostly soft-bodied but a few with hard parts. A trophic web has been reconstructed (Conway Morris 1986; Burzin et al., this volume: figure 10.2.2).

It is considered that many other open-shelf communities that lacked the circumstances for exceptional preservation originally had a rich soft-part component broadly similar to the Burgess Shale at this time (Conway Morris 1992:631). The Burgess Shale fauna is, however, conservative (Conway Morris 1989b), suggesting that colonization of even the "deeper-water" parts of the oceans was not yet far advanced. However, the apparently greater abundance and diversity of trace fossils than of body fossils in many deep-water Early and Middle Cambrian sequences may indicate that, with increased dispersal pressures and the development of hard skeletons in shallow-water environments, soft-bodied animals were forced out into deeper water, where they were probably responsible for most of the traces. Simplification, including diminution of hard skeletal elements, is a common feature of present-day abyssal animals, because of scarcity of food resources (Hessler and Wilson 1983; Zezina 1989).

Indeed, the Burgess Shale fauna includes several forms of Ediacara-type fossils; and

Conway Morris (1993) considers that *Thaumaptilon* may be comparable to frondlike fossils such as *Charniodiscus,* that *Mackenzia* may have affinities to *Inaria, Protechiurus,* and *Platypholinia,* and that *Emmonsaspis* could be related to *Pteridinium.* These possibilities may suggest that the Ediacara fauna survived beyond the Precambrian-Cambrian boundary in part by moving into deeper water. The discovery of a rigid-bodied but nonskeletal biota of Ediacaran affinities in a thick (thousands of meters), deep-water, turbidite sequence of Late Cambrian age in Eire suggests colonization not just of "deeper water" but also of the deep ocean. The biota comprises large (50–200 mm in diameter) disks referred to *Ediacaria* and common discoidal forms included in *Nimbia* (Crimes et al. 1995). However, there is evidence that other "medusoid"-like forms continue as rare occurrences in shallow water. Pickerill (1982) described disk-like "medusoid" fossils, showing concentric and widely spaced radial ornament, from the Late Cambrian *Agnostus* Cove Formation of the St. John Group in New Brunswick. "Medusoids" were also recorded from sediments deposited above storm wave base in the late Middle to early Late Cambrian King's Square Formation, also of the St. John Group in New Brunswick (Tanoli and Pickerill 1989). Finally, Late Cambrian to Early Ordovician intertidal and shallow subtidal sediments of the Kelly's Island and Beach formations of the Bell Island Group of East Newfoundland have yielded "medusoid impressions" that show prominent radial structures likened to those of *Cyclomedusa* (Nautiyal 1973). It is possible therefore that the Ediacara-type fauna may have expanded into deep water during the Cambrian, rather than retreated.

FAUNAL CHANGES ACROSS THE CAMBRO-ORDOVICIAN BOUNDARY AND MAJOR COLONIZATION OF THE DEEP SEA

The Cambrian evolutionary fauna started a long, gradual decline as the end of the Cambrian approached (Sepkoski 1990). This was accompanied by the migration of many of its component forms into deeper water. Data assembled by Sepkoski and Miller (1985: figure 2) as a time-environment diagram show that there was only minimal colonization of outer-shelf and slope environments even by the Late Cambrian, but that there was major progressive colonization of the outer shelf by the Early Ordovician, and of the slope toward the end of the Middle Ordovician. Even trilobite-rich communities, which dominated Cambrian shelf seas, penetrated into deep-water environments (Berry 1972, 1974; Sepkoski and Miller 1985).

In the Late Cambrian of Wales, the olenid community existed in stagnant bottoms, and agnostids replaced earlier eodiscids, while olenids replaced earlier protolenids, but no burrows occur (McKerrow 1978). Some groups of polymeroid trilobites bear features of adaption to low oxygen tension (Fortey and Wilmot 1991). By the Middle to Late Ordovician they had colonized deep-water basinal marine habitats, as is shown by the occurrence of the trilobite traces *Cruziana* and *Rusophycus* in flysch sediments of the Lotbinière Formation of Quebec (Pickerill 1995). *Rusophycus* also occurs in Late Ordovician turbidites at Llangranog in central Wales, where the body form of the resting traces indicates production by trinucleids. They are accompanied by walking traces of *Diplichnites*-type (pers. obs.).

An interesting fauna occurs in the Late Cambrian to Early Ordovician Hales Limestone of Nevada, which consists of dark-colored lime mudstone and wackestone, interbedded with coarse-textured allochthonous gravity-flow and slump deposits interpreted as deposited in deep-water slope environments (Cooke and Taylor 1975, 1977). The indigenous assemblage occurs in the lime mudstones and wackestones and includes spiculate sponges and trilobites, such as agnostids, olenids, ceratopygids, and papyriaspidids, which may have lived below the thermocline. A similar fauna occurs in the Chopko Formation of the Siberian Platform, where the trilobite fauna is not so diverse and the burrowing and infauna are restricted, but the presence of palaeoscolecidans, benthic bradoriids, conodonts, and abundant sponge spicules suggests filter and suspension feeding and carnivorous activity (Barskov and Zhuravlev 1988; Varlamov and Pak 1993).

Trace fossils, however, show a significant rise in diversity in deep water rather earlier than body fossils, and Crimes and Crossley (1991: figure 18) show these changes commencing at the Cambro-Ordovician boundary. Indeed, a deep-water flysch sequence of Tremadoc to Lower Arenig age within the Ribband Group of Wexford County, Eire, contains abundant trace fossils representing 14 ichnogenera, testifying to significant colonization of the deep sea (figure 13.2C). The Early Ordovician Skiddaw Group of the Lake District, England, has also yielded 14 ichnogenera, including 5 not recorded in the Ribband Group (Orr 1996).

All the main lineages of typical deep-sea trace fossils, including rosette, patterned, meandering, and spiral forms evolved in shallow water during the Vendian or Early Cambrian but migrated into the deep sea during the Ordovician (Crimes et al. 1992). It seems that animals producing trace fossils penetrated into deep water first and were followed by skeletal animals, which produced body fossils.

CONCLUSIONS

The main conclusions that can be drawn from this review are as follows:

1. The earliest megafossils were stromatolites, which appeared around 3.5 Ga. The growth of large cratonic masses of continental lithosphere during the Archean-Proterozoic interval created carbonate platforms with extensive ecospace, which may have contributed to their dramatic increase in abundance and diversity. They extended their habitat to well below fair-weather wave base but were unable to penetrate below the photic zone. This may have contributed to their decline in the late Proterozoic, because following the advent of Metazoa, which may have been able to graze on them, and competitive exclusion by higher algae, they may have had no more available ecospace.

2. Colonization of deeper water commenced in the Vendian, when Ediacara fau-

nas and their coeval ichnofaunas, which dominated shallow water, gradually spread to the continental slope, and some were preserved within turbidites commonly associated with slumps. Most of the body fossils were from sessile animals, whereas surface-grazing, deposit-feeding animals provided most of the traces. Any bioturbation was concentrated in the top few millimeters, with deeper penetration not occurring until the Early Cambrian. Fossils and trace fossils have not been found in truly bathyal sediments in rocks of this age.

3. There may have been a small decrease in body fossil diversity at about the Precambrian-Cambrian boundary, but the same is not true of trace fossils. There were a significant number of unusual short-ranging forms, many of which died out in the late Vendian, but this loss was largely balanced by new appearances, most of which were long-ranging. There does not, therefore, appear to have been a mass extinction at the boundary, but there may have been a biotically controlled change from "evolutionary failure" in the Vendian to "evolutionary success" in the Phanerozoic.

4. The Ediacara fauna that dominated shallow, clastic shelf seas in much of the Vendian may have spread to slope environments by the Middle Cambrian (Burgess Shale) and to the deep sea by the Late Cambrian (Ribband Group).

5. Trilobites may have extended into deeper water by the Botoman, but in general there was minimal colonization of even the outer shelf by skeletal forms by the Late Cambrian. However, exceptionally preserved, dominantly soft-bodied fossils occurred in slope environments during the Early to Middle Cambrian but provided only conservative faunas.

6. In contrast, low-diversity ichnofaunas, commonly including *Oldhamia*, were widespread in Early and Middle Cambrian deep-sea turbidite sequences and testify to an earlier and more successful colonization of the deep sea by soft-bodied forms.

7. Trace fossils were abundant and diverse in the deep sea by the Early Ordovician, but the skeletal elements of the Cambrian evolutionary fauna began a slow migration into deeper water during the Ordovician. Significant colonization of the outer shelf took place in the Early Ordovician, followed by migration to the slope by the Middle Ordovician. Trilobites did, however, manage to penetrate at least to the foot of the slope by the Middle to Late Ordovician.

It seems that during the progressive colonization of the deep sea, soft-bodied faunas, including those forms responsible for producing traces, migrated first and were generally well established before the arrival of faunas rich in skeletal forms. The earliest forms in the deep sea were mostly surface grazers, well adapted to life on muddy substrates.

Acknowledgments. I am exceedingly grateful to Dr. A. Yu. Zhuravlev for providing me with much helpful information, particularly concerning the Russian literature and to Dr. P. J. Brenchley for critical reading of the manuscript. This paper is a contribution to IGCP Projects 319, 320, and 366.

REFERENCES

Aceñolaza, F. G. 1978. El Paleozoico inferior de Argentina según sus trazas fósiles. *Ameghiniana* 15:12, 15–64.

Aceñolaza, F. G. and F. Durand. 1973. Trazas fósiles del basamento cristalino del Noroeste Argentino. *Boletín de la Asociación Geológica de Córdoba* 2:45–55.

Aceñolaza, F. G. and A. J. Toselli. 1981. *Geología del Noroeste Argentino*. Universidad Nacional de Tucumán Publicación 1287.

Anderson, M. M. and S. Conway Morris. 1982. A review, with descriptions of four unusual forms, of the soft-bodied fauna of the Conception and St. John's groups (late-Precambrian), Avalon Peninsula, Newfoundland. In B. Mamet and M. J. Copeland, eds., *Proceedings of the Third North American Palaeontological Convention* 1: 1–8.

Anderson, M. M. and S. B. Misra. 1968. Fossils found in the Precambrian Conception Group in southeastern Newfoundland. *Nature* 220:680–681.

Astashkin, V. A., T. V. Pegel', L. N. Repina, A. Yu. Rozanov, Yu. Ya. Shabanov, A. Yu. Zhuravlev, S. S. Sukhov, and V. M. Sundukov. 1991. *The Cambrian System on the Siberian Platform*. International Union of Geological Sciences, Publication 27.

Awramik, S. M. 1971. Precambrian columnar stromatolite diversity: Reflection of metazoan appearance. *Science* 174:825–827.

Bambach, R. K. 1977. Species richness in marine benthic habitats through the Phanerozoic. *Paleobiology* 3:152–167.

Banks, N. L. 1973. Innerelv Member: Late Precambrian marine shelf deposits, east Finnmark. *Norges geologiske undersøkelse* 288: 7–25.

Barskov, I. S. and A. Yu. Zhuravlev. 1988. Myagkotelye organizmy kembriya Sibirskoy platformy [Soft-bodied organisms from the Cambrian of the Siberian Platform]. *Paleontologischeskiy zhurnal* 1988 (1):3–9.

Bassett, D. A. and E. K. Walton. 1960. The Hell's Mouth Grits: Cambrian greywackes in St. Tudwal's Peninsula, North Wales. *Quarterly Journal of the Geological Society of London* 116:85–110.

Bekker, Yu. R. and N. V. Kishka. 1989. Discovery of Ediacara biota in the southern Urals. In T. N. Bogdanov and L. I. Khozatsky, eds., *Proceedings of the All-Union Palaeontological Society*, pp. 109–120. Leningrad: Nauka.

Benus, A. P. 1988. Sedimentological context of a deep-water Ediacaran fauna (Mistaken Point Formation, Avalon Zone, eastern Newfoundland). *Bulletin New York State Museum* 463:8–9.

Berry, W. B. N. 1972. Early Ordovician bathyurid province lithofacies, biofacies, and correlations—their relationship to a proto-Atlantic Ocean. *Lethaia* 5:69–84.

Berry, W. B. N. 1974. Types of early Paleozoic faunal replacements in North America: Their relationship to environmental change. *Journal of Geology* 82:371–382.

Bottjer, D. J., M. L. Droser, and D. Jablonski. 1988. Palaeoenvironmental trends in the history of trace fossils. *Nature* 333:252–255.

Boynton, H. E. and T. D. Ford. 1995. Ediacaran fossils from the Precambrian (Charnian Supergroup) of Charnwood Forest, Leicestershire, England. *Mercian Geologist* 13: 165–182.

Brasier, M. D. 1989. On mass extinction and faunal turnover near the end of the Precambrian. In S. K. Donovan, ed., *Mass Extinctions, Processes, and Evidence*, pp. 73–88. New York: Columbia University Press.

Butterfield, N. J., A. H. Knoll, and K. Swett.

1988. Exceptional preservation of fossils in an upper Proterozoic shale. *Nature* 334: 424–427.

Byerly, G. R., D. R. Lowe, and M. M. Walsh. 1986. Stromatolites from the 3,300–3,500 Myr Swaziland Supergroup, Barberton Mountain Land, South Africa. *Nature* 319:489–491.

Chang, S. 1994. The planetary setting of prebiotic evolution. In S. Bengtson, ed., *Early Life on Earth*, pp. 10–23. Nobel Symposium no. 84. New York: Columbia University Press.

Conway Morris, S. 1985. Cambrian Lagerstätten: Their distribution and significance. *Philosophical Transactions of the Royal Society of London B* 311:49–65.

Conway Morris, S. 1986. The community structure of the Middle Cambrian Phyllopod Bed (Burgess Shale). *Palaeontology* 29: 423–467.

Conway Morris, S. 1989a. South-eastern Newfoundland and adjacent areas (Avalon Zone). In J. W. Cowie and M. D. Brasier, eds., *The Precambrian-Cambrian Boundary*, pp. 7–39. Oxford: Clarendon Press.

Conway Morris, S. 1989b. The persistence of Burgess Shale–type faunas: Implications for the evolution of deeper-water faunas. *Transactions of the Royal Society of Edinburgh (Earth Sciences)* 80:271–283.

Conway Morris, S. 1992. Burgess Shale–type faunas in the context of the "Cambrian explosion": A review. *Journal of the Geological Society of London* 149:631–636.

Conway Morris, S. 1993. Ediacaran-like fossils in Cambrian Burgess Shale–type faunas of North America. *Palaeontology* 36: 593–635.

Cooke, H. E. and M. E. Taylor. 1975. Early Paleozoic continental margin sedimentation, trilobite biofacies, and the thermocline, western United States. *Geology* 3:559–562.

Cooke, H. E. and M. E. Taylor. 1977. Comparison of continental slope and shelf environments in the Upper Cambrian and lowest Ordovician of Nevada. In H. E. Cook and P. Enos, eds., *Deep-Water Carbonate Environments*, pp. 51–81. *Society of Economic Paleontologists and Mineralogists Special Publication* 25.

Crimes, T. P. 1970. A facies analysis of the Cambrian of Wales. *Palaeogeography, Palaeoclimatology, Palaeoecology* 7:113–170.

Crimes, T. P. 1974. Colonization of the early ocean floor. *Nature* 248:328–330.

Crimes, T. P. 1992. Changes in trace fossil biota across the Proterozoic-Phanerozoic boundary. *Journal of the Geological Society* 149:637–646.

Crimes, T. P. 1994. The period of early evolutionary failure and the dawn of evolutionary success: The record of biotic changes across the Precambrian-Cambrian boundary. In S. K. Donovan, ed., *The Palaeobiology of Trace Fossils*, pp. 105–133. England: John Wiley and Sons.

Crimes, T. P. and M. M. Anderson. 1985. Trace fossils from late Precambrian–Early Cambrian strata of southeastern Newfoundland (Canada): Temporal and environmental implications. *Journal of Paleontology* 50:310–343.

Crimes, T. P. and J. D. Crossley. 1968. The stratigraphy, sedimentology, ichnology, and structure of the lower Palaeozoic rocks of part of northeastern Co. Wexford. *Proceedings of the Royal Irish Academy* 67B: 185–215.

Crimes, T. P. and J. D. Crossley. 1991. A diverse ichnofauna from Silurian flysch of the Aberystwyth Grits Formation. *Geological Journal* 26:27–64.

Crimes, T. P. and M. L. Droser. 1992. Trace fossils and bioturbation: The other fossil record. *Annual Review of Ecology and Systematics* 23:339–360.

Crimes, T. P. and M. A. Fedonkin. 1994. Evolution and dispersal of deepsea traces. *Palaios* 9:74–93.

Crimes, T. P. and M. A. Fedonkin. 1996. Biotic changes in platform communities across the Precambrian-Phanerozoic boundary. *Revista Italiana di Paleontologia e Stratigrafia* 102:317–332.

Crimes, T. P., J. F. Garcia Hidalgo, and D. G. Poire. 1992. Trace fossils from Arenig flysch sediments of Eire and their bearing on the early colonisation of the deep seas. *Ichnos* 2:61–77.

Crimes, T. P., A. Insole, and B. P. J. Williams. 1995. A rigid-bodied Ediacaran biota from Upper Cambrian strata in Co. Wexford, Eire. *Geological Journal* 30:89–109.

Dhonau, N. B. and C. H. Holland. 1974. The Cambrian of Ireland. In C. H. Holland, ed., *Cambrian of the British Isles, Norden, and Spitsbergen*, pp. 157–176. London: John Wiley and Sons.

Durham, P. N. and N. V. Sennikov. 1993. A new rhabdopleurid hemichordate from the Middle Cambrian of Siberia. *Palaeontology* 36:283–296.

Farmer, J., G. Vidal, H. Moczydłowska, H. Strauss, P. Ahlberg, and A. Siedlecka. 1992. Ediacaran fossils from the Innerelv Member (Late Proterozoic) of the Tanafjorden area, northeastern Finnmark. *Geological Magazine* 129:181–195.

Fedorov, A. B., L. I. Egorova, T. V. Pegel', and I. P. Popov. 1986. Analiz stroeniya i opyt stratigraficheskogo raschleneniya rifovykh kompleksov na primere Anabarskogo rifovogo srednekembriyskogo massiva [Analysis of structure and essay on stratigraphic subdivision of reef complexes from the example of the Middle Cambrian Anabar Reef Massif]. In F. G. Gurari and V. I. Krasnov, eds., *Regional'nye i mestnye stratigrafischeskie podrazdeleniya dlya krupnomasshtabnogo geologicheskogo kartirovaniya Sibiri* [Regional and local stratigraphic subdivisions for a large-scale geological mapping of Siberia], pp. 48–58. Novosibirsk: Siberian Scientific-Research Institute of Geology, Geophysics, and Mineral Resources.

Fedonkin, M. A. 1992. Vendian faunas and the early evolution of Metazoa. In J. H. Lipps and P. W. Signor, eds., *Origin and Early Evolution of the Metazoa*, pp. 87–129. New York: Plenum Press.

Fortey, R. A. and N. V. Wilmot. 1991. Trilobite cuticle thickness in relation to palaeoenvironment. *Paläontologische Zeitschrift* 65:141–51.

Garrett, P. 1970. Phanerozoic stromatolites: Non-competitive ecologic restriction by grazing and burrowing animals. *Science* 169:171–173.

Gehling, J. G. 1991. The case for Ediacaran roots to the Metazoan tree. *Geological Society of India, Memoir* 20:181–224.

Gibson, G. G. 1989. Trace fossils from late Precambrian Carolina Slate Belt, south-central North Carolina. *Journal of Paleontology* 63:1–10.

Gibson, G. G., S. A. Teeter, and M. A. Fedonkin. 1984. Ediacaran fossils from the Carolina Slate Belt, Stanly County, North Carolina. *Geology* 12:387–390.

Glaessner, M. F. 1984. *The Dawn of Animal Life: A Biohistorical Study.* Cambridge: Cambridge University Press.

Goldring, R. and S. Jensen. 1996. Trace fossils and biofabrics at the Precambrian-Cambrian boundary interval in western Mongolia. *Geological Magazine* 133:403–415.

Greenly, E. 1919. *Geology of Anglesey.* Memoir of the Geological Survey of Great Britain. London: His Majesty's Stationery Office.

Grotzinger, J. P. 1990. Geochemical model for Proterozoic stromatolite decline. *American Journal of Science* 290:80–103.

Grotzinger, J. P. 1994. Trends in Precambrian carbonate sediments and their implication for understanding evolution. In S. Bengtson, ed., *Early Life on Earth,* pp. 245–258. *Nobel Symposium no. 84.* New York: Columbia University Press.

Grotzinger, J. P., S. A. Bowring, B. Z. Saylor, and A. J. Kaufman. 1995. Biostratigraphic and geochronologic constraints on early animal evolution. *Science* 270:598–604.

Groves, D. I., J. S. R. Dunlop, and R. Buick. 1981. An early habitat of life. *Scientific American* 245:64–73.

Hessler, R. R. and H. L. Sanders. 1967. Faunal diversity in the deep-sea. *Deep-Sea Research* 14:65–78.

Hessler, R. R. and G. D. F. Wilson. 1983. The origin and biogeography of malacostracan crustaceans in the deep sea. In R. M. Sims, J. H. Price, and P. E. S. Whalley, eds., *Evolution, Time, and Space: The Emergence of the Biosphere,* pp. 227–254. The Systematics Association Special Volume 23. London: Academic Press.

Hofmann, H. J. 1985. The mid-Proterozoic Little Dal macrobiota, Mackenzie Mountains, north-west Canada. *Palaeontology* 28:331–354.

Hofmann, H. J. and J. D. Aitken. 1979. Precambrian biota from the Little Dal Group, Mackenzie Mountains, northwestern Canada. *Canadian Journal of Earth Sciences* 16:150–166.

Hofmann, H. J. and I. M. Patel. 1989. Trace fossils from the type "Etcheminian Series" (Lower Cambrian Ratcliffe Brook Formation), Saint John area, New Brunswick, Canada. *Geological Magazine* 126:138–157.

Hofmann, H. J., G. M. Narbonne, and J. D. Aitken. 1990. Ediacaran remains from intertillite beds in northwestern Canada. *Geology* 18:1199–1202.

Hofmann, H. J., M. P. Cecile, and L. S. Lane. 1994. New occurrences of *Oldhamia* and other trace fossils in the Cambrian of the Yukon and Ellesmere Island, arctic Canada. *Canadian Journal of Earth Sciences* 31:767–782.

Hou, X.-G. and W.-G. Sun. 1988. Discovery of Chengjiang fauna at Meishucun, Jinning, Yunnan. *Acta Palaeontologica Sinica* 27:1–12.

Jablonski, D. 1995. Extinctions in the fossil record. In J. H. Lawton and R. M. May, eds., *Extinction Rates,* pp. 25–44. Oxford: Oxford University Press.

Jablonski, D. and D. J. Bottjer. 1983. Soft-bottom epifaunal suspension-feeding assemblages in the Late Cretaceous: Implications for the evolution of benthic palaeo-communities. In M. J. S. Tevesz and P. L. McCall, eds., *Biotic Interactions in Recent and Fossil Benthic Communities,* pp. 747–812. New York: Plenum Press.

Jenkins, R. F. 1992. Functional and ecological aspects of Ediacaran assemblages. In J. H. Lipps and P. W. Signor, eds., *Origin and Early Evolution of the Metazoa,* pp. 131–176. New York: Plenum Press.

Jenkins, R. J. F. and P. Hasenohr. 1989. Trilobites and their trails in a black shale: Early Cambrian of the Fleurieu Peninsula, South Australia. *Royal Society of South Australia, Transactions* 113:195–203.

Kuznetsov, A. P. 1989. Glubokovodnaya fauna. Osnovy adaptatsiy k glubokovodnomu obrazu zhizni. Istoriya formirovaniya [Deep-sea fauna. Its adaptive principles. The history of formation]. *Trudy, Institut okeanologii, Akademiya nauk SSSR* 123:7–22.

Kuznetsov, A. P., V. P. Strizhov, and S. V. Galkin. 1994. Pozdenepaleozoyskie (karbonovye) vestimentifery (Vestimentifera, Obturata, Pogonophora) severo-vostochnoy Azii [Late Paleozoic (Carboniferous) vestimentiferans (Vestimentifera, Obturata, Pogonophora) of north-eastern Asia]. *Izve-*

stiya Akademii nauk SSSR, Seriya biologiche-skaya 1994 (6):898–906.

Lazarenko, N. P. and N. I. Nikiforov. 1972. Sredniy i verkhniy kembriy severa Sibirskoy platformy i prilegayushchikh skladchatysk oblastey [Middle and Upper Cambrian of the north of the Siberian Platform and adjacent fold regions]. In V. Ya. Ban'kov, ed., *Stratigrafiya, paleogeografiya i poleznye iskopaemye Sovetskoy Arktiki* [Stratigraphy, palaeogeography, and mineral resources of the Soviet Arctic], pp. 4–9. Leningrad: Scientific-Research Institute of Geology of Arctic.

McKerrow, W. S., ed. 1978. *The Ecology of Fossils.* Cambridge: MIT Press.

Menzies, R. J., R. Y. George, and Q. T. Rowe. 1973. *Abyssal Environment and Ecology of the World Oceans.* New York: John Wiley and Sons.

Misra, S. B. 1969. Late Precambrian (?) fossils from southeastern Newfoundland. *Geological Society of America Bulletin* 80:2133–2139.

Mount, J. F. and P. W. Signor. 1985. Early Cambrian innovation in shallow subtidal environments: Paleoenvironments of Early Cambrian shelly fossils. *Geology* 13:730–733.

Myrow, P. M. 1995. Neoproterozoic rocks of the Newfoundland Avalon zone. *Precambrian Research* 73:123–136.

Narbonne, G. M. 1994. New Ediacaran fossils from the Mackenzie Mountains, Northwestern Canada. *Journal of Paleontology* 68:411–416.

Narbonne, G. M. and J. D. Aitken. 1990. Ediacaran fossils from the Sekwi Brook area, Mackenzie Mountains, northwestern Canada. *Palaeontology* 33:945–980.

Narbonne, G. M. and H. J. Hofmann. 1987. Ediacaran biota of the Wernecke Mountains, Yukon, Canada. *Palaeontology* 30:647–676.

Nautiyal, A. C. 1973. Medusoid impressions in the lower Paleozoic rocks of eastern Newfoundland. *Canadian Journal of Earth Sciences* 10:1016–1020.

Orłowski, S. 1989. Trace fossils in the Lower Cambrian sequence in the Świętokrzyskie Mountains, central Poland. *Acta Palaeontologica Polonica* 34:211–231.

Orr, P. J. 1996. The ichnofauna of the Skiddaw Group (Early Ordovician) of the Lake District, England. *Geological Magazine* 133:193–216.

Pacześna, J. 1985. Ichnorodzaj *Paleodictyon* Meneghini z dolnego kambru Zbilutki (Góry Świętokrzyskie). *Kwartalnik Geologiczny* 29:589–596.

Pel'man, Yu. L., N. A. Aksarina, S. P. Koneva, L. Ye. Popov, L. P. Sobolev, and G. T. Ushatinskaya. 1992. *Drevneyshie brakhiopody territorii Severnoy Evrazii* [Oldest brachiopods from the territory of northern Eurasia]. Novosibirsk: United Institute of Geology, Geophysics, and Mineralogy, Siberian Branch, Russian Academy of Sciences.

Pickerill, R. K. 1982. Cambrian medusoids from the St. John Group, southern New Brunswick. *Geological Survey of Canada, Current Research* 82-1B:71–76.

Pickerill, R. K. 1995. Deep-water marine *Rusophycus* and *Cruziana* from the Ordovician Lotbinière Formation of Quebec. *Atlantic Geology* 31:103–108.

Pickerill, R. K., J. M. Hurst, and F. Surlyk. 1982. Notes on Lower Palaeozoic flysch trace fossils from Hall Land and Peary Land, North Greenland. *Grønlands Geologiske Undersøgelse, Bulletin* 108:25–29.

Rex, M. A. 1973. Deep-sea species diversity: Decreased gastropod diversity at abyssal depths. *Science* 181:1051–1053.

Rex, M. A. 1976. Biological accommodation in the deep-sea benthos: Comparative evidence on the importance of predation and productivity. *Deep Sea Research* 23:975–987.

Rex, M. A. 1981. Community structure in deep-sea benthos. *Annual Review of Ecology and Systematics* 12:331–353.

Rickards, R. B., P. W. Baillie, and J. B. Jago. 1990. An Upper Cambrian (Idamean) dendroid assemblage from near Smithton, northwestern Tasmania. *Alcheringa* 14:207–232.

Runnegar, B. N. 1992. Evolution of the earliest animals. In J. W. Schopf, ed., *Major Events in the History of Life*, pp. 65–93. Boston: Jones and Bartlett.

Sanders, H. L. 1968. Marine benthic diversity: A comparative study. *American Naturalist* 102:243–282.

Sanders, H. L. and R. R. Hessler. 1969. Ecology of the deep-sea benthos. *Science* 163:1419–1424.

Schopf, J. W. 1994. The oldest known records of life: Early Archean stromatolites, microfossils, and organic matter. In S. Bengtson, ed., *Early Life on Earth*, pp. 10–23. Nobel Symposium no. 84. New York: Columbia University Press.

Sdzuy, K. 1974. Mittelkambrische Graptolithen aus NW-Spanien. *Paläontologische Zeitschrift* 48:110–139.

Seilacher, A. 1984. Late Precambrian and Early Cambrian Metazoa: Preservational or real extinctions? In H. D. Holland and A. F. Trendall, eds., *Patterns of Change in Earth Evolution*, pp. 159–168. Berlin: Springer Verlag.

Seilacher, A. 1989. Organismic construction in the Proterozoic biosphere. *Lethaia* 22:229–239.

Seilacher, A. 1992. Vendobionta and Psammocorallia: Lost constructions of Precambrian evolution. *Journal of the Geological Society of London* 149:607–613.

Sepkoski, J. J., Jr. 1990. Evolutionary faunas. In D. E. G. Briggs and P. R. Crowther, eds., *Palaeobiology: A Synthesis*, pp. 37–41. Oxford: Blackwell Scientific Publishers.

Sepkoski, J. J., Jr. 1991. A model of onshore-offshore change in faunal diversity. *Paleobiology* 17:58–77.

Sepkoski, J. J., Jr. 1992. Proterozoic–Early Cambrian diversification of metazoans and metaphytes. In J. W. Schopf and C. Klein, eds., *The Proterozoic Biosphere: A Multidisciplinary Study*, pp. 553–561. Cambridge: Cambridge University Press.

Sepkoski, J. J., Jr. and A. I. Miller. 1985. Evolutionary faunas and the distribution of Palaeozoic marine communities in space and time. In J. J. Valentine, ed., *Phanerozoic Diversity Patterns*, pp. 153–190. Princeton, N.J.: Princeton University Press.

Sepkoski, J. J., Jr. and P. M. Sheehan. 1983. Diversification, faunal change, and community replacement during the Ordovician radiations. In M. J. S. Tevesz and P. L. McCall, eds., *Biotic Interactions in Recent and Fossil Benthic Communities*, pp. 673–717. New York: Plenum Press.

Shackleton, R. M. 1969. The Pre-Cambrian of North Wales. In A. Wood, ed., *The Pre-Cambrian and Lower Palaeozoic Rocks of Wales*, pp. 1–22. Cardiff: University of Wales Press.

Signor, P. W. 1990. The geologic history of diversity. *Annual Review of Ecology and Systematics* 21:509–539.

Sokolov, B. S. and A. B. Iwanowski, eds. 1985. *The Vendian System*. Berlin: Springer Verlag.

Sokolova, M. N. 1989. Usloviya pitaniya i razmernaya kharakteristika glubokovodnogo makrobentosa [Feeding conditions and size characteristics of the deep-sea macrobenthos]. *Trudy, Institut okeanologii, Akademiya nauk SSSR* 123:23–34.

Steele-Petrovic, H. M. 1979. The physiological differences between articulate brachiopods and filter-feeding bivalves as a factor in the evolution of marine level-bottom communities. *Palaeontology* 22:101–134.

Sweet, N. L. and G. M. Narbonne. 1993. Occurrence of the Cambrian trace fossil *Oldhamia* in southern Quebec. *Atlantic Geology* 29:69–73.

Tanoli, S. K. and R. K. Pickerill. 1989. Cambrian shelf deposits of the King Square Formation, Saint John Group, southern New Brunswick. *Atlantic Geology* 25:129–141.

Theokritoff, G. 1985. Early Cambrian biogeography in the North Atlantic region. *Lethaia* 18:283–293.

Valentine, J. W. and D. Jablonski. 1983. Speciation in the shallow sea: General patterns and biogeographic controls. In R. W. Simms, J. H. Price, and P. E. S. Whalley, eds., *Evolution, Time, and Space: The Emergence of the Biosphere,* pp. 203–228. London: Academic Press.

Varlamov, A. I. and K. L. Pak. 1993. Soobshchestva trilobitov i fatsii verkhnekembriyskikh otlozheniy severo-zapada Sibirskoy platformy [Upper Cambrian trilobite communities and facies on the northwest of the Siberian Platform]. *Stratigrafiya, Geologicheskaya korrelyatsiya* 1:104–10.

Veizer, J. and W. Compston. 1976. $^{87}Sr/^{86}Sr$ in Precambrian carbonates as an index of crustal evolution. *Geochimica et Cosmochimica Acta* 40:905–914.

Walter, M. R. 1983. Archean stromatolites: Evidence of the earth's earliest benthos. In J. W. Schopf, ed., *Earth's Earliest Biosphere: It's Origin and Evolution,* pp. 187–213. Princeton, N.J.: Princeton University Press.

Walter, M. R. 1994. Stromatolites: The main geological source of information on the evolution of the early benthos. In S. Bengtson, ed., *Early Life on Earth,* pp. 270–286. Nobel Symposium no. 84. New York: Columbia University Press.

Walter, M. R. and G. R. Heys. 1985. Links between the rise of the Metazoa and the decline of stromatolites. *Precambrian Research* 29:149–174.

Woese, C. R. 1987. Bacterial evolution. *Microbiological Reviews* 51:221–271.

Wood, D. 1969. The base and correlation of the Cambrian rocks of North Wales. In A. Wood, ed., *The Pre-Cambrian and Lower Palaeozoic Rocks of Wales,* pp. 47–66. Cardiff: University of Wales Press.

Zezina, O. H. 1989. O gipomorfnykh priznakakh u glubokovodnykh donnykh zhivotnykh [On hypomorphic features of deepsea bottom animals]. *Trudy, Institut okeanologii, Akademiya nauk SSSR* 123:35–48.

PART III

Ecologic Radiation of Major Groups of Organisms

CHAPTER FOURTEEN

Françoise Debrenne and Joachim Reitner

Sponges, Cnidarians, and Ctenophores

Sponges and coralomorphs were sessile epibenthic suspension feeders living in normal marine environments. Sponges with calcified skeletons, including archaeocyaths, mainly inhabited shallow to subtidal and intertidal domains, while other sponges occupied a variety of depths, including slopes. The high diversity of sponges in many Cambrian Lagerstätten suggests that complex tiering and niche partitioning were established early in the Cambrian. Hexactinellida were widespread in shallow-water conditions from the Tommotian; some of them may have been restricted to deepwater environments later in the Cambrian. Calcareans (pharetronids), together with solitary coralomorphs, thrived in reef environments, mostly in cryptic niches protected from very agitated waters. Rigid demosponges (anthaspidellids and possible axinellids) appeared by the end of the Early Cambrian and inhabited hardgrounds and reefs from the Middle Cambrian. The overall diversity of sponge and coralomorph types indicates that during the Cambrian these groups, like other metazoans, evolved a variety of architectural forms not observed in subsequent periods.

RAPID DIVERSIFICATION near the Proterozoic-Phanerozoic boundary implies the mutual interactions of ecosystems and biotas. One of the most striking features in the distribution of Early Paleozoic sessile benthos is the poor Middle–Late Cambrian record (Webby 1984).

The present contribution deals with the ecologic radiation of sponges and cnidarians.

SPONGES

Earliest Metazoans?

Sponges are a monophyletic metazoan group characterized by choanoflagellate cells (choanocytes). Based on studies made by Mehl and Reiswig (1991), Reitner (1992),

Müller et al. (1994), and Reitner and Mehl (1995), the first sponges originated in the Proterozoic from a choanoflagellate ancestor. The ancestral sponge was probably an aggregate of choanoflagellates, closely associated with various microbial communities. Important data are given by the analysis of metazoan β-galactose–binding lectins (S-type lectins) in sponges, hitherto analyzed only from vertebrates and one species of nematode (Müller et al. 1994). The development of this sponge lectin may have occurred before 800 Ma (Hirabayashi and Kasai 1993). Also remarkable are biomarker analyses made by McCaffrey et al. (1994), who detected C_{30} sterane, which is characteristic for demosponges, in 1.8-Ma-old black shales. This biochemical argument that sponges are Proterozoic metazoans is proven by new finds of undoubted sponge spicules and even entire phosphatized juvenile sponges with well-preserved sclerocytes (spicule-forming cells) from the late Sinian Doushantuo Formation of China (Ding et al. 1985; Li et al. 1998). Gehling and Rigby (1996) illustrate a nearly complete hexactinellid sponge from the Ediacarian Rawnsley Quartzite from South Australia. Additional specimens were described by them, but not all exhibit sponge affinities. The most convincing is *Paleophragmodictya,* which exhibits hexactinellid spicule patterns. Nevertheless, most previous records of Precambrian sponge spicules have proven upon examination either not to be sponges or not to be of Precambrian age (Rigby 1986a).

Sponges are represented in the fossil record as disarticulated spicules, soft-body casts, spicular networks, and spicular or calcareous skeletons. Since the review of Finks (1970), there has been a considerable number of new discoveries, but the ecologic history of sponges has yet to be revised.

Spicule Record

The oldest isolated spicules belong to the hexactinellids: stauractines, pentactines, and hexactines, in the Nemakit-Daldynian of Mongolia, Tommotian of Siberia, and Meishucunian of South China (Fedorov in Pel'man et al. 1990; Brasier et al. 1997). The Tindir Group (now dated by carbon isotopic correlation as Riphean—Kaufman et al. 1992) in Alaska contains possible hexactinellid spicules. Rare hexactine occurrences are found in pretrilobitic sequences, but hexactines become more numerous and widespread in the Atdabanian.

Genuine demosponge spicules are present in the upper quarter of the Atdabanian as tetractines with various additional elements that show much higher diversity than previously recognized (Bengtson et al. 1990).

By the Atdabanian, demosponges and hexactinellids seem to have become widespread in low-energy, offshore marine environments in Siberia and Australia (James and Gravestock 1990; Debrenne and Zhuravlev 1996), suggesting deeper-water occurrence.

In the Botoman, some microscleres are recognized, autapomorphic of the Tetractinellida (Reitner and Mehl 1995). Spongoliths of pentactines and hexactines are known from the Sinsk and Kuonamka formations (Botoman of Siberia—Fedorov and Pereladov 1987; Rozanov and Zhuravlev 1992). In addition, these formations contain a large number of inflated pillowlike stauractines (e.g., *Cjulankella*), which may compose dermal armoring layers of hexactinellids (Rozanov and Zhuravlev 1992; Reitner and Mehl 1995). Armoring probably reflects development of protective structures against predators.

In the Ordian (late Early Cambrian) of the Georgina Basin, Australia, Kruse (in Kruse and West 1980) found sigmata microscleres, autapomorphic of the ceractinomorph demosponges (Reitner and Mehl 1995).

Most tetractine spicules exhibiting diagenetic features have previously been recorded from Mesozoic siliceous sponges. In contrast, regular triaene spicules of the Calcarea are represented by a single crystal (Reitner and Mehl 1995). Among demosponges, the tetractines are restricted to the Tetractinellida. Additionally, typically modified dermal spicules (nail-type), monaxons (large tylostyles), and large aster microscleres (sterraster autapomorphic of the Geodiidae) have been found in the Early Cambrian, demonstrating the advanced state of tetractinellid evolution since that time. The rapid diversification of demosponges with clearly differentiated spicules occurred only in the Middle Cambrian.

The first calcarean spicules (Tommotian Pestrotsvet Formation, Siberian Platform —Kruse et al. 1995) have a triradiate symmetry. Their systematic position among the Calcarea is under discussion (Bengtson et al. 1990) (figures 14.1C,D). Previously known regular calcitic triaene spicules were Mesozoic. The Heteractinida, with multirayed spicules or characteristic octactines, are typical Paleozoic Calcarea. Regular triaene spicules of the Polyactinellida are common in early Paleozoic strata (Mostler 1985). The observed calcarean spicules have affinities with those of modern Calcaronea; spicules with calcinean affinities (regular triaenes) are rare in the Cambrian.

Sponge spicule assemblages are abundant in the Early Cambrian. In the lower Middle Cambrian of the Iberian Chains (Spain), spicules are so common with echinoderm ossicles that eocrinoid-sponge meadows are inferred for low-energy shallow subtidal environments (Alvaro and Vennin 1997). In general, spicule assemblages display high morphologic diversity, with many spicule types unknown in living sponges (Mostler 1985; Bengtson 1986; Fedorov and Pereladov 1987; Fedorov in Shabanov et al. 1987; Zhang and Pratt 1994; Dong and Knoll 1996; Mehl 1998). Their composition indicates the early appearance of hexactinellid, and possible calcarean, sponges in shallow-water archaeocyath-calcimicrobial mounds and the dominance of these sponges over archaeocyaths in deeper-water mounds. Relatively deep environments yield only demosponge and hexactinellid spicules, with the latter being prevalent (Fedorov and Pereladov 1987; James and Gravestock 1990; Zhang and Pratt 1994; Debrenne and Zhuravlev 1996; Dong and Knoll 1996).

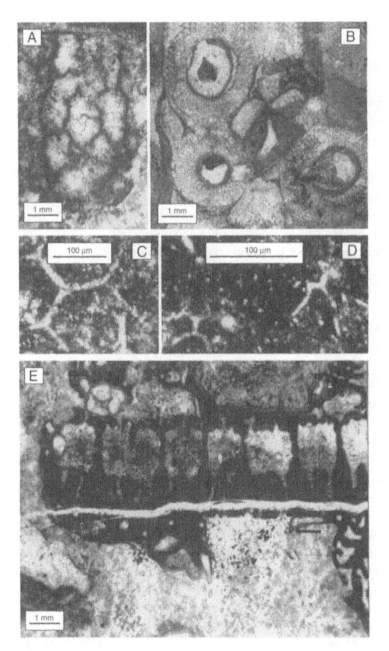

Figure 14.1 Thin sections. *A,* Cryptic thalamid sponge *Tanchocyathus amgaensis* (Vologdin 1963) PIN, Middle Cambrian, Mayan Tangha Formation (Amga River, Siberian Platform, Russia). *B,* Frame-building anthaspidellid demosponge *Rankenella* ex gr. *mors* (Gatehouse), IGS, Middle Cambrian, Kushanian Mila Formation (Elburz Mountains, Iran). *C and D,* Remains of modified tetractines (dodecaactinellids) described as Calcarea, Lower Cambrian, Atdabanian Wilkawillina Limestone (Arrowie Basin, Australia). *E,* Cryptic pharetronid *Gravestockia pharetronensis* Reitner anchored on the inner wall of an archaeocyath cup and partially overgrown by its secondary skeleton, Lower Cambrian, Atdabanian Wilkawillina Limestone (Arrowie Basin, Australia). *Source:* Photographs A and B courtesy of Andrey Zhuravlev.

Soft-Bottom Communities of Sponges

Most sponges are soft-bodied animals, which means that their preservation potential is poor. Entirely preserved sponges are the exception. Sponges, such as coralline sponges, with a rigid skeleton do exist and include archaeocyaths and lithistid demosponges, which are characterized by a rigid framework of choanosomal spicules.

Preserved soft sponges are now recorded from the southern China Nuititang Formation at Sansha (Steiner et al. 1993), first attributed to Tommotian, since co-occurrence of the associated bivalved arthropod *Perspicaris* favors a younger age. A nearly complete hexactinellid spicule cluster of protospongid character has been found at the base of the formation (basal chert) (Steiner et al. 1993). The middle part of the formation bears a diverse fauna of complete specimens of hexactinellids, together with one doubtful demosponge taxon (*Saetaspongia*). The gray pelitic rocks, completely free of carbonate, probably correspond to a typical soft substrate under low-energy marine conditions; the sponges were morphofunctionally adapted to this environment. The hexactinellids demonstrate two main types of spicule architecture: rosselleid type (*Solactinella*) (figure 14.2B) and hyalonemid-like spicule root tufts (*Hyalosinica*) (figure 14.2A). Thin spicule mats have also been identified, on which grow numerous young hexactinellids, a strategy similar to the one observed on the top of the Recent Vesterisbanken Seamount in the Greenland Sea (Henrich et al. 1992).

Atdabanian rocks of northern Greenland (Sirius Passet) have yielded two genera of demosponges (Rigby 1986b) that are also known with a similar preservation in the younger Burgess Shale fauna. This soft-bodied fauna was deposited in deep-water shales on the margin of the outer detrital belt, on shelves facing the open ocean (Conway Morris et al. 1987; Conway Morris 1989). The forms noted as Paleozoic Dictyospongiidae are hexactinellids with bundles of long and large diactines (Mehl 1996).

After arthropods, Botoman sponges represent the most diverse metazoan group in the Chengjiang fauna, with at least 11 genera and 20 species (Chen et al. 1989, 1990; Chen and Erdtmann 1991; Rigby and Hou 1995). Those described by Chen et al. (1989, 1990) are hexactinellids and not demosponges. The spicule arrangement of the so-called leptomitid sponges has nothing in common with that of demosponges. The simple diactine spicules are very long (several mm to 1 cm), with a rectangular arrangement more characteristic of lyssacine hexactinellids. Some hexactinellids bear diactine spicules, which are actually reduced hexactines, with the typical hexactine cross in the center of the axial canal (Mehl 1992). For example, the modern Euplectellidae and most of lyssakiin hexactinellids exhibit this structure.

The Chengjiang sponges, embedded in mudstones of a low-energy environment, represent a sessile, suspension-feeding epifauna. Evidence of niche partitioning among them is visualized from their tiering complexity: choiids mostly occupying a lower-level epifaunal tier (<2 cm) or even being infaunal, and leptomitids feeding at the

Figure 14.2 *A, Hyalosinica archaica* Mehl and Reitner with long spicule root tuft with small isolated hexactine on top, holotype SAN 109ab, Lower Cambrian, Qiongzhusian Niutitang Formation (Sansha, China). *B,* Hexactinellid sponge with strong lyssacyne character, *Solactinella plumata* Mehl and Reitner, holotype SAN 107ab, Lower Cambrian, Qiongzhusian Niutitang Formation (Sansha, China). *C,* Encrusting anthaspidellid *Rankenella mors* (Gatehouse), weathered out and etched specimens, AGSO CPC 21244, Lower Cambrian, Ordian Arthur Creek Formation (Georgina Basin, Australia). *D,* Heteractinid *Eiffelia globosa* Walcott, USNM 66521, Middle Cambrian Burgess Shale (British Columbia, Canada).

intermediate level (5–15 cm), with a higher tier represented by a new globular sponge exhibiting a four-layered skeleton.

Early Cambrian articulated sponges have been recorded in Laurentia from Vermont (*Leptomitus*) and Pennsylvania (*Hazelia*), indicating that these two lineages had diverged by the end of the Early Cambrian (Rigby 1987).

Sponges constitute the most important Burgess Shale group in terms of number of specimens (Walcott 1920; Rigby 1986a; Ushatinskaya, this volume: figure 16.6), with at least 15 genera represented. The majority of these are hexactinellids resembling *Protospongia:* they consist of a single layer of parallel stauractines with rare pentactines, organized as a vasiform sheet. There are demosponges among them: *Choia, Hazelia,* and a probable keratose sponge, *Vauxia.* The calcareous heteractinid genus *Eiffelia* (figure 14.2D) has a thin-walled subspherical skeleton, with three ranks of oriented sexiradiate spicules. Most of these sponges are endemic, except for *Eiffelia* and *Choia,* the latter having also been reported from other localities of Laurentia, Europe, and possibly from South America and Australia (Rigby 1983).

More-complex complete bodies of spicular sponges have been found only in Laurentia: *Hintzespongia,* occurring in slightly younger rocks than the Burgess Shale, and thin-walled *Ratcliffespongia.* These sponges have, beneath an outer (dermal) layer of stauract spicules, an inner (endosomal) layer of stauractines and hexactines in a non-parallel arrangement, surrounding numerous circular aporhyses, covered externally by the outer layer (Finks 1983). Sponges of these lineages appear to have had their origin in the moderate deep shelf, in relatively constant temperatures and similar-chemistry waters of the shelf and outer margin of the continents (Rigby 1986a). The early hexactinellid sponges seem to have lived in warm shallow-water and high-energy environments and in rather deep and quiet water, on muddy sea floors, and colonizing sandy limestone substrates by the end of the Cambrian.

These sponges were sessile epibenthic suspension feeders on picoplankton and/or dissolved organic matter. Detailed investigations of the Chengjiang and Burgess faunas suggest that various niches existed: nutrients differing in type and size were ingested by different species at different heights (tiering), showing that the fundamental trophic structure of marine metazoan life was established very early in metazoan evolution (Conway Morris 1986) and that the maximum height of the community above the sediment-water interface was greater than suggested in the tiering model of Ausich and Bottjer (1982).

Reefal and Hardground Sponges

In addition to the secretion of siliceous and calcareous spicules, nonspicular calcareous skeletons have been independently acquired at different times, both in Demospongea and Calcarea.

Archaeocyaths

Functional and constructional analyses of archaeocyaths support a poriferan affinity for the group (Debrenne and Vacelet 1984; Kruse 1990; Zhuravlev 1990; Debrenne and Zhuravlev 1992), possibly with demosponges (Debrenne and Zhuravlev 1994). As sessile benthic filter-feeding organisms, archaeocyaths appeared in the Tommotian, progressively colonizing Atdabanian carbonate platforms, reaching their acme of development in the Botoman, and then declining in the Toyonian. Only a few forms persisted into the Middle and Late Cambrian.

Archaeocyaths are divided into two groups, according to the reconstructed position of their soft tissues: the Ajacicyathida (Regulares) and the Archaeocyathida (Irregulares). In the Regulares (Debrenne et al. 1990b), soft tissue filled the entire body and nutrient flows circulated through a complex aquiferous system corresponding to the different types of skeletal porosity. In the Irregulares (Debrenne and Zhuravlev 1992), the living tissue was restricted to the upper part of the cup, and a secondary skeleton developed that separated dead from living parts; thus nutrient flows in the Irregulares were less dependent on skeletal porosity, which is not as diverse as it is in the Regulares. The respective position of the living tissue in both groups also influenced their ecologic responses (figure 14.3A).

Archaeocyaths are associated with calcimicrobes but commonly play a subordinate role in reef building (Wood et al. 1992; Kruse et al. 1995; Pratt et al., this volume). Regulares were mainly solitary, with a high degree of individualization and thus with limited possibilities of being efficient frame builders. They tended to settle on soft bottoms in environments with low energy and low sedimentation rate, commonly at reef peripheries. Irregulares had a higher degree of integration that was propitious for modularization and for tolerance of associations with other species; they produced abundant secondary skeletal links between adjacent cups (figure 14.4A). All these features enhanced frame-building ability. They settled on stable substrates, after stabilization of the soft bottom, and were supported by cement and calcimicrobes—the principal reef builders (Pratt et al., this volume: figures 12.1A and 12.2A). Archaeocyaths differentiated from the late Tommotian into distinct open-surface and crypt dwellers (Zhuravlev and Wood 1995). Solitary ajacicyathids and modular branching archaeocyathids dominated open-surface assemblages, while solitary archaeocyathids and solitary chambered forms (capsulocyathids and kazachstanicyathids) were preferentially housed in crypts. Some species of *Dictyofavus, Altaicyathus,* and *Polythalamia* were obligate cryptobionts (figure 14.4B; Pratt et al., this volume: figure 12.1B).

Overall, archaeocyaths were adapted to restricted conditions of temperature, salinity, and depth. They were limited to tropical seas, as confirmed by paleomagnetic continental reconstructions (McKerrow et al. 1992; Debrenne and Courjault-Radé 1994). Under conditions of increased salinity, archaeocyath assemblages became depleted, and they were represented by the simplest forms (Debrenne and Zhuravlev 1996).

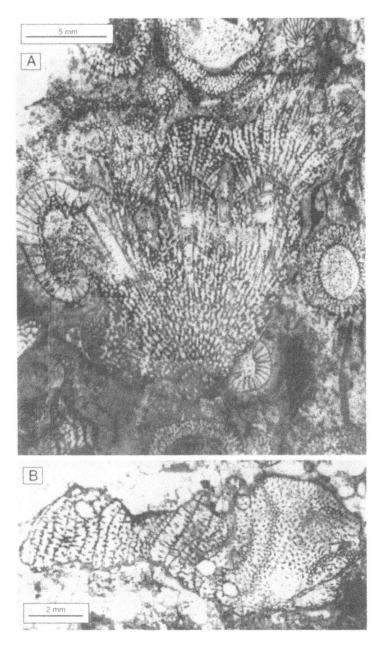

Figure 14.3 Archaeocyaths in thin section. *A,* Modular Archaeocyathida (*Archaeocyathus arborensis* Okulitch and *Arrythmocricus macdamensis* [Handfield]) and solitary Ajacicyathida (*Robustocyathellus pusillus* [Debrenne] and *Palmericyathus americanus* [Okulitch]), MNHN M83075, Lower Cambrian, Botoman Puerto Blanco Formation (Cerro Rajón, Mexico). *B,* Stromatoporoid *Korovinella sajanica* (Yaworsky), MNHN M81017, Lower Cambrian, Botoman Verkhnemonok Formation (Karakol River, Western Sayan, Russia).

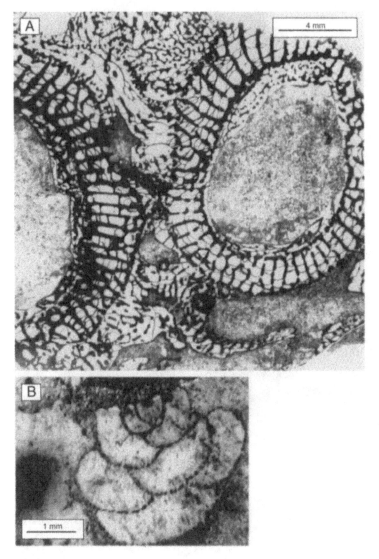

Figure 14.4 Archaeocyaths in thin section. A, Modular *Metaldetes profundus* (Billings), GSC 62113, Lower Cambrian, Botoman Forteau Formation (Labrador, Canada). B, Cryptic thalamid *Polythalamia americana* Debrenne and Wood, anchored to cyanobacterial crust-forming crypt, USNM 443584, Lower Cambrian, Botoman Scott Canyon Formation (Battle Mountain, Nevada, USA).

Archaeocyaths occupied the intertidal to subtidal zones. Basinward, the communities became impoverished and commonly were associated with hexactinellid sponge spicules, suggesting that with increasing depth, spicular sponges came to dominate sponges with a calcified skeleton (e.g., the Atdabanian of the Lena River—Debrenne and Zhuravlev 1996; Pratt et al., this volume: figure 12.2). Deeper-water bioherms (e.g., Sellick Hill Formation, Australia) contain oligotypic assemblages of archaeocyaths developing exocyathoid buttresses, interpreted as a response to higher water pressure (Debrenne and Zhuravlev 1996). Erosional features may also be observed in some places (e.g., Khara Ulakh, Siberian Platform, and Sardinia) that are indicative of peritidal conditions in which some archaeocyaths existed.

As filter feeders, archaeocyaths were better adapted to environments with sufficient current activity to transport nutrients. Complex outer walls promoted inhalant-exhalant flow through the cup, while annular inner walls accelerated the initial speed of the exhalant current (Debrenne and Zhuravlev 1996). Metallic models in fume tanks have shown that porous septa are better adapted to low-energy currents and aporous septa to high-energy environments (Savarese 1992); these conclusions are in accordance with the observations of Zhuravlev (1986) of an archaeocyath reef facies assemblage where genera have mostly aporose septa, whereas in back-reef facies their analogs have porous septa.

In conclusion, archaeocyaths were stenothermal, stenohaline, stenobathic marine sessile filter-feeding organisms, employing both active and passive current flow to move water through their systems. The nature of their food remains uncertain (Signor and Vermeij 1994); like their modern poriferan relatives, they probably fed primarily on bacteria and similarly sized particles. Whether some archaeocyaths possessed photosymbionts remains controversial (Camoin et al. 1989; Wood et al. 1992; Surge et al. 1997; Riding and Andrews 1998), but if photosymbionts were associated with archaeocyaths, they were rare, as in Recent marine sponges.

Thalamid Coralline Sponges ("Sphinctozoans")

A thalamid grade of organization is recognized in various classes of calcified sponge (Archaeocyatha [Capsulocyathida], Demospongea, and Calcarea) and in one species of Hexactinellida that lacks a calcareous skeleton. This type of skeleton is thus polyphyletic (Vacelet 1985; Reitner 1990), and the term *sphinctozoan* is only morphologic and without systematic significance.

Apart from archaeocyaths (see above), sphinctozoans of Early Cambrian age described from Australia either are not sponges or lack a sphinctozoan grade of organization. Simple sebargasiids have been found in marginal shelf deposits of New South Wales (Pickett and Jell 1983). Some of these are of doubtful affinity: single-chambered *Blastulospongia*, considered as a possible ancestor for the whole group, has been reinterpreted as a radiolarian (Bengtson 1986). Nonetheless, its large size and apparent attachment to the substrate do not fit closely to the radiolarian model of the type *Blastulospongia* species. As for the multichambered and cateniform *Nucha* and *Amblysiphonella?*, reexamination of the holotypes (Reitner and Pickett, unpubl. data) suggests that they might not be sponges.

Coeval "sphinctozoans" *Jawonya* and *Wagima* (Kruse 1987) have been found in platform deposits (Tindall Limestone) of northern Australia. Upon reexamination, Wood (in Kruse 1990; Rigby 1991) noted the presence in these of spicules. They are modi-fied octactines, confirming *Jawonya* as a heteractinid sponge (Wewokellidae). Kruse (1996) has recently demonstrated that *Jawonya* is in fact two-walled, with a compli-cated exopore architecture. The related genus *Wagima* is also considered to be two-walled. They lived in a low-energy, open-shelf environment on the muddy substrate, stabilized by calcimicrobes (Kruse 1996).

Questionable *Jawonya,* from older Atdabanian strata of South Australia (Kruse 1987), is a rimmed single-chambered form (not with "sphinctozoan" grade of organ-ization). It differs from contemporaneous one-walled archaeocyaths in its size and inferred microstructure; its affinity remains uncertain. This form is intimately associ-ated with reefal facies (in this case, calcimicrobial-archaeocyath mounds).

Tanchocyathus amgaensis (Vologdin 1963), from the Middle Cambrian of Siberia, is probably a thalamid, nonarchaeocyathan sponge that lived in cryptic communities (Zhuravlev 1996) (see figure 14.1A).

Stromatoporoid Coralline Sponges

Forms exhibiting a stromatoporoid grade of organization have been noted from the Botoman. The archaeocyath order Kazachstanicyathida (Debrenne and Zhuravlev 1992) has the thalamid type of cup development and a stromatoporoid growth pat-tern, even with astrorhizae (figure 14.3B). They are associated with calcimicrobial-archaeocyathan reefs.

Calcarea with a Rigid Skeleton ("Pharetronida")

Apart from isolated regular calcitic spicules, one articulated taxon is known from the Flinders Ranges, South Australia, in beds of Atdabanian equivalent age: *Graves-tockia pharetronensis* Reitner (Reitner 1992). This is a pharetronid sponge with a rigid skeleton of cemented choanosomal simple tetractine calcareous spicules and diac-tine free dermal spicules. It is anchored on an archaeocyath inner wall in a cryptic niche (figure 14.1E) and may in turn have been locally overgrown by the archaeocy-ath's secondary skeleton. *Gravestockia* is associated with calcimicrobial-archaeocyath bioconstructions.

Bottonaecyathus, from the Botoman of the Altay Sayan Foldbelt, Tuva, Morocco, and Mongolia, was originally described as an archaeocyath. It is now considered a prob-able sponge with a calcified skeleton. It lived together with archaeocyaths in reefal en-vironments (Kruse et al. 1996).

Demosponges with Desma-Type Spicules ("Lithistida")

The "Lithistida" are a highly polyphyletic group of demosponges, including taxa of both Tetractinellida and Ceractinomorpha (Reiner 1992). The oldest known (Ordian to early Templetonian) desma-bearing sponge, the anthaspidellid *Rankenella,* inhabited a low-energy, shallow subtidal marine environment, with abundant mud and high productivity (Ranken Limestone) (see figure 14.2C), and also even anaerobic low-energy shelf areas of limited circulation (Arthur Creek Formation) (Kruse 1996). A similar sponge has been identified from the late Early Cambrian to early Middle Cambrian Dedebulak Formation of Kyrgyzstan (Teslenko et al. 1983). Such sponges are restricted to a stable soft bottom and are presumed to be encrusting forms. From the late Middle Cambrian, anthaspidellid and axinellid demosponges became ubiquitous elements of fossil assemblages in Laurentia, Altay Sayan, and Iran (Wilson 1950; Okulitch and Bell 1955; Zhuravleva 1960). They encrusted hardgrounds (Brett et al. 1983; Zhuravlev et al. 1996) and even built their own reefs—Mila Formation, Iran (Hamdi et al. 1995; see also figure 14.1B) and Wilberns Formation, USA (Wood 1999; Pratt et al., this volume: figure 12.2C).

COELENTERATA

Soft-Bodied Cnidaria and Ctenophores

In contrast with the Precambrian Ediacara fauna, which is dominated by medusoids, representatives of the soft-bodied cnidaria and ctenophores are relatively poorly represented in the Cambrian. A great number of Cambrian forms have been assigned to Cnidaria with varying degrees of uncertainty. Impressions of putative jellyfish have been reported in Cambrian rocks since Walcott (1911), but most of them have been reinterpreted as trace fossils, sponges, echinoderms, arthropod appendages, or worms; others have been designated as incertae sedis or are unrecognizable forms (Harrington and Moore 1956; Conway Morris 1993a). The discovery of annulated disks alone is insufficient to place them in the chondrophores. The Tommotian records are still doubtful. Associated with *Lapworthella,* 50 m above the Cadomian peneplain, forms provisionally attributed to scyphozoans have been recorded (Doré 1985) (figure 14.5B).

Other discoidal fossils have been described in Europe but have not recently been reinvestigated, so their possible attribution to cnidarians remains uncertain. *Ichnusina cocozzai* (nom. correct. herein) (figure 14.5A)—from Sardinia, Italy, at the base of the "Arenarie di San Vito" (Middle-Upper Cambrian)—is one of these. It consists of a hemispheric body with undifferentiated center, dichotomized radial lobes and peripheral tentacles. If considered as a possible cnidarian, then this organism would have had a swimming or floating lifestyle.

Within the Middle Cambrian Burgess Shale–type fauna, some specimens resembling elements of the Ediacara fauna have a cnidarian affinity (Conway Morris 1993b). *Thaumaptilon* is a bilaterally symmetrical foliate animal with a holdfast and is related to pennatulaceans. It was benthic, and its mode of feeding rather conjectural, probably trapping the food particles by means of small tentacles of putative zooids. *Ge-*

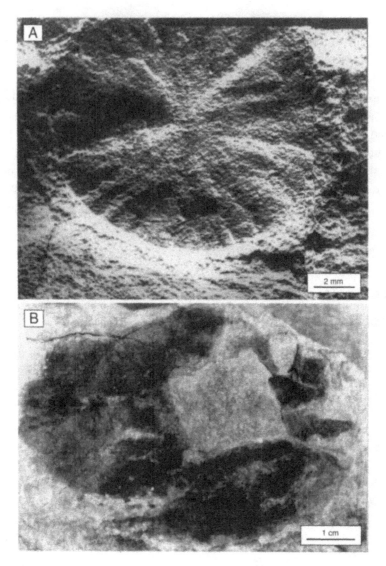

Figure 14.5 A, Disk of a possible chondro-phore cast of *Ichnusina cocozzai* (Debrenne), MNHN M84160, Middle-Upper Cambrian (Sardinia, Italy). B, Cubic medusoid with square central part (gastrogenital cavity?), with a tentacle springing from the lower right angle of the manubrium (?), surrounded by a dark organic circle (umbrella?), N 1368A Caen University, Lower Cambrian "Schistes et calcaires" Formation (Normandy, Val de May, Normandy, France). Source: Photograph courtesy of Francis Doré.

lenoptron is tentatively assigned to chondrophorines (Conway Morris 1993b), together with some undetermined disks with spaced annulations and tentacles. *Emmonsaspis* from the Early Cambrian Parker Slate of the Appalachians is tentatively interpreted as a benthic suspension feeder or microcarnivore (Conway Morris 1993b).

The trace fossil *Dolopichnus* is interpreted as a possible cnidarian burrow (Alpert and Moore 1975; Birkenmajer 1977). It contains trilobite debris, indicating a carnivorous diet. Such trace fossils might be produced by animals similar to the Early Cambrian *Xianguangia* or Middle Cambrian *Mackenzia*. *Xianguangia* from Chengjiang is interpreted as an anthozoan-like cnidarian on account of a basal disk, a polyp-like body with possible septal impressions, and a distal crown of tentacles bearing closely spaced pinnules (Chen and Erdtmann 1991). *Mackenzia costalis* Walcott, having a baglike body with possible internal partitions, is compared with some putative actinians (Conway Morris 1993b).

Ctenophores, representatives of another branch of the coelenteratan grade, were active swimmers that combed the pelagic realm in search of tiny metazoans and larvae (Conway Morris and Collins 1996; Chen and Zhou 1997).

Coralomorphs

The mass radiation of Metazoans included mineralized skeletons of solitary calcium carbonate cups and, later, slender irregular cerioid polygonal tubes, near the beginning of the Cambrian. These were originally grouped as coralomorphs because of their probable cnidarian affinities (Jell 1984). New descriptions of Early Cambrian coralomorphs, including studies of the biocrystals characteristic of their microstructure, their systematic position, and their stratigraphic distribution, have recently been made (Zhuravlev et al. 1993; Sorauf and Savarese 1995).

The oldest coralomorph, *Cysticyathus* (figure 14.6B), occurs in middle Tommotian calcimicrobial-archaeocyath bioherms of Siberia. It was previously included in archaeocyaths, despite its aporous skeleton. Tannuolaiids (=khasaktiids) (figure 14.6A) appeared in the Atdabanian of Siberia, diversifying as they migrated throughout the Ural-Mongolian Belt, and are always associated with reefs.

Hydroconozoa began with the Atdabanian but are not known later than the Botoman, when modular ramose forms developed. The skeletal microstructure of *Hydroconus* is most likely similar to that of genuine corals (Lafuste et al. 1990).

The Botoman was the acme for all Cambrian coralomorphs. In addition to tannuolaiids and hydroconozoans, which are characteristic of Siberia, one of the most convincing cnidarians, *Tabulaconus* (low modular) (Debrenne et al. 1987) (figure 14.6C), also appeared in Laurentia, along with the solitary *Aploconus* (Debrenne et al. 1990a) and the high modular *Rosellatana* (Kobluk 1984). In Australia, *Flindersipora* occurs. It was thought to comprise the oldest tabulate corals (Lafuste et al. 1991) (figure 14.6D) but is considered by Scrutton (1992) to be an unassigned early skeletonized anthozoan lacking linear descent to any major coral group. The newly discovered *Moorowipora* and *Arrowipora,* with their cerioid coral forms and typical coralline wall structure, short septal spines, and tabulae, suggest an assignment with Tabulata (Fuller and Jenkins 1994, 1995; Sorauf and Savarese 1995). The latter authors also propose inclusion of *Tabulaconus* in the Tabulata, thereby greatly extending the stratigraphic range of the group. Scrutton (1997), however, prefers to classify Cambrian

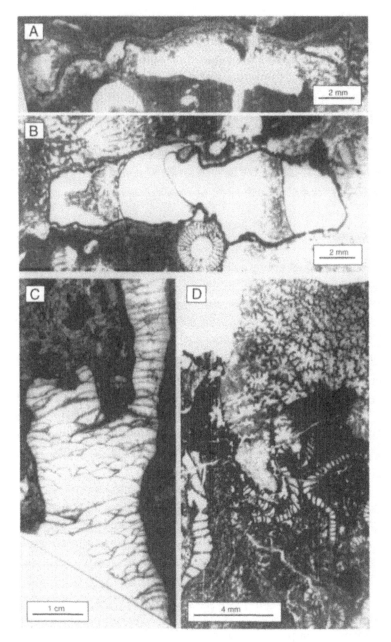

Figure 14.6 Coralomorphs in thin section. A, Encrusting *Khasaktia vesicularis* Sayutina, PIN, Lower Cambrian, Atdabanian Pestrotsvet Formation (middle Lena River, Siberian Platform, Russia). B, Branching *Cysticyathus tunicatus* Zhuravleva, MNHN M81016, Lower Cambrian, Tommotian Pestrotsvet Formation (middle Lena River, Siberian Platform, Russia). C, Branching *Tabulaconus kordae* Handfield, UA 2526, Lower Cambrian, Botoman Adams Argillite (Tatonduk area, Alaska, USA). D, Association of archaeocyaths (*Ajacicyathus aequitriens* [Bedford and Bedford]) and tabulate *Flindersipora bowmanni* Lafuste, MNHN M42048, Lower Cambrian, Botoman Moorowie Limestone (Arrowie Basin, Australia).

zoantharian corals as a separate order Tabulaconida without an assignment to other Paleozoic coral clades.

All Atdabanian and Botoman coralomorphs are associated with calcimicrobial-archaeocyath reef environments, with *Flindersipora* and *Yaworipora* even participating in bioconstruction (Zhuravlev 1999). *Khasaktia* and *Rosellatana* can be cryptobionts in calcimicrobial-archaeocyathan reef cavities.

The modular Laurentian *Labyrinthus* is known from the late Botoman Forteau Formation of Labrador. Colonies are often attached to archaeocyath skeletons, indicating a preference for hard substrates. They are found in the "upper biostrome complex," which underlies and interfingers with ooid beds containing oncoids and diverse skeletal fragments. This implies shallow, agitated water conditions in the vicinity of a bioconstruction (Kobluk 1979).

Lipopora and *Cothonion*, from New South Wales, Australia, are the latest Early Cambrian (Ordian) coralomorphs (Jell and Jell 1976). Solitary or modular, they occur in carbonate beds, associated with *Girvanella* oncoids and a rich fauna of trilobites, brachiopods, mollusks, and sponges. The high faunal diversity, the predominance of cyanobacteria, and the carbonate petrology suggest a warm shallow-water carbonate bank environment.

Other proposed Early Cambrian cnidarians have doubtful records (inorganic concretions, algae, bryozoans, or synonyms of already described tannuolaiids or hydroconozoans) and consequently are not considered here.

A Middle Cambrian (Floran-Undillan) coralomorph *Tretocylichne* is found in reworked clasts within inner submarine fan deposits of northeastern New South Wales (Engelbretsen 1993). The single example of a possible Late Cambrian coral is found in Montana (Fritz and Howell 1955).

Coralomorphs were suspension feeders living in warm waters and generally associated with calcimicrobial-archaeocyath bioherms as coconstructors or cryptobionts. Some lived in agitated waters near biostromes or carbonate banks.

Other Possible Skeletal Cnidarians

Among Cambrian small shelly fossils, a number of tiny, often septate, conoidal tubes have been suggested to be of cnidarian affinity, namely, paiutiids, quadriradial carinachitiids and hexangulaconulariids, triradial anabaritids, and byroniids (for reviews, see Conway Morris and Chen 1989, 1992; Bengtson et al. 1990; Rozanov and Zhuravlev 1992). Except for byroniids, these animals are restricted to the Early Cambrian. Most of them are suggested to be sessile forms. Tentacle-bearing *Cambrorhytium* might be a cnidarian possessing an organic-walled tube (Conway Morris and Robison 1988). It is worth noting that phosphatized spheroids, in Nemakit-Daldynian strata containing anabaritids, resemble nonplanktotrophic cnidarian actinula larvae (Kouchinsky et al. 1999).

A

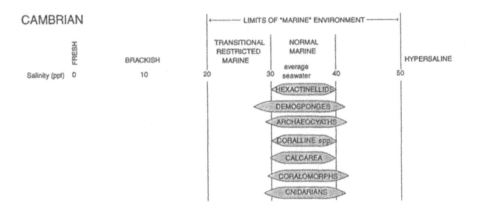

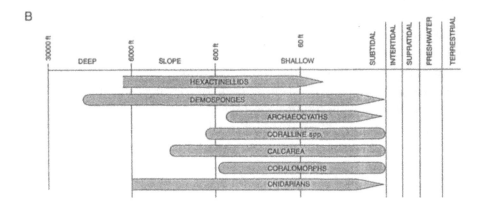

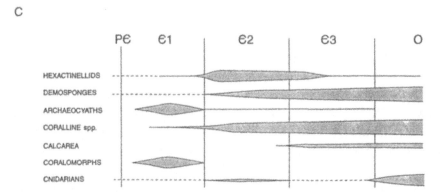

Figure 14.7 Distribution of sponges and cnidarians in relation to salinity (A), in relation to depth (B), and in relation to time (C).

CONCLUSIONS

Siliceous sponges, either as spicules or complete bodies, are known since the Ediacarian. From the Atdabanian and later, they were widespread in low-energy offshore marine environments (figure 14.7), suggesting a deep-water origin on open ocean-facing shelves. Ceractinomorphs are found only from the Middle Cambrian; they appear to have occupied shallow waters.

Calcified skeletons occur in different groups: archaeocyaths, pharetronids, and wewokellids. Archaeocyaths (first appearance in the Tommotian) occupied intertropical, intertidal to subtidal environments of low to normal salinity (figure 14.7), in well-agitated waters associated with reefs. Archaeocyaths with a stromatoporoid grade of organization were present in reefs, whereas the chambered forms ("sphinctozoans") were crypt dwellers.

Calcareous spicules are rare in the Early Cambrian. The known pharetronids grew on Atdabanian archaeocyath-calcimicrobe reefs, whereas late Early Cambrian heteractinids (Wewokellidae) were level-bottom dwellers.

The Middle Cambrian Burgess Shale fauna contains possible chondrophores and pennatulaceans. If the interpretation of forms unrecognizable and/or difficult to interpret as chondrophores is correct, they would have had a pelagic mode of life, because frondlike organisms were sessile organisms. In general, fossils of free-swimming cnidarians are rare in the Cambrian.

All Atdabanian and Botoman coralomorphs (Siberia, Australia, Laurentia) were associated with calcimicrobial-archaeocyath Tommotian to Botoman reefs, as open-surface and crypt dwellers. Late Early Cambrian coralomorphs from Australia were probably dwellers of warm agitated water with carbonate banks.

REFERENCES

Alpert, S. P. and J. N. Moore. 1975. Lower Cambrian trace fossil evidence for predation on trilobites. *Lethaia* 8:223–230.

Alvaro, J. J. and E. Vennin. 1997. Episodic development of Cambrian eocrinoid-sponge meadows in the Iberian Chains (NE Spain). *Facies* 37:49–64.

Ausich, W. I. and D. J. Bottjer. 1982. Tiering in suspension-feeding communities on soft substrata throughout the Phanerozoic. *Science* 216:173–174.

Bengtson, S. 1986. Siliceous microfossils from the Upper Cambrian of Queensland. *Alcheringa* 10:195–216.

Bengtson, S., S. Conway Morris, B. J. Cooper, P. A. Jell, and B. N. Runnegar. 1990. Early Cambrian fossils from South Australia. *Association of Australasian Palaeontologists, Memoir* 9:1–364.

Birkenmajer, K. 1977. Trace fossil evidence for predation on trilobites from Lower Cambrian of South Spitsbergen. *Norsk Polarinstitut Årsbok* 1976:187–195.

Brasier, M. D., O. Green, and G. Shields. 1997. Ediacarian sponge spicule clusters from southwestern Mongolia and the origins of the Cambrian fauna. *Geology* 25:303–306.

Brett, C. E., W. D. Liddell, and K. L. Derstler.

1983. Late Cambrian hard substrate communities from Montana/Wyoming: The oldest known hardground encrusters. *Lethaia* 16:281–289.

Camoin, G., F. Debrenne, and A. Gandin. 1989. Premières images des communautés microbiennes dans les écosystèmes cambriens. *Comptes rendus sommaires de l'Académie des Sciences, Paris,* 2d ser., 308: 1451–1458.

Chen, J. and G. Zhou. 1997. Biology of Chengjiang fauna. *Bulletin of the National Museum of Natural Science, Taichung, Taiwan* 10:11–105.

Chen, J.-Y. and B. D. Erdtmann. 1991. Lower Cambrian fossil Lagerstätte from Yunnan, China: Insights for reconstructing early metazoan life. In A. M. Simonetta and S. Conway Morris, eds., *The Early Evolution of Metazoa and the Significance of Problematic Taxa,* pp. 57–76. Cambridge: Cambridge University Press.

Chen, J.-Y., X.-G. Hou, and H.-Z. Lu. 1989. Lower Cambrian leptomitids (Demospongea), Chengjiang, Yunnan. *Acta Palaeontologica Sinica* 28:17–30.

Chen, J.-Y., X.-G. Hou, and G.-X. Li. 1990. New Lower Cambrian demosponges— *Quadrolaminella* gen. nov. from Chengjiang, Yunnan. *Acta Palaeontologica Sinica* 29:402–413.

Conway Morris, S. 1986. The community structure of the Middle Cambrian Phyllopod Bed (Burgess Shale). *Palaeontology* 29: 423–467.

Conway Morris, S. 1989. The persistence of Burgess Shale–type faunas: Implications for the evolution of deeper-water faunas. *Transactions of the Royal Society of Edinburgh (Earth Sciences)* 80:271–283.

Conway Morris, S. 1993a. The fossil record and the evolution of the Metazoa. *Nature* 361:219–225.

Conway Morris, S. 1993b. Ediacaran-like fossils in Cambrian Burgess Shale type–fauna of North America. *Palaeontology* 36:593–635.

Conway Morris, S. and M. Chen. 1989. Lower Cambrian anabaritids from South China. *Geological Magazine* 126:615–632.

Conway Morris, S. and M. Chen. 1992. Carinachitiids, hexangulaconulariids, and *Punctatus:* Problematic metazoans from the Early Cambrian of South China. *Journal of Paleontology* 66:384–406.

Conway Morris, S. and D. H. Collins. 1996. Middle Cambrian ctenophores from the Stephen Formation, British Columbia, Canada. *Philosophical Transactions of the Royal Society of London B* 351:279–308.

Conway Morris, S. and R. A. Robison. 1988. More soft-bodied animals and algae from the Middle Cambrian of Utah and British Columbia. *University of Kansas Paleontological Contributions* 122:1–48.

Conway Morris, S., J. S. Peel, A. K. Higgins, N. J. J. Soper, and N. C. Davis. 1987. A Burgess Shale–like fauna from the Lower Cambrian of North Greenland. *Nature* 326:181–183.

Debrenne, F. and P. Courjault-Radé. 1994. Répartition paléogéographique des archéocyathes et délimitation des zones intertropicales au cambrien inférieur. *Bulletin de la Société géologique de France* 165:459–467.

Debrenne, F. and J. Vacelet. 1984. Archaeocyatha: Is the sponge model consistent with their structural organization? *Palaeontographica Americana* 54:358–369.

Debrenne, F. and A. Yu. Zhuravlev. 1992. *Irregular Archaeocyaths.* Paris: Cahiers de Paléontologie, Éditions du Centre National de la Recherche Scientifique.

Debrenne, F. and A. Yu. Zhuravlev. 1994. Archaeocyathan affinities: How deep can we go into the systematic affiliation of an extinct group? In R. W. M. Van Soest, T. M. G.

Van Kempen, and J. C. Braekman, eds., *Sponges in Time and Space*, pp. 3–12. Rotterdam: Balkema.

Debrenne, F. and A. Yu. Zhuravlev. 1996. Archaeocyatha, palaeoecology: A Cambrian sessile fauna. In A. Cherchi, ed., *Autoecology of Selected Fossil Organisms: Achievement and Problems. Bollettino della Società Paleontologica Italiana, Special Volume* 3: 77–85.

Debrenne, F., M. Debrenne, R. A. Gangloff, and J. G. Lafuste. 1987. *Tabulaconus* Handfield: Microstructure and its implication in the taxonomy of primitive corals. *Journal of Paleontology* 61:1–9.

Debrenne F., A. Gandin, and R. A. Gangloff. 1990a. Analyse sédimentologique et paléontologique de calcaires organogènes du Cambrien inférieur de Battle Mountain (Nevada, USA). *Annales de Paléontologie* 76:73–119.

Debrenne, F., A. Yu. Rozanov, and A. Yu. Zhuravlev. 1990b. *Regular Archaeocyaths.* Paris: Cahiers de Paléontologie, Éditions du Centre National de la Recherche Scientifique.

Ding, Q.-X., Y.-S. Xing, and Y.-Y. Chen. 1985. Metazoa and trace fossils. In *Biostratigraphy of the Yangtze Gorge Area. 1: Sinian*, pp. 115–119. Beijing: Geological Publishing House.

Dong, X.-P. and A. H. Knoll. 1996. Middle and Late Cambrian sponge spicules from Hunnan, China. *Journal of Paleontology* 70: 173–184.

Doré, F. 1985. Premières méduses et premières faunes à squelette dans le Massif Armoricain: Problème de la limite Précambrien-Cambrien. *Terra Cognita* 5:2–3, 235.

Engelbretsen, M. J. 1993. A Middle Cambrian possible cnidarian from the Murrawong Creek Formation, NE New South Wales. *Association of Australasian Palaeontologists, Memoir* 15:51–56.

Fedorov, A. B. and V. S. Pereladov. 1987. Kremnevye spikuly gubok iz Kuonamskoy svity severo-zapada Sibirskoy platformy [Siliceous sponge spicules from the Kuonamka Formation of the northeastern Siberian Platform]. In S. P. Bulynnikova and I. G. Klimova, eds., *Novye vidy drevneishikh bespozvonochnykh i rasteniy iz fanerozoya Sibiri* [New species of ancient invertebrates and plants of the Phanerozoic of Siberia], pp. 36–46. Novosibirsk: Siberian Scientific-Research Institute of Geology, Geophysics, and Mineral Resources.

Finks, R. M. 1970. The evolution and ecologic history of sponges during Palaeozoic times. In W. G. Frey, ed., *The Biology of the Porifera*, pp. 3–22. New York: Academic Press.

Finks, R. M. 1983. Fossil Hexactinellida. In T. W. Broadhead, ed., *Sponges and Spongiomorphs: Notes for a Short Course*, pp. 101–115. University of Tennessee, Department of Geological Sciences, Studies in Geology 7.

Fritz, M. A. and B. F. Howell. 1955. An Upper Cambrian coral from Montana. *Journal of Paleontology* 29:181–183.

Fuller, M. K. and R. J. F. Jenkins. 1994. *Moorowipora chamberensis*, a coral from the Early Cambrian Moorowie Formation, Flinders Ranges, South Australia. *Royal Society of South Australia, Transactions* 118: 227–235.

Fuller, M. K. and R. J. F. Jenkins. 1995. *Arrowipora fromensis*, a new genus and species of tabulate-like coral from the Early Cambrian Moorowie Formation, Flinders Ranges, South Australia. *Royal Society of South Australia, Transactions* 119:75–82.

Gehling J. G. and J. K. Rigby. 1996. Long expected sponge from the Neoproterozoic Ediacara Fauna of South Australia. *Journal of Paleontology* 70:185–195.

Hamdi, B., A. Yu. Rozanov, and A. Yu. Zhuravlev. 1995. Latest Middle Cambrian metazoan reef from northern Iran. *Geological Magazine* 132 : 367–373.

Harrington, H. J. and R. C. Moore. 1956. Medusae incertae sedis and unrecognizable forms. In R. C. Moore, ed., *Treatise on Invertebrate Paleontology, Part F: Coelenterata,* pp. F153–F161. Boulder, Colo.: Geological Society of America.

Henrich, R., M. Hartmann, J. Reitner, P. Schäfer, A. Freiwald, P. Dietrich, and J. Thiede. 1992. Facies belts and communities of the Arctic Vesterisbanken Seamount (central Greenland Sea). *Facies* 27 : 1–352.

Hirabayashi, J. and K. Kasai. 1993. The family of metazoan metal-independent β-galactose-binding lectins: Structure, function, and molecular evolution. *Glycobiology* 3 : 297–304.

James, N. P. and D. I. Gravestock. 1990. Lower Cambrian shelf and shelf-margin buildups, Flinders Ranges, South Australia. *Sedimentology* 37 : 455–480.

Jell, J. S. 1984. Cambrian cnidarians with mineralized skeletons. *Palaeontographica Americana* 54 : 105–109.

Jell, P. A. and J. S. Jell. 1976. Early Middle Cambrian corals from western New South Wales. *Alcheringa* 1 : 181–195.

Kaufman, A. J., A. H. Knoll, and S. M. Awramik. 1992. Biostratigraphic and chemostratigraphic correlation of Neoproterozoic sedimentary successions: Upper Tindir Group, northwestern Canada, as a test case. *Geology* 20 : 181–185.

Kobluk, D. R. 1979. A new and unusual skeletal organism from the Lower Cambrian of Labrador. *Canadian Journal of Earth Sciences* 16 : 2040–2045.

Kobluk, D. R. 1984. A new compound skeletal organism from the Rosella Formation (Lower Cambrian), Atan Group, Cassiar Mountains, British Columbia. *Journal of Paleontology* 58 : 703–708.

Kouchinsky, A., S. Bengtson, and L.-A. Gershwin. 1999. Cnidarian-like embryos associated with the first shelly fossils in Siberia. *Geology* 27 : 609–612.

Kruse, P. D. 1987. Further Australian Cambrian sphinctozoans. *Geological Magazine* 124 : 543–553.

Kruse, P. D. 1990. Are archaeocyaths sponges, or are sponges archaeocyaths? *Geological Society of Australia, Special Publication* 16 : 310–323.

Kruse, P. D. 1996. Update on the northern Australian Cambrian sponges *Rankenella, Jawonya,* and *Wagima. Alcheringa* 20 : 161–178.

Kruse, P. D. and P. W. West. 1980. Archaeocyatha of the Amadeus and Georgina basins. *BMR Journal of Australian Geology and Geophysics* 5 : 165–181.

Kruse, P. D., A. Yu. Zhuravlev, and J. P. James. 1995. Primordial metazoan-calcimicrobial reefs: Tommotian (Early Cambrian) of the Siberian Platform. *Palaios* 10 : 291–321.

Kruse, P. D., A. Gandin, F. Debrenne, and R. Wood. 1996. Early Cambrian bioconstructions in the Zavkhan Basin of western Mongolia. *Geological Magazine* 133 : 429–444.

Lafuste, J., F. Debrenne, and A. Yu. Zhuravlev. 1990. Les fuscinules, type nouveau de biocristaux dans le squelette d'*Hydroconus* Korde 1963, coralomorphe du Cambrien inférieur. *Comptes rendus sommaires de l'Académie des Sciences, Paris,* 2d ser., 310 : 1553–1559.

Lafuste, J., F. Debrenne, A. Gandin, and D. Gravestock. 1991. The oldest tabulate coral and the associated Archaeocyatha, Lower Cambrian, Flinders Ranges, South Australia. *Géobios* 24 : 697–718.

Li, C.-W., J.-Y. Chen, and T.-E. Hua. 1998. Precambrian sponges with cellular structures. *Science* 279 : 879–882.

McCaffrey, M. A., J. M. Moldowan, P. A. Lipton, R. E. Summons, K. E. Peters, A. Jegenathan, and D. S. Watt. 1994. Paleoenvironmental implications of novel C_{30} steranes in Precambrian to Cenozoic age petroleum and bitumen. *Geochimica et Cosmochimica Acta* 58:529–532.

McKerrow, W. S., C. R. Scotese, and M. D. Brasier. 1992. Early Cambrian continental reconstructions. *Journal of the Geological Society, London* 149:599–606.

Mehl, D. 1992. Die Entwicklung der Hexactinellidae seit dem Mesozoikum: Paläobiologie, Phylogenie, und Evolutionsökologie. *Berliner geowissenschaftliche Abhandlungen, Reihe E* 2:1–164.

Mehl, D. 1996. Phylogenie und Evolutionsökologie der Hexactinellida (Porifera) im Paläozoikum. *Geologische Paläontologische Mitteilungen der Universitat Innsbruck, Sonderband* 4:1–55.

Mehl, D. 1998. Porifera and Chancelloriidae from the Middle Cambrian of the Georgina Basin, Australia. *Palaeontology* 41:1153–1182.

Mehl, D. and H. L. Reiswig. 1991. The presence of flagellar vanes in choanomeres of Porifera and their possible phylogenetic implications. *Zeitschrift für Zoologische Systematik und Evolutionforschung* 28:312–319.

Mostler, H. 1985. Neue heteractinide Spongien (Calcispongea) aus dem Unter- und Mittelcambrium Sudwestsardiniens. *Berichte des Naturwissenschaftlich-medizinichen Vereins Innsbruck* 72:7–32.

Müller, W. E. G., H. C. Schröder, and V. Gamulin. 1994. Phylogenetic relationship of ubiquitin repeats in the polyubiquitin gene from the marine sponge *Geodia cyonium*. *Journal of Molecular Evolution* 39:369–377.

Okulitch, V. J. and W. G. Bell. 1955. *Gallatinospongia*, a new siliceous sponge from the Upper Cambrian of Wyoming. *Journal of Paleontology* 29:460–461.

Pel'man, Yu. L., V. V. Ermak, A. B. Fedorov, V. A. Luchinina, I. T. Zhuravleva, L. N. Repina, V. I. Bondarev, and Z. V. Borodaevskaya. 1990. Novye dannye po stratigrafii i paleontologii nizhnego kembriya r. Dzhandy (pravyy pritok r. Aldan) [New data on the Lower Cambrian stratigraphy and paleontology on the Dzhanda River (Aldan River right tributary)]. *Trudy, Institut geologii i geofiziki, Sibirskoe otdelenie, Akademiya nauk SSSR* 765:3–32.

Pickett, J. and P. A. Jell. 1983. Middle Cambrian Sphinctozoa (Porifera) from New South Wales. *Association of Australasian Palaeontologists, Memoir* 1:85–92.

Reitner, J. 1990. Polyphyletic origin of the sphinctozoans. In K. Rützler, ed., *New Perspective in Sponge Biology: Third International Sponge Conference 1985*, pp. 33–42. Washington, D.C.: Smithsonian Institution Press.

Reitner, J. 1992. "Coralline Spongien" der Versuch einer phylogenetisch-taxonomischen Analyse. *Berliner geowissenschaftliche Abhandlungen, Reihe E* 1:1–352.

Reitner, J. and D. Mehl. 1995. Early Paleozoic diversification of sponges: New data and evidences. *Geologische Paläontologische Mitteilungen der Universitat Innsbruck, Sonderband* 20:335–347.

Riding, R. and J. E. Andrews. 1998. Carbon isotopic evidence for photosynthesis in Early Cambrian oceans: Comment. *Geology* 26:191.

Rigby, J. K. 1983. Fossil Demospongia. In T. W. Broadhead, ed., *Sponges and Spongiomorphs: Notes for a Short Course*, pp. 12–39. University of Tennessee, Department of Geological Sciences, Studies in Geology 7.

Rigby, J. K. 1986a. Sponges of the Burgess Shale (Middle Cambrian), British Columbia. *Palaeontographica Canadiana* 2:1–105.

Rigby, J. K. 1986b. Cambrian and Silurian sponges from North Greenland. *Rapport*

Grønlands Geologiske Undersøgelse 132:51–63.

Rigby, J. K. 1987. Early Cambrian sponges from Vermont and Pennsylvania, the only ones described from North America. *Journal of Paleontology* 61:451–461.

Rigby, J. K. 1991. Evolution of Paleozoic heteractinid calcareous sponges and demosponges—Patterns and records. In J. Reitner and H. Keupp, eds., *Fossil and Recent Sponges*, pp. 83–101. Berlin: Springer Verlag.

Rigby, J. K. and X.-G. Hou. 1995. Lower Cambrian demosponges and hexactinellid sponges from Yunnan, China. *Journal of Paleontology* 69:1009–1019.

Rozanov, A. Yu. and A. Yu. Zhuravlev. 1992. The Lower Cambrian fossil record of the Soviet Union. In J. H. Lipps and P. W. Signor, eds., *Origin and Early Evolution of the Metazoa*, pp. 205–282. New York: Plenum Press.

Savarese, M. 1992. Functional analysis of archaeocyathan skeletal morphology and its paleobiological implications. *Paleobiology* 18:464–480.

Scrutton, C. T. 1992. *Flindersipora bowmani* Lafuste and the early evolution of tabulate corals. *Fossil Cnidaria and Porifera Newsletter* 21:29–33.

Scrutton, C. T. 1997. The Palaeozoic corals. 1: Origins and relationships. *Proceedings of the Yorkshire Geological Society* 51:177–208.

Shabanov, Yu. Ya., V. A. Astashkin, T. V. Pegel', L. I. Egorova, I. T. Zhuravleva, Yu. L. Pel'man, V. M. Sundukov, M. V. Stepanova, S. S. Sukhov, A. B. Fedorov, B. B. Shishkin, N. V. Vaganova, V. I. Ermak, K. V. Ryabukha, A. G. Yadrenkina, G. P. Abaimova, T. V. Lopushinskaya, O. V. Sychev, and T. A. Moskalenko. 1987. *Nizhniy paleozoy yugozapadnogo sklona Anabarskoy anteklizy (po*

dannym bureniya) [Lower Paleozoic of the southwestern slope of the Anabar Anteclise (according to boring data)]. Novosibirsk: Nauka.

Signor, P. W. and G. J. Vermeij. 1994. The plankton and the benthos: Origins and early history of an evolving relationship. *Paleobiology* 20:297–319.

Sorauf, J. E. and M. Savarese. 1995. A Lower Cambrian coral from South Australia. *Palaeontology* 38:757–770.

Steiner, M., D. Mehl, J. Reitner, and B. D. Erdtmann. 1993. Oldest entirely preserved sponges and other fossils from the lowermost Cambrian and a new facies reconstruction of the Yangtse Platform (China). *Berliner geowissenschaftliche Abhandlungen, Reihe E* 9:293–329.

Surge, D. M., M. Savarese, J. R. Dodd, and K. C. Lohmann. 1997. Carbon isotopic evidence for photosynthesis in Early Cambrian oceans. *Geology* 25:503–506.

Teslenko, I. L., A. M. Mambetov, I. T. Zhuravleva, E. I. Myagkova, and N. P. Meshkova. 1983. Dedebulakskaya biogermnaya gryada i istoriya ee razvitiya [The Dedebulak Bioherm Belt and the history of its development]. *Trudy, Institut geologii i geofiziki, Sibirskoe otdelenie, Akademiya nauk SSSR* 569:124–138.

Vacelet, J. 1985. Coralline sponges and the evolution of Porifera. In S. Conway Morris, J. D. George, R. Gibson, and H. M. Platt, eds., *The Origins and Relationships of Lower Invertebrates*, pp. 1–13. Systematics Association Special Publication 28. Oxford: Clarendon Press.

Vologdin A. G. 1963. Pozdne-srednekembriyskie arkheotsiaty basseyna reki Amgi (Sibirskaya platforma) [Late Middle Cambrian archaeocyaths from the Amga River basin (Siberian Platform)]. *Doklady Akademii nauk SSSR* 151:946–949.

Walcott, C. D. 1911. Cambrian geology and paleontology 2: Middle Cambrian holothurians and medusae. *Smithsonian Miscellaneous Collections* 57:145–228.

Walcott, C. D. 1920. Cambrian geology and paleontology 4: Middle Cambrian Spongiae. *Smithsonian Miscellaneous Collections* 85:1–46.

Webby, B. D. 1984. Early Phanerozoic distribution pattern of some major groups of sessile organisms. In *Palaeontology*, vol. 2 of *Proceedings of the 27th International Geological Congress, Moscow, 1984*, pp. 193–208. Utrecht: VNU Science Press.

Wilson, J. A. 1950. Upper Cambrian pleospongiid (?). *Journal of Paleontology* 24:460–461.

Wood, R. 1999. *Reef Evolution*. Oxford: Oxford University Press.

Wood, R., A. Yu. Zhuravlev, and F. Debrenne. 1992. Functional biology and ecology of Archaeocyatha. *Palaios* 7:131–156.

Zhang, X. G. and B. R. Pratt. 1994. New and extraordinary Early Cambrian sponge spicule assemblage from China. *Geology* 22:43–46.

Zhuravlev, A. Yu. 1986. Evolution of archaeocyaths and palaeobiogeography of the Early Cambrian. *Geological Magazine* 123:377–385.

Zhuravlev, A. Yu. 1990. Sistema arkheotsiat [Systematics of archaeocyaths]. In V. V. Menner, ed., *Sistematika i filogeniya bes-pozvonochnykh: Kriterii vydeleniya vysshikh taksonov* [Systematics and phylogeny of invertebrates: Criteria of high taxa establishing], pp. 28–54. Moscow: Nauka.

Zhuravlev, A. Yu. 1996. Reef ecosystem recovery after the Early Cambrian extinction. In M. B. Hart, ed., *Biotic Recovery from Mass Extinction Events*, pp. 79–96. Geological Society Special Publication 102.

Zhuravlev, A. Yu. 1999. A new coral from the Lower Cambrian of Siberia. *Paleontologicheskiy zhurnal* 1999 (5):27–33.

Zhuravlev, A. Yu. and R. Wood. 1995. Lower Cambrian reefal cryptic communities. *Palaeontology* 18:443–470.

Zhuravlev, A. Yu., F. Debrenne, and J. Lafuste. 1993. Early Cambrian microstructural diversification of Cnidaria. *Courier Forschungsinstitut Senckenberg* 164:365–372.

Zhuravlev, A. Yu., B. Hamdi, and P. D. Kruse. 1996. IGCP 366: Ecological aspects of the Cambrian radiation—field meeting. *Episodes* 19:136–137.

Zhuravleva, I. T. 1960. Tip Porifera: Gubki [Phylum Porifera: Sponges]. In L. L. Khalfin, ed., *Biostratigrafiya paleozoya Sayano-Altayskoy gornoy oblasti. Tom 1, Nizhniy paleozoy* [Paleozoic biostratigraphy of the Sayan Altay Mountain region. Vol. 1, Lower Paleozoic], pp. 140–141. Novosibirsk: Siberian Scientific-Research Institute of Geology, Geophysics, and Mineral Resources.

CHAPTER FIFTEEN

Artem V. Kouchinsky

Mollusks, Hyoliths, Stenothecoids, and Coeloscleritophorans

Molluskan diversification was a result of the adaptation of skeletonized forms to various habitats. The ecologic radiation of Cambrian skeletonized mollusks and their possible relatives led to the appearance of all trophic groups, many of them during the Cambrian: deposit feeders (orthothecimorphs, low-spired helcionelloids, and tergomyans), scrapers and grazers (multiplated mollusks, some gastropods), suspension feeders (stenothecoids, chancelloriids, hyolithomorphs, and some macluritid gastropods, orthothecimorphs, Yochelcionella-like helcionelloids), predators, and scavengers (halkieriids and cephalopods). The distinction between suspension and deposit feeding, as well as that of semi-infaunal versus epifaunal habitats, may be meaningless for such small animals approaching interstitial sizes, as the majority of the Early Cambrian mollusks were. Size increase in Cambrian mollusks might have resulted from the invasion of shallow-water high-energy environments. Significant changes in life-cycles could have followed, one of the most important of which was possibly the appearance of the planktotrophic veliger larva.

THE CONTINUOUS Phanerozoic history of marine mollusks that bore mineralized skeletons began in the Early Cambrian. Molluskan remains constitute an important part of the earliest skeletal assemblages (Bengtson and Conway Morris 1992; Dzik 1994). In the present chapter, hyoliths, stenothecoids, and coeloscleritophorans are treated together because even now most of these are considered to be mollusks (Marek and Yochelson 1976; Bengtson 1992; Starobogatov and Ivanov 1996). Nonetheless, the systematic position among the class Mollusca of many of these Cambrian groups is still disputed (Runnegar and Pojeta 1974, 1985; Yochelson 1978; Linsley and Kier 1984; Missarzhevsky 1989; Peel 1991; Geyer 1994; Runnegar 1996). In this chapter, I follow the systematics of the principal groups of Early Paleozoic mollusks developed by Peel (1991), which is supported by morphologic-functional analyses as well as by the observed diversification pattern (Wagner 1996; Zhuravlev, this volume).

This pattern displays the highest diversity of helcionelloids, as well as some minor groups treated as Early Cambrian paragastropods, pelecypods, and rostroconchs, in the Tommotian, followed by continued steady decline during the Cambrian, interrupted by almost complete elimination during the early Botoman Sinsk event (Zhuravlev and Wood 1996). In contrast, indisputable rostroconchs, gastropods, tergomyans, polyplacophorans, and cephalopods started to diversify at the end of the Cambrian and achieved their first peak of diversification in the latest Sunwaptan (Zhuravlev, this volume). True pelecypods diversified even later on, in the Ordovician. Discussion of the paleoecology of these Cambrian groups (figures 15.1 and 15.2) will focus on their morphologic adaptations, possible trophic orientations, and organism-substrate relationships.

MOLLUSKS

Polyplacophorans

The first probable multiplated mollusks appeared during the latest Late Cambrian (Bergenhayan 1960; Stinchcomb and Darrough 1995). Early Cambrian *Triplicatella*, previously interpreted as the earliest chiton (Yates et al. 1992), is an operculum (Conway Morris and Peel 1995). The morphology of the Late Cambrian multiplated mollusks, probable members of the class Polyplacophora, is the subject of some debate. They may be reconstructed as metamerized sluglike animals bearing about eight mid-dorsal plates (Pojeta 1980; Stinchcomb and Darrough 1995). A Late Cambrian multiplated mollusk, *Matthevia*, has been described in detail, based on co-occurrence of three morphologic types of matthevian shells (valves) (Runnegar et al. 1979). Each shell possesses two large ventral holes; no multiple muscle scars were found. All the valves, when clustered in situ, are of essentially the same shape. The armor might have consisted of more or less than eight shells. *Hemithecella* and *Elongata*, which were described by Stinchcomb and Darrough (1995), differ from representatives of the post-Cambrian order Paleoloricata (class Polyplacophora) and *Matthevia*. The assignment of such forms to the Polyplacophora is questionable because the number and arrangement of scars are similar to those of monoplacophorans.

Conical shells of the multiplated mollusks were robust enough to withstand storm-wave activity. Like Recent chitons, the Late Cambrian multiplated mollusks possibly were scrapers or grazers that fed on algal and bacterial mats (figure 15.2:9, 10) (Taylor and Halley 1974; Runnegar et al. 1979). Shells of multiplated mollusks are associated with stromatolite cores that show little abrasion and rarely breakage. This suggests that they occupied stromatolitic reef areas and may well have lived on firm substrates of stromatolitic buildups (Runnegar et al. 1979; Stinchcomb and Darrough 1995). Like Recent chitons, they possibly lived in intertidal and shallow subtidal environments.

Figure 15.1 Generalized reconstruction of the Early Cambrian community of mollusks, hyoliths, stenothecoids, and coeloscleritophorans (background = calcimicrobial-archaeocyathan mounds). Helcionelloids: *1, Oelandiella; 2, Anabarella; 4, Yochelcionella; 5, Ilsanella.* Paragastropod: *3, Aldanella.* Stenothecoid: *6, Stenothecoides.* Rostroconch: *7, Watsonella.* Pelecypod: *8, Fordilla.* Orthothecimorph hyoliths: *9, Ladatheca; 10, Conotheca.* Hyolithomorph hyolith: *11, Burithes.* Coeloscleritophorans: *12, Chancelloria. 13, Halkieria.*

Figure 15.2 Generalized reconstruction of the Late Cambrian community of mollusks and hyoliths (background = stromatolithic mounds). Gastropods: *1, Sinuopea; 2, Strepsodiscus; 3, Matherella; 4, Spirodentalium.* Tergomyans: *5, Proplina; 7, Hypseloconus.* Helcionelloid: *6, Scenella.* Cephalopod: *8, Plectronoceras.* Polyplacophorans: *9, Matthevia; 10, Hemithecella.* Rostroconchs: *11, Pleuropegma; 12, Oepikila; 13, Ribeiria.* Orthothecimorph hyolith: *14, Tcharatheca.* Hyolithomorph hyolith: *15, Linevitus.*

Helcionelloids and Paragastropods

The majority of Cambrian univalves (helcionelloids) fall into three main morphologic categories. These reflect adaptive strategies but are also important evolutionarily, giving rise to pelecypods, rostroconchs, and, subsequently, scaphopods.

The earliest helcionelloid, *Bemella*, is a small caplike shell, with the apex usually lying outside by a slightly elongate apertural ring. Planispirally coiled *Latouchella*-like and *Bemella*-like shells, with relatively broad apertures, are abundant and diverse in the lowermost Lower Cambrian and also subsequently. They exhibit a compromise between a flattened shell with broad aperture and a tightly coiled shell with a small aperture (Runnegar and Pojeta 1985). They occur in various facies worldwide and, based on their low-spired and widely expanded shell (Linsley 1978), would have had a broad foot, which characterizes sluggish epifaunal deposit feeders (figure 15.1 : 1, 5) (Kruse et al. 1995; Gubanov and Peel 1999). They were probably an ancestor for other morphologic-adaptive lineages of helcionelloids.

The principal morphologic trend among helcionelloids is lateral compression of the shell and aperture and loss of strong comarginal ornamentation, often followed by the development of emarginations such as sinus, internal ridges, and snorkel. Such shells have an elongate narrow aperture and a high rate of expansion, with rather smooth but often plicate walls (e.g., *Anabarella, Stenotheca*). Peel (1991) has reconstructed *Eotebenna* as a transitional range of forms from sinus-bearing to elongated with snorkel. These emarginations are assumed to have had an exhalant function (sometimes both exhalant and inhalant) and were oriented posteriorly (Peel 1991). Some reconstructions place them anteriorly (Runnegar and Jell 1980), but the small cross-sectional area of the snorkel in *Yochelcionella*, and the development of the snorkel in *Eotebenna* and *Oelandia*, suggest its posterior direction and exhalant function (Peel 1991). Lateral compression of the shells may be consistent with a vagrant semi-infaunal living mode and with suspension or detritus feeding (Runnegar and Pojeta 1985). Using the criteria of Linsley (1978) and McNair et al. (1981), laterally compressed and widely umbilical helcionelloids with a long aperture, such as *Bemella, Anabarella,* and *Yochelcionella,* are inferred to have been actively mobile on soft substrate in low-energy conditions and thus to have been semi-infaunal filter feeders (Peel 1991; Kruse et al. 1995; Gubanov and Peel 1999) (figure 15.1 : 2, 4).

However, the distinction between suspension and deposit feeding, as well as between semi-infaunal and epifaunal habitats, may be meaningless in such small animals, approaching interstitial sizes. Among modern macrofauna, deposit-feeding invertebrates feed principally upon bacteria, whereas suspension feeders ingest phytoplankton (Levinton 1974). For diminutive Early-Middle Cambrian mollusks, such a distinctive difference might be inappropriate.

Another main adaptive strategy of helcionelloids is shell elongation and subsequent compaction by means of coiling into a bilaterally symmetric, or dissymmetric, spiral. This mode of development is seen in low-spired bilaterally symmetric *Latouchella*-like forms when the beak deviates to the left (e.g., *Pseudoyangtzespira*) or to the right (e.g., *Archaeospira*), giving rise to dextrally or sinistrally coiled forms, respectively (Qian and Bengtson 1989). Together with increase in the number of revolutions, sculptural relief becomes lower in a succession of dextral forms: *Aldanella crassa–A. operosa– Paraaldanella* (Golubev 1976). The shell becomes involute or tightly coiled evolute, with more revolutions in groups of sinistral mollusks (*Barskovia hemisymmetrica–B. rotunda; Beshtashella–Yuwenia–Kistasella*) (Missarzhevsky 1989; Bengtson et al. 1990) and planispiral forms (*Khairkhania* n.sp.–*K. evoluta–K. rotata*) (Esakova and Zhegallo 1996). Hook-shaped forms (e.g., *Ceratoconus*) probably often precede loosely and tightly coiled symmetric or asymmetric conchs with low rates of expansion. Uncoiled, tall, small-apertured shells have a high pressure point and center of gravity (Linsley 1978). To balance such a shell when moving, it is necessary to obtain a lower center of gravity and pressure point and to minimize the frontal cross-sectional area. Curvature and coiling enable a shell held by a snail to be balanced, because movement with a tall or loosely coiled shell is difficult in agitated water. Achievement of a proportionately small cross-sectional area, low pressure point, and low center of gravity favors active locomotion.

Because they were compact, strong, and able to contain a relatively voluminous body, tightly coiled shells could successfully compete with other forms and invade various ecologic niches. Detritus-feeding or grazing is usually assigned to the Cambrian coiled mollusks (Runnegar and Pojeta 1985). Minute shell size, especially in the Early Cambrian, suggests that many Cambrian paragastropods may well have used algae as substrates. Peel (1991) concluded this for Recent and Silurian gastropods of 1–2 mm in size. On the other hand, small paragastropods, with their elongated tangential aperture, have also been inferred to have been mobile epifaunal deposit feeders on soft substrates (Linsley and Kier 1984) (figure 15.1 : 3).

It is possible that small or large individuals of the same species occurred in different environments in the Early Cambrian, depending on water energy. A large helcionelloid, *Randomia aurorae,* was common in microbial mud mounds of the Fosters Point Formation (Landing 1992). Another large helcionelloid, described as *Bemella jacutica,* was recovered in the vicinity of calcimicrobial-archaeocyath reefs of the Pestrotsvet Formation (Dzik 1991). Peribiohermal facies of the Selinde River calcimicrobial-archaeocyath reefs are surrounded by limestones with abundant *Helcionella* with diameters up to 1.5 cm. These occur with their apex upright, which is suggestive of their in situ life position (Repina and Zhuravleva 1977). *Tannuella elata,* a large (2–3 cm) Atdabanian helcionelloid, occurs in interbiohermal and peribiohermal facies of the Medvezh'ya River archaeocyathan reefs (Sundukov and Fedorov 1986). Shallow subtidal wackestones of the Medvezh'ya Formation abound with *Aldanella costata* (pers. obs.). This organism probably dominated subtidal muddy soft substrates of the Tommotian Yudoma-Olenek Basin, Siberian Platform (Vasil'eva and Rudavskaya 1989). Very shallow level-bottom environments are indicated by condensed peritidal limestones that form, for example, the tops of shoaling cycles in Member 4 of the Early Cambrian Chapel Island Formation (Myrow and Landing 1992), with numerous firm surfaces containing abundant but small helcionelloids. In general, mollusks that inhabited reefal areas were relatively sizable forms with robust conchs.

Rostroconchs

Primitive riberiid rostroconchs of the Cambrian had laterally compressed bivalved shells with a univalved protoconch. The development of this morphology was probably the result of a change in living habit, from mainly epifaunal to semi-infaunal suspension/detritus feeding of an *Anabarella*-like ancestor (Runnegar 1978). *Watsonella* (?=*Heraultipegma*) is the earliest-known rostroconch. Landing (1989) noted that although some small (<1 cm) shelly organisms (cf. ostracodes) could be epifaunal crawlers in spite of their laterally compressed condition, a quarter of *Watsonella* specimens collected were oriented vertically in situ; a position more compatible with an infaunal or semi-infaunal habit. Kruse et al. (1995) suggested that *Watsonella*, based on its morphologic similarity to *Anabarella*-like helcionelloids, was more probably a semi-infaunal suspension feeder (figure 15.1:7).

Rostroconchs occurred throughout the Cambrian and constitute an important part of latest Late Cambrian fossil assemblages from China and Australia (Pojeta et al. 1977; Druce et al. 1982). Rostroconchs became diverse and abundant at the very end of the Late Cambrian (latest Sunwaptan-Datsonian), before the first diversity explosion of bivalves. This time interval was previously placed in the Early Ordovician, and the major rostroconch diversification was therefore assigned to that epoch (Runnegar 1978; Pojeta 1979). A variety of life habits, ranging from epifaunal seston feeding (*Euchasma*) to infaunal seston or deposit feeding (*Ptychopegma*), appeared by the end of the Late Cambrian, but semi-infaunal deposit feeding or suspension feeding was probably the most common life strategy until the Permian, when rostroconchs died out (figure 15.2:11–13) (Runnegar and Pojeta 1985).

The paleoecology of Early Cambrian *Watsonella crosbyi* is relatively well known. It has been recovered from various lithofacies, including subtidal siliciclastic mudstones, intertidal stromatolites, and peritidal wacke-packstones of warm- and coolwater environments of various depths and probably of normal salinity (Landing 1989). Late Cambrian rostroconchs are known from warm-water environments, where they seem to have preferred quiet conditions in offshore muds and carbonates (Pojeta and Runnegar 1976; Runnegar 1978).

Pelecypods

A massive radiation of the Bivalvia, which effectively competed with rostroconchs, occurred in the Ordovician. They evolved from mostly polar onshore infaunal deposit feeders (nuculoids) and suspension feeders (conocardiids, babinkiids, cycloconchids) in the Early Ordovician to principally epifaunal suspension feeders in the Middle Ordovician (Pojeta 1971; Babin 1995). A byssus was a key adaptation for elaboration of sessile modes of life among Ordovician pelecypods, both infaunal and epifaunal (Stanley 1972).

Several genera of Cambrian bivalved mollusks have been described (Pojeta et al. 1973; Jell 1980; MacKinnon 1982; Shu 1986; Krasilova 1987; Hinz-Schallreuter 1995; Geyer and Streng 1998). It seems possible that the bivalved condition appeared independently several times within the Cambrian. According to Runnegar and Pojeta (1985), a ligament was the critical point in the origin of the Bivalvia.

The Early Cambrian *Pojetaia* and *Fordilla* (and their numerous synonyms) are usually referred to as the Bivalvia (divided valves, adductor muscles and ligament), even though there are no intermediates between them and Ordovician clams. *Pojetaia* and *Fordilla* occur, with rare exceptions, in articulate closed mode, which may well signify infaunal habitation (Runnegar and Bentley 1983; Ermak 1986, 1988). Otherwise, valves would be disarticulated because of bottom current action. The author has about a hundred specimens of *Fordilla* sp. from the Siberian Platform and of *Pojetaia runnegari* from Australia, entirely in closed mode. However, a shell hash in thin section may well correspond to their detritus, and it is likely that there has been dissolution of disarticulated carbonate valves and selective preservation of phosphatized internal molds in residues. Even if the valves were in closed condition, it does not appear to show convincingly their infaunal lifestyle; some small (<5 mm) recent clams do not spring open after their death on the sediment surface. The disarticulation process depends on decay rate of the adductor muscles and ligament and on intensity of sediment disturbance (Tevesz and McCall 1985). The growth lines on the ligamental area of *Pojetaia,* arranged parallel to the hinge, indicate that the ligament was composed of multiple layers and was probably very weak. Therefore the valves would not necessarily have sprung open after death. A weak ligament is additional evidence that the clams perhaps did not burrow at all, because an elastic ligament is essential for burrowing.

Recent infaunal clams possess a deep pallial sinus, which is lacking among their Cambrian relatives. This seems to be in agreement with a supposed absence of siphuncles due to the unfused mantle of the earliest pelecypods (Stanley 1975). The beaks of most burrowing forms are directed forward (prosogyrous). Such an adaptation increases burrowing efficiency (Stanley 1975) and might explain the prosogyrate shape of *Fordilla*-like mollusks, but the lack of a blunt anterior contradicts this interpretation. Thus, their size and morphology are not incompatible with an epifaunal mode of life (Tevesz and McCall 1985). Again, the distinction between epifaunal and infaunal life modes is difficult to make, given that the size of the animal approaches that of sediment grains (MacKinnon 1982, 1985). Recent juvenile and adult bivalves, less than 3 mm in size, pick up individual food particles with the foot but do not filter water for food (Reid et al. 1992). Supposedly, these bivalves were either inhalant deposit feeders, using ciliated body and mantle surfaces to collect and sort particles of food (Runnegar and Bentley 1983), or epifaunal suspension feeders (Tevesz and McCall 1976, 1985) (figure 15.1:8).

Tergomyans

Another adaptive lineage of univalves is represented by bilaterally symmetric orthoconic or cyrtoconic tergomyans, more or less flattened or tall, with the apex inside the apertural ring. Most of these have a rather large whorl expansion rate and relatively isometric broad aperture, providing stability to the shell on the substrate. In this case, the substrate functioned as a "ventral valve" to protect the animal. Linsley (1978) noted that shell shape is significantly correlated with rate of locomotion. Flattened shells, like *Proplina* and *Kalbiella,* had low position of both pressure point and center of gravity (figure 15.2 : 5, 6). Recent tergomyans with flat shells inhabit quieter environments feeding on detritus. Well-developed radular and pedal muscular scars in the Late Cambrian tergomyan *Pilina* from North China indicate that it was indeed a clamping and crawling grazer (Yu and Yochelson 1999).

Gastropods

Torsion may be regarded as an initial adaptation for living in an elongate coiled shell with a rather narrow aperture. In this case, Cambrian coiled forms could be torted, partially torted, or untorted. Asymmetric Early-Middle Cambrian gastropod-like mollusks may well have been incompletely torted and thus were not gastropods but paragastropods (Runnegar 1981a; Linsley and Kier 1984), the taxonomic rank of which, however, is relatively low. The global lack of predominance of dextral over sinistral forms in the Early Cambrian raises even more suspicion about their gastropod affinity. If it is admitted that the exogastric shell might have had an adaptive significance for the planktotrophic larva, when torsion occurred at the end of the veliger stage, then torsion would be merely an aftereffect of size increase. Late Cambrian gastropods were indeed quite sizable animals in comparison with other Early Cambrian coiled mollusks, including even the largest representatives quoted by Dzik (1991).

True archaeogastropods with a deep sinus and slit appeared in the Late Cambrian and include the orders Pleurotomariida, Bellerophontida, and Macluritida (figure 15.2 : 1–3). The first probable gastropod was Middle Cambrian *Protowenella,* with an ultradextral shell coiling and bellerophontid muscle scar position. Judging from the scar position, deep inside the conch on the umbilical shoulder, Brock (1998) suggested that the animal was capable of retracting in the shell as gastropods do. Based on spire heights and apertural inclinations, the Late Cambrian Macluritacea and Pleurotomariacea were restricted to clear water and hard substrates, since a large amount of suspended fine sediment would have easily fouled the complex aspidobranchial gill (Vermeij 1971). On the other hand, poorly balanced shells, a diffuse nervous system, and weak radulae would have restricted Cambrian archaeogastropods, slow and unstreamlined animals, to a diet of mud (Yochelson 1978; Hickman 1988). Throughout the entire Paleozoic, archaeogastropods were indeed confined to soft sediment environments (Peel 1985). Nonetheless, filter feeding is postulated for sinistral open-coiled macluritid gastropods reported from the late Late Cambrian (Yochelson 1987; Yochelson and Stinchcomb 1987) and even from the late Middle Cambrian (Peel 1988). Their open-coiled shape is not compatible with movement (figure 15.2 : 4) (Yochelson and Stinchcomb 1987). There is a similarity in form and morphologic gradation between Cambrian apparently sinistral (ultradextral?) forms (e.g., *Scaevogyra, Matherella, Kobayashella*) and operculate hyperstrophic macluritacean gastropods of the Ordovician (*Palliseria, Teiichispira, Maclurites*). The Ordovician *Maclurites* has also been interpreted as immobile filter feeders living on reef flats (Webers et al. 1992). Among gastropods, gross shell morphology often reflects basic trophic strategy and function. Thus, the concentration of such a large number of major transitions per time interval in the Late Cambrian–Middle Tremadoc (Wagner 1995) may indicate that the principal trophic groups had already evolved by then.

Late Cambrian assemblages include abundant and relatively large hypseloconids and macluritids, predominantly in high-energy bioclastic carbonates deposited in nearshore medium to high-energy environments, often on reef flats (Webers et al. 1992).

Cephalopods

Forms with tall, cyrtoconic, slightly coiled septate shells of the order Hypseloconida (*Knightoconus, Hypseloconus, Shelbyoceras*) could be ancestors of the first cephalopods (Teichert 1988) (figure 15.2:7). Early cephalopods, such as *Plectronoceras*, were mainly endogastric and rarely exogastric. The direction of coiling does not appear to be of high taxonomic value but might have had an adaptive significance, because endogastric shells could be more suitable for benthic forms (figure 15.2:8). A large number of cephalopods have been described from the Late Cambrian of North China (Chen et al. 1979a,b; Chen and Qi 1982), and about 150 species, 40 genera, 8 families, and 4 orders were recognized (Chen and Teichert 1983). An additional but much less diverse cephalopod fauna is known from Kazakhstan, Siberia, and Laurentia. This surprising diversity of early cephalopods, most of which had become extinct by the end of the Late Cambrian, is consistent with the explosive record of other Late Cambrian mollusks. However, the data are restricted to mostly Chinese localities and need more investigation.

Early cephalopods might have been carnivores, although, this has not been adequately demonstrated. The earliest forms were basically benthic, with their shells vertical in life. The elaboration of a regulatory mechanism controlling buoyancy made it possible to inhabit an ecologic niche with very good prospects. Further evolution led to increase of mobility: "From what is known of the early straight shelled cephalopods, . . . they were not restricted to a benthonic mode of life. It is far more likely that hard parts supplied a balancing mechanism which permitted active swimming, leading to nektonic existence and in rare cases even perhaps a planktonic mode of life" (Yochelson et al. 1973:296).

Cambrian cephalopods favored warm-water environments but do not seem to have been restricted to a single type of substrate. They occurred from shallow water to the outer shelf and continental slope. The first cephalopod fauna, which includes only species of *Plectronoceras*, inhabited well-oxygenated, more or less turbulent shallow-water environments. Cephalopods then became dominants and occupied all available ecologic niches in the inner and outer shelf and the continental slope (Chen and

Teichert 1983). They occurred in turbulent water on the seaward side of stromatolite reefs and in quieter waters of level-bottom environments. Nektonic cephalopods of low diversity occurred in deposits of stagnant basins with euxinic bottoms, which is not compatible with a benthic cephalopod fauna.

COELOSCLERITOPHORANS

Metazoan communities during the Early-Middle Cambrian abounded in problematic organisms called coeloscleritophorans (Bengtson and Missarzhevsky 1981), bearing calcareous sclerites of various size, shape, and degree of mineralization. Sluglike coeloscleritophorans, so far as is known from scleritomes of *Wiwaxia corrugata* and *Halkieria evangelista,* were bilaterally symmetric and probably metameric forms (Conway Morris and Peel 1995). Paired arrangements of elongate sclerites might correspond to an eight- or nine-segmented body (Dzik 1986). There is a certain analogy between wiwaxiidan weakly mineralized leaflike scales and the elitra of segmented annelids (Butterfield 1990). The group might be closely related to the Annelida, but similarities with Mollusca and Brachiopoda also exist (Conway Morris and Peel 1995). On the other hand, Starobogatov and Ivanov (1996) consider that differentiation of the body into a dorsal surface with "metameric" organization of transverse sclerite rows, and a ventral surface without cuticularization, as well as the presence of a cuticular radula-like apparatus, still allow *Wiwaxia* to be ascribed to the subphylum Aculifera (Mollusca). Furthermore, the anterior and posterior shells of *Halkieria* may be homologous with the first and last plates of chitons, and consequently, Starobogatov and Ivanov (1996) assign *Halkieria* to the class Polyplacophora.

Comparative functional morphology of sluglike coeloscleritophorans, such as *Wiwaxia* and *Halkieria,* bears on their ecology. According to reconstructions (Conway Morris 1985), *Wiwaxia* had a slightly elongate, almost isometric body covered with imbricated rows of flattened sclerites, and additionally carrying two sets of elongate spinose sclerites. *Halkieria* had a more elongate and flattened form. Its scleritome included about 2,000 imbricate sclerites, which are smaller than those of *Wiwaxia* (Conway Morris and Peel 1990, 1995). Elongate spinose sclerites are believed to have served in defense, judging from their upright position (Conway Morris 1985). Imbricate sclerites and two terminal shells of *Halkieria evangelista* possibly had a protective function. Wiwaxiids were probably able to shed their sclerites (Conway Morris 1985), although halkieriids may have grown without molting, because the two terminal shells grew accretionally, and there were several zones where new sclerites were generated.

Locomotion of *Halkieria* and *Wiwaxia* could have been effected by locomotory waves along the muscular sole, rimmed by lateral sclerites. No discrete locomotory appendages have been observed. The halkieriid body was very flexible, could shorten, and possibly enroll (Conway Morris and Peel 1995). A vagrant epifaunal lifestyle has been suggested for both genera (figure 15.1:13).

Two or three rows of posteriorly directed teeth are recognized within the anterior part of *Wiwaxia* (Conway Morris 1985). A similar feeding apparatus might have been concealed under the anterior shell of *Halkieria* but has not yet been identified (Conway Morris and Peel 1995). Such a position of the mouth is typical for deposit feeders, scavengers, and grazers. Surprisingly, *Halkieria* bodies show bradoriids stacked in the stomach.

Scaly structures of various shapes (from small groups of fused sclerites and scaly plates to relatively high scaly and tuberculate cones) are common in Early Cambrian strata (Qian and Bengtson 1989; Bengtson et al. 1990). They can all be referred to scleritophoran animals called siphogonuchitids (Bengtson 1992). Apart from an unrestricted and fairly large basal opening, siphogonuchitid sclerites are superficially similar to those of halkieriids. Siphogonuchitids presumably lacked terminal shells with growth lines, but some bilaterally symmetric caplike shells, often confused with monoplacophorans sensu lato, might belong to siphogonuchitid sclerotomes (e.g., *Maikhanella, Purella*). However, none has yet been found.

Spongelike chancelloriids are referred to as coeloscleritophorans, because they bear sclerotomes consisting of hollow calcareous sclerites (Bengtson and Missarzhevsky 1981). Recent investigations of chancelloriids suggest that they were attached to the substrate (A. Zhuravlev, pers. comm., 1996) and that sclerites were hardly used to provide grip on the sediment (cf. Bengtson 1994). The most completely preserved specimens referable to chancelloriids recovered in the Burgess Shale are of *Allonnia* sp. and are up to 20 cm high. *Chancelloria eros* (up to 10 cm high) and *C. pentactina* were also described as bag-shaped forms from the Middle Cambrian Wheeler Shale of Laurentia (Rigby 1978). They show that the sclerotome consists of a different type of rosettes. These animals have funnel-shaped bodies covered with sclerites and a thin skinlike layer between sclerites (Mehl 1996). A group of small sclerites located at one end might represent a growth zone and/or mouth. Rare specimens form groups of individuals of various growth stages. Mehl (1996) inferred from the type of biomineralization and growth that chancelloriids represent an early branch of the Deuterostomia rather than mollusks, but Butterfield and Nicholas (1996) compare the microstructure of organic-walled chancelloriid sclerites with that of the fibers in certain horny sponges. However, sessile conical pelecypods, rudists, should be kept in mind.

Even if chancelloriids were not sponges, in spite of their overall body shape and the presence of spiny corolla (similar to that of hexactinellids and archaeocyaths) surrounding an osculum-like opening, they would still most probably be sessile suspension feeders (figure 15.1:12).

Coeloscleritophorans inhabited various environments. Complete specimens of *Halkieria evangelista* have been found in relatively deep subtidal deposits. This is suggestive of in situ preservation or minimal transport from an adjoining shallow subtidal setting of a carbonate platform reminiscent of many other Burgess Shale–type faunas (Conway Morris and Peel 1995; Butterfield 1995). The earliest diverse shelly assemblage containing siphogonuchitids (Bokova 1985; Rozanov and Zhuravlev

1992) has been reported and described from the mouth of the Kotuykan River (Anabar Uplift, Siberian Platform) by many authors (Rozanov et al. 1969; Missarzhevsky 1989; Khomentovsky and Karlova 1993). This unit occurs between two calcimicrobial biostromes in the uppermost Manykay Formation and includes rare calcimicrobial mounds (Luchinina 1985). Mounds and adjacent rocks contain scaly caps of *Purella* and disarticulated siphogonuchitid sclerites.

STENOTHECOIDS

Stenothecoids represent a group of enigmatic Early to Middle Cambrian bivalved organisms. The overall shape of the skeleton resembles that of brachiopods or pelecypods, and even hinge teeth are found (Pel'man 1985). Their adult shells are usually slightly inequivalve, and individual valves are often curved or have asymmetric margins. Stenothecoids are, to some extent, similar to monoplacophorans sensu lato, considering the valve shape and metameric paired possible muscle scars on the internal surface. Based on this, Runnegar and Pojeta (1974) regarded them as possible bivalved monoplacophorans.

Stenothecoid soft-body anatomy is unknown, and the plane of symmetry is questionable. In this chapter the viewpoint of Yochelson (1969) is accepted—that is, that stenothecoids are a distinctive class of brachiopod-like animals with the plane of symmetry crossing the valves. Nonetheless, Aksarina (1968) has proposed a pelecypod-like bilateral symmetry, and Rozov (1984) established a new phylum Stenothecata based on the two planes of symmetry.

Stenothecoids (*Stenothecoides? kundatensis*) are first reported from the late Tommotian of the Altay Sayan Foldbelt (Pel'man et al. 1992) but definitely occur in the Atdabanian (Rozanov and Zhuravlev 1992). Similar shells were found in a sub-Tommotian part of the Manykay Formation (Missarzhevsky 1989), where they were described as the earliest stenothecoids by Bokova (1985). Their steinkerns are slightly asymmetric and bear concentric plication, growth lines, and small tubercles. No teeth have been found. There is a similarity in shape to smooth steinkerns of *Purella* cf. *antiqua* from the same sample, which normally has scaly walls and co-occurs with siphogonuchitidan sclerites.

It can be speculated that stenothecoids might have evolved from a halkieriid-like ancestor with two terminal shells, by reduction of intermediate sclerites. This process may have been accompanied by transition from vagrant to sessile lifestyle. Stenothecoids probably were suspension feeders inhabiting calm silty interreef environments (Spencer 1981; Zhuravlev 1996) (figure 15.1:6).

HYOLITHS

Cambrian elongate cone-shaped calcareous conchs are ascribed to hyoliths. An operculum (or the second valve) sealed the conch in true hyoliths. Two taxonomic subdivisions are recognized among hyoliths (Sysoev 1976): hyolithomorphs, which possess a ventral projecting ligula, a pair of extensible, curved, strutlike helens, and a complex muscle system; and orthothecimorphs, which lack ligula and helens and have a simpler muscle system.

Hyolithan soft parts have been reconstructed using paleoecologic and morphologic data (Runnegar et al. 1975). The body must have been curved toward the apex in the conical shell. A mud-filled folded intestine, forming a bend and preserved in some shells, corroborates this suggestion. It consists of a tightly folded ventral part and a dorsally situated straight structure (rectum). No mouth or gullet is observed, but an anus may well be present at the end of the rectum. The configuration of the digestive tract makes it difficult to reconstruct a metameric body and serially arranged dorsoventral muscles at the rear part of the body. Serial muscle scars may be the result of migration of muscle attachment areas to the morphologically anterior part of the growing body. However, this reconstruction, based on the inner structure of orthothecimorphs, has been used for hyolithomorphs in which such an intestine has never been found.

Hyoliths have been variously considered to have been pelagic (plankton and nekton) (Dzik 1981; Sysoev 1984), vagrant deposit feeders (Missarzhevsky 1989), free-lying benthic suspension feeders (Sysoev 1984), and semi-infaunal suspension feeders (Landing 1993). Based on the larval shells of hyoliths, Dzik (1978) noted that they are closely similar to molluskan (especially gastropodan) protoconchs. When smooth, the protoconch probably developed within the egg covers. Otherwise, when ornamented by growth lines, it may correspond to a free-living planktotrophic larva. At the early ontogenic stage, some hyoliths supposedly were planktic. Hyolith larval conchs are usually separated from the adult part by a septum. Some hyoliths kept on secreting septa during the postlarval stages. Unperforated septa are common in the apical parts of the mature conchs. According to Dzik (1981) and Sysoev (1984), hyoliths used a "gas camera" to swim and remained pelagic even when mature. It seems unlikely, however, that unperforated septa could regulate buoyancy without a siphuncle maintaining gas exchange. Unperforated septa are known from shells of Recent benthic mollusks (Yochelson et al. 1973), in which they serve as adaptations to border abandoned space in elongate conchs. Hence, life habits of the youngest planktic and adult hyoliths may have been different.

A benthic habit is generally accepted for mature hyoliths (Runnegar et al. 1975; Marek and Yochelson 1976). The hyolithomorph ventral lip and flattened or concave ventral side are regarded as adaptations for crawling (Missarzhevsky 1989). However, the common presence of strong longitudinal sculpture on very robust conchs is difficult to reconcile with this suggestion, unless they had a powerful apparatus to move, for which no evidence exists (Fisher 1962). Studies of muscle scars on hyolith shells (Yochelson 1974; Sysoev 1976) have led to the conclusion that the operculum could not have opened widely and that hyoliths would have had difficulties in moving their shells. Relative sizes of helens and their curvature in some hyoliths may negate the idea that they could be withdrawn into the shell (Yochelson 1974). These discrepancies cast some doubt on their proposed homology with the brachiopod lophophore (Sysoev 1981). Marek (1963) suggested that they might have had the function of oars, but Yochelson (1974) rejected this speculation. Babcock and Robison (1988), Zhou et al. (1996), and Butterfield and Nicholas (1996) described and figured several hyolithomorph specimens from the Middle Cambrian of Laurentia and South China, with preserved helens at different stages of withdrawal, including the complete placing of them inside the conch. In any case the helens would have been inefficient in assisting in locomotion. The appendages were very thin and would have been readily broken. Besides, small gaps between the operculum and the conch would not have been sufficient to allow them to move extensively. Despite the abundance of hyoliths, trails ascribed to them have not been described.

The earliest hyoliths (e.g., circothecid orthothecimorphs) have rounded cross sections and poorly ornamented shells with almost exclusively transverse sculpture. Some orthothecimorphs may well have been semi-infaunal suspension feeders. Occurrences of in situ vertically embedded "*Ladatheca*" *cylindrica* (Landing 1993) in the Nemakit-Daldynian of Avalonia imply that some orthothecimorphs favored a suspension-feeding mode of life under slightly higher bottom-current velocities (figure 15.1:9). Similar forms, which are smooth and commonly laterally curved, with rounded cross sections, are occasionally observed in vertical orientation within early Tommotian Siberian reefs (Riding and Zhuravlev 1995: figure 3B). They occurred in a wide range of environments.

In the middle Tommotian, hyolithomorphs appeared that were epifaunal suspension feeders; helens and prominent dorsal and lateral keels might have enhanced orientation capability relative to bottom currents but allowed them only limited movement in maintaining a rheophile posture (Sysoev 1984; Yochelson 1984; Kruse et al. 1995; Marek et al. 1997) (figures 15.1:11 and 15.2:15). The reinterpretation of the hyolithomorph anatomical dorsum and ventrum further supports a strategy of suspension feeding rather than deposit feeding for them (Kruse 1997). Helens might have provided stability for the conch. Meshkova (1973) reported *Doliutus* sp. and *Trapezovitus* sp., from the Erkeket Formation of the Olenek River in Siberia, oriented on the lithified bottom surface. This suggests that hyoliths could orient their shells along suitable water currents that possibly provided food particles. Flume studies on scale models of hyoliths confirmed that the ligula accelerated currents over the conch in accordance with Bernoulli's principle facilitating suspension feeding (Marek et al. 1997). Tabulate epizoans were described on Devonian hyoliths (Marek and Galle 1976). Individual corallites were oriented longitudinally on the dorsal surface of hyolithid shells, but no epizoans have been found associated with orthothecid hyoliths. This observation further supports the suggestion that hyolithomorphs and orthothecimorphs differed ecologically.

Preserved gut fillings, which led Runnegar et al. (1975) to suggest that hyoliths were semisessile epifaunal deposit feeders, are in fact restricted to orthothecimorphs with a dorsoventrally compressed conch, bearing a differentiated dorsoventral sculpture and cardial processes on the operculum (Meshkova and Sysoev 1981). Thus, this model should be limited to that group. If the constraints for shell-bearing mollusks posed by Linsley (1978) are also applicable to orthothecimorphs, then they should be slow shell draggers (figures 15.1 : 10 and 15.2 : 14).

Hyolithomorphs and orthothecimorphs have been reported from deep to shallow subtidal and reefal environments (Conway Morris 1986; Kruse et al. 1995). Hyoliths occur in Late Cambrian shallow-water sandstones of Laurentia, deposited in a wide spectrum of water energies (Marek and Yochelson 1976). In general, hyoliths are much more common in fine-grained than in coarse-grained sediments (Fisher 1962; Marek 1967; Marek and Yochelson 1976). Orthothecimorph "*Ladatheca*" *cylindrica* ranges from deep subtidal mudstones to peritidal limestones (Landing 1993). "*Ladatheca*" itself formed buildups in shallow subtidal environments. Hyolithomorphs reached their maximum diversity in intrabiohermal facies, such as the Medvezh'ya Formation (Sundukov and Fedorov 1986). The vicinity of calcimicrobial reefs was characterized by fine carbonate detritus and various bottom currents. These facies were likely favored by hyoliths, and this is in accordance with their presumed suspension-feeding mode of life.

CONCLUSIONS

The first occurrences of mollusks were concentrated in shallow subtidal level-bottom open-shelf environments (Mount and Signor 1985, 1992), representing siliciclastic and carbonate units deposited below fair-weather but above storm wave base. There, long periods of relatively quiet conditions, dominated by the accumulation of mudstones, lime mudstones, and wackestones, were interrupted by episodic storm wave erosional events, reflected by rare lenticular packstones, grainstones, and fine sand interbeds. Cambrian mollusks occupied cool- and warm-water environments of various salinities and depths. By the Late Cambrian, a variety of mollusks had invaded the stromatolitic littoral zone of clastic shorelines and offshore carbonate banks (Runnegar 1981b). They became larger and more diverse. Higher endemism of the Early and Middle Cambrian mollusks, in contrast with globally distributed Late Cambrian–Ordovician taxa (Signor 1992; Webers et al. 1992), cannot simply be explained by environmental restriction of specialized forms. One possible explanation is the increase in body size that occurred in the Late Cambrian, possibly accompanied by the appearance of planktotrophic veliger larvae (Chaffee and Lindberg 1986; Signor and Vermeij 1994).

All existing classes of shelled mollusks had appeared by the end of the Tremadoc. This diversification might have taken no more than 50 Ma (Bowring et al. 1993) and was caused by the adaptation of skeletonized forms to various habitats. The ecologic radiation of Cambrian skeletonized mollusks and their possible relatives led to the appearance of all trophic groups among them during the Cambrian. Detritus feeders included likely epifaunal or semi-infaunal orthothecimorphs, low-spired helcionelloids, and tergomyans, at least. Scrapers and grazers were represented by multiplated mollusks, some gastropods, and possibly cephalopods. Suspension feeders (sessile and vagile) might be common among Cambrian mollusks and related animals. Stenothecoids, chancelloriids, hyolithomorphs, and some macluritid gastropods and orthothecimorphs were apparent filter feeders. Perhaps some *Yochelcionella*-like helcionelloids, rostroconchs, and pelecypods were facultatively vagrant suspension feeders. Predators and scavengers, supposedly existed among halkieriids and cephalopods.

In the modern macrofaunal world, deposit-feeding invertebrates feed principally upon bacteria, whereas suspension feeders feed upon phytoplankton (Levinton 1974). Pectin is not digested by mollusks, and consequently algae with pectin envelopes pass intact through the alimentary channel; the diatom diet of Recent mollusks is a Mesozoic innovation (Starobogatov 1988). Thus, the principal diet of filter-feeding Cambrian mollusks had to be bacterioplankton rather than phytoplankton. Based on the same reasoning, Recent primitive mollusks (Aplacophora) have a microcarnivorous mode of feeding, and molluskan sister groups (Turbellaria, Nemertini, Echiurida, Sipunculida, Annelida) never show primary herbivorous conditions. Therefore, carnivory or microcarnivory, given the small size of most early mollusks, might have been the original molluskan mode of nourishment (Haszprunar 1992). In this case, predators were much more common among the earliest mollusks. However, the distinction between suspension and deposit feeding and between microcarnivorous and bacteriovorous diet may be meaningless in such small animals approaching interstitial sizes, as the Early Cambrian mollusks were.

Acknowledgments. John Peel is thanked for helpful review of the manuscript, Andrey Zhuravlev for sharing ideas and data, and Mary Droser for correcting the English. This paper is a contribution to IGCP Project 366.

REFERENCES

Aksarina, N. A. 1968. Probivalvia—novyy klass drevneyshikh mollyuskov [Probivalvia—A new class of ancient mollusks]. In G. A. Selyatitskiy, ed., *Novye dannye po geologii i poleznym iskopaemym Zapadnoy Sibiri* [New data on geology and natural resources of western Siberia], vol. 3, pp. 77–86. Tomsk, Russia: Tomsk University Press.

Babcock, L. E. and R. A. Robison. 1988. Taxonomy and paleobiology of some Middle Cambrian *Scenella* (Cnidaria) and hyolithids (Mollusca) from western North America. *University of Kansas Paleontological Contributions* 121:1–22.

Babin, C. 1995. The initial Ordovician bivalve mollusc radiations on the western Gondwanan shelves. In J. D. Cooper, M. L. Droser, and S. C. Finney, eds., *Ordovician*

Odyssey: Short Papers for the Seventh Annual Symposium on the Ordovician System (Las Vegas, Nevada, USA, June 1995), pp. 491–498. Fullerton, Calif.: Pacific Section Society for Sedimentary Geology.

Bengtson, S. 1992. The cap-shaped Cambrian fossil *Maikhanella* and the relationship between coeloscleritophorans and molluscs. *Lethaia* 25:401–420.

Bengtson, S. 1994. Functional morphology of non-interlocking scleritomes: Implications for autecology. *Terra Nova* 6 (Abstract Supplement 3):1.

Bengtson, S. and S. Conway Morris. 1992. Early radiation of biomineralizing phyla. In J. H. Lipps and P. W. Signor, eds., *Origin and Early Evolution of the Metazoa*, pp. 447–481. New York: Plenum Press.

Bengtson, S. and V. V. Missarzhevsky. 1981. Coeloscleritophora—a major group of enigmatic Cambrian metazoans. In M. E. Taylor, ed., *Short Papers for the Second International Symposium on the Cambrian System 1981*, pp. 19–21. U.S. Geological Survey Open-File Report 81-743.

Bengtson, S., S. Conway Morris, B. J. Cooper, P. A. Jell, and B. N. Runnegar. 1990. Early Cambrian fossils from South Australia. *Association of the Australasian Paleontologists, Memoir* 9:1–364.

Bergenhayan, J. B. M. 1960. Cambrian and Ordovician loricates from North America. *Journal of Paleontology* 34:168–178.

Bokova, A. R. 1985. Drevneyshiy kompleks organizmov kembriya Zapadnogo Prianabar'ya [The oldest assemblage of Cambrian organisms from western Prianabar'e]. In V. V. Khomentovsky, ed., *Stratigrafiya pozdnego dokembriya i rannego paleozoya Sibiri: Vend i rifey* [Late Precambrian and Early Paleozoic stratigraphy of Siberia: Vendian and Riphean], pp. 13–28. Novosibirsk: Institute of Geology and Geophysics, Siberian Branch, USSR Academy of Sciences.

Bowring, S. A., J. P. Grotzinger, C. E. Isachsen, A. H. Knoll, S. M. Pelechaty, and P. Kolosov. 1993. Calibrating rates of Early Cambrian evolution. *Science* 261:1293–1298.

Brock, G. A. 1998. Middle Cambrian molluscs from the southern New England Fold Belt, New South Wales, Australia. *Géobios* 31:571–586.

Butterfield, N. J. 1990. A reassessment of the enigmatic Burgess Shale fossil *Wiwaxia corrugata* (Matthew) and its relationship to the polychaete *Canadia spinosa* Walcott. *Paleobiology* 16:287–303.

Butterfield, N. J. 1995. Secular distribution of Burgess Shale–type preservation. *Lethaia* 28:1–13.

Butterfield, N. J. and C. J. Nicholas. 1996. Burgess Shale–type preservation of both non-mineralizing and "shelly" Cambrian organisms from the Mackenzie Mountains, northwestern Canada. *Journal of Paleontology* 70:893–899.

Chaffee, C. and D. R. Lindberg. 1986. Larval biology of Early Cambrian molluscs: The implications of small body size. *Bulletin of Marine Science* 39:536–549.

Chen, J.-Y. and D. L. Qi. 1982. Upper Cambrian Cephalopoda from Suxian of Anhui Province. *Acta Palaeontologica Sinica* 21:392–403.

Chen, J.-Y. and C. Teichert. 1983. Cambrian cephalopods. *Geology* 11:647–650.

Chen, J.-Y., S.-P. Tsou, T.-E. Chen, and D.-L. Qi. 1979a. Late Cambrian cephalopods of North China—Plectronocerida, Protactinocerida (ord. nov.), and Yanhecerida (ord. nov.). *Acta Palaeontologica Sinica* 18:1–24.

Chen, J.-Y., X.-P. Zou, and D. L. Qi. 1979b. Late Cambrian Ellesmerocerida (Cephalopoda) of North China. *Acta Palaeontologica Sinica* 18:103–124.

Conway Morris, S. 1985. The Middle Cambrian metazoan *Wiwaxia corrugata* (Matthew) from the Burgess Shale and *Ogygopsis* Shale, British Columbia, Canada. *Philosophical Transactions of the Royal Society of London B* 307:507–586.

Conway Morris, S. 1986. The community structure of the Middle Cambrian Phyllopod Bed (Burgess Shale). *Palaeontology* 29:423–467.

Conway Morris, S. and J. S. Peel. 1990. Articulated halkeriids from the Lower Cambrian of North Greenland. *Nature* 345:802–805.

Conway Morris, S. and J. S. Peel. 1995. Articulated halkeriids from the Lower Cambrian of North Greenland and their role in early protostomate evolution. *Philosophical Transactions of the Royal Society of London B* 347:305–358.

Druce, E. C., J. H. Shergold, and B. M. Radke. 1982. A reassessment of the Cambrian-Ordovician boundary section at Black Mountain, western Queensland, Australia. In M. G. Bassett and W. T. Dean, eds., *The Cambrian-Ordovician Boundary: Sections, Fossil Distribution, and Correlations,* pp. 193–209. Geological Series 3. Cardiff: National Museum of Wales.

Dzik, J. 1978. Larval development of hyoliths. *Lethaia* 11:293–299.

Dzik, J. 1981. Origin of the Cephalopoda. *Acta Palaeontologica Polonica* 26:161–169.

Dzik, J. 1986. Turrilepadida and other Machaeridia. In A. Hoffman and M. H. Nitecki, eds., *Problematic Fossil Taxa,* pp. 116–134. New York: Oxford University Press.

Dzik, J. 1991. Is fossil evidence consistent with traditional views of the early metazoan phylogeny? In A. Simonetta and S. Conway Morris, eds., *The Early Evolution of Metazoa and the Significance of Problematic Taxa,* pp. 47–56. Cambridge: Cambridge University Press.

Dzik, J. 1994. Evolution of "small shelly fossils" assemblages of the Early Paleozoic. *Acta Palaeontologica Polonica* 39:247–313.

Ermak, V. V. 1986. Rannekembriyskie fordillidy (Bivalvia) severa Sibirskoy platformy [Early Cambrian fordillids (Bivalvia) from the north of the Siberian Platform]. *Trudy, Institut geologii i geofiziki, Sibirskoe otdelenie, Akademiya nauk SSSR* 669:183–188.

Ermak, V. V. 1988. Stroenie zamochnogo apparata, mikrostruktura rakoviny i obraz zhizni rannekembriyskikh fordillid (Bivalvia) [Hinge construction, shell microstructure, and life strategy of Early Cambrian fordillids (Bivalvia)]. *Trudy, Institut geologii i geofiziki, Sibirskoe otdelenie, Akademiya nauk SSSR* 720:179–184.

Esakova, N. V. and E. A. Zhegallo. 1996. *Stratigrafiya i fauna nizhnego kembriya Mongolii* [Lower Cambrian stratigraphy and fauna of Mongolia]. *Trudy, Sovmestnaya Rossiysko-Mongol'skaya paleontologicheskaya ekspeditsiya* 46:1–208.

Fisher, D. W. 1962. Small conoidal shells of uncertain affinities. In R. C. Moore, ed., *Treatise on Invertebrate Paleontology. Part W: Miscellanea,* pp. W98–W143. Lawrence, Kans.: Geological Society of America and University of Kansas Press.

Geyer, G. 1994. Middle Cambrian mollusks from Idaho and early conchiferan evolution. *New York State Museum Bulletin* 481:69–86.

Geyer, G. and M. Streng. 1998. Middle Cambrian pelecypods from the Anti-Atlas, Morocco. *Revista Española de Paleontología, no. extraordinario, Homenaje al Prof. Gonzalo Vidal,* 83–96.

Golubev, S. N. 1976. Ontogeticheskie izmeneniya i evolyutsionnye tendentsii rannekembriyskikh gastropod [Ontogenetic changes and evolutionary trends in Early Cambrian gastropods]. *Paleontologicheskiy zhurnal* 1976 (2):34–40.

Gubanov, A. and J. S. Peel. 1999. Oelandiella, the earliest Cambrian helcionelloid mollusc from Siberia. *Palaeontology* 42:211–222.

Haszprunar, G. 1992. The first mollusks—small animals. *Bolletino Zoologico* 59:1–16.

Hickman, C. S. 1988. Archaeogastropod evolution, phylogeny, and systematics: A reevaluation. *Malacological Review, Supplement* 4:17–34.

Hinz-Schallreuter, I. 1995. Muscheln (Pelecypoda) aus dem Mittelkambrium von Bornholm. *Geschiebekunde aktuel* 11:71–84.

Jell, P. A. 1980. Earliest known pelecypod on Earth—a new Early Cambrian genus from South Australia. *Alcheringa* 4:233–239.

Khomentovsky, V. V. and G. A. Karlova. 1993. Biostratigraphy of the Vendian-Cambrian beds and the Lower Cambrian boundary in Siberia. *Geological Magazine* 130:29–45.

Krasilova, I. N. 1987. Pervye predstaviteli dvustvorchatykh mollyuskov [The earliest representatives of bivalved mollusks]. *Paleontologicheskiy zhurnal* 1987 (4):24–30.

Kruse, P. D. 1997. Hyolith guts in the Cambrian of northern Australia—turning hyolithomorph upside down. *Lethaia* 29:213–217.

Kruse, P. D., A. Yu. Zhuravlev, and N. P. James. 1995. Primordial metazoan-calcimicrobial reefs: Tommotian (Early Cambrian) of the Siberian Platform. *Palaios* 10:291–321.

Landing, E. 1989. Paleoecology and distribution of the Early Cambrian rostroconch *Watsonella crosbyi* Grabau. *Journal of Paleontology* 63:566–573.

Landing, E. 1992. Lower Cambrian of southern Newfoundland: Epeirogeny and Laza-

rus faunas, lithofacies-biofacies linkages, and the myth of a global chronostratigraphy. In J. H. Lipps and P. W. Signor, eds., *Origin and Early Evolution of the Metazoa*, pp. 283–309. New York: Plenum Press.

Landing, E. 1993. In situ earliest Cambrian tube worms and the oldest metazoan-constructed biostrome (Placentian series, southeastern Newfoundland). *Journal of Paleontology* 67:333–342.

Levinton, J. S. 1974. Trophic group and evolution in bivalve molluscs. *Palaeontology* 17:579–585.

Linsley, R. M. 1978. Locomotion rates and shell form in the Gastropoda. *Malacologia* 17:193–206.

Linsley, R. M. and W. M. Kier. 1984. The Paragastropoda: A proposal for a new class of Paleozoic Mollusca. *Malacologia* 25:241–254.

Luchinina, V. A. 1985. Vodoroslevye postroyki rannego paleozoya severa Sibirskoy platformy [Early Paleozoic algal buildups on the north of the Siberian Platform]. *Trudy, Institut geologii i geofiziki, Sibirskoe otdelenie, Akademiya nauk SSSR* 628:45–50.

MacKinnon, D. I. 1982. *Tuarangia paparua* n.gen. and n.sp.: A late Middle Cambrian pelecypod from New Zealand. *Journal of Paleontology* 56:58–98.

MacKinnon, D. I. 1985. New Zealand late Middle Cambrian molluscs and the origin of Rostroconchia and Bivalvia. *Alcheringa* 9:65–81.

Marek, L. 1963. New knowledge on the morphology of *Hyolithes*. *Sbornik geologickikh Věd, Paleontologie P* 1:5–73.

Marek, L. 1967. The class Hyolitha in the Caradoc of Bohemia. *Sbornik geologickikh Věd, Paleontologie P* 9:51–114.

Marek, L. and A. Galle. 1976. The tabulate coral *Hyostragulum*: An epizoan with bearing on hyolithid ecology and systematics. *Lethaia* 9:51–64.

Marek, L. and E. L. Yochelson. 1976. Aspects of the biology of Hyolitha (Mollusca). *Lethaia* 9:65–82.

Marek, L., R. L. Parsley, and A. Galle. 1997. Functional morphology of hyoliths based on flume studies. *Věstník Českeho Geologickěho Ústavu* 72:351–358.

McNair, C. G., W. M. Kier, P. D. LaCroix, and R. M. Lindsay. 1981. The functional significance of aperture form in gastropods. *Lethaia* 14:63–70.

Mehl, D. 1996. Organization and microstructure of the chancelloriid skeleton: Implications for the biomineralization of the Chancelloriidae. *Bulletin de l'Institut océanographique, Monaco*, no. spécial 14 (4): 377–385.

Meshkova, N. P. 1973. Nekotorye voprosy tafonomii i ekologii rannekembriyskikh khiolitov [Some questions on taphonomy and ecology of the Early Cambrian hyoliths]. *Trudy, Institut geologii i geofiziki, Sibirskoe otdelenie, Akademiya nauk SSSR* 169:124–126.

Meshkova, N. P. and V. A. Sysoev. 1981. Nakhodka slepkov pishchevaritel'nogo apparata nizhnekembriyskikh khiolitov [A find of intestine casts of the Early Cambrian hyoliths]. *Trudy, Institut geologii i geofiziki, Sibirskoe otdelenie, Akademiya nauk SSSR* 481:82–85.

Missarzhevsky, V. V. 1989. *Drevneyshie skeletnye okamenelosti i stratigrafiya pogranichnykh tolshch dokembriya i kembriya* [The oldest skeletal fossils and stratigraphy of the Precambrian-Cambrian boundary strata]. *Trudy, Geologicheskiy institut, Akademiya nauk SSSR* 443:1–237.

Mount, J. F. and P. W. Signor. 1985. Early Cambrian innovation in shallow subtidal environments: Paleoenvironments of Early

Cambrian shelly fossils. *Geology* 13:730–733.

Mount, J. F. and P. W. Signor. 1992. Faunas and facies—fact and artifact: Paleoenvironmental controls on the distribution of Early Cambrian faunas. In J. H. Lipps and P. W. Signor, eds., *Origin and Early Evolution of the Metazoa*, pp. 27–51. New York: Plenum Press.

Myrow, P. M. and E. Landing. 1992. Mixed siliciclastic-carbonate deposition in an Early Cambrian oxygen-stratified basin, Chapel Island Formation, southeastern Newfoundland. *Journal of Sedimentary Petrology* 62:455–473.

Peel, J. S. 1985. Autecology of Silurian gastropods and monoplacophorans. *Special Papers in Palaeontology* 32:165–182.

Peel, J. S. 1988. Molluscs of the Holm Dal Formation (late Middle Cambrian), central North Greenland. *Meddelelser om Grønland, Geoscience* 20:145–168.

Peel, J. S. 1991. Functional morphology, evolution, and systematics of Early Palaeozoic univalved molluscs. *Grønlands Geologiske Undersøgelse* 161:1–116.

Pel'man, Yu. L. 1985. Novye stenotekoidy iz nizhnego kembriya Zapadnoy Mongolii [New stenothecoids from the Lower Cambrian of western Mongolia]. *Trudy, Institut geologii i geofiziki, Sibirskoe otdelenie, Akademiya nauk SSSR* 632:103–114.

Pel'man, Yu. L., N. A. Aksarina, S. P. Koneva, L. Ye. Popov, L. P. Sobolev, and G. T. Ushatinskaya. 1992. *Drevneyshie brakhiopody territorii severnoy Evrazii* [The oldest brachiopods from the territory of northern Eurasia]. Novosibirsk: United Institute of Geology, Geophysics, and Mineralogy, Siberian Branch, Russian Academy of Sciences.

Pojeta, J., Jr. 1971. Review of Ordovician pelecypods. *United States Geological Survey Professional Paper* 695:1–46.

Pojeta, J., Jr. 1979. Geographic distribution of Cambrian and Ordovician rostroconch molluscs. In J. Gray and A. J. Boucot, eds., *Historical Biogeography, Plate Tectonics, and the Changing Environment*, pp. 27–36. Eugene, Oreg.: Oregon State University Press.

Pojeta, J., Jr. 1980. Molluscan phylogeny. *Tulane Studies in Geology and Paleontology* 16:55–80.

Pojeta, J., Jr., and B. Runnegar. 1976. The paleontology of rostroconch molluscs and the early history of the phylum Mollusca. *United States Geological Survey Professional Paper* 968:1–88.

Pojeta, J., Jr., B. Runnegar, and J. Kriz. 1973. *Fordilla troyensis* Barrande: The oldest known pelecypod. *Science* 180:866–868.

Pojeta, J., Jr., J. Gilbert-Tomlison, and J. H. Shergold. 1977. Cambrian and Ordovician rostroconch molluscs from northern Australia. *United States Geological Survey, Bulletin* 171:1–54.

Qian, Y. and S. Bengtson. 1989. Paleontology and biostratigraphy of the Early Cambrian Meishucunian Stage in Yunnan Province, South China. *Fossils and Strata* 24:1–156.

Reid, R. G. B., R. F. McMahon, D. O'Foighil, and R. Finnigan. 1992. Anterior inhalent currents and pedal feeding in bivalves. *Veliger* 35:93–104.

Repina, L. N. and I. T. Zhuravleva. 1977. Novoe mestonakhozhdenie bigogermov s arkheotsiatami [A new locality of bioherms with archaeocyaths]. *Trudy, Institut geologii i geofiziki, Sibirskoe otdelenie, Akademiya nauk SSSR* 302:134–136.

Riding, R. and A. Yu. Zhuravlev. 1995. Structure and diversity of oldest sponge-microbe reefs: Lower Cambrian, Aldan River, Siberia. *Geology* 23:649–652.

Rigby, J. K. 1978. Porifera of the Middle Cambrian Wheeler Shale from the Wheeler Amphitheater, House Range, in Western

Utah. *Journal of Paleontology* 52:1325–1345.

Rozanov, A. Yu. and A. Yu. Zhuravlev. 1992. The Lower Cambrian fossil record of the Soviet Union. In J. H. Lipps and P. W. Signor, eds., *Origin and Early Evolution of the Metazoa*, pp. 205–282. New York: Plenum Press.

Rozanov, A. Yu., V. V. Missarzhevsky, N. A. Volkova, L. G. Voronova, I. N. Krylov, B. M. Keller, I. K. Korolyuk, K. Lendzion, R. Michniak, N. G. Pykhova, and A. D. Sidorov. 1969. *Tommotskiy yarus i problema nizhney granitsy kembriya* [The Tommotian stage and the Lower Cambrian boundary problem]. *Trudy, Geologicheskiy institut, Akademiya nauk SSSR* 241:1–380.

Rozov, S. N. 1984. Morfologiya, terminologiya, i sistematicheskoe polozhenie stenotekoid [Morphology, terminology, and systematic affinity of stenothecoids]. *Trudy, Institut geologii i geofiziki, Sibirskoe otdelenie, Akademiya nauk SSSR* 597:117–133.

Runnegar, B. 1978. Origin and evolution of the Class Rostroconchia. *Philosophical Transactions of the Royal Society of London B* 284:319–333.

Runnegar, B. 1981a. Muscle scars, shell form, and torsion in Cambrian and Ordovician univalved molluscs. *Lethaia* 14:311–322.

Runnegar, B. 1981b. Biostratigraphy of Cambrian molluscs. In M. E. Taylor, ed., *Short Papers for the Second International Symposium on the Cambrian System 1981*, pp. 198–202. United States Geological Survey Open-File Report 81-743.

Runnegar, B. 1996. Early evolution of the Mollusca: The fossil record. In J. D. Taylor, ed., *Origin and Evolutionary Radiation of the Mollusca*, pp. 77–87. Oxford: Oxford University Press.

Runnegar, B. and C. Bentley. 1983. Anatomy, ecology, and affinities of the Australian

Early Cambrian bivalve *Pojetaia runnegari* Jell. *Journal of Paleontology* 57:73–92.

Runnegar, B. and P. A. Jell. 1980. Australian Middle Cambrian molluscs: Corrections and additions. *Alcheringa* 4:111–113.

Runnegar, B. and J. Pojeta, Jr. 1974. Molluscan phylogeny: The paleontological viewpoint. *Science* 186:311–317.

Runnegar, B. and J. Pojeta, Jr. 1985. Origin and diversification of the Mollusca. In E. R. Trueman and M. R. Clarke, eds., *Evolution*, vol. 10 of *The Mollusca*, pp. 1–57. Orlando, Fla.: Academic Press.

Runnegar, B., J. Pojeta, Jr., N. J. Morris, J. D. Taylor, M. E. Taylor, and G. McClung. 1975. Biology of the Hyolitha. *Lethaia* 8: 181–191.

Runnegar, B., J. Pojeta, Jr., M. E. Taylor, and D. Collins. 1979. New species of the Cambrian and Ordovician chitons *Matthevia* and *Chelodes* from Wisconsin and Queensland: Evidence for the early history of polyplacophoran molluscs. *Journal of Paleontology* 53:1374–1394.

Shu, D.-G. 1986. Notes on the oldest fossil bivalves from the Niutitang Formation of Fuquan, Guizhou. *Acta Palaeontologica Sinica* 25:219–222.

Signor, P. W. 1992. Evolutionary and tectonic implications of Early Cambrian faunal endemism. In C. A. Hall, Jr., V. Doyle-Jones, and B. Widawski, eds., *The History of Water: Eastern Sierra Nevada, Owens Valley, White-Inyo Mountains*, pp. 1–13. White Mountain Research Station Symposium, vol. 4. Los Angeles: Regents of the University of California.

Signor, P. W. and G. J. Vermeij. 1994. The plankton and the benthos: Origins and early history of an evolving relationship. *Paleobiology* 20:297–319.

Spencer, L. M. 1981. Palaeoecology of a Lower Cambrian archaeocyathid inter-reef fauna

from southern Labrador. In M. E. Taylor, ed., *Short Papers for the Second International Symposium on the Cambrian System 1981*, pp. 215–218. United States Geological Survey Open-File Report 81-743.

Stanley, S. M. 1972. Functional morphology and evolution of byssally attached bivalve molluscs. *Journal of Paleontology* 46:165.

Stanley, S. M. 1975. Adaptive themes in the evolution of Bivalvia (Mollusca). *Annual Review of Earth and Planetary Sciences* 3:361–385.

Starobogatov, Ya. I. 1988. O sootnoshenii mezhdu mikro- i makroevolyutsiey [On the correlation between micro- and macro-evolution]. In E. I. Kolchinskiy and Yu. I. Polyanskiy, eds., *Darvinizm: Istoriya i sovremennost'* [Darwinism: History and present], pp. 293–302. Leningrad: Nauka.

Starobogatov, Ya. I. and D. L. Ivanov. 1996. O problematichnykh iskopaemykh ostatkakh, otnosimykh k mollyuskam [On the problematic fossils assigned to mollusks]. In V. N. Shimanskiy, A. Yu. Zhuravlev, and An. F. Veis, eds., *Vserossiyskiy simpozium: Zagadochnye organizmy v evolyutsii i filogenii, tezisy dokladov* [All-Russian symposium: Enigmatic organisms in evolution and phylogeny, abstracts], pp. 85–87. Moscow: Paleontological Institute, Russian Academy of Sciences.

Stinchcomb, B. L. and G. Darrough. 1995. Some molluscan Problematica from the Upper Cambrian–Lower Ordovician of the Ozark Uplift. *Journal of Paleontology* 69: 52–65.

Sundukov, V. M. and A. B. Fedorov. 1986. Paleontologicheskaya kharakteristika i vozrast sloev s vodoroslevo-arkheotsiatovymi biogermami r. Medvezh'ey [Paleontological characteristics and the age of the strata containing algal-archaeocyathan bioherms from the Medvezh'ya River]. *Trudy, Institut geologii i geofiziki, Sibirskoe otdelenie, Akademiya nauk SSSR* 669:108–118.

Sysoev, V. A. 1976. O sistematike i sistematicheskoy prinadlezhnosti khiolitov [About the systematics and systematic affinity of hyoliths]. In V. N. Shimanskiy, ed., *Osnovnye problemy sistematiki zhivotnykh* [The chief problems of animal systematics], pp. 28–34. Moscow: Paleontological Institute, USSR Academy of Sciences.

Sysoev, V. A. 1981. Iz istorii izucheniya hiolitov (K diskussii o sistematicheskom polozhenii) [On the history of hyolith study (To the discussion on a systematic position)]. *Trudy, Institut geologii i geofiziki, Sibirskoe otdelenie, Akademiya nauk SSSR* 481:85–92.

Sysoev, V. A. 1984. Morfologiya i sistematicheskaya prinadlezhnost' khiolitov [Morphology and systematic affinity of hyoliths]. *Paleontologicheskiy zhurnal* 1984 (2): 3–14.

Taylor, M. E. and R. B. Halley. 1974. Systematics, environment, and biogeography of some Late Cambrian and Early Ordovician trilobites from eastern New York State. *United States Geological Survey Professional Paper* 834:1–38.

Teichert, C. 1988. Main features of cephalopod evolution. In M. R. Clarke and E. R. Trueman, eds., *Paleontology and Neontology of Cephalopods*, vol. 12 of *The Mollusca*, pp. 11–75. Orlando, Fla.: Academic Press.

Tevesz, M. J. S. and P. L. McCall. 1976. Primitive life habitats and adaptive significance of the pelecypod form. *Paleobiology* 2: 183–190.

Tevesz, M. J. S. and P. L. McCall. 1985. Primitive life habit of Bivalvia reconsidered. *Journal of Paleontology* 59:1326–1330.

Vasil'eva, N. I. and N. V. Rudavskaya. 1989. Zakonomernosti rasprostraneniya fauny i fitoplanktonnykh soobshchestv v pogranichnykh otlozheniyakh venda i nizhnego kembriya na Sibirskoy platforme [Regularities in the distribution of the fauna and phytoplanktonic associations from the Vendian and Lower Cambrian boundary strata on the Siberian Platform]. In M. S. Mesezhnikov and S. A. Chirva, eds., *Metodicheskie aspekty stratigraficheskikh issledovaniy v neftegazonosnykh basseynakh* [Methodic aspects of stratigraphic researches in oil- and gas-bearing basins], pp. 69–79. Leningrad: All-Union Scientific-Research Geological-Exploring Institute.

Vermeij, G. J. 1971. Gastropod evolution and morphological diversity in relation to shell geometry. *Journal of Zoology* 16:15–23.

Wagner, P. J. 1995. Testing evolutionary constraint hypotheses with early Paleozoic gastropods. *Paleobiology* 21:248–272.

Wagner, P. J. 1996. Morphologic diversification of early Paleozoic "archaeogastropods." In J. D. Taylor, ed., *Origin and Evolutionary Radiation of the Mollusca*, pp. 161–169. Oxford: Oxford University Press.

Webers, G. F., J. Pojeta, Jr., and E. L. Yochelson. 1992. Cambrian mollusca from the Minaret Formation, Ellsworth Mountains, West Antarctica. In G. F. Webers, C. Craddock, and J. F. Splettstoesser, eds., *Geology and Paleontology of the Ellsworth Mountains, West Antarctica*, pp. 181–248. Geological Society of America, Memoir 170.

Yates, A. M., K. L. Gowlett-Holmes, and B. J. McHenry. 1992. *Triplicatella disdoma* reinterpreted as the earliest known polyplacophoran. *Journal of the Malacological Society of Australia* 13:71.

Yochelson, E. L. 1969. Stenothecoida, a proposed new class of Cambrian Mollusca. *Lethaia* 2:49–62.

Yochelson, E. L. 1974. Redescription of the Early Cambrian *Helenia bella* Walcott, an appendage of Hyolithes. *United States Geological Survey Journal of Research* 2:717–722.

Yochelson, E. L. 1978. An alternative approach to the interpretation of the phylogeny of ancient mollusks. *Malacologia* 17:165–191.

Yochelson, E. L. 1984. Speculative functional morphology and morphology that could not function: The example of *Hyolithes* and *Biconulithes*. *Malacologia* 25:255–264.

Yochelson, E. L. 1987. Redescription of *Spirodentalium* Walcott (Gastropoda: Late Cambrian) from Wisconsin. *Journal of Paleontology* 61:66–69.

Yochelson, E. L. and B. L. Stinchcomb. 1987. Recognition of *Macluritella* (Gastropoda) from the Upper Cambrian of Missouri and Nevada. *Journal of Paleontology* 61:56–61.

Yochelson, E. L., R. H. Flower, and G. F. Webers. 1973. The bearing of the new Late Cambrian monoplacophoran genus *Knightoconus* upon the origin of the Cephalopoda. *Lethaia* 6:275–310.

Yu, W. and E. I. Yochelson. 1999. Some Late Cambrian molluscs from Liaoning Province, China. *Records of the Western Australian Museum* 19:379–389.

Zhou, Y., J. Yuan, Z. Zhang, Y. Huang, X. Chen, and Z. Zhou. 1996. Composition and significance of the Middle Cambrian Kaili Lagerstätte in Taijiang County, Guizhou Province, China: A new Burgess Shale type Lagerstätte. *Guizhou Geology* 13:7–14.

Zhuravlev, A. Yu. 1996. Reef ecosystem recovery after the Early Cambrian extinction. In M. B. Hart, ed., *Biotic Recovery from Mass Extinction Events,* pp. 79–96. Geological Society Special Publication, London 102.

Zhuravlev, A. Yu. and R. A. Wood. 1996. Anoxia as the cause of the mid–Early Cambrian (Botomian) extinction event. *Geology* 24:311–314.

CHAPTER SIXTEEN

Galina T. Ushatinskaya

Brachiopods

*All brachiopods are sessile benthic organisms; in feeding style they are ciliary sus-
pension feeders. Cambrian brachiopods show several types of substrate relationships:
pedicle-anchoring, free-lying, cemented epifaunal, infaunal, quasi-infaunal, and interstitial, as well as possibly pseudoplanktic. The earliest brachiopods are known
from Early Cambrian carbonates of the Siberian Platform. Lingulates appeared at
the beginning of the Tommotian, and calciates arose in the middle Tommotian. Additional lingulate orders appeared during the late Atdabanian in siliciclastic sediments
in northern European areas. The acquisition of a mineralized skeleton by brachiopods at the beginning of the Cambrian may have been connected with changes in
ocean water chemistry. Differences in diet probably defined distinctions in skeletal
composition: lingulates could consume phytoplankton, but calciates preferred animal
proteins.*

BRACHIOPODS BELONG to the subkingdom Eumetazoa and are characterized by two
unique features. First, they have an intermediate protostomian-deuterostomian embryology. It is likely that brachiopods separated from other Bilateralia prior to protostomian-deuterostomian differentiation (Malakhov 1976, 1983). Second, the Brachiopoda is the only phylum that produces both calcium carbonate and phosphatic shells.

By the second half of the twentieth century, the brachiopod systematics developed
by Huxley (1869) became widely accepted (Sarycheva 1960; Williams and Rowell
1965). This subdivided brachiopods into two classes, Articulata and Inarticulata,
based on presence or absence of valve articulation, respectively. Therefore, the Inarticulata included brachiopods with calcium carbonate shells, as well as those with
phosphatic shells. During the last decades, studies of Recent brachiopods have revealed that forms possessing shells of different composition and microstructure are
also distinguished in their embryology and molecular phylogeny (e.g., Williams 1968;

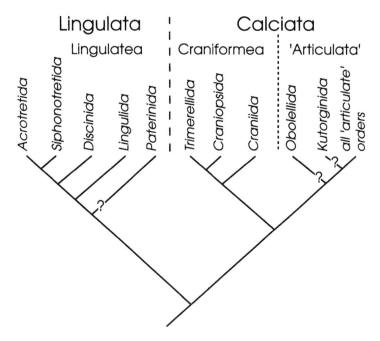

Figure 16.1 Subdivision of the phylum Brachiopoda into classes Lingulata and Calciata and orders, according to Popov et al. (1993).

Malakhov 1976; Jope 1977, 1986; Ushatinskaya 1990; Nielsen 1991; Williams and Holmer 1992). They therefore appear to belong to different lineages. As a result, a new systematic scheme was proposed by Gorjansky and Popov (1985) and later modified by Popov et al. (1993). According to this scheme, the phylum Brachiopoda includes the classes Lingulata and Calciata (figure 16.1). The former contains brachiopods with phosphatic shells only, and the latter those with calcareous shells. Recently, this supra-ordinal classification has been developed by Williams et al. (1996).

CAMBRIAN BRACHIOPOD RADIATION AND DIVERSITY

Both classes are known from the Early Cambrian. In the Cambrian, Calciata were represented by the orders Obolellida, Kutorginida, Craniopsida, Naukatida, and Chileida, in addition to traditional articulates (Orthida and Pentamerida orders). However, their diversity was low, and they became abundant only during the Ordovician to Devonian. Lingulates were much more diverse, and all five orders were present in the Cambrian (figure 16.2). Three of these—Paterinida, Lingulida, and Acrotretida—appeared during the Early Cambrian, the Siphonotretida appeared in the Middle Cambrian, and the Discinida arose at the end of the Cambrian.

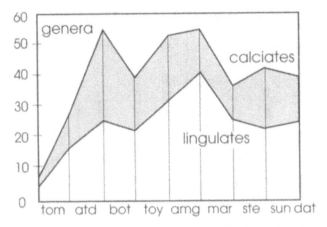

Figure 16.2 Generic diversity of calciate and lingulate brachiopods in the Cambrian. Stages: *tom* = Tommotian; *atd* = Atdabanian; *bot* = Botoman; *toy* = Toyonian; *amg* = Amgan; *mar* = Marjuman; *ste* = Steptoan; *sun* = Sunwaptan; *dat* = Datsonian.

LARVAL ECOLOGY AND BRACHIOPOD DISTRIBUTION

General brachiopod paleoecology has often been discussed (e.g., Lochman 1956; Sarycheva 1960; Ivanova 1962; Rudwick 1965, 1970; Gorjansky 1969; Rowell and Krause 1973; McKerrow 1978; Percival 1978; Williams and Lockley 1983; Bassett 1984; Holmer 1989; Popov et al. 1989; Wright and McClean 1991; Popov 1995), and the following information is based on this research.

All brachiopods are passive suspension feeders that use the lophophore, a variously looped or coiled extension of the mesocoelom, for water and food uptake. Chuang (1959) showed that the alimentary system of *Lingula unguis* contains fermenters that allow the animal to digest phytoplankton. The remaining Recent lingulates do not differ from it in this respect. In contrast, Recent calciates lack such fermenters and feed, mainly, on bacterial aggregates and dissolved nutrient matter (Atkins 1960; McCammon 1969; Rhodes and Thompson 1993).

All brachiopods belonging to sessile benthos spread at the larval stage. Recent calciates have a lecithotrophic larva that is free-living from several hours to 1–2 days. After that, the larva settles on a substrate. Recent lingulates possess a planktotrophic larva that floats in the water column for from several days to a month (Malakhov 1976). In some cases, under unfavorable conditions, a complete morphogenesis is observed, and a pedicle and a lophophore, bearing numerous cirri, are formed that appear just like those in anchored animals (Zezina 1976). These features allow lingulates exclusive facilities for migration. Many Recent lingulates are widespread in shallow waters of the Indian and western Pacific oceans, and *Pelagodiscus atlanticus* (order Discinida) is a cosmopolitan species. Some Cambrian species of the Paterinida,

Acrotretida, and Lingulida orders have a wide, sometimes global distribution. This phenomenon may be due to a protracted pelagic phase (Jablonski 1986; Rowell 1986). Indirect evidence for this is that larval shell sizes do not exceed one-third to one-fifth of the whole shell (figures 16.3A,B,D). In addition, the larval shell surface of all acrotretids and most Cambrian lingulids bears numerous minute pits (figures 16.3C,G). Many researchers ascribe this feature to a vesicular structure of the periostracum or to an entirely organic larval shell (Biernat and Williams 1970; Williams and Curry 1991). Popov et al. (1982) regarded the pitted microornament of the umbonal area in acrotretids as a negative impression of the entirely organic larval shell and suggested that the acrotretid shell acquired mineralization only after settlement. In any case, such a structure may increase buoyancy and probably is an adaptation to a pelagic lifestyle.

Prolongation of the pelagic larval stage might serve to enhance the distribution of some acrotretids. The acceleration in foramen development might indirectly substantiate such a suggestion (Popov and Ushatinskaya 1992). Early Cambrian *Linnarssonia* possessed only a vestigial delthyrium on the posterior margin of the larval shell. Development of this delthyrial opening into a foramen occurred after the settlement of the animal. Later genera, *Homotreta* and *Hadrotreta,* had a well-defined delthyrium that turned into a foramen soon after settlement. Several genera (*Neotreta, Quadrisonia, Angulotreta, Rhondellina*) had larval shells with well-developed foramens (figures 16.3I,J). Some Middle Cambrian paterinids (*Paterina, Micromitra, Dictyonina*) from the Siberian Platform had larval valves of about one-half to one-third of the entire shell size, although usually the size of larval valves was about one-fifth to one-tenth of the adult shell. This phenomenon probably also indicates prolongation of the larval stage.

SETTLEMENT AND INTERACTION WITH THE SUBSTRATE

Although overall brachiopod diversity was low during the Cambrian, most of the ecologic types already existed. These included epifaunal anchored, cemented, and free-lying forms, as well as infaunal, quasi-infaunal, interstitial, and possibly pseudo-planktic brachiopods.

Epifaunal Anchored Brachiopods

During the Early Cambrian, brachiopods inhabited shallow subtidal environments. The earliest brachiopods (family Cryptotretidae, order Paterinida) occur in the Tommotian and Atdabanian of the Anabar-Sinsk Basin of the Siberian Platform and the Zavkhan Province of Mongolia. Cryptotretids occurred on calcareous-argillaceous interreefal substrates of shallow warm epicontinental seas, often under high-energy

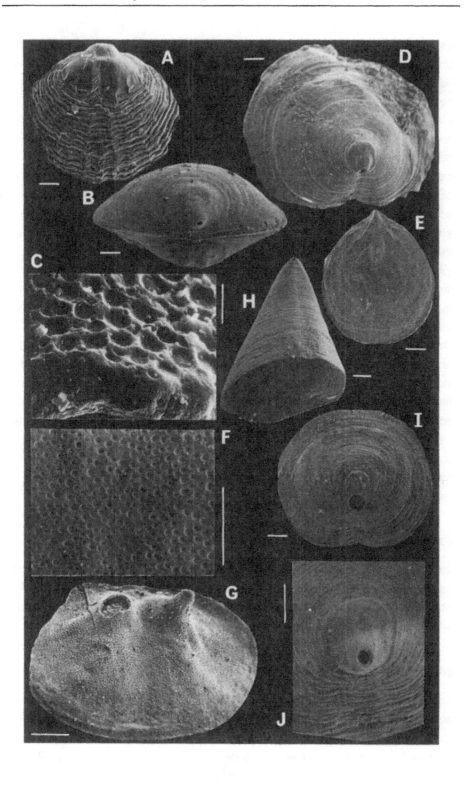

Figure 16.3 A, *Dictyonina* sp., PIN 4290/206, ventral valve, Middle Cambrian, Marjuman stage (Kotuy River, Siberian Platform), scale bar 1 mm. B and C, *Stilpnotreta inaequalis* Ushatinskaya, pitted larval shell surface, Middle Cambrian, Marjuman stage (Siberian Platform, Kotuy River); B, PIN 4290/141, scale bar 0.5 mm; C, PIN 4511/76, scale bar 0.2 mm. D, *Linnarssonia rowelli* Pel'man, PIN 3848/3001, ventral valve, Early Cambrian, Botoman Perekhod Formation (Ulakhan-Kyyry-Taas, middle Lena River, Siberian Platform), scale bar 1 mm. E, *Eoobolus* sp., PIN 4290/252, ventral valve interior, Early Cambrian, Atdabanian Krasnyy Porog Formation (Sukharikha River, Siberian Platform), scale bar 0.5 mm. F, *Fossuliella linguata* (Pel'man), PIN 4290/252, adult shell surface, Middle Cambrian, Amgan Stage (Siberian Platform, Olenek River), scale bar 0.1 mm. G, *Acrothele* sp., PIN 4290/253, posterior part of dorsal valve exterior, thin spines on the larval shell serving as adaptation to soft substrate, Middle Cambrian, Marjuman stage (Olenek River, Siberian Platform), scale bar 0.1 mm. H, *Semitreta* sp., PIN 4511/41, iceberg-type brachiopod shell of quasi-infaunal lingulate with highly conical ventral valve, inhabiting soft muddy substrate, Middle Cambrian, Marjuman stage (Kotuy River, Siberian Platform), scale bar 1 mm. I, *Batenevotreta formosa* Ushatinskaya, holotype PIN 4377/124, ventral valve, Middle Cambrian, Amgan Sladkie Koren'ya Formation (Batenevsky Ridge, Altay-Sayan Foldbelt), scale bar 1 mm. J, *Quadrisonia simplex* Koneva, Popov, and Ushatinskaya, PIN 4321/1, posterior part of ventral valve, Late Cambrian, Steptoean stage (Olenty-Shiderty Province, northeastern Kazakhstan), scale bar 0.5 mm.

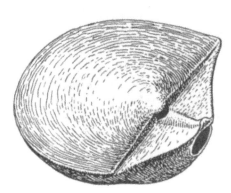

Figure 16.4 Reconstruction of the shell of the lingulate *Salanygolina*, possessing a large pedicle opening and high flattened pseudoarea on the ventral valve, Early Cambrian, Atdabanian Stage (Zavkhan Province, Mongolia).

conditions (Zhuravleva 1966; Wood et al. 1993). These genera, *Aldanotreta, Cryptotreta*, and *Dzunarzina*, together with some later *Salanygolina*, possessed a large pedicle opening and flattened high pseudoarea on the ventral valve. The latter feature might have provided additional support on substrates (figure 16.4). These forms were probably anchored to small pebbles and shell fragments, which were abundant in the vicinity of reefs.

Obolellids appeared in the Anabar-Sinsk Basin during the late Tommotian, and nisusiids developed there during the middle Atdabanian. Both groups retained an opening throughout life and were typical anchored forms. Obolellids were confined to interreef and reef habitats. They were common elements of reefal cryptic communities from the early Atdabanian (Kobluk and James 1979; Kobluk 1985). *Obolella* and other brachiopods attached at their posterior margin, with the shell opening into the cavity. Attachment was effected by a short stout pedicle, probably bearing papillae

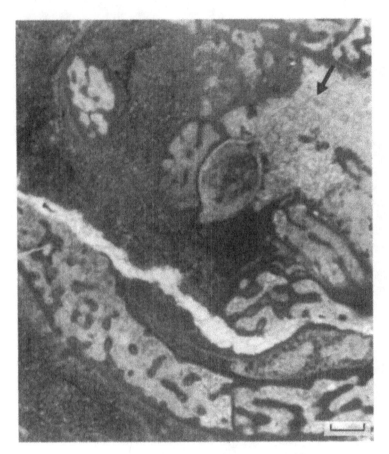

Figure 16.5 Cryptic cavity formed by archaeo-cyathan secondary skeleton in a calcimicrobial-archaeocyathan reef containing the calciate *Obolella* sp. attached by its rear, with the aperture opening into the cavity, × 10, thin section PIN 3848/710, Early Cambrian, Atdabanian Pestrotsvet Formation (Oi-Muran village, middle Lena River, Siberian Platform). The carbonate substrate is dissolved slightly where papillae rooted into it (*arrow*). *Source:* Photograph courtesy of Andrey Zhuravlev.

on its surface. Carbonate substrates are dissolved slightly where papillae rooted (figure 16.5). A similar feature is common among Recent attached brachiopods (Bromley and Surlyk 1973). Jackson et al. (1971) described Recent *Thecidellina* and *Aegyrotheca,* which are similar in size to Cambrian cryptic brachiopods, from coral reefs. Many of these forms anchor on the upper surface of reefal caves but are absent from the floors. Thus, the animals occurred well above the water-sediment interface, minimizing occlusion by mud. Such a strategy was perhaps exploited by Early Cambrian obolellids that inhabited calcimicrobial-archaeocyath reefs. *Obolella* shell pavements several meters long and 0.5–1.0 cm thick are preserved in middle Atdabanian calcareous-argillaceous mudstones. This might have resulted from local transport and redeposition of valves adjacent to a mass settlement of brachiopods. In the late Middle

Cambrian Bol'shoy Kitat Formation of Kuznetsky Alatau (Altay-Sayan Foldbelt), banks formed by densely spaced in situ *Diraphora* occur (Aksarina 1983). The brachiopods were anchored at their posterior margin to the originally muddy carbonate substrate.

Initially, Lingulida and Acrotretida were mainly restricted to Atdabanian siliciclastic sediments in higher latitudes. The earliest of them are found in thin limestones at Comley (British part of Avalonia) and from siltstones in the western Baltica (Keller and Rozanov 1979; Hinz 1987; Jendryka-Fuglewicz 1992). Simultaneously, or shortly after, during the late Atdabanian–early Botoman, lingulids and acrotretids appeared in argillaceous carbonate facies on the Siberian Platform (Perekhod and Krasnyy Porog formations) and Laurentia (Sekwi Formation) (Voronova et al. 1987; Astashkin et al. 1991) These regions were shallow basins with low water energies and slow submergence (Rushton 1974; Rozanov and Łydka 1987). The brachiopods were characterized by small (<5 mm), thin shells. An opening for the pedicle, by which the animals appeared to be anchored, had the form of either a foramen or a groove and was located near the apex (see figures 16.3D,E).

Clustered accumulations of very small (1–2 mm) lingulates occur in the late Early to early Middle Cambrian calcareous-argillaceous Kuonamka facies of the Siberian Platform. A single 200–300 g sample contains up to 200–300 well-preserved valves, and sometimes complete shells. In these clusters, species number is generally two or three and up to five. Acrotretids dominate and lingulids are less common. Extremely fine grain and homogeneous structure of the Kuonamka facies indicate calm conditions (Bakhturov et al. 1988). The abundance of cyanobacterial fossils indicates a relatively shallow depth within the photic zone, approximately 50–100 m, for the basin (Zhegallo et al. 1994). Dominantly soft silty substrate, and the presence in both the acrotretids and lingulids confined to this facies of a pedicle opening that functioned throughout life, suggest restriction of these brachiopods to algal thickets. These algae could be *Margaretia,* comprising abundant carbonaceous beds of the Kuonamka facies (Barskov and Zhuravlev 1988). The clustered distribution of brachiopod settlements might be related to sporadic occurrence of algal thickets. The Kuonamka facies (Sinsk and Kuonamka formations) accumulated in anoxic conditions (Zhuravlev and Wood 1996). Thus, attachment of lingulates to benthic algae allowed them to rise above the anaerobic bottom water layer. Such attached shells have been discovered in situ recently on *Margaretia* thalli from the Sinsk Formation (Ivantsov et al. 2000). On the other hand, hemerythrin molecules that are responsible for oxygen transport in brachiopod blood impart relatively low oxygen requirements. Some Recent brachiopods can survive periods of anoxia and are capable of both aerobic and anaerobic metabolism (Brunton 1982). In the Early Triassic, for instance, lingulids were typical of lower dysaerobic assemblages (Hallam 1994). Thus, abundant lingulate assemblages confined to anoxic strata of the Kuonamka facies were probably well adapted to dysaerobic-anaerobic conditions. Similarly, lingulates were common elements of the Late Cambrian Olenid community that existed in stagnant bottom conditions in

Figure 16.6 Calciate *Nisusia* sp. attached to the spicular sponge *Pirania muricata* Walcott, lectotype USNM 66459, Middle Cambrian, Amgan Stephen Formation, Burgess Shale (Mount Stephen, British Columbia, North America). Scale bar equals 1 cm. *Source:* Reprinted with permission from Rigby 1986: plate 20, figure 1.

Wales (McKerrow 1978). Relatively low metabolic rates in brachiopods were also significant for survival in such conditions (James et al. 1992).

The Middle Cambrian lingulate *Dictyonina* and the calciate *Nisusia* may have attached to the large spicular sponge *Pirania* from the Burgess Shale of Laurentia (Walcott 1920; Whittington 1980; Conway Morris 1982; Rigby 1986) (figure 16.6). The sponge skeletons were complete, and the sponges probably alive, when brachiopods attached to their spicules in order to capture higher, and thus stronger, currents. Conway Morris (1986) suggested brachiopod-sponge commensalism.

Free-Lying Forms

Some paterinids, *Micromitra* and *Paterina*, inhabiting both siliciclastic and siliciclastic-carbonate Early Cambrian substrates, did not have a separate pedicle opening. They possessed slightly convexo-convex shells with gaps between the valves in the area of the pseudodelthyrium. The possibility cannot be excluded that these brachiopods were attached only when young, as in later strophomenids. Kutorginids may also have been free-lying forms restricted to carbonate substrates. They had large shells (up to 2–3 cm) bearing concentric wrinkling and thickened posterior margins that allowed them to maintain a stable position on the sea floor.

Infaunal Brachiopods

For long it was accepted that the majority of Cambrian lingulids were infaunal forms inhabiting shallow nearshore conditions similar to their Recent representatives (e.g., Pel'man 1982). Indeed, both Recent genera, *Lingula* and *Glottidia,* are burrowing animals adapted to unstable intertidal environments. They have a set of features that provide good adaptation to life on shifting sands-silts in shallow conditions. First, they anchor by a pedicle whose distal end produces a sticky substance that binds loose substrate (Thayer and Steele-Petrovic 1975). Second, they have developed a more effective mechanism for protecting and cleaning the mantle cavity and lophophore from foreign particles than have other brachiopods (Chuang 1961). Nonetheless, Rudwick (1965), Gorjansky (1969), and Krause and Rowell (1975) doubted whether all Early Paleozoic lingulids were infaunal. Analysis of umbonal morphology in middle Late Cambrian lingulids from the Leningrad region (Baltica) revealed that they were typical epifaunal forms inhabiting the entire shelf (Popov et al. 1989). Most of them had pseudoareas projecting far from the rear margin of the valves. Such projections may have prevented movement of valves from causing dipping into the substrate. In addition, complete closing of the pedicle groove was sometimes observed in adolescent and gerontic forms. The very small size and thin shells of many Early Cambrian lingulids suggest that they were unable to burrow.

Recently, however, Jin et al. (1993) described Botoman lingulids in the Chengjiang fauna from fine siltstone in southern China. The posterior parts of shells and very long pedicles are preserved in these brachiopods (Burzin et al., this volume: figure 10.1:9). In Recent brachiopods, such a pedicle serves for attachment in deep burrowings. Nonetheless, L. Ye. Popov (pers. comm., 1996) believes that these lingulids were epibenthic, with a shell supported by a long pedicle, for these reasons: (1) the shell and the main part of the pedicle are invariably preserved on a bedding surface, and only the distal end of the pedicle is embedded in the sediment; (2) there are no bioturbation features in the rock; and (3) soft-bodied preservation suggests anaerobic conditions within the sediment and possibly in the lower part of the water column. On the contrary, Erdtmann et al. (1990) developed a scenario in which infaunal elements of the Chengjiang fauna migrated to the sediment surface during short temporal anoxic events. Some Cambrian lingulate burrows from the Botoman Bradore Formation (Labrador) were evidently from infaunal forms (Pemberton and Kobluk 1978).

Quasi-Infaunal Adaptations

The abundance of soft muddy substrates during the Middle and Late Cambrian possibly fostered the appearance of the quasi-infaunal lifestyle among brachiopods (Henderson 1974; Percival 1978). Brachiopods that lived in such conditions developed highly conical ventral valves turned to the substrate. As a result, the commissure was above the substrate surface during the entire life of the brachiopod, resembling an iceberg (see figure 16.3H). Thin spines on the acrothelid larval shell served as another soft substrate adaptation. Once settled, the larva used spines to keep its anterior and lateral margins slightly above the surface. The presence of spines on both valves allowed freshly settled animals to survive even being toppled onto their dorsal valves (see figure 16.3G) (Henderson 1974). During subsequent holoperipherical growth, the shell became subconical, and its anterior and lateral commissures were kept raised even when the shell was covered by a thin layer of mud. The spines might atrophy, only the tubercles were preserved on the surface of the larval shell, and the shell itself evolved a snowshoe morphology, to use a phrase from Rudwick (1970), for support.

Interstitial Lifestyle

Another infaunal adaptation, interstitial lifestyle, was common in some minute acrotretids inhabiting shallow turbulent conditions (Swedmark 1964; Bassett 1984). Acrotretids 1–2 mm or even less in length are often present in Middle Cambrian sandstones of Kazakhstan and the Altay Sayan Foldbelt. These dimensions are smaller than those of the sand grains around them, yet the shells are well preserved. It seems likely that these brachiopods were attached in shelters between sand grains or beneath larger shell fragments, as the Recent terebratulid *Gwinia capsula* is (Bassett 1984). Such habitats protected brachiopods from storm action. Through being buried by sediment, the brachiopods could be locked in their shelters, and the shells preserved.

Cemented Brachiopods

The formation of hardgrounds, which began at the end of the Middle Cambrian, provided new opportunities for brachiopod ecologic radiation. The earliest forms cemented to rigid substrates are observed among orthids in the late Middle Cambrian Mila Formation of northern Iran (Zhuravlev et al. 1996) and in the Late Cambrian Snowy Range Formation of Montana and Wyoming (Brett et al. 1983).

Possible Pseudoplanktic Brachiopods

The question of possible pseudoplanktic brachiopods has been discussed a number of times (e.g., Schuchert 1911; Ager 1962; Popov 1981, 1995; Pel'man 1982; Williams and Lockley 1983; Bassett 1984; Holmer 1989). It has been suggested that brachiopods might have attached to floating sargasso-type algae or, rarely, to other floating organisms. Discovery of clusters of minute shells in Ordovician organic-rich shales, sometimes together with graptolites, and the cosmopolitan occurrence of some brachiopods support such a view. Holmer (1989) scrutinized all arguments in favor of the existence of pseudoplanktic brachiopods in the Early Paleozoic and concluded that such a lifestyle was an exception to the rule. In most cases, small thin-shelled lingulates were anchored by a pedicle. Nonetheless, the pitted microsculpture of larval shells might increase buoyancy. Popov et al. (1982) discovered a pitted surface on the entire valves of the Late Ordovician thin-shelled genus *Paterula* and suggested that this genus had completely lost a benthic stage in its life cycle and had become planktic. However, Lenz (1993) described and illustrated several shells of *Paterula* (erroneously assigned to *Craniops*) clustered along the oscular margin of the distal end of sponges. Among Middle and Late Cambrian lingulids, both small thin-shelled *Fossuliella* and relatively large thick-shelled *Zhanatella* possessed shell surfaces bearing pitted microsculpture (see figure 16.3F). In addition, the latter has a well-developed pedicle groove and a thickened posterior (Pel'man et al. 1992). Thus, a feature such as pitted sculpture is not a good indicator of planktic life, but it is not necessary to overlook this feature completely.

MAJOR BRACHIOPOD BIOFACIES AND COMMUNITIES

Intense diversification of brachiopods occurred during the late Middle to early Late Cambrian; the radiation of the Acrotretida and Lingulida orders was especially pronounced. Lingulids were mainly restricted to siliciclastic facies. Recent features of these brachiopods, such as ability to bind loose substrate and presence of a lophophore protection mechanism, probably started to develop in the early stages of the evolution of the order. These features allowed them to occupy wide belts of mobile sandy substrate of Baltica, from which the first abundant occurrences of lingulids are known (Middle Cambrian Sablinka Formation). Such occurrences are dominated by one or two species (Popov et al. 1989). Popov and Khazanovich (1988) ascribed the growth of lingulate biomass at the end of the Cambrian to increased primary productivity in this region, which led to the formation of the first significant shell beds and, consequently, phosphorite deposits. Brachiopods continued to inhabit the inner shelf. Sometimes their shells are preserved in high-energy carbonate-arenaceous facies. Acrotretids in these conditions possessed a relatively large foramen, which probably provided space for a thick pedicle, and a posterior margin thickened by secondary shell lamellae. *Batenevotreta, Prototreta,* and *Erbotreta* exemplify such forms in the Amgan Stage of the Batenevsky Ridge, Altay Sayan Foldbelt (see figure 16.3I). At the same time, brachiopods were dispersed throughout relatively deep-water habitats, including the outer shelf and upper continental slope. Such deeper communities were widespread on the northern Siberian Platform (Tyussala and Eyra formations), in the

Maly Karatau Ridge (Aktas and Zhumabay formations), and in Greenland (Holm Dal Formation) (Zell and Rowell 1988; Pel'man et al. 1992; Ushatinskaya 1994). High diversity (seven to nine genera) and widespread acrotretid assemblages dominated these communities.

ENVIRONMENT, DIET, AND SKELETAL BIOMINERALIZATION

In contrast to other phyla, the Brachiopoda have two skeletal mineralogies, calcium carbonate and phosphate. Both of these elements used in skeleton formation, calcium and phosphorus, are required for many functions and were used by organisms long before mineralized skeletons appeared. The study of phosphatic *Glottidia* by Pan and Watabe (1988) revealed that ambient seawater is the principal source of Ca^{2+} in a brachiopod shell. Ca^{2+} ions enter the lophophore from the surroundings by diffusion. Phosphorus is transported with the food, which is the major source of PO_4^{3-} ions for shell formation.

The appearance of mineralized skeletons at the advent of the Cambrian was due to a unique combination of environmental conditions and the stage of development of the organic world (e.g., Lowenstam 1984). Supercontinent disruption and subsequent dispersal of its fragments allowed the formation of warm epicontinental seas in low latitudes and a new circulation pattern, which in turn influenced the redistribution of nutrients and other chemical substances (e.g., Cook and Shergold 1984; Rozanov 1984; Donnelly et al. 1990; Brasier 1991; Brasier and Lindsay, this volume). The central Siberian Platform was just such a warm shallow basin rich in diverse organisms (Rozanov et al. 1969; Rozanov and Zhuravlev 1992), among which, probably, were the ancestors of brachiopods. Increased phosphate concentrations, observed in the lower Tommotian of the Siberian Platform (Rozanov 1979), might influence phytoplankton productivity. Experiments with Recent cyanobacteria indicate that they store phosphate in volutin granules even under conditions of insignificant phosphate content in the ambient water (Gerasimenko et al. 1994). When the phosphate content increases, the cyanobacterial cells are almost completely infilled by it. It is possible to suggest that, at the beginning of the Cambrian, animals that consumed phytoplankton had their food enriched in phosphate. An improvement in PO_4^{3-} ion balance regulation became necessary and led to discharge of the surplus into the epithelium in a process similar to that of shell formation in Recent brachiopods. Paterinids were the first brachiopods with phosphatic skeletons. The second maximum in the diversification of brachiopods with phosphatic shells occurred in the late Atdabanian, when sediments also are characterized by increased phosphate content (Cook 1992). That was the very time when both acrotretids and lingulids appeared.

The same factors—increased nutrient content and related plankton productivity—may have caused lingulates to thrive during the Cambrian–Early Ordovician and may have delayed the diversification of calciates (see figure 16.2). Calciates are restricted to areas of limited food supply and stop feeding when phytoplankton concentrations are high (Thayer 1986; Rhodes and Thompson 1993). In addition, spirolophous lingulates can filter more effectively under higher particle concentrations than can plectolophous species (Rhodes and Thompson 1993). At the same time, more-turbid conditions probably favored brachiopods over bivalves (Steele-Petrovic 1975). If the plot of relative phosphorite abundance by Cook (1992; see Zhuravlev, this volume: figure 8.1E) to some extent reflects nutrient availability, then nutrient-rich conditions during the Cambrian–Early Ordovician would be more suitable for lingulates than for calciates. Decrease in nutrient levels in the Middle Ordovician provided better conditions for the diversification of calciates.

Acknowledgments. This paper is a contribution to IGCP Project 366, and I am grateful to its leaders for translation and editing of the manuscript. L. Ye. Popov is thanked for helpful review comments. A. D. McCracken (editor, *Palaeontographica Canadiana*), Geological Survey of Canada, and J. K. Rigby are acknowledged for kind permission to reproduce their illustration. This work was supported by the Russian Foundation for Basic Research Projects 00-04-484099 and 00-15-97764.

REFERENCES

Ager, D. V. 1962. The occurrence of pediculate brachiopods in soft sediments. *Geological Magazine* 99:184–186.

Aksarina, N. A. 1983. Nekotorye paleoekologo-tafonomicheskie nablyudeniya nad kembriyskimi brakhiopodami zapadnoy chasti Altae-Sayanskoy oblasti [Some paleoecologic-taphonomic observations on Cambrian brachiopods in the western part of the Altay Sayan Foldbelt]. In V. P. Udodov, ed., *Priroda i ekonomika Kuzbassa (Tezisy dokladov k predstoyashchey nauchnoy konferentsii)* [The nature and economics of Kuzbass (Abstracts of papers for the foregoing scientific conference)], pp. 95–97. Novokuznetsk: USSR Geographic Society.

Astashkin, V. A., T. V. Pegel', L. N. Repina, A. Yu. Rozanov, Yu. Ya. Shabanov, A. Yu. Zhuravlev, S. S. Sukhov, and V. M. Sundukov. 1991. *Cambrian System on the Siberian Platform.* International Union of Geological Sciences, Publication 27.

Atkins, D. 1960. The ciliary feeding mechanism of the Megathyridae (Brachiopoda) and the growth stages of the lophophore. *Journal of the Marine Biological Association of the United Kingdom* 39:459–479.

Bakhturov, S. F., V. M. Evtushenko, and V. S. Pereladov. 1988. *Kuonamskaya bituminoznaya karbonatno-slantsevaya formatsiya* [Kuonamka bituminous carbonate-shelly formation]. *Trudy, Institut geologii i geofiziki, Sibirskoe otdelenie, Akademiya nauk SSSR* 671:1–152.

Barskov, I. S. and A. Yu. Zhuravlev. 1988. Myagkotelye organizmy kembriya Sibirskoy platformy [Soft-bodied organisms in the Cambrian of the Siberian Platform]. *Paleontologicheskiy zhurnal* 1988 (1):3–9.

Bassett, M. G. 1984. Life strategies of Silurian brachiopods. *Special Papers in Palaeontology* 32:237–263.

Biernat, G. and A. Williams. 1970. Ultrastructure of the protegulum of some acrotretide brachiopods. *Palaeontology* 13:491–502.

Brasier, M. D. 1991. Nutrient flux and the evolutionary explosion across the Precambrian-Cambrian boundary interval. *Historical Biology* 5:85–93.

Brett, C. E., W. D. Liddell, and K. L. Derstler. 1983. Late Cambrian hard substrate communities from Montana/Wyoming: The oldest known hardground encrusters. *Lethaia* 16:281–289.

Bromley, R. G. and F. Surlyk. 1973. Borings produced by brachiopod pedicles, fossil and Recent. *Lethaia* 6:349–365.

Brunton, C. H. C. 1982. The functional morphology and paleoecology of the Dinantian brachiopod *Levitusia*. *Lethaia* 15:149–167.

Chuang, S. H. 1959. The structure and function of the alimentary canal in *Lingula unguis* (L.) (Brachiopoda). *Proceedings of the Zoological Society of London* 132:283–311.

Chuang, S. H. 1961. Description of *Lingula parva* Smith (Brachiopoda) from the coast of tropical West Africa. *Atlantida Reports* 6:161–168.

Conway Morris, S., ed. 1982. *Atlas of the Burgess Shale*. London: Palaeontological Association.

Conway Morris, S. 1986. The community structure of the Middle Cambrian Phyllopod Bed (Burgess Shale). *Palaeontology* 29:423–467.

Cook, P. J. 1992. Phosphogenesis around the Proterozoic-Phanerozoic transition. *Journal of the Geological Society, London* 149:615–620.

Cook, P. J. and J. H. Shergold. 1984. Phosphorus, phosphorites, and skeletal evolution at the Precambrian-Cambrian boundary. *Nature* 308:231–236.

Donnelly, T. H., J. H. Shergold, P. N. Southgate, and C. J. Barnes. 1990. Events lead-ing to global phosphogenesis around the Proterozoic/Cambrian boundary. In A. J. Notholt and J. Jarvis, eds., *Phosphorite Research and Development*, pp. 273–287. Geological Society Special Publication 52.

Erdtmann, B.-D., P. Huttel, and J. Chen. 1990. Depositional environment and taphonomy of the Lower Cambrian soft-bodied fauna at Chengjiang, Yunnan Province, China. *Third International Symposium on the Cambrian System, Novosibirsk, USSR, Abstracts*, p. 91. Novosibirsk: Institute of Geology and Geophysics, Siberian Branch, USSR Academy of Sciences.

Gerasimenko, L. M., I. V. Goncharova, G. A. Zavarzin, E. A. Zhegallo, I. V. Pochtareva, A. Yu. Rozanov, and G. T. Ushatinskaya. 1994. Dinamika vysvobozhdeniya i pogloshcheniya phosphora cyanobacteriyami [Dynamics of release and absorption of phosphorus by cyanobacteria]. In A. Yu. Rozanov and M. A. Semikhatov, eds., *Ekosistemnye perestroyki i evolyutsiya biosfery, vypusk 1* [Ecosystem restructures and the evolution of biosphere, issue 1], pp. 348–353. Moscow: Nedra.

Gorjansky, V. Yu. 1969. *Bezzamkovye brakhiopody kembriyskikh i ordovikskikh otlozheniy Russkoy platformy* [Inarticulate brachiopods of the Cambrian and Ordovician strata on the Russian Platform]. Leningrad: Nedra.

Gorjansky, V. Yu. and L. Ye. Popov. 1985. Morfologiya, systematicheskoe polozhenie, i proiskhozhdenie bezzamkovykh brakhiopod s karbonatnoy rakovinoy [Morphology, systematic position, and origin of inarticulate brachiopods with a calcareous shell]. *Paleontologicheskiy zhurnal* 1985 (3):3–14.

Hallam, A. 1994. The earliest Triassic as an anoxic event, and its relationship to the end-Palaeozoic mass extinction. *Canadian Society of the Petroleum Geologists, Memoir* 17, pp. 797–804.

Henderson, R. A. 1974. Shell adaptation in

acrothelid brachiopods to settlement on a soft substrate. *Lethaia* 7:57–62.

Hinz, I. 1987. The Lower Cambrian microfauna of Comley and Rushton, Shropshire/England. *Palaeontographica A* 196: 1–100.

Holmer, L. E. 1989. Middle Ordovician phosphatic brachiopods from Västergötland and Dalarma, Sweden. *Fossils and Strata* 26:1–172.

Huxley, T. H. 1869. *An Introduction to the Classification of Animals.* London: John Churchill and Sons.

Ivanova, E. A. 1962. *Ekologiya i razvitie brakhiopod silura i devona Kuznetskogo, Minusinskogo i Tuvinskogo basseynov* [Ecology and development of the Silurian and Devonian brachiopod in the Kuznetsk, Minusinsk, and Tuva basins]. *Trudy, Paleontologicheskiy institut, Akademiya nauk SSSR* 87:1–150.

Ivantsov, A. Yu., A. Yu. Zhuravlev, V. A. Krassilov, A. V. Leguta, L. M. Melnikova, L. N. Repina, A. Urbanek, G. T. Ushatinskaya, and Ya. E. Malakhovskaya. 2000. *Unikal'nye sinskie mestonakhozhdeniya rannekembriyskikh organizmov (Sibirskaya platforma)* [Extraordinary Sinsk localities of Early Cambrian organisms (Siberian Platform)]. Moscow: Paleontological Institute.

Jablonski, D. 1986. Larval ecology and macroevolution in marine invertebrates. *Bulletin of Marine Science* 39:565–587.

Jackson, J. B. C., T. F. Goreau, and W. D. Hartman. 1971. Recent brachiopod-coralline-sponge communities and their paleoecological significance. *Science* 173:623–625.

James, M. A., A. D. Ansell, M. J. Collins, G. B. Curry, L. S. Peck, and M. C. Rhodes. 1992. Biology of living brachiopods. *Advances in Marine Biology* 28:175–387.

Jendryka-Fugłewicz, B. 1992. Analiza porównawcza ramienionogów z utworów kambru Gór Świętokrzyskich i platformy pre-kambryjskiej w Polsce. *Przegląd Geologiczny* 3:150–155.

Jin, Y., X. Hou, and H. Wang. 1993. Lower Cambrian pediculate lingulids from Yunnan, China. *Journal of Paleontology* 67: 788–798.

Jope, M. 1977. Brachiopod shell proteins: Their functions and taxonomic significance. *American Zoologist* 17:133–140.

Jope, M. 1986. Evolution of the Brachiopoda: The molecular approach. In P. B. R. Racheboeuf and C. C. Emig, eds., *Les Brachiopod fossiles et actuels,* pp. 103–111. Biostratigraphie du Paléozoique 4.

Keller, B. M. and A. Yu. Rozanov, eds. 1979. *Stratigrafiya verkhnedokembriyskikh i kembriyskikh otlozheniy zapada Vostochno-Evropeyskoy platformy* [Stratigraphy of the upper Precambrian and Cambrian strata on the East European Platform]. Moscow: Nauka.

Kobluk, D. R. 1985. Biota preserved within cavities in Cambrian *Epiphyton* mounds, upper Shady Dolomite, southern Virginia. *Journal of Paleontology* 59:1158–1172.

Kobluk, D. R. and N. P. James. 1979. Cavity-dwelling organisms in Lower Cambrian patch reefs from southern Labrador. *Lethaia* 12:193–218.

Krause, F. F. and A. J. Rowell. 1975. Distribution and systematics of the inarticulate brachiopods of the Ordovician carbonate mud mound of Meiklejohn Peak, Nevada. *University of Kansas Paleontological Contributions* 61:1–74.

Lenz, A. C. 1993. A Silurian sponge-inarticulate brachiopod life? association. *Journal of Paleontology* 67:138–139.

Lochman, C. 1956. Stratigraphy, paleontology, and paleogeography of the *Elliptocephala asaphoides* strata in Cambridge and Hoosick quadrangles, New York. *Bulletin of the Geological Society of America* 67:1331–1396.

Lowenstam, H. A. 1984. Biomineralization processes and products and the evolution of biomineralization. In *Palaeontology*, vol. 2 of *Proceedings of the 27th International Geological Congress, Moscow, 1984*, pp. 79–95. Utrecht: VNU Science Press.

Malakhov, V. V. 1976. Nekotorye stadii razvitiya zamkovoy brachiopody *Cnismatocentrum sakhalinensis parvum* i problema evolyutsii sposoba zakladki celomitcheskoy mesodermy [Some ontogenic stages of the articulate brachiopod *Cnismatocentrum sakhalinensis parvum* and the problem of evolution in the coelomic mesoderm insertion]. *Zoologicheskiy zhurnal* 55:66–75.

Malakhov, V. V. 1983. Stroenie lichinok zamkovoy brachiopody *Cnismatocentrum sakhalinensis parvum* [The larval structure of the articulate brachiopod *Cnismatocentrum sakhalinensis parvum*]. In A. V. Ivanov, ed., *Evolyutsionnaya morfologiya bespozvonochnykh* [Evolutionary morphology of invertebrates], pp. 147–155. Leningrad: Nauka.

McCammon, H. M. 1969. The food of articulate brachiopods. *Journal of Paleontology* 43:976–985.

McKerrow, W. S., ed. 1978. *The Ecology of Fossils*. Cambridge: MIT Press.

Nielsen, C. 1991. The development of the brachiopod *Crania* (*Neocrania*) *anomala* (O. F. Müller) and its phylogenetic significance. *Acta Zoologica, Stockholm* 33:187–213.

Pan, C.-M. and N. Watabe. 1988. Uptake and transport of shell minerals in *Glottidia pyramidata* Stimpson (Brachiopoda; Inarticulata). *Journal of Experimental Marine Biology and Ecology* 118:257–268.

Pel'man, Yu. L. 1982. Paleobiotsenoticheskie gruppirovki fauny v srednekembriyskikh domanikoidnikh otlozheniyakh kuonamskoy svity (Sibirskaya platforma, r. Muna) [Paleobiocoenotic faunal groups in the Middle Cambrian domanicoid sediments of the Kuonamka Formation (Siberian Platform, Muna River)]. *Trudy, Institut geologii i geofiziki, Sibirskoe otdelenie, Akademiya nauk SSSR* 510:60–74.

Pel'man, Yu. L., N. A. Aksarina, S. P. Koneva, L. Ye. Popov, L. P. Sobolev, and G. T. Ushatinskaya. 1992. *Drevneyshie brakhiopody territorii Severnoy Evrazii* [Oldest brachiopods from the territory of northern Eurasia]. Novosibirsk: United Institute of Geology, Geophysics, and Mineralogy, Siberian Branch, Russian Academy of Sciences.

Pemberton, S. G. and D. R. Kobluk. 1978. Oldest known brachiopod burrow: The Lower Cambrian of Labrador. *Canadian Journal of Earth Sciences* 15:1385–1389.

Percival, I. G. 1978. Inarticulate brachiopods from the Late Ordovician of New South Wales, and their palaeoecological significance. *Alcheringa* 2:117–142.

Popov, L. Ye. 1981. Pervaya nakhodka mikroskopicheskikh brakhiopod semeystva Acrotretidae v silure Estonii [The first find of microscopic brachiopods of the family Acrotretidae in the Silurian of Estonia]. *Eesti NSV Teaduste Akademia Toimetised, Geologia* 30:34–41.

Popov, L. Ye. 1995. Major events in the evolution of the Ordovician linguloids. *Paleontologicheskiy Journal* 29 (4):134–141.

Popov, L. Ye. and K. K. Khazanovich. 1988. Znachenie paleoekologii i tafonomii bezzamkovykh brakhiopod dlya rasshifrovki zakonomernostey formirovaniya zalezhey biogennykh fosforitov Pribaltiki [Significance of the inarticulate brachiopod paleoecology and taphonomy for the deciphering of regularities in the formation of biogenic phosphorite deposits in the Baltic region]. In D. L. Kaljo, ed., *Problemy geologii fosforitov: Tezisy dokladov i putevoditel' VI Vsesoyuznogo soveshchaniya, Tallin, 18–21 aprelya 1988 g.* [Problems in phosphorite geology: Abstracts of papers and guidebook of the sixth All-Union Meeting, Tallin, 18–21 April 1988], p. 88. Tallin: Estonian SSR Academy of Sciences.

Popov, L. Ye. and G. T. Ushatinskaya. 1992. Uskorenie razvitiya foramena v ontogeneze u kembriyskikh akrotretid (brakhiopody) [Acceleration of the foramen development in the Cambrian acrotretid (brachiopod) ontogeny]. *Paleontologicheskiy zhurnal* 1992 (4):76–78.

Popov, L. Ye., O. N. Zezina, and J. Nolvak. 1982. Mikrostruktura apikal'noy chasti rakoviny bezzamkovykh brakhiopod i ee ekologicheskoe znachenie [Microstructure of the apical part in the inarticulate brachiopod shell and its ecological significance]. *Byulleten' Moskovskogo obshchestva ispytateley prirody, Otdel biologicheskiy* 87: 94–104.

Popov, L. Ye., K. K. Khazanovich, N. G. Borovko, S. P. Sergeeva, and R. F. Sobolevskaya. 1989. *Opornye razrezy i stratigrafiya kembro-ordovikskoy fosforitonosnoy obolovoy tolshchi na severo-zapade Russkoy platformy* [Reference sections and stratigraphy of the Cambrian-Ordovician phosphate-bearing *Obolus* Unit on the northwest of the Russian Platform]. Leningrad: Nauka.

Popov, L. Ye., M. C. Bassett, L. E. Holmer, and J. Laurie. 1993. Phylogenetic analysis of higher taxa of Brachiopoda. *Lethaia* 26: 1–5.

Rhodes, M. C. and R. J. Thompson. 1993. Comparative physiology of suspension-feeding in living brachiopods and bivalves: Evolutionary implications. *Paleobiology* 19:322–334.

Rigby, J. K. 1986. Sponges of the Burgess Shale (Middle Cambrian), British Columbia. *Palaeontographica Canadiana* 2:1–105.

Rowell, A. J. 1986. The distribution and inferred larval dispersal of *Rhondellina dorei*: A new Cambrian brachiopod (Acrotretida). *Journal of Paleontology* 60:1056–1065.

Rowell, A. J. and F. F. Krause. 1973. Habitat diversity in the Acrotretacea (Brachiopoda, Inarticulata). *Journal of Paleontology* 47:791–800.

Rozanov, A. Yu. 1979. Nekotorye problemy izucheniya drevneyshikh skeletnykh organizmov [Some problems in the study of the oldest skeletal organisms]. *Byulleten' Moskovskogo obshchestva ispytateley prirody, Otdel geologicheskiy* 54 (3):62–9.

Rozanov, A. Yu. 1984. Some aspects of studies on bio- and paleogeography of the Cambrian. In *Palaeontology*, vol. 2 of *Proceedings of the 27th International Geological Congress, Moscow, 1984*, pp. 143–157. Utrecht: VNU Science Press.

Rozanov, A. Yu. and K. Łydka, eds. 1987. *Palaeogeography and Lithology of the Vendian and Cambrian of the Western East-European Platform*. Warsaw: Wydawnictwa Geologiczne.

Rozanov, A. Yu. and A. Yu. Zhuravlev. 1992. Lower Cambrian fossil records of the Soviet Union. In J. H. Lipps and P. W. Signor, eds., *Origin and Early Evolution of the Metazoa*, pp. 205–281. New York: Plenum Press.

Rozanov, A. Yu., V. V. Missarzhevskiy, N. A. Volkova, L. G. Voronova, I. N. Krylov, B. M. Keller, I. K. Korolyuk, K. Lendzion, R. Michniak, N. G. Pykhova, and A. D. Sidorov. 1969. *Tommotskiy yarus i problema nizhney granitsy kembriya* [The Tommotian Stage and the Cambrian lower boundary problem]. *Trudy, Geologitcheskiy institit, Akademiya nauk SSSR* 206:1–380.

Rudwick, M. J. S. 1965. Ecology and paleoecology. In R. C. Moore, ed., *Brachiopoda*, part H of *Treatise on Invertebrate Palaeontology*, pp. H199–H214. Lawrence, Kans.: University of Kansas Press and Geological Society of America.

Rudwick, M. J. S. 1970. *Living and Fossil Brachiopods*. London: Hutchinson.

Rushton, A. W. A. 1974. The Cambrian of Wales and England. In C. H. Holland, ed., *The Cambrian of the British Islands, Norden and Spitsbergen,* pp. 43–123. London: Wiley Interscience.

Sarycheva, T. G., ed. 1960. Tip Brachiopoda. Brakhiopody [Phylum Brachiopoda. Brachiopods]. In Yu. A. Orlov, ed., *Osnovy paleontologii: Mshanki, brakhiopody. Prilozhenie: foronidy* [Principles of paleontology: Bryozoans, brachiopods. Supplement: Phoronids], pp. 115–324. Moscow: Izdatel'stvo Akademii nauk SSSR.

Schuchert, C. 1911. Paleogeographic and geologic significance of Recent Brachiopoda. *Geological Society of America Bulletin* 22: 258–275.

Steele-Petrovic, U. M. 1975. An explanation for the tolerance of brachiopods and relative intolerance of filter-feeding bivalves for soft muddy bottoms. *Journal of Paleontology* 49:552–556.

Swedmark, B. 1964. The interstitial fauna in marine sand. *Biological Reviews* 39:1–42.

Thayer, C. W. 1986. Are brachiopods better than bivalves? Mechanisms of turbidity tolerance and their interactions with feeding in articulates. *Paleobiology* 12:161–174.

Thayer, C. W. and H. M. Steele-Petrovic. 1975. Burrowing of the lingulid brachiopod *Glottidia piramidata:* Its ecologic and paleoecologic significance. *Lethaia* 8:209–221.

Ushatinskaya, G. T. 1990. Mikrostruktura i sekretsiya rakovini u brakhiopod otryada Acrotretida [Shell microstructure and secretion among brachiopods of the Acrotretida order]. *Paleontologicheskiy zhurnal* 1990 (1):55–65.

Ushatinskaya, G. T. 1994. Novye sredneverkhnekembriyskie akrotretidy (brakhiopody) severa Sibirskoy platformy [New Middle-Late Cambrian acrotretids (brachiopods) from the north of the Siberian Platform]. *Paleontologicheskiy zhurnal* 1994 (4):38–54.

Voronova, L. G., N. A. Drozdova, N. V. Esakova, E. A. Zhegallo, A. Yu. Zhuravlev, A. Yu. Rozanov, T. A. Sayutina, and G. T. Ushatinskaya. 1987. Iskopaemye nizhnego kembriya gor Makkenzi (Kanada) [Lower Cambrian fossils of the Mackenzie Mountains (Canada)]. *Trudy, Paleontologicheskiy institut, Akademiya nauk SSSR* 224: 1–88.

Walcott, C. D. 1920. Middle Cambrian Spongiae. *Smithsonian Miscellaneous Collections* 67:261–364.

Whittington, H. B. 1980. The significance of the fauna of the Burgess Shale, Middle Cambrian, British Columbia. *Proceedings of the Geological Association* 91:127–148.

Williams, A. 1968. A history of skeletal secretion among articulate brachiopods. *Lethaia* 1:268–287.

Williams, A. and G. B. Curry. 1991. The microarchitecture of some Acrotretidae brachiopods. In D. I. MacKinnon, D. E. Lee, and J. D. Campbell, eds., *Brachiopods Through Time,* pp. 133–145. Rotterdam: Balkema.

Williams, A. and L. E. Holmer. 1992. Ornamentation and shell structure of acrotretoid brachiopods. *Palaeontology* 35:657–692.

Williams, A. and A. J. Rowell. 1965. Classification. In R. C. Moore, ed., *Brachiopoda,* part H of *Treatise on Invertebrate Palaeontology,* pp. H214–H234. Lawrence, Kansas: University of Kansas Press and Geological Society of America.

Williams, A., S. J. Carlson, C. H. C. Brunton, L. E. Holmer, and L. Popov. 1996. A supraordinal classification of the Brachiopoda. *Philosophical Transactions of the Royal Society of London B* 351:1171–1193.

Williams, S. H. and C. M. Lockley. 1983. Ordovician inarticulate brachiopods from graptolitic shales at Dob's Linn, Scotland; Their morphology and significance. *Journal of Paleontology* 57:391–400.

Wright, A. D. and A. E. McClean. 1991. Microbrachiopods and the end-Ordovician event. *Historical Biology* 5:123–129.

Wood, R., A. Yu. Zhuravlev, and A. Chimed Tseren. 1993. The ecology of Lower Cambrian buildups from Zuune Arts, Mongolia: Implications for early metazoan reef evolution. *Sedimentology* 40:829–858.

Zell, M. G. and A. J. Rowell. 1988. Brachiopods of the Holm Dal Formation (late Middle Cambrian), central North Greenland. In J. S. Peel, ed., *Stratigraphy and Palaeontology of the Holm Dal Formation (late Middle Cambrian), central North Greenland. Meddelelser om Grønland, Geoscience* 20: 119–144.

Zezina, O. N. 1976. *Ekologiya i rasprostranenie sovremennykh brakhiopod* [Ecology and distribution of Recent brachiopods]. Moscow: Nauka.

Zhegallo, E. A., A. G. Zamiraylova, and Yu. N. Zanin. 1994. Mikroorganizmy v sostave porod kuonamskoy svity nizhnegosrednego kembriya Sibirskoy platformy (r. Molodo) [Microorganisms in the Lower-Middle Cambrian Kuonamka Formation rocks on the Siberian Platform (Molodo River)]. *Litologiya i poleznye iskopaemye* 1994 (5):123–127.

Zhuravlev, A. Yu. and R. A. Wood. 1996. Anoxia as the case of the mid-Early Cambrian (Botomian) extinction event. *Geology* 24:311–314.

Zhuravlev, A. Yu., B. Hamdi, and P. D. Kruse. 1996. IGCP 366: Ecological aspects of the Cambrian radiation—field meeting. *Episodes* 19:136–137.

Zhuravleva, I. T. 1966. Rannekembriyskie organogennye postroyki na territorii Sibirskoy platformy (Early Cambrian organogenous buildups on the territory of the Siberian Platform). In R. T. Hecker, ed., *Organizm i sreda v geologicheskom proshlom* [Organism and the environment in the geological past], pp. 61–84. Moscow: Nauka.

CHAPTER SEVENTEEN

Nigel C. Hughes

Ecologic Evolution of Cambrian Trilobites

*Skeletonized Cambrian trilobites are both varied and abundant and provide poten-
tial proxies for understanding the evolution of nonskeletonized arthropod groups.
Soft- and hard-part morphology suggests that Cambrian Trilobita pursued a variety
of feeding habits, ranging from predator-scavenger activity to sediment ingesting and
suspension feeding. They occupied habitats ranging from infaunal to probably pelagic
and lived in ecosystems that were structured in a manner comparable to those of
marine habitats today. The range of ecologic diversity among skeletonized Cambrian
trilobites is similar to that exhibited by nonskeletonized Cambrian arthropods. Data
on taxonomic, morphologic, and size diversity, in combination with information
about abundance and occurrence, suggest that considerable ecologic diversity was established by the appearance of trilobites in the fossil record. Species richness and the
absolute abundance of individuals increased during the remainder of the Cambrian,
but in at least some biogeographic provinces the rate of morphologic diversification
was constrained after the Early Cambrian. This constraint may have been related to
the demise of carnivorous redlichiid trilobites and the radiation of primitive libristomate trilobites with a primary consumption feeding mode. Many of the phylogenetic
and ecologic components of Ordovician trilobite communities appeared no later than
the Middle Cambrian but did not rise to dominance until the establishment of the
Paleozoic fauna.*

THE BIOMASS OF TRILOBITES in scientific collections far exceeds that of all other
Cambrian metazoans put together. This fact reflects the volumetric and taxonomic
abundance of trilobites in a wide range of Cambrian sediments, their intricate and la-
bile morphology, and their occurrence throughout the majority of Cambrian time.
These attributes have given the group unrivaled utility as zonal fossils in Cambrian
strata, and as the principal faunal element used to assess Cambrian paleobiogeogra-

phy. Paradoxically, while trilobites serve as the timekeepers by which we gauge the *ecologic* evolution of other Cambrian metazoans, the ecology of Cambrian trilobites remains poorly resolved. This chapter summarizes current knowledge of the ecology of Cambrian trilobite species and their place in Cambrian communities, outlines the difficulties in making paleoecologic inferences in this group, explores a number of indirect measures of ecologic diversity, and presents an overview of the *ecologic* evolution of these fossils.

Recent interest in the Cambrian radiation has been fueled by the redescription of Cambrian soft-bodied organisms and by the discovery of new ones. Advances in arthropod systematics have constrained the taxonomic position of the Trilobita (e.g., Wheeler et al. 1993; Wills et al. 1994). The trilobites are a monophyletic constituent (Fortey and Whittington 1989) of a larger clade of arachnate arthropods that were common in Cambrian marine environments and that exceeded other Cambrian arthropods clades in terms of taxic diversity (at least within individual Burgess Shale–type Lagerstätten). Furthermore, schizoramid arthropods (arachnates + crustaceanomorphs + marrellomorphs) apparently dominated Cambrian communities in terms of numbers of taxa, individuals, and biovolume (Conway Morris 1986). Trilobites are thus important not only in their own right, but also as possible proxies for understanding patterns of *ecologic* evolution in other soft-bodied Cambrian arthropods, which played a dominant role in Cambrian ecologies.

Despite the good fossil record of trilobites, interpretation of their life habits is often difficult. We are unable to use modern representatives for direct insights into the ecology of Cambrian relatives because trilobites are extinct. Although extant arachnate horseshoe crabs can provide some pointers about possible trilobite lifestyles, this information does little to resolve the *ecologic* significance of particular trilobite morphotypes or characteristic features. Hence knowledge of Cambrian trilobite autecology is based on case studies of particularly well-preserved or morphologically distinctive trilobites.

INSIGHTS INTO THE AUTECOLOGY OF CAMBRIAN TRILOBITES

Although trilobite exoskeletal morphology is not intimately linked to feeding strategy, as in some Paleozoic groups (e.g., Wagner 1995), major *morphologic* differences likely imply different ecologies, and many aspects of trilobite form and habits bear on the ecologic evolution of the group. These include sensory systems (e.g., Clarkson 1973), locomotion (e.g., Whittington 1980), molting behaviors (e.g., McNamara 1986; Whittington 1990), and reproductive strategies (e.g., Hughes and Fortey 1995). Of the "economic" aspects of ecology (sensu Eldredge 1989), inferences on feeding behavior are most important because these may indicate the role of trilobites in the trophic structure of Cambrian marine communities and the habitats that they occupied. Direct evidence for feeding strategies comes from appendage morphology, known in some exceptionally preserved faunas. Indirect indicators such as exoskeletal shape, trace fossils, and functional modeling provide additional information.

Direct Evidence for Feeding: Exceptionally Preserved Material

Soft-part preservation in Cambrian deposits has permitted reconstructions of the principal external features of several taxa, including (1) Early Cambrian redlichiids *Eoredlichia intermedia* and *Yunnanocephalus yunnanensis* (Shu et al. 1995; Ramsköld and Edgecombe 1996); (2) Middle Cambrian corynexochides *Olenoides serratus* and *Kootenia burgessensis* (Whittington 1975; Whittington 1980) and soft-bodied nectaspid trilobites *Naraoia compacta* (Whittington 1977) and *Tegopelte gigas* (Whittington 1985); and (3) the Late Cambrian agnostid *Agnostus pisiformis* (Müller and Walossek 1987). These studies, and others of post-Cambrian trilobites, suggest that trilobites lacked specialized feeding appendages. All trilobites apparently fed by passing food to the midline and then moving it forward to the mouth, which in *A. pisiformis* was posteriorly directed. This movement was achieved by rotating the basis, the plate to which both endopodites and exopodites are attached, in the horizontal plane. Hence, locomotion and feeding were combined processes, as in other arachnomorphs (Müller and Walossek 1987).

Naraoia compacta and *O. serratus* possessed spinose gnathobases on the basis that likely shredded food. These trilobites also had spinose endopods and are interpreted as predators or scavengers on benthic organisms (Whittington 1975; Whittington 1980; Briggs and Whittington 1985). *Agnostus pisiformis* also possessed a spinose gnathobase (Müller and Walossek 1987), but because adult *Agnostus* was so much smaller than *Naraoia* or *Olenoides,* the type of food particles macerated by *Agnostus* must have differed. Based on the structure of the thorax and of the appendages, Müller and Walossek (1987) concluded that *A. pisiformis* lived partially enrolled and fed by collecting suspended detrital particles while actively swimming or by processing material at the sea floor.

Exceptional preservation of gut morphology in Late Cambrian *Pterocephalia* from British Columbia (Chatterton et al. 1994) provides details of both the alimentary canal and the food source. The composition of the gut contents suggests that this trilobite was a deposit feeder and that it ingested fine-grained sediment. Similar structures have been reported in *Eoredlichia,* and the putative absence of spines on the endopods of this animal (Shu et al. 1995) might suggest that the food particles ingested were small. Further investigations of the limb structure in *Eoredlichia,* however, suggest a level of endopod spinosity comparable to that of *Naraoia* (Ramsköld and Edgecombe 1996). This at least suggests that food particles handled by *Eoredlichia* were larger. Negative allometry of the hypostome in *Eoredlichia* with respect to overall size supports the idea of a relatively small food particle size throughout growth. In conclusion, exceptionally preserved material indicates a variety of feeding strategies among Cambrian trilobites ranging from predator-scavenger activity to sediment ingesting and suspension feeding.

Recent analyses have shown that the soft-bodied forms *Naraoia* and *Tegopelte* may not be the closest relatives of skeletonized trilobites or of each other (Edgecombe and Ramsköld 1999). Additional discoveries of anatomically disparate Early Cambrian trilobite-like arachnates (e.g., Ivantsov 1999) further strengthen the impression of broad morphologic and, by proxy, ecologic diversity among Early Cambrian arachnates.

Indirect Evidence for Feeding

The Generalized Trilobite Body Plan

Although Cambrian trilobites displayed a wide variety of form, features general to their morphology provide broad indicators of life habits. On the basis of functional design, analogy with living arthropods, and homology with extant arachnates, the generalized body plan common to most trilobites, consisting of a rigid dorsal exoskeleton with eyes perched on the dorsal surface and homopodous walking legs, suggests a vagile benthic or nektobenthic life. Marked departures from this basic morphology suggest alternative lifestyles.

Specialized Morphologies and "Morphotypes"

In some cases, more-detailed inferences on ecology can be deduced from exoskeletal morphology. Fortey (1985), using explicit criteria based on occurrence, analogy, and functional morphology, presented strong arguments for pelagic life habits among some Ordovician trilobites. The convergence of a set of *morphologic* and occurrence features (particularly related to the form of the eye) among members of several different clades permitted the recognition of a generalized pelagic trilobite morphotype and the recognition of specializations within this broad habit. Fortey (1985:227) also suggested a candidate Cambrian pelagic morphotype, exemplified by the Late Cambrian primitive libristomate *Irvingella*. This trilobite had elongated eyes, a relatively wide axis (permitting the attachment of large muscles), spinose posterior thoracic pleurae, and a distribution spanning a wide range of lithofacies and paleocontinents. This generalized morphotype and a similarly widespread distribution were found in the Middle Cambrian redlichiid *Centropleura* and the latest Cambrian olenid *Jujuyaspis,* and each of these forms may have been pelagic in adult life. However, as the eye structure of these animals is poorly known, and the functional significance of the extended pleurae unclear, the case for a pelagic habit remains incomplete.

An alternative example of a derived, specialized morphology is found in several Late Cambrian trilobites that are characterized by an inflated and effaced cephalon with small eyes, angular articulation of cephalon and thorax, and postcephalic segments with wide axes. This morphotype is epitomized by *Stenopilus pronus* and is interpreted to be the result of a shallow infaunal habit (Stitt 1976). Both the overall form of the animal, and minor modifications such as the surface sculpture, suggest that *Stenopilus* occupied the sediment by adopting the bumastoid stance (Fortey 1986), with the cephalon resting horizontally on the sediment surface, and the thorax and pygidium extending vertically down. Trilobites adopting this morphology are thought to have been suspension feeders (Stitt 1976; Westrop 1983), although there is no appendage evidence to support this feeding mode. This is a case in which the morphology of the animal is modified such that its life mode can be directly inferred from functional morphology. Unfortunately, such cases are rare among Cambrian trilobites.

The nature of attachment of the hypostome to the remainder of the cephalon may provide a feature of importance in interpreting broad feeding habits for many trilobites (Fortey 1990). Natant or "floating" hypostomes, which are not attached by calcified exoskeleton to the remainder of the dorsal shield, show *morphologic* conservatism through the Cambrian and beyond. Based on the style of attachment, small size, and evolutionary conservatism, Fortey (1990:553) suggested that natant hypostomes characterize trilobites that consumed small organic particles extracted by the gnathobases or that directly ingested sediment. Conterminant trilobites, with hypostomes attached to the remainder of the exoskeleton, display a wider variety of hypostomal forms, some of which may have been specialized for processing larger food items, including prey. Evidence for this interpretation includes the greater strength of the buttressed hypostome in conterminant forms, and the presence of special adaptations such as posterior forks on the hypostomes (in post-Cambrian forms) that may have assisted in food maceration. The recognition of these two basic feeding types among Cambrian trilobites is important because it links the feeding habits deduced from exceptionally preserved taxa to *morphologic* characters that can be recognized in the majority of Cambrian trilobites.

Fortey and Hughes (1998) argued that a sagittal swelling anterior to the glabella in some primitive libristomate trilobites, most common in the Cambrian, may represent a brood pouch. The ideas of Fortey (1990) on broader aspects of trilobite feeding ecology have been significantly expanded, notably providing stronger support for filter feeding in post-Cambrian trilobites (Fortey and Owens 1999).

Major "morphotypes" have been recognized on the basis of the form of the dorsal shield (Jell 1981; Repina 1982; Fortey and Owens 1990a), and attempts have been made to link these morphologies to particular *ecologic* strategies. The morphotypes of Fortey and Owens (1990a) were defined as similar morphologies that arose convergently among different clades of trilobite (figure 17.1). They suggested that convergence on a common morphology argued for a common *ecologic* strategy, even if the nature of that strategy remained unresolved. Recognition of these morphotypes is based on either the overall form of the animal (such as the miniaturized morphotype)

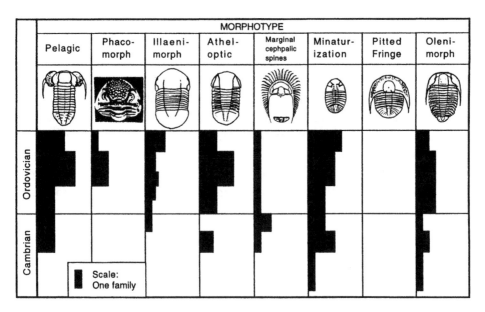

Figure 17.1 Cambrian and Ordovician occurrences of eight common trilobite morphotypes plotted against time. Note that most morphotypes are represented in the Cambrian. Modified from Fortey and Owens (1990a: figures 5.4–5.6).

or a specific character state (such as the atheloptic morphotype, which had reduced eyes). This approach provides a way of assessing the *ecologic* diversification of trilobites that is partially independent both of taxonomy and of the need to identify specific niches for each form. The results of this approach are discussed below.

Trace Fossils

The direct association of trilobites with trace fossils proves that they were the makers of some lower Paleozoic traces (e.g., Osgood 1970; Draper 1980; Geyer et al. 1995). Such direct associations are unknown in the Cambrian. Nontrilobite arthropods are known to have produced *Cruziana*-like tracks (Seilacher 1985), and given the diversity of Cambrian homopodous arachnomorphs, many of these traces could have been made by organisms other than trilobites. The common occurrence of *Rusophycus avalonensis* in the pretrilobitic Cambrian suggests that organisms making *Rusophycus* were not always preservable as body fossils. Nevertheless, *Cruziana/Rusophycus* makers likely occupied niches similar to those of trilobites, and the supposed parallel trends in size and abundance of *Cruziana/Rusophycus* and trilobites argue that trilobites were the principal architects of these traces (Seilacher 1985; but see also Whittington 1980). An alternative interpretation is that the evolutionary history of trilobites was mirrored by that of other cruzianaeform trace producers, but in either case the evolutionary history of trilobites is likely representative of that of the trace maker.

The case for an association of the Late Cambrian trace fossil *Cruziana semiplicata* and the trilobite *Maladioidella* cf. *colcheni,* found in adjacent beds, was made recently by Fortey and Seilacher (1997), but no direct association was observed.

Despite the abundance of cruzianaeform trace fossils, there is little strong evidence as to their function. An exception is the association between *Rusophycus* and teich-ichnian burrows in the Early Cambrian of Sweden (Bergström 1973; Jensen 1990), which provides evidence that burrowing arthropods preyed on infaunal worms. The large size of the burrows and the form of a cephalic impression are consistent with the makers' being olenelloid trilobites, which are associated with these deposits. *Cruziana* and *Rusophycus* provide unequivocal evidence of infaunal activity; some formed interstratally (Goldring 1985), while others suggest surficial burrowing (Droser et al. 1994).

Functional Modeling

Experiments with models of trilobites have provided insights into the hydrodynamics of Ordovician trilobites (Fortey 1985). Cambrian trilobites with morphologies similar to those modeled presumably behaved in similar fashions, and on this basis Hughes (1993) suggested a bottom-hugging life mode of the Late Cambrian asaphide *Dikelocephalus.*

INSIGHTS INTO CAMBRIAN TRILOBITE SYNECOLOGY

The Burgess Shale fauna provides the clearest evidence of the role of trilobites in Cambrian marine communities (Briggs and Whittington 1985; Conway Morris 1986). Trilobites from that assemblage include free-swimming suspension feeders (e.g., *Ptychagnostus*), benthic primary consumers (e.g., *Elrathina*), and carnivores (e.g., *Naraoia* and *Olenoides*). These broad lifestyles were shared with a wide variety of other schizoramid arthropods. Hence, trilobite morphology did not constrain the group to a limited range of *ecologic* opportunities; rather the group exploited the same broad range of niches available to other arthropods. The presence of benthic primary consumers (e.g., *Eoredlichia*), carnivores (e.g., *Naraoia*), and a possible free-swimming eodiscid from the Early Cambrian Chengjiang fauna (Shu et al. 1995), suggests that, minimally, this pattern was in place shortly after the advent of skeletonization, and possibly prior to that time. Identifying specific synecologic relationships within "normal" assemblages of Cambrian trilobites is more difficult, but specific size and habitat partitioning relationships have been suggested for Early and Middle Cambrian agnostid trilobites (Robison 1975), based on differences in maximum sizes of individual taxa and their relationship to lithofacies.

Abnormalities of various kinds also provide direct evidence of Cambrian trilobite synecology. A variety of skeletal abnormalities have been described in trilobites, resulting either from developmental anomalies, disease, infestation, or injury (e.g.,

Figure 17.2 Abnormalities in a Cambrian trilobite, possibly related to parasitism (see Hughes 1993 : 15). Divisions on scale bars in millimeters; arrows mark positions of structures of interest. A, Swelling on glabella of *Dikelocephalus minnesotensis*, UW 4006-70. B, Tunnels and ridges on composite mold of *D. minnesotensis* pygidium, presumed to be related to boring of the internal surface of the exoskeleton, UW 4006-90a.

Owen 1985; Jell 1989). Infestation by both microscopic and macroscopic organisms (figures 17.2A,B) indicates host-infester relationships among Cambrian trilobites. Healed injuries in many Cambrian trilobites (e.g., Conway Morris and Jenkins 1985; Babcock 1993) demonstrate the presence of macrophagous predators in the Early Cambrian, sophisticated repair mechanisms within the Trilobita, and possible behavioral styles within the group. The presence of macerated trilobite fragments within the gut contents of other Cambrian arthropods (e.g., Robison 1991 : 91) confirms that trilobites served as food sources. Arcuate bite marks are consistent with the mouthpart morphology of large Cambrian soft-bodied predators (e.g., Whittington and Briggs 1985 on *Anomalocaris*) and may even occur on large trilobites (Hughes 1993 : plate 7, figure 8), which were themselves likely predators.

Pratt (1998) has argued that extinction of a major predator on trilobites occurred during the Late Cambrian, based on changes in sclerite fracture in the lower Rabbitkettle Formation. Although imaginative, it remains unclear why the putative predator should have actively fractured exuvae, which likely formed the large majority of species examined. Furthermore, no candidate predator capable of smashing calcified exoskeletons in the manner envisaged by Pratt (1998) has yet been identified among the Burgess Shale–type faunas. A nonbiological explanation for the change in fracturing, such as a longer time interval prior to shell bed cementation, remains a viable alternative.

The discussion above indicates that Cambrian trilobites likely occupied a range of habitats from infaunal to probably pelagic realms. Indirect and direct evidence consistently suggests a number of feeding strategies among Cambrian trilobites, including sediment ingestion, suspension feeding, and active predation. Other feeding strategies, such as filter feeding, have been proposed (e.g., Bergström 1973; Stitt 1983) but are less firmly established. Evidence that a wide variety of trilobites were hosts for parasites, and prey for other organisms, suggests that they lived in ecosystems that are at least comparable to those found in marine habitats today. Trilobites apparently exploited a range of *ecologic* strategies similar to those employed by other Cambrian arthropods.

LIMITS ON ECOLOGIC RESOLUTION IN CAMBRIAN TRILOBITES

Despite progress toward understanding feeding and habitats of Cambrian trilobites, several major problems remain unsolved. The inability to infer specific life habits and niches for the majority of Cambrian trilobites presents the greatest challenge to understanding the ecologic evolution of these forms. Even though it is obvious that distinctive morphotypes must have had specific functional constraints, we are often at a loss to identify these constraints. An example is the multisegmented *Cermatops*-like pygidium. This morphotype is characterized by reduced propleurae and a wide doublure (Hughes and Rushton 1990; Rushton and Hughes 1996) and evolved independently in peri-Gondwanan early Late Cambrian iwayaspinids and idahoiids and in latest Cambrian dikelocephalids from Laurentia. Pygidia are indistinguishable among certain species belonging to distantly related groups. Repeated convergence on this morphology suggests a specific function for this pygidium, but that function remains unknown. Paleoenvironmental distributions offer no clues: taxa bearing the *Cermatops*-like pygidium appear in a wide variety of lithofacies, ranging from carbonate shelf environments to clastic submarine fan deposits and deeper-water dysaerobic environments. They also occur at a wide range of paleolatitudes and around several Cambrian landmasses (Rushton and Hughes 1996). The same difficulty extends across a wide variety of morphologies. For example, the distinctive catillicephalid morphotype (Jell 1981), consisting of a bulbous glabella, a small pygidium, and a small number of segments, was almost certainly related to a specific feeding habit—yet, beyond a general resemblance to *Stenopilus*, that habit is unknown (see the different interpretations offered by Stitt [1975] and Ludvigsen and Westrop [1983]).

The *Cermatops*-like pygidium reflects another broad difficulty in studies of Cambrian trilobites: rampant convergent evolution. Although convergent structures may indicate functional constraints, they can also confound attempts to assess phylogenetic relationships. Some, but not all, Cambrian trilobites show marked intraspecific variation (e.g., Westergård 1936; Rasetti 1948; Hughes 1994) and mosaic patterns of variation among related species (Kiaer 1917; Whittington 1989). This plasticity presents problems for systematics because of the difficulty of recognizing discrete taxa, distinguished by stable character sets. The "ptychopariid problem" (e.g., Lochman 1947; Schwimmer 1975; Ahlberg and Bergström 1978; Palmer and Halley 1979; Blaker 1986), which is the seemingly intractable systematics of a paraphyletic group of primitive libristomates, is an expression of this phenomenon. Reasons for the high levels of homoplasy among Cambrian trilobites are poorly known. They may reflect procedural or preservational artifacts, such as the desire to recognize stratigraphically diagnostic species (Hughes and Labandeira 1995), or greater absolute abundance of trilobites during the Cambrian than at later times (Li and Droser 1997). These factors could increase the range of intermediate morphotypes relative to units that are poorly studied or sampled. Alternatively, high levels of homoplasy may reflect a developmental or ecologic constraint that reduced the numbers of viable character states among primitive libristomate trilobites (see the section "A History of Cambrian Trilobite Ecology" below).

TRILOBITE DIVERSITY, ABUNDANCE, AND OCCURRENCE AS TOOLS FOR ECOLOGIC ANALYSES

Given the ignorance of the specifics of trilobite ecology, we must find alternative ways of estimating ecologic diversity. A comparative approach can provide useful information on the ecologic evolution of the group. Estimates of taxic and morphologic diversity, and patterns of trilobite occurrence and abundance, can serve to indicate aspects of the ecologic structure of the group. By assessing these parameters through Cambrian time, the comparative ecologic evolution of the group can be charted, even though we lack details of the role of each form within its own community. The skeletonized Trilobita are the only Cambrian clade sufficiently common to permit this kind of broad-scale analysis, and hence the group provides a unique perspective on Cambrian ecologic evolution. Furthermore, Burgess Shale–type faunas suggest that Cambrian trilobites occupied a range of niches similar to those of other Cambrian arthropods. Hence it is possible that the evolutionary history of the trilobites may be representative of the history of schizoramid arthropods as a whole. A discussion of measures of ecological diversity follows.

Taxonomic Diversity

Taxonomic diversity provides a rough measure of morphologic variety. It is approximate because it is impossible to standardize systematic judgments in groups with divergent morphologies, patterns of variation, and preservational styles (see Lochman 1947; Rasetti 1948) and because other factors, such as stratigraphic position, paleogeography, and taxonomic philosophy, have influenced systematic placement (Fortey 1990; Hughes and Labandeira 1995). Given that morphologic variety reflects ecologic diversity, the taxonomic history of trilobites provides insights into their ecologic evolution. The diversity of trilobites increased through the Cambrian at all taxonomic levels, and Cambrian ordinal-level diversity is likely to increase further as systematic studies are refined and additional basal sister taxa of post-Cambrian clades are identified (see Fortey and Owens 1990b). Generic and species-level diversity increased

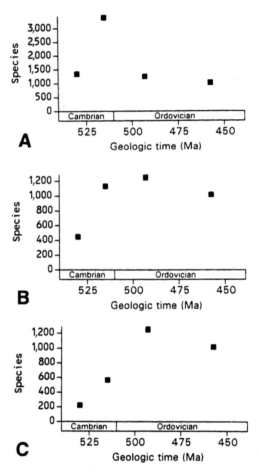

Figure 17.3 Species diversity in Cambrian and Ordovician trilobites based on the compilation of Foote (1993). A, Raw diversity data for the intervals earlier and later Cambrian, and earlier and later Ordovician. Standard error bars are smaller than the symbols. Note the sharp peak in species diversity in the later Cambrian. B and C, Estimates of trilobite standing taxonomic diversity (i.e., those alive at any one time). B, Cambrian species richness di-vided by 3 to account for Cambrian trilobite species turnover rates being estimated at 3 times greater in the Cambrian than in the Ordovician (Foote 1988). C, Cambrian species richness divided by 6. This figure was chosen because of updated estimates of the duration of the Cambrian, which is now thought to occupy a shorter span than used in Foote (1988). *Source:* Figures computed by Mike Foote.

dramatically through the Cambrian (e.g., Foote 1993: figure 5; Zhuravlev and Wood 1996: figure 2), reaching higher levels in the later Cambrian than at any other time in trilobite history (figure 17.3A). This sharp increase in later Cambrian diversity partly reflects high species turnover rates in the Late Cambrian (Foote 1988). Estimates of the number of taxa alive at any one time can be computed by calculating species turnover rates. Foote (1988) calculated that turnover rates were three times higher in the Cambrian than in the Ordovician, but revised estimates of Cambrian duration suggest that the average turnover rate could have been up to six times that of the Ordovician (M. Foote, pers. comm., 1997). These results confirm that although trilobite species diversity was greatest in the later Cambrian (figure 17.3A), the standing diversity of species at any one time may have been similar to (figure 17.3B) or significantly lower than (figure 17.3C) that during Ordovician times.

Despite the overall increase in taxonomic diversity during the Cambrian, the Early Cambrian contained a wide diversity of trilobite forms, typified by olenelloids, other "redlichiids," agnostids, corynexochids, primitive libristomates, and possibly also odontopleurid trilobites. Although the earliest collections of Cambrian trilobites at various sections worldwide usually contain only a single taxon (e.g., Brasier 1989a,b), the global diversity of the earliest trilobites and their well-established provincialism suggest that the appearance of trilobites in the rock record was not congruent with the earliest evolution of the group (Fortey and Owens 1990b; Fortey et al. 1996). While trilobite taxonomic diversity continued to increase rapidly in the Cambrian at a variety of taxonomic levels, many of the most distinctive Cambrian trilobite morphotypes were established by the close of Early Cambrian time. Trilobite higher taxa of Middle and Late Cambrian age have commonly been erected on the basis of numbers of constituent lower taxa rather than specified quanta of morphologic variation (Hughes and Labandeira 1995). Hence it is unclear whether the increase in taxonomic diversity in the later Cambrian is related to the abundance of trilobites in the rock record (and its relationship to taxonomic practice) or to continued rapid *morphologic* diversification.

Trilobite taxonomic diversity peaked in the Ordovician (Stubblefield 1959; Foote 1993), suggesting that trilobites reached their maximal ecologic diversity at that time. This argument is strengthened by (1) "morphospace" analyses, which assess aspects of trilobite diversity independently of taxonomy (Foote 1991, 1992); and (2) "morphotype" approaches, which estimate the numbers of clades contributing to distinctive recurrent morphotypes that presumably shared common life habits (Fortey and Owens 1990a) (see figure 17.1). Both these approaches indicate that the maximum diversity of trilobites occurred during the Ordovician and that it was coupled with the radiation of clades of distinctive and disparate trilobites. Morphospace approaches to the diversification of trilobites have proved particularly instructive in this regard and are discussed further below.

Morphospace Analyses

Morphospace studies abstract and simplify information on morphologic variation, using a variety of mathematic algorithms. The principal advantage of this approach is that, provided that the same information is abstracted for each specimen analyzed, relationships among taxa can be evaluated within a uniform reference frame. Hence the user can be confident that like is being compared with like, largely independently of taxonomy. Here the term *morphospace* rather than *morphologic* is used here, because these studies consider the relative placement of individuals within a space that is defined by the same set of individuals. Because morphospaces are based on a sample of the overall morphology, the extent to which they summarize the group's morphologic variation depends on the degree to which the sample is representative of total morphology. Morphospace approaches have been used to address a variety of questions within Cambrian trilobites (e.g., Ashton and Rowell 1975; Schwimmer 1975), but the most relevant application to the ecologic evolution of the Trilobita has been attempts to assess the morphologic diversification of trilobites throughout their evolutionary history (Foote 1990, 1991, 1992, 1993).

Using an analysis of the outline of the cranidium in trilobites that have a dorsal suture and the outline of the cephalon in forms that do not, Foote (1989) suggested that the morphologic diversity of polymerid trilobites increased from the earlier to later Cambrian, followed by a sharp increase in diversity in the later Ordovician (Foote 1991) (figure 17.4A). Although the earlier Cambrian shows the lowest overall diversity, the transition to the later Cambrian is not marked by a significant jump in area of occupied morphospace, despite the large increase in numbers of species sampled. Furthermore, the variance of earlier Cambrian trilobites apparently exceeds that of later Cambrian forms (Foote 1993) (figure 17.4B). The transition from Cambrian to later Ordovician was marked by the appearance of several distinct trilobite morphotypes, which went on to dominate the remainder of trilobite evolutionary history. Given the roughly similar volume of morphologic space occupied by earlier and later Cambrian trilobites, the greater variance of earlier forms, and the profound difference in numbers of species in each interval, estimates of diversity must be corrected to assess the effects of differing sample sizes. These analyses showed that earlier Cambrian morphologic diversity might actually have been higher than that of the later Cambrian (Foote 1992). Alternatively, if the increased sample size of later Cambrian trilobites reflects an absolute increase in taxic diversity during that time, it may suggest that the later Cambrian diversification of trilobites was morphologically constrained.

Foote's work permits an improved understanding of the morphologic diversification of trilobites, but interpretation of his data is complicated by the fact that the cephalic structures studied were not homologous among all the trilobites surveyed (Foote 1991). Early Cambrian olenelloids lacked a dorsal facial suture, and so the outline of the cephalon was used as a proxy for cranidial form. The cephalic outline includes the genal spine, a character showing considerable variation, and the presence of this spine contributes to the high disparity among olenelloid taxa (Foote 1991: text-figure 3) relative to forms in which cranidial outline was used. The argument that inclusion of the genal spine increases intragroup variability is supported by the pattern shown in the Ordovician cheirurids, which also occupied a larger proportion of morphospace than other groups and show a greater intrataxon disparity. Cheirurids had a proparian facial suture, with the result that their genal spines were also in-

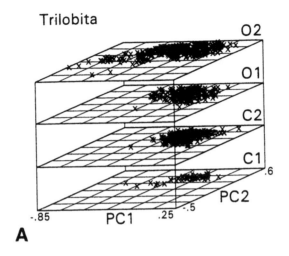

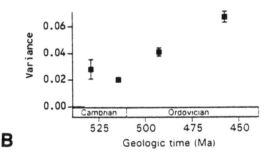

Figure 17.4 *A,* Morphological diversification of Cambrian and Ordovician trilobites as expressed by the first two principal components of Fourier coefficients of the outlines of cephalic structures (Foote 1989, 1993). Note the relatively constant area of morphological space occupied from the earlier Cambrian through earlier Ordovician. *B,* Morphological variance of trilobites. Note that the variance of earlier Cambrian trilobites is slightly higher and shows greater error estimates than that of the later Cambrian. *Source:* Figures computed by Mike Foote.

cluded in the data set. Whether this anomaly explains the relatively high variance in Early Cambrian trilobites as revealed by rarefaction analysis (Foote 1992) is unclear. Nevertheless, high variation in the Early Cambrian is consistent with the appearance of five trilobite orders during that time, each with a distinctive morphology.

Size Ranges

Estimates of the range of maximum sizes within a group provide a measure of ecologic diversity, because maximal body size is directly related to ecologic activity (McKinney 1990). Analysis of the size ranges of Cambrian trilobites was attempted using data on

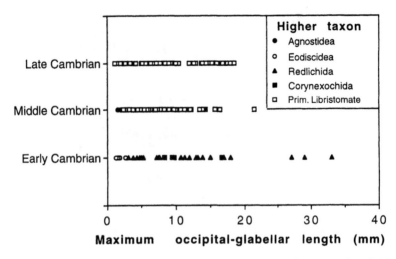

Figure 17.5 Maximum glabellar lengths of 253 species of skeletonized trilobites from China and Australia, as measured from systematic illustrations (see text for discussion). Note that maximal size diversity is found during the Early Cambrian, and the dominance of primitive libristomate trilobites in Middle and Late Cambrian faunas.

253 Gondwanan Cambrian trilobite species from China and Australia (figure 17.5). Maximum occipital-glabellar lengths were calculated using the largest cranidia of each species, illustrated in two extensive monographs (Zhang and Jell 1987; Bengtson et al. 1990), and a supplementary paper (Zhu and Jiang 1981). Trilobites from each of the Chinese Cambrian stages (or its correlatives) were sampled and assigned to the following age classes: Early, Middle, and Late Cambrian. This area was chosen because the large monograph by Zhang and Jell (1987) includes trilobites from each Cambrian epoch, and because Jell was also coauthor of the paper on Australian Early Cambrian forms (Bengtson et al. 1990). By limiting the sources to comprehensive works with a common author, I have attempted to maximize the consistency of the sample analyzed.

Results indicate that maximum size diversity occurred in the Early Cambrian ($n = 36$, mean = 10.7 mm, standard deviation [SD] = 7.9 mm). This finding is due to the presence of several large redlichiid trilobites in the Early Cambrian data set. The Middle Cambrian shows reduced size ranges, despite having by far the largest number of species sampled ($n = 153$, mean = 6.6 mm, SD = 3.8 mm). The Late Cambrian shows a slight increase in the numbers of larger trilobites ($n = 65$, mean = 8.9 mm, SD = 4.8 mm). Several biases affect this data set, including different numbers of taxa sampled within each time interval, variable durations among the time intervals, differences of paleoenvironment both within and between time intervals, variation in glabellar structure among the taxa sampled, and inconsistent underestimation of maximal glabellar lengths of the taxa analyzed. Nevertheless, the overall pattern demonstrates that trilobites achieved a broad distribution of maximal sizes during the Early Cambrian, with many large redlichiid trilobites present at that time. In this data set the Middle Cambrian shows a relatively restricted range of sizes, related to the decline of redlichiids and dominance of primitive libristomate forms. The Late Cambrian shows a slight expansion in the number of larger libristomate trilobites.

The results accord with data from the taxonomic diversification of trilobites and with the morphospace analyses of Foote (1991, 1992, 1993). Small agnostid and eodiscid trilobites were both present during the Early Cambrian, as were large olenelloids, many of which are significantly larger than the largest Early Cambrian trilobites in the data set above (e.g., Geyer and Palmer 1995). Some paleogeographic areas, such as the Mediterranean sector of peri-Gondwanaland, have large redlichiid trilobites persisting well into the Middle Cambrian, with an expansion in the range of maximal sizes at that time. This finding is due to the large Middle Cambrian paradoxidids (e.g., Bergström and Levi-Setti 1978). In other areas, this pattern would be mirrored by large Middle Cambrian forms such as the xystridurids (e.g., Öpik 1975). However, all these forms were redlichiids, a group with attached hypostomes that appeared in the Early Cambrian and had become extinct by Late Cambrian time. Faunal provinces may differ in patterns of maximal size distribution, but in at least some regions, Early Cambrian size distribution was more diverse than at later Cambrian times. The decline in size-range diversity in the Middle Cambrian in the data set presented (figure 17.5) relates to the rise to prominence of primitive libristomate trilobites (commonly called "ptychoparioids"), which were generally quite small. This result is consistent with Foote's suggestion of a morphologically constrained diversification in later Cambrian times, which was based on data from Laurentia (Foote 1992). After the extinction of redlichiids, the increased number of larger trilobites in the Late Cambrian was related to the advent of advanced trilobite groups with attached hypostomes such as the asaphids, some of whose latest Cambrian members are among the largest of all Cambrian trilobites (e.g., Hughes 1994).

The analysis of trilobite size diversity presented herein contrasts with the results of an analysis of a Treatise-based Cambrian Ptychopariina and Asaphina (Trammer and Kain 1997), in which greatest size diversity was found late in the Cambrian. The contrast is explained by the fact that large Lower and Middle Cambrian trilobites considered in my analysis were excluded from that of Trammer and Kaim (1997) because these species are members of other higher taxa. Neither their nor my analyses are comprehensive, and both should be viewed as exploratory.

Abundance

Few data exist on the abundance of trilobites during Cambrian time, but analyses of the nature and frequency of shell accumulations through the Cambrian of the Great Basin provide insight in this regard (Li and Droser 1997). The overall thickness and abundance of shell beds increased during the trilobite-bearing Cambrian, as did the phylum-level diversity of these concentrations. After attempting to assess the influence of the depositional history on these trends, Li and Droser (1997) conclude that increase in the absolute abundance of skeletonized animals was the major influence responsible for this trend. Increased abundance is compatible with the rise in trilobite taxonomic diversity during the same time period, because larger species numbers are likely to produce more fossils. However, it is also possible that the numbers of individuals within species also increased. The relationship between specimen abundance and taxic diversity will repay further study, because it is important to assess whether the numbers of specimens analyzed per species show variation through the Cambrian (Hughes and Labandeira 1995). Late Cambrian species turnover rates are three times higher than in the Early Ordovician (Foote 1988). This may reflect a fundamental difference in speciation patterns during the two periods, or alternatively it may be an artifact of different taxonomic practices applied during the two intervals. Trilobites are the most abundant macroinvertebrates found in Upper Cambrian rocks, and consequently there may have been a tendency to proliferate the numbers of species for the purpose of biostratigraphic resolution.

Linkages between numbers of specimens preserved in sedimentary rocks and recovered for analysis, and the numbers of described taxa and absolute abundances of individuals within species, are incompletely understood. Extracting detailed information on relative abundance of individuals from a myriad of taphonomic and preservational influences could be intractable (Westrop and Adrain 1998), but at the biofacies level at least it appears that specimen abundance contains information of biological import.

Occurrence

Trilobite taxa differ in their temporal and geographic distributions. The study of trilobite distributions provides information on the ecologic evolution of the group, even if this information cannot be directly related to specific niches. Occurrence, along with analogy and functional considerations, can constrain hypotheses about trilobite life habits (Fortey 1985). For example, the widespread geographic occurrence of agnostid trilobite species, upon which much of intercontinental Cambrian biocorrelation rests, supports morphology-based arguments that these animals were free-swimming and possibly pelagic (see the section "Specialized Morphologies and 'Morphotypes'" above). Pioneering studies of the global distribution of Cambrian trilobites (e.g., Richter and Richter 1941; Repina 1968, 1985; Cowie 1971; Jell 1974; Taylor 1977; Shergold 1988) indicate broad faunal provinces during Cambrian time. Faunal data are broadly consistent with other indicators of Cambrian global paleogeography. Laurentian shelf faunas are apparently the most distinctive, a characteristic consistent with the notion that Laurentia was geographically isolated during Cambrian times. A widespread shelf fauna occurs about the peri-Gondwanan margin, although the restriction of many elements to specific regions suggests some paleolatitudinal control of faunal distribution. Faunas adapted to cooler waters had more widespread occur-

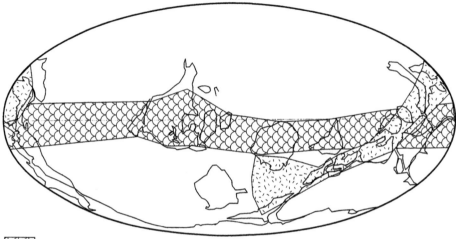

 Erixanium occurrence

Maladioidella occurrence

Figure 17.6 Contrasting biogeographic distributions of two Late Cambrian trilobite genera. The distribution of *Erixanium*, which shows a worldwide equatorial distribution (Stitt et al. 1994), was apparently constrained by factors related to latitude. In contrast, *Maladioidella* shows a widespread peri-Gondwanan distribution (Rushton and Hughes 1996). The greatest constraint on its distribution was crossing oceanic basins. The *Maladioidella* distribution tract is superimposed over the *Erixanium* tract for graphic clarity only. *Source:* Base map provided by Chris Scotese.

rence than did faunas restricted to equatorial shelf environments (e.g., Wilson 1957; Taylor 1977). Cambrian trilobite faunas have now been described from most parts of the world, but a great deal of phylogenetic analysis is necessary before the potential of Cambrian trilobites for assessing paleogeography is fully realized. Studies of the global distributions of genera or species suggest that ecologic factors controlling distributions differ markedly, even among taxa with broadly similar morphologies (figure 17.6).

In spite of these problems, the distributions of distinctive morphotypes suggest consistent broad patterns of trilobite distribution among paleocontinents. For example, distinctive oryctocephalid, olenid, eodiscid, and agnostid trilobites are common in slope facies and had wide geographic distributions during life. These forms are characterized by thin cuticles, which is common also in Ordovician trilobites inhabiting deeper water (Fortey and Wilmot 1991). Offshore benthic polymerids commonly share the "olenimorphic" morphotype, consisting of multisegmented thoraces with narrow axes and wide pleurae (Fortey and Owens 1990a). A comparison of Laurentian and Siberian faunas suggests morphologic and distributional similarities among shelf faunas, even though phylogenetic relationships among these faunas remain unclear. For example, diverse assemblages of Late Cambrian trilobites are known

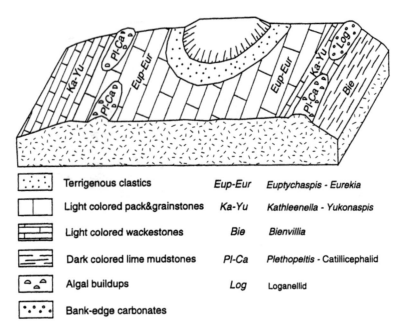

Terrigenous clastics	*Eup-Eur*	*Euptychaspis - Eurekia*
Light colored pack&grainstones	*Ka-Yu*	*Kathleenella - Yukonaspis*
Light colored wackestones	*Bie*	*Bienvillia*
Dark colored lime mudstones	*Pl-Ca*	*Plethopeltis - Catillicephalid*
Algal buildups	*Log*	*Loganellid*
Bank-edge carbonates		

Figure 17.7 Cartoon cross section of northern North America during the Late Cambrian serotina interval, showing lithofacies and associated trilobite biofacies. The biofacies represent distinguishable faunal associations. *Eup-Eur,* *Ka-Yu,* and *Bie* represent level-bottom biofacies; *Pl-Ca* and *Log* are outer platform bankedge biofacies. *Source:* Modified from Ludvigsen and Westrop (1983: figure 17).

from outer shelf facies in both paleocontinents (Pegel' 1982; Ludvigsen and Westrop 1983). Similarly, Late Cambrian nearshore assemblages show relatively reduced diversity (Hughes 1993), although unusual morphotypes can be common in these facies (Hughes et al. 1997).

At a smaller geographic scale, repeated associations of taxa have been recognized as biofacies (e.g., Lochman-Balk and Wilson 1958). In many cases, biofacies can be related to specific paleoenvironments, and a variety of statistical methods have been used in their definition (e.g., Ludvigsen and Westrop 1983; Pratt 1992; Babcock 1994; Westrop 1995). Detailed biofacies analyses have been undertaken in North America for portions of Cambrian time. In many cases these biofacies possess distinctive suites of trilobites and can be related to specific lithofacies. For example, the Sunwaptan *Euptychaspis-Eurekia* biofacies consistently shows a similar array of taxa in approximately constant proportions and is always associated with light-colored shelf packstones and grainstones (Ludvigsen and Westrop 1983; Westrop 1986) (figure 17.7). However, other biofacies, such as the *Kathleenella-Yukonaspis* biofacies, which occurs in light-colored wackestones from ramp settings, show considerable variation in constituent taxa and their relative abundances (e.g., Ludvigsen and Westrop 1983; Westrop 1995). Different biofacies vary markedly in the numbers of constituent taxa, their taxonomic integrity, and the range of lithofacies that they occupy. Hence, tracking the temporal and geographic establishment and demise of biofacies, and their constituent taxa, can provide insight into the ecologic evolution of Cambrian communities.

THE ECOLOGIC EVOLUTION OF CAMBRIAN TRILOBITES

Based on the information and approaches presented above, some inferences on the Cambrian ecologic history of trilobites can now be presented. First, the Upper Cambrian biomeres of Laurentia are discussed because they provide insight into the dynamics of the ecologic evolution of trilobites at low taxonomic levels. Second, comments on the overall ecologic evolution of Cambrian trilobites, with speculations on the significance of trilobites for understanding the Cambrian evolutionary radiation, serve to summarize current knowledge of the broad history of the group as a whole and to outline directions for future research.

Laurentian Upper Cambrian Biomeres

Studies of stratigraphically thick and richly fossiliferous Upper Cambrian deposits in Laurentia suggest cyclic changes in trilobite diversity during the Late Cambrian and provide the basis for recognition of a series of biostratigraphic stages commonly known as biomeres (Palmer 1965a). Biomeres are interpreted as continental-scale episodes of evolutionary radiation that are bounded by episodes of mass extinction. A broad pattern of taxonomic, morphologic, and ecologic diversification within biomeres has been recognized, and authors agree that boundaries between biomeres are related to drastic environmental changes. Patterns of ecologic change associated with biomeres offer important insights into the ecologic evolution of Cambrian trilobites, particularly at the family level and below.

Phylogenetic Basis for Biomere Evolution

Biomeres are named after individual trilobite families that characterize them. Central to the biomere model is the notion that evolutionary radiation within each biomere took place in a closed system, seeded by immigration at the base of each biomere but thereafter evolving in isolation until complete faunal decimation by a terminal extinction. The phylogenetic implication of this model is that all taxa within a biomere are closely related and that these clades are exclusive to the biomere in which they occur. Cambrian trilobite taxonomy is incompletely resolved (see the section "Limits on Ecologic Resolution in Cambrian Trilobites" above), and it has been suggested that some major groups may transgress biomere boundaries, with close relatives on either sides of the boundary known by different names (see Fortey 1983; Briggs et al. 1988; Fortey and Owens 1990b; Edgecombe 1992). Using arguments derived from phylogenetic relationships, Edgecombe (1992:148) suggested that many Cambrian trilobite taxa must have evolved substantially earlier than their age of first appearance in the fossil record. Assuming that these phylogenies are correct, the whereabouts of these "ghost lineages" are uncertain. They may have been either extremely rare and as yet unrecovered from known faunas, or they may have occupied areas or environments with a poorly known record. Alternatively, sister taxa occurring in different biomeres may simply have been mistaken to be more distantly related (Fortey 1983). Examples of groups characterized by long temporal or geographic gaps give credence to the idea of ghost lineages. For example, saukiid trilobites, which have earliest stratigraphic records in Australia during the time of the first Late Cambrian biomere, first appear in Laurentian strata in the middle portion of the third and last Late Cambrian biomere. Given these phylogenetic uncertainties, this discussion of biomeres will concentrate on those aspects of biomere development that are largely independent of phylogeny. Hence, when increasing taxonomic diversity within biomeres is mentioned below, it refers to the pattern of increasing number of taxa per collection within a given biomere and does not necessarily imply evolution of these forms in situ.

Ecologic Evolution Within Biomeres

The basal parts of biomeres are characterized by faunas of low diversity, the genera and species of which tend to have short stratigraphic occurrence ranges. These are replaced by faunas of higher diversity whose constituent taxa commonly occur over longer stratigraphic intervals (Palmer 1965a,b; Stitt 1971; but also see Westrop 1996). Ecologic arguments have been used to explain this trend. Stitt (1975, 1977) saw a direct coupling of taxonomic radiation and ecologic diversification. He proposed that low diversity faunas composed of morphologically variable species with broad niches were gradually replaced by morphologically discrete species adapted to specialized niches. These species formed stable ecologic communities in which new innovation was constrained by niche incumbency. In this model, the ecologic evolution of the biomere is driven by such biological factors as niche partitioning and community structure.

Patterns of diversification can be examined in the context of paleoenvironmental conditions, some of which are independent from biological systems (e.g., sea level, lithofacies distribution, and shelf area). Westrop (1988, 1996) noted that the patterns of diversification in biomeres differ among paleoenvironments. High rates of diversification characterize shelf margin settings throughout biomere evolution. This may be due to the combined effects of local speciations plus species immigration from slope environments. In subtidal shelf carbonates and storm deposits, which are distant from the shelf margin, diversification rates slacken toward the top of biomeres. This may reflect an overall decline in the rate of within-habitat speciation. Despite these environmental differences, faunas from regions throughout Laurentia confirm an overall rise in within-collection species diversity (alpha diversity) during biomere evolution, suggesting that the taxonomic diversification is not simply a consequence of environmental diversity.

During the recovery from a mass extinction at the base of the ptychaspid biomere (Sunwaptan stage), diversity in terms of both within-habitat species richness and biofacies differentiation increased (Westrop 1988, 1996). Initial stages of recovery were dominated by a small number of widespread biofacies with low taxonomic diversity but relatively high environmental diversity. These were later replaced by a larger number of biofacies with higher taxonomic variety and taxonomic integrity that track lithofacies shifts. The change from a few environmentally widespread biofacies of low diversity to a larger number of higher-diversity, environmentally circumscribed biofacies is one aspect of ecologic evolution during biomere development (Westrop 1988, 1996).

Morphometric analysis of biomere faunas indicates that taxonomic diversification is coupled with increased morphologic diversity (Sundberg 1996), suggesting that late-stage biomeres contained an increased variety of trilobite niches at all spatial levels from within individual collections to the number of biofacies present in Laurentia. New insights into the ecologic evolution of biomeres and associated extinctions will come from improved understanding of trilobite phylogeny and from further integration of diversification patterns with other types of geologic information, such as lithofacies, sea level, and isotopic data (e.g., Osleger and Read 1993; Saltzman et al. 1995).

Data on specimen abundance per collection (see Palmer 1979, 1984) suggest that the numbers of individual trilobites remained relatively constant across biomere boundaries (or even slightly increased), despite the sharp drops in taxonomic diversity. Hence, early species are represented by large numbers of specimens. Stitt (e.g., 1971) suggested that the level of intraspecific variation in trilobites declined during biomere evolution, and he argued that as niches became more finely divided, intraspecific variation was constrained. Morphometric analysis by Ashton and Rowell (1975) (mistakenly criticized by Hughes 1994:54) of approximately equal-size samples of species from throughout a biomere failed to find the predicted decrease in levels of intraspecific variation. A possible explanation of Stitt's comments relates to the abundance of large numbers of individuals belonging to a small variety of species early within biomeres. Because patterns of variation can depend on the numbers of specimens analyzed, marked variability in these early forms might be a function of their high abundance (Hughes and Labandeira 1995).

Ecologic Changes at Biomere Boundaries

Biomere boundaries are defined by marked changes in trilobites at species and higher taxonomic levels, and the rapid decline of diversity across these boundaries suggests periods of accelerated extinction rate. These faunal changes have been attributed to major changes in paleoenvironment, but the nature of these changes is disputed. Authors who emphasize the apparently synchronous extinction of many taxa prefer to invoke geologically rapid changes, such as major changes in ocean temperature (Lochman and Duncan 1944; Palmer 1965a,b; Stitt 1975, 1977; Loch et al. 1993) or levels of dissolved oxygen in sea water (Palmer 1979, 1984; Saltzman et al. 1995). Others, who recognize paleoenvironmentally selective extinctions that occur over longer stratigraphic intervals, favor gradual collapse of biofacies structure due to widespread marine regression (Ludvigsen 1982; Ludvigsen and Westrop 1983; Westrop and Ludvigsen 1987). Regardless of the causes of extinction, it appears that family-level survivorship across episodes of mass extinction was strongly influenced by geographic range. Families confined to narrow geographic or paleoenvironmental ranges suffered significantly greater extinction than widespread families (Westrop and Ludvigsen 1987; Westrop 1989, 1991). Because clades that are confined to shelf facies tend to be both geographically restricted and endemic, biomere extinctions strongly select shelf faunas. Although there is good evidence for selective extinction at biomere boundaries, it is worth pointing out that the extinction patterns presented rely on taxonomic data, which may be subject to revision.

Biomeres and Other Elements of the Cambrian Fauna

Other skeletonized elements of the Cambrian fauna that are volumetrically subordinate to trilobites have received less usage in Cambrian biostratigraphy and commonly contain fewer morphologic components. Nevertheless, marked faunal turnovers in nontrilobite members of the Cambrian fauna, including both inarticulate and articulate brachiopods and conodonts, coincide with biomere extinctions (Palmer 1984; Westrop 1996). Hence, the faunal patterns seen in trilobites during biomeres may have parallels in these groups (see also Zhuravlev, this volume). Trilobites occupied an ecologic spectrum comparable to that of other arachnomorphic arthropods (see the section "Direct Evidence for Feeding" above), so it is likely that this broader clade experienced similar diversity fluctuations. Soft-bodied trilobite genera, such as *Naraoia* and *Tegopelte,* are known to have relatively long stratigraphic ranges (Robison 1991), a finding that may be related to a preference for deeper-water habitats, which promoted greater taxonomic longevity (Conway Morris 1989: figure 2). This pattern may be consistent with the recognition of conservative, slowly evolving skeletonized trilobite groups in slope environments (Palmer 1965a; Stitt 1977).

In summary, biomeres apparently represent cycles of evolutionary and ecologic diversification and collapse. Diversifications appear to have been linked not to key innovations within individual clades but to new environmental opportunities or biologic interactions exploited synchronously by several clades. In these respects they resemble "economic" evolutionary radiations (Erwin 1992).

A History of Cambrian Trilobite Ecology

The presence of *Rusophycus* in pretrilobitic Cambrian rocks, indicating organisms of comparable organization and behavior, and the presence of a variety of clades among the oldest collections containing trilobites suggest a Cambrian prehistory for the clade that is not yet known from the fossil record. Arguably, the oldest beds containing Cambrian trilobites yield more than a single species (see Brasier 1989b); thus niche partitioning among trilobites apparently existed from their earliest appearance in the fossil record. Fortey and Owens (1990b) suggested that the oldest trilobites occur nearshore and that this may reflect the site of their evolutionary origin. This suggestion may be misleading, because trilobite origins remain cryptic (Fortey et al. 1996) and also because much of the Cambrian worldwide was characterized by passive margin subsidence following major Neoproterozoic rifting (Burrett and Richardson 1978; Bond et al. 1984). A result of this pattern is that much of the early Cambrian record was deposited in shallow water.

The Early Cambrian is characterized by a variety of distinctive trilobite clades that differed greatly in their morphology, size, ranges of geographic distribution, and feeding habits. Carnivores and both benthic and nektobenthic primary consumers were represented, suggesting that the major trophic levels were already occupied during the earliest trilobite-bearing Cambrian. Trilobite morphologic diversity may even have been higher in the Early Cambrian than later in Cambrian time (Foote 1992; see also Budd 1995). Nevertheless, trilobite species richness was reduced, and the absolute abundance of trilobites was likely lower in the Early Cambrian (Li and Droser 1997). The close of the Early Cambrian was marked by the demise of a major group of carnivorous trilobites, the olenelloids, and a change in the ecologic structure of the group.

Proliferation of trilobite species in the Middle and Late Cambrian was accompanied by an increased abundance of trilobite fossils (Li and Droser 1997). Despite these increases, the diversification was morphologically constrained, and the morphologic distinction between taxa limited (Foote 1990). Groups radiating at this time were mostly primitive libristomates with natant hypostomes. An ecologic constraint related to primary consumption feeding mode, coupled with extinction or decline of carnivorous forms, might explain the slackening of morphologic diversification. Primitive sister taxa of advanced trilobite clades, which rose to dominance in the post-Cambrian, first appeared at this time and evolved from primitive libristomates (Fortey and Owens 1990b). Similarly, some of the major post-Cambrian iterative morphotypes also first appeared at this time (Fortey and Owens 1990a: figure 5.5). For example, of the eight major iterative morphotypes of trilobite evolution recognized by Fortey and Owens (1990a), two were present in the Early Cambrian and six were present in the Middle and Late Cambrian. All of these morphotypes are postulated to be primary consumers. Hence, many of the basic phylogenetic and ecologic components of Ordovician trilobite communities were in place as early as the Middle Cambrian but did not rise to dominance until the establishment of the Paleozoic fauna (see also Droser et al. 1996).

The transition between the low-diversity, morphologically disparate Early Cambrian and the high-diversity, morphologically constrained later Cambrian suggests an important ecologic restructuring that likely affected other elements of the Cambrian arachnomorphic fauna. Although the morphologically constrained later Cambrian diversification may be related to feeding mode, several major questions remain. Species turnover rates within Late Cambrian biomeres are among the highest known in the fossil record (Foote 1988; Westrop 1996). Phylogenetic concepts (see Edgecombe 1992) and the desire to recognize stratigraphically diagnostic taxa (Hughes and Labandeira 1995) may have artificially inflated species turnover rates at this time, but other evidence suggests that it might be related to a distinctive pattern of variation within and among basal libristomate species. Evolutionary "lability" in the Late Cambrian trilobites has been attributed to structural and behavioral complexity (Westrop 1996), although why such lability is temporally restricted (Foote 1988) remains unclear. Some intraspecific studies of these trilobites indicate unusual variability in holaspid segment numbers, in other meristic characters, and in body proportions (McNamara 1983; Hughes 1994; Hughes and Chapman 1995), and the morphometric distances between basal libristomate species are, on average, shorter than between Ordovician species (Foote 1990). The "ptychopariid problem" (see the section "Limits on Ecological Resolution in Cambrian Trilobites" above) also attests to the high levels of homoplasy and convergence within this group. The relationship between morphologic plasticity and repeated biomere mass extinctions is unclear, but evidence now suggests (Hughes and Chapman 1995) that elevated degrees of intraspecific variation in some Cambrian trilobites were related to environmental or ecologic constraints, rather than to phylogenetic ones (contra Hughes 1991). The nature of those constraints remains unknown.

The validity of the "Cambrian Evolutionary Fauna" as a distinct ecologic entity has been questioned on the grounds that the structure of Cambrian and Ordovician trilobite biofacies are comparable (Ludvigsen and Westrop 1983:315). Although this may be true, major changes in trilobite phylogeny (Stubblefield 1959), the morphologic distinctness of species and clades (Foote 1990, 1991), and the array of iterative morphotypes (Fortey and Owens 1990a) all suggest that trilobites underwent a major ecologic transition in the Ordovician that was coincident with the establishment of the Paleozoic fauna. The rise of trilobite clades in the Paleozoic fauna continued through the Lower Ordovician, with their first widespread occurrence coincident with the base of the Middle Ordovician (Droser et al. 1996). In some cases, trilobite species diversity in particular paleoenvironments remained constant during this transition (Westrop et al. 1995), but trilobite individuals were apparently less abundant than in the later Cambrian (Li and Droser 1997).

The role of global-scale tectonic events in governing the evolution of Cambrian trilobites is being investigated using a phylogenetic approach (Leiberman 1997, 1999). Results suggest that slight elevation of speciation rates relative to Paleozoic norms may have been related to continental fragmentation during the Cambrian. Further studies of the role of constraints in trilobite evolution suggest that phylogenetic constraints may, in some cases, be less important than ecologic ones (Hughes et al. 1999).

Future Research Directions

The fossil record of trilobites offers a unique database for understanding aspects of the Cambrian radiation. Nevertheless, the potential of this database remains far from being fully realized.

Improved phylogenetic analyses are essential for attempts to interpret processes that governed the evolution of Cambrian trilobites. Even though high levels of homoplasy may hinder the resolution of well-supported groups, it is important to dissect patterns and levels of character support for phylogenetic hypotheses in order to investigate controls underpinning the radiation of the group. These data are critical for determining whether the morphologic constraint in Middle and Late Cambrian trilobites is linked to specific characters or character arrays. Until phylogenies are better resolved, current interpretations of biomere evolution remain questionable, because it is possible that diversifications were strongly affected by survivorship across biomere extinctions or by immigrations from other areas during biomere development.

With improved phylogenies it will be possible to document more accurately the relationship between morphologic and phylogenetic diversification, as has recently been shown in Cambrian gastropods (Wagner 1995). This information could then be integrated with environmental data to constrain environmental controls on diversification patterns and to test the relationships between speciation rates and biologic and environmental diversity. For example, it would be interesting to determine the phylogenetic and morphometric evolution of biofacies within biomeres and to compare these patterns across biomeres to determine the extent to which biomere diversification events are iterative.

Finally, although attempts have been made to document and relate morphologic and phylogenetic attributes to "economic" aspects of trilobite lifestyles (principally feeding strategy), almost nothing is yet known of the reproductive strategies of Cambrian trilobites, which are a critical aspect of their ecology. More-detailed studies of life history schedules of trilobite species in articulated assemblages may provide important information in this regard.

Progress in understanding the evolution of Cambrian trilobites will be facilitated by the establishment of a comprehensive electronic database that is accessible to a large number of scientists. Such a database, including images of specimens, data on biostratigraphic and lithofacies occurrence, geographic distribution data, and synonymy information, would permit ready evaluation of phylogenetic and ecologic relationships.

Acknowledgments. My thanks to Mary Droser, Doug Erwin, Mike Foote, Richard Fortey, Chris Scotese, Steve Westrop, and other colleagues for information and discussion of topics presented in this chapter. Mike Foote kindly generated the graphs in figures 17.3 and 17.4. Mary Droser, Andrey Zhuravlev, Richard Fortey, Gerd Geyer, Robert Riding, and Mark Webster provided helpful comments on the manuscript. This paper is a contribution to IGCP Project 366.

REFERENCES

Ahlberg, P. and J. Bergström. 1978. Lower Cambrian pytchopariid trilobites from Scandinavia. *Sveriges Geologiska Undersökning* 49:1–34.

Ashton, J. H. and A. J. Rowell. 1975. Environmental stability and species proliferation in Late Cambrian trilobite faunas: a test of the niche-variation hypothesis. *Paleobiology* 1:161–174.

Babcock, L. E. 1993. Trilobite malformations and the fossil record of behavioral asymmetry. *Journal of Paleontology* 67:217–229.

Babcock, L. E. 1994. Biogeography and biofacies patterns of polymeroid trilobites from North Greenland: Palaeogeographic and palaeo-oceanographic implications. *Grønlands Geologiske Undersøgelse, Bulletin* 169:129–147.

Bengtson, S., S. Conway Morris, B. J. Cooper, P. A. Jell, and B. N. Runnegar. 1990. Early Cambrian fossils from South Australia. *Association of Australasian Palaeontologists, Memoir* 9:1–364.

Bergström, J. 1973. Organisation, life, and systematics of trilobites. *Fossils and Strata* 2:1–69.

Bergström, J. and R. Levi-Setti. 1978. Phenotypic variation in the Middle Cambrian trilobite *Paradoxides davidis* Salter at Manuels, S.E. Newfoundland. *Geologica et Palaeontologica* 12:1–40.

Blaker, M. R. 1986. Notes on the trilobite faunas of Henson Gletscher Formation (Lower and Middle Cambrian) of central North Greenland. *Rapport Grønlands Geologiske Undersøgelse* 132:65–73.

Bond, G. C., P. A. Nickeson, and M. A. Kominz. 1984. Breakup of a supercontinent between 625 Ma and 555 Ma: New evidence and implications for continental histories. *Earth and Planetary Science Letters* 70:325–345.

Brasier, M. D. 1989a. China and the palaeotethyan belt (India, Pakistan, Iran, Kazakhstan, and Mongolia). In J. W. Cowie and M. D. Brasier, eds., *The Precambrian-Cambrian Boundary,* Oxford Monographs on Geology and Geophysics 12, pp. 40–74. Oxford: Clarendon Press.

Brasier, M. D. 1989b. Towards a biostratigraphy of the earliest skeletal biotas. In J. W. Cowie and M. D. Brasier, eds., *The Precambrian-Cambrian Boundary,* Oxford Monographs on Geology and Geophysics 12, pp. 117–165. Oxford: Clarendon Press.

Briggs, D. E. G. and H. B. Whittington. 1985. Modes of life of arthropods from the Burgess Shale, British Columbia. *Transactions of the Royal Society of Edinburgh (Earth Sciences)* 76:149–160.

Briggs, D. E. G., R. A. Fortey, and E. N. K. Clarkson. 1988. Extinction and the fossil record of the arthropods. In G. P. Larwood, ed., *Extinction and Survival in the Fossil Record,* pp. 171–209. Oxford: Systematic Association and Clarendon Press.

Budd, G. E. 1995. *Kleptothule rasmusseni* gen. et sp. nov.: An ?olenellinid-like trilobite from the Sirius Passet fauna (Buen Formation, Lower Cambrian, North Greenland). *Transactions of the Royal Society of Edinburgh (Earth Sciences)* 86:1–12.

Burrett, C. F. and R. G. Richardson. 1978. Cambrian trilobite diversity related to cratonic flooding. *Nature* 272:717–719.

Chatterton, B. D. E., Z. Johanson, and G. Sutherland. 1994. Form of the trilobite digestive system: Alimentary structures in Pterocephalia. *Journal of Paleontology* 68:294–305.

Clarkson, E. N. K. 1973. Morphology and evolution of the eye in Upper Cambrian Olenidae (Trilobita). *Palaeontology* 16:735–763.

Conway Morris, S. 1986. The community structure of the Middle Cambrian Phyllopod Bed (Burgess Shale). *Palaeontology* 29:423–467.

Conway Morris, S. 1989. The persistence of Burgess Shale–type faunas: Implications for the evolution of deeper-water faunas. *Transactions of the Royal Society of Edinburgh (Earth Sciences)* 80:271–283.

Conway Morris, S. and R. J. F. Jenkins. 1985. Healed injuries in Early Cambrian trilobites from South Australia. *Alcheringa* 9:167–177.

Cowie, J. W. 1971. Lower Cambrian faunal provinces. In F. A. Middlemiss, P. F. Rawson, and G. Newall, eds., *Faunal Provinces in Space and Time,* pp. 31–46. Geological Journal, Special Issue, London 4.

Draper, J. J. 1980. *Rusophycus* (Early Ordovician ichnofossil) from the Mithaka Formation, Georgian Basin. *BMR Journal of Australian Geology and Geophysics* 5:57–61.

Droser, M. L., N. C. Hughes, and P. A. Jell. 1994. Infaunal communities and tiering in Early Palaeozoic nearshore clastic environments: Trace fossil evidence from the Cambro-Ordovician of New South Wales. *Lethaia* 27:273–283.

Droser, M. L., R. A. Fortey, and X. Li. 1996. The Ordovician radiation. *American Scientist* 84:122–131.

Edgecombe, G. D. 1992. Trilobite phylogeny and the Cambrian-Ordovician "Event": Cladistic reappraisal. In M. J. Novacek and Q. D. Wheeler, eds., *Extinction and Phylogeny,* pp. 144–77. New York: Columbia University Press.

Edgecombe, G. D. and L Ramsköld. 1999. Relationships of Cambrian Arachnata and the systematic position of Trilobita. *Journal of Paleontology* 73:263–287.

Eldredge, N. 1989. *Marcoevolutionary Dynamics.* New York: McGraw-Hill.

Erwin, D. H. 1992. A preliminary classification of evolutionary radiations. *Historical Biology* 6:133–147.

Foote, M. 1988. Survivorship analysis of Cambrian and Ordovician trilobites. *Paleobiology* 14:258–171.

Foote, M. 1989. Perimeter-based Fourier analysis: A new morphometric approach applied to the trilobite cranidium. *Journal of Paleontology* 63:880–885.

Foote, M. 1990. Nearest-neighbor analysis of trilobite morphospace. *Systematic Zoology* 39:371–382.

Foote, M. 1991. Morphologic patterns of diversification: Examples from trilobites. *Palaeontology* 34:461–485.

Foote, M. 1992. Rarefaction analysis of morphological and taxonomic diversity. *Paleobiology* 18:1–16.

Foote, M. 1993. Discordance and concordance between morphological and taxonomic diversity. *Paleobiology* 19:185–204.

Fortey, R. A. 1983. Cambrian-Ordovician trilobites from the boundary beds in western Newfoundland and their phylogenetic significance. *Special Papers in Palaeontology* 30:179–211.

Fortey, R. A. 1985. Pelagic trilobites as an example of deducting life habits in extinct arthropods. *Transactions of the Royal Society of Edinburgh (Earth Sciences)* 76:219–230.

Fortey, R. A. 1986. The type species of the Ordovician trilobite *Symphysurus:* Systematics, functional morphology, and terrace ridges. *Paläontologische Zeitschrift* 60:255–275.

Fortey, R. A. 1990. Ontogeny, hypostome attachment, and trilobite classification. *Palaeontology* 33:529–576.

Fortey, R. A. and N. C. Hughes. 1998. Brood pouches in trilobites. *Journal of Paleontology* 72:638–649.

Fortey, R. A. and R. M. Owens. 1990a. Trilobites. In K. J. McNamara, ed., *Evolutionary Trends,* pp. 121–42. London: Belhaven Press.

Fortey, R. A. and R. M. Owens. 1990b. Evolutionary radiations in the Trilobita. In P. D. Taylor and G. P. Larwood, eds., *Major Evolutionary Radiations,* Systematics Association 42, pp. 139–164. Oxford: Clarendon Press.

Fortey, R. A. and R. M. Owens. 1999. Feeding habits in trilobites. *Palaeontology* 42:429–465.

Fortey, R. A. and A. Seilacher. 1997. The trace fossil *Cruziana semiplicata* and the trilobite that made it. *Lethaia* 30:105–112.

Fortey, R. A. and H. B. Whittington. 1989. The Trilobita as a natural group. *Historical Biology* 2:125–138.

Fortey, R. A. and N. V. Wilmot. 1991. Trilobite cuticle thickness in relation to palaeoenvironment. *Paläontologische Zeitschrift* 65:141–151.

Fortey, R. A., D. E. G. Briggs, and M. A. Wills. 1996. The Cambrian evolutionary "explosion": Decoupling cladogenesis from morphological disparity. *Biological Journal of the Linnean Society* 57:13–33.

Geyer, G. and A. R. Palmer. 1995. Neltneriidae and Holmiidae (Trilobita) from Morocco and the problem of Early Cambrian intercontinental correlation. *Journal of Paleontology* 69:459–474.

Geyer, G., W. Heldmaier, and E. Landing. 1995. Arthropod traces in the Middle Cambrian of Morocco. *Beringeria, Special Issue* 2:254.

Goldring, R. 1985. The formation of the trace fossil *Cruziana. Geological Magazine* 122:65–72.

Hughes, N. C. 1991. Morphological plasticity and genetic flexibility in a Cambrian trilobite. *Geology* 19:913–916.

Hughes, N. C. 1993. Distribution, taphonomy, and functional morphology of the Upper Cambrian trilobite *Dikelocephalus. Milwaukee Public Museum Contributions in Biology and Geology* 84:1–49.

Hughes, N. C. 1994. Ontogeny, intraspecific variation, and systematics of the Late Cambrian trilobite *Dikelocephalus. Smithsonian Contributions to Paleobiology* 79:1–89.

Hughes, N. C. and R. E. Chapman. 1995. Growth and variation in the Silurian proetide trilobite *Aulacopleura konincki* and its implications for trilobite palaeobiology. *Lethaia* 28:333–353.

Hughes, N. C. and R. A. Fortey. 1995. Sexual dimorphism in trilobites, with an Ordovician case study. In J. C. Cooper, M. L. Droser, and S. C. Finney, eds., *Ordovician Odyssey,* pp. 419–421. Los Angeles: Society of Economic Paleontologists and Mineralogists, Pacific Section.

Hughes, N. C. and C. C. Labandeira. 1995. The stability of species in taxonomy. *Paleobiology* 21:401–403.

Hughes, N. C. and A. W. A. Rushton. 1990. Computer-aided restoration of a Late Cambrian ceratopygid trilobite from Wales, and its phylogenetic implications. *Palaeontology* 33:429–445.

Hughes, N. C., G. O. Gunderson, and M. J. Weedon. 1997. Circumocular suture and visual surface of *"Cedaria" woosteri* (Trilobita, Late Cambrian) from the Eau Claire Formation, Wisconsin. *Journal of Paleontology* 71:103–107.

Hughes, N. C., R. E. Chapman, and J. M. Adrain. 1999. The stability of thoracic segmentation in trilobites: A case study in developmental and ecological constraints. *Evolution and Development* 1:24–35.

Ivantsov, A. Y. 1999. Trilobite-like arthropod from the Lower Cambrian of the Siberian Platform. *Acta Palaeontologica Polonica* 44: 455–466.

Jell, P. A. 1974. Faunal provinces and possible planetary reconstruction of the Middle Cambrian. *Journal of Geology* 82:319–350.

Jell, P. A. 1981. Trends and problems in Cambrian trilobite evolution. In M. E. Taylor, ed., *Short Papers for the Second International Symposium on the Cambrian System*, p. 91. U.S. Geological Survey Open-File Report 81-743.

Jell, P. A. 1989. Some aberrant exoskeletons from fossil and living arthropods. *Memoirs of the Queensland Museum* 27:491–498.

Jensen, S. 1990. Predation by Early Cambrian trilobites on infaunal worms—evidence from the Swedish *Mickwitzia* Sandstone. *Lethaia* 23:29–42.

Kiaer, J. 1917. The Lower Cambrian *Holmia* fauna at Tømten in Norway. *Norske Videnskaps Academi i Oslo Skrifter, Matematisk-videnskalelig Klasse* 10:1–141.

Li, X. and M. L. Droser. 1997. Nature and distribution of Cambrian shell concentrations: Evidence from the Basin and Range Province of the western United States (California, Nevada, and Utah). *Palaios* 12: 111–126.

Lieberman, B. S. 1997. Early Cambrian paleogeography and tectonic history: A bio-geographic approach. *Geology* 25:1039–1042.

Lieberman, B. S. 1999. Testing the Darwinian legacy of the Cambrian radiation using trilobite phylogeny and biogeography. *Journal of Paleontology* 73:176–181.

Loch, J. D., J. H. Stitt, and J. R. Derby. 1993. Cambro-Ordovician boundary interval extinctions: Implications of revised trilobite and brachiopod data from Mount Wilson, Alberta, Canada. *Journal of Paleontology* 67:497–517.

Lochman, C. 1947. Analysis and revision of eleven Lower Cambrian trilobites. *Journal of Paleontology* 21:59–71.

Lochman, C. and D. Duncan. 1944. Early Upper Cambrian faunas from central Montana. *Geological Society of America Special Papers* 54:1–181.

Lochman-Balk, C. and J. L. Wilson. 1958. Cambrian biostratigraphy in North America. *Journal of Paleontology* 32:312–350.

Ludvigsen, R. 1982. Upper Cambrian and Lower Ordovician trilobite biostratigraphy of the Rabbitkettle Formation, western District of Mackenzie. *Royal Ontario Museum Life Sciences Contributions* 134:1–188.

Ludvigsen, R. and S. R. Westrop. 1983. Trilobite biofacies of the Cambrian-Ordovician boundary interval in northern North America. *Alcheringa* 7:301–319.

McKinney, M. L. 1990. Trends in body-size evolution. In K. J. McNamara, ed., *Evolutionary Trends*, pp. 75–118. London: Belhaven Press.

McNamara, K. J. 1983. Progenesis in trilobites. *Special Papers in Palaeontology* 30: 59–68.

McNamara, K. J. 1986. Techniques of exuviation in Australian species of the Cambrian trilobite *Redlichia*. *Alcheringa* 10: 403–412.

Müller, K. J. and D. Walossek. 1987. Morphology, ontogeny, and life habit of *Agnostus pisiformis* from the Upper Cambrian of Sweden. *Fossils and Strata* 19:1–124.

Öpik, A. A. 1975. Templetonian and Ordian xystridurid trilobites of Australia. *Bureau of Mineral Resources, Australia, Bulletin* 121: 1–84.

Osgood, R. G., Jr. 1970. Trace fossils of the Cincinnati area. *Palaeontographica Americana* 41:281–439.

Osleger, D. and J. F. Read. 1993. Comparative analysis of methods used to define eustatic variations in outcrop: Late Cambrian interbasinal sequence development. *American Journal of Science* 293:157–216.

Owen, A. W. 1985. Trilobite abnormalities. *Transactions of the Royal Society of Edinburgh (Earth Sciences)* 76:255–272.

Palmer, A. R. 1965a. Biomere—a new kind of biostratigraphic unit. *Journal of Paleontology* 39:149–153.

Palmer, A. R. 1965b. Trilobites of the Late Cambrian Pterocephaliid biomere in the Great Basin, United States. *United States Geological Survey Professional Paper* 493: 1–105.

Palmer, A. R. 1979. Biomere boundaries revisited. *Alcheringa* 3:33–41.

Palmer, A. R. 1984. The biomere problem: Evolution of an idea. *Journal of Paleontology* 58:599–611.

Palmer, A. R. and R. B. Halley. 1979. Physical stratigraphy and trilobite biostratigraphy of the Carrara Formation (Lower and Middle Cambrian) in the southeastern Great Basin. *United States Geological Survey Professional Paper* 1047:1–131.

Pegel', T. V. 1982. Kharakter raspredeleniya trilobitovykh soobshchestv v Diringdinskom rifovom komplekse (kembriy yugo-zapadnogo Prianabartya) [Character of the distribution of trilobite communities in the Diringde reef complex (Cambrian of the southeastern Anabar area)]. In V. A. Astashkin, ed., *Stratigrafiya i fatsii osadochnykh basseynov Sibiri* [Stratigraphy and facies of sedimentary basins in Siberia], pp. 82–99. Novosibirsk: Siberian Scientific-Research Institute of Geology, Geophysics, and Mineral Resources.

Pratt, B. R. 1992. Trilobites of the Marjuman and Steptoean stages (Upper Cambrian), Rabbitkettle Formation, southern Mackenzie Mountains, northwest Canada. *Palaeontographica Canadiana* 9:1–179.

Pratt, B. R. 1998. Probable predation on Upper Cambrian trilobites and its relevance for the extinction of soft-bodied Burgess Shale–type animals. *Lethaia* 31:73–88.

Ramsköld, L. and G. D. Edgecombe. 1996. Trilobite appendage structure—*Eoredlichia* reconsidered. *Alcheringa* 20:269–276.

Rasetti, F. 1948. Lower Cambrian trilobites from the conglomerates of Quebec (exclusive of the Ptychopariidae). *Journal of Paleontology* 22:1–24.

Repina, L. N. 1968. Biogeografiya rannego kembriya Sibiri po trilobitam [Biogeography of the Early Cambrian of Siberia according to trilobites]. In R. T. Hecker, ed., *Problemy paleontologii: Doklady sovetskikh geologov na XXIII sessii Mezhdunarodnogo geologicheskogo kongressa* [Problems of paleontology: Papers of Soviet geologists on the 23rd session of the International Geological Congress], pp. 46–56. Moscow: Nauka.

Repina, L. N. 1982. Ekotipy olenelloidnykh trilobitov i ikh rasprostranenie v perekhodnom tipe razreza (Siberskaya platforma) [Olenelloid trilobite ecotypes and their distribution in the transitional type sequence (Siberian Platform)]. *Trudy, Institut geologii i geofiziki, Sibirskoe otdelenie, Akademiya nauk SSSR* 510:46–60.

Repina, L. N. 1985. Rannekembriyskie morya

zemnogo shara i paleobiogeograficheskie podrazdeleniya po trilobitam [Early Cambrian global seas and their paleobiogeographic subdivision by trilobites]. *Trudy, Institut geologii i geofiziki, Sibirskoe otdelenie, Akademiya nauk SSSR* 626: 5–17.

Richter, R. and E. Richter. 1941. Das Kambrium am Toten Meer und die altese Tethys. *Abhanglungen von der Senkenbergischen Naturforschenden Gesellschaft* 460: 1–50.

Robison, R. A. 1975. Species diversity among agnostid trilobites. *Fossils and Strata* 4: 219–226.

Robison, R. A. 1991. Middle Cambrian biotic diversity: Examples from four Utah Lagerstätten. In A. Simonetta and S. Conway Morris, eds., *The Early Evolution of Metazoa and the Significance of Problematic Taxa,* pp. 77–98. Cambridge: Cambridge University Press.

Rushton, A. W. A. and N. C. Hughes. 1996. Biometry, systematics, and biogeography of the Late Cambrian trilobite *Maladioidella abdita. Transactions of the Royal Society of Edinburgh (Earth Sciences)* 86: 247–256.

Saltzman, M. R., J. P. Davidson, P. Holden, B. N. Runnegar, and K. C. Lohmann. 1995. Sea-level driven changes in ocean chemistry at an Upper Cambrian extinction horizon. *Geology* 23: 893–896.

Schwimmer, D. R. 1975. Quantitative taxonomy and biostratigraphy of Middle Cambrian trilobites from Montana and Wyoming. *Mathematical Geology* 7: 149–166.

Seilacher, A. 1985. Trilobite palaeobiology and substrate relationships. *Transactions of the Royal Society of Edinburgh (Earth Sciences)* 76: 231–337.

Shergold, J. H. 1988. Review of trilobite biofacies distributions at the Cambrian-Ordovician boundary. *Geological Magazine* 125: 363–380.

Shu, D., G. Geyer, L. Chen, and X. Zhang. 1995. Redlichiacean trilobites with preserved soft-parts from the Lower Cambrian Chengjiang fauna (South China). *Beringeria Special Issue* 2: 203–241.

Stitt, J. H. 1971. Late Cambrian and earliest Ordovician trilobites, Timbered Hills and Lower Arbuckle groups, Western Arbuckle Mountains, Murray County, Oklahoma. *Oklahoma Geological Survey, Bulletin* 110: 1–80.

Stitt, J. H. 1975. Adaptive radiation, trilobite paleoecology, and extinction, Ptychaspidid biomere, Late Cambrian of Oklahoma. *Fossils and Strata* 4: 381–390.

Stitt, J. H. 1976. Functional morphology and life habits of the Late Cambrian trilobite *Stenopilus pronus* Raymond. *Journal of Paleontology* 50: 561–756.

Stitt, J. H. 1977. Late Cambrian and earliest Ordovician trilobites, Wichita Mountains area, Oklahoma. *Oklahoma Geological Survey, Bulletin* 124: 1–75.

Stitt, J. H. 1983. Enrolled Late Cambrian trilobites from the Davis Formation, southeast Missouri. *Journal of Paleontology* 57: 93–105.

Stitt, J. H., J. D. Rucker, N. D. Boyer, and W. D. Hart. 1994. New *Elvinia* Zone (Upper Cambrian) trilobites from new localities in the Collier Shale, Ouachita Mountains, Arkansas. *Journal of Paleontology* 68: 518–523.

Stubblefield, C. J. 1959. Evolution in trilobites. *Journal of the Geological Society, London* 115: 145–162.

Sundberg, F. A. 1996. Morphological diversification of Ptychopariida (Trilobita) from the Marjumiid biomere (Middle and Upper Cambrian). *Paleobiology* 22: 49–65.

Taylor, M. E. 1977. Late Cambrian of western North America: Trilobite biofacies, environmental significance, and biostratigraphic implications. In E. G. Kauffman and J. E. Hazel, eds., *Concepts and Methods of Biostratigraphy,* pp. 397–425. Strouds-burg: Dowden, Hutchinson and Ross.

Trammer, J. and A. Kain. 1997. Body size and diversity exemplified by three trilobite clades. *Acta Palaeontologica Polonica* 42:1–12.

Wagner, P. J. 1995. Testing evolutionary con-straint hypotheses: Examples with early Paleozoic gastropods. *Paleobiology* 21:248–272.

Westergård, A. H. 1936. *Paradoxides oelandicus* beds of Öland. *Sveriges Geologiska Undersökning* C394:1–66.

Westrop, S. R. 1983. The life habits of the Or-dovician illaenine trilobite *Bumastoides.* *Lethaia* 16:15–24.

Westrop, S. R. 1986. Trilobites of the Up-per Cambrian Sunwaptan Stage, southern Canadian Rocky Mountains, Alberta. *Palaeontographica Canadiana* 3:1–179.

Westrop, S. R. 1988. Trilobite diversity pat-terns in an Upper Cambrian Stage. *Paleobiology* 14:401–409.

Westrop, S. R. 1989. Macroevolutionary im-plications of mass extinction—evidence from an Upper Cambrian Stage boundary. *Paleobiology* 15:46–52.

Westrop, S. R. 1991. Intercontinental varia-tion in mass extinction patterns: Influ-ence of biogeographic structure. *Paleobiol-ogy* 17:363–368.

Westrop, S. R. 1995. Sunwaptan and Ibex-ian (Upper Cambrian–Lower Ordovician) trilobites of the Rabbitkettle Formation, Mountain River region, northern Macken-zie Mountains, northwest Canada. *Palae-ontographica Canadiana* 12:1–75.

Westrop, S. R. 1996. Temporal persistence and stability of Cambrian biofacies: Sun-waptan (Upper Cambrian) trilobite faunas of North America. *Palaeogeography, Palaeo-climatology, Palaeoecology* 127:33–46.

Westrop, S. R. and J. M. Adrain. 1998. Trilo-bite alpha diversity and the reorganization of Ordovician benthic marine communi-ties. *Paleobiology* 24:1–16.

Westrop, S. R. and R. Ludvigsen. 1987. Bio-geographic control of trilobite mass ex-tinction at an Upper Cambrian "biomere" boundary. *Paleobiology* 13:84–99.

Westrop, S. R., J. V. Tremblay, and E. Landing. 1995. Declining importance of trilobites in Ordovician nearshore paleocommunities: Dilution or displacement? *Palaios* 10:75–79.

Wheeler, W. C., P. Cartwright, and C. Y. Ha-yashi. 1993. Arthropod phylogeny: A com-bined approach. *Cladistics* 9:1–39.

Whittington, H. B. 1975. Trilobites with ap-pendages from the Middle Cambrian Bur-gess Shale, British Columbia. *Fossils and Strata* 4:97–136.

Whittington, H. B. 1977. The Middle Cam-brian trilobite *Naraoia,* Burgess Shale, Brit-ish Columbia. *Philosophical Transactions of the Royal Society of London B* 280:409–443.

Whittington, H. B. 1980. Exoskeleton, moult stage, appendage morphology, and habits of the Middle Cambrian trilobite *Olenoides serratus. Palaeontology* 23:171–204.

Whittington, H. B. 1985. *Tegopelte gigas,* a sec-ond soft-bodied trilobite from the Burgess Shale, Middle Cambrian, British Colum-bia. *Journal of Paleontology* 59:1251–1274.

Whittington, H. B. 1989. Olenelloid trilobites: Type species, functional morphology, and higher classification. *Philosophical Transac-tions of the Royal Society of London B* 324:111–147.

Whittington, H. B. 1990. Articulation and ex-uviation in Cambrian trilobites. *Philosoph-ical Transactions of the Royal Society of Lon-don B* 329:27–49.

Whittington, H. B. and D. E. G. Briggs. 1985.

The largest Cambrian animal, *Anomalocaris*, Burgess Shale, British Columbia. *Philosophical Transactions of the Royal Society of London B* 309:569–609.

Wills, M. A., D. E. G. Briggs, and R. A. Fortey. 1994. Disparity as an evolutionary index: A comparison of Cambrian and Recent arthropods. *Paleobiology* 20:93–130.

Wills, M. A., Briggs, D. E. G., Fortey, R. A. and M. Wilkinson. 1995. The significance of fossils in understanding arthropod evolution. *Verhandlungen der Deutschen Zoologischen Gesellschaft* 88:203–215.

Wilson, J. L. 1957. Geography of olenid trilobite distribution and its influence on Cambro-Ordovician correlation. *American Journal of Science* 255:321–340.

Zhang, W.-T. and P. A. Jell. 1987. *Cambrian Trilobites of North China.* Beijing: Science Press.

Zhu, Z.-L. and L.-F. Jiang. 1981. An Early Cambrian trilobite fanule from Yeshan, Luhe district, Jiangsu. *Geological Society of America Special Paper* 187:153–159.

Zhuravlev, A. Yu. and R. A. Wood. 1996. Anoxia as the cause of the mid-Early Cambrian (Botomian) extinction event. *Geology* 24:311–314.

CHAPTER EIGHTEEN

Graham E. Budd

Ecology of Nontrilobite Arthropods and Lobopods in the Cambrian

Arthropods and lobopods first appear for certain in the body fossil record in the Atdabanian and, at the time of this appearance, already exhibit a wide spread of ecologic strategies. Investigation of Cambrian arthropod ecology is hampered, however, by three factors: the paucity of authentic nontrilobite trace fossils; the restriction of the wide variety of poorly sclerotized taxa to the principal Cambrian Lagerstätten, which may not necessarily provide a representative aliquot of Cambrian environments; and the continuing lack of firm consensus over the systematics of nontrilobite forms. Cambrian arthropod ecology is thus still largely based on functional morphology, with as yet only a poor understanding of ecologic interactions and trophic webs. In recent years several promising areas for research into early arthropod ecologies have emerged, including the study of previously unsuspected miniature taxa from Swedish orsten and the Canadian Mount Cap Formation. Such discoveries have demonstrated that Cambrian arthropods played a critical role at all levels of the trophic web, as indeed they continue to do today. However, a few strategies (e.g., sessile filter feeding, mineralization of limbs) are probably not present in the Cambrian. Moreover, the ecologic sophistication of Cambrian arthropods was limited by their relatively simple body plans, involving a small number of tagmata, as defined with reference to their segment types. This simplicity, which reflects a primitive deployment of homeotic genes rather than the much more complex patterns seen in advanced arthropods, may have been an important factor in distinguishing Cambrian from Recent ecologies.

The recent recognition of the "lobopods" as an important morphologic grouping in the Cambrian was entirely unexpected. Although some distance must be covered before a full understanding of their systematics is attained, they appear to form a paraphyletic grade, out of which the arthropods emerged, probably via the Anomalocaris-like taxa (Anomalocaris, Opabinia, and Kerygmachela, plus related forms). As such, they constitute the stem group to the arthropods, but with the Onychophora, Tardigrada, and perhaps the Pentastomida as extant representatives.

They exhibit an astonishing variety of ecologies, including ecto- and endoparasitism, predation, miniaturization, and scavenging. The range of ecologic strategies seen in the lobopods may be allied directly to the development of arthropodization; several key morphologic innovations may be identified.

The evolution of arthropod ecology is hard to track, but one possibility is that the euarthropods are primitively predatory, with more derived taxa radiating to fill lower ecologic niches previously occupied by lobopods.

THE ARTHROPODS TODAY make up perhaps 80 percent of animals, and their dominance was scarcely less in the historic record: indeed, their importance in the *marine* realm is likely to have been even greater in the past than it is today. On the basis of trace fossils (*Rusophycus*), arthropods are known from at least the Tommotian onward. They have certainly been important contributors to ecologic webs and hierarchies throughout the Cambrian. Discerning ecologic paths and strategies of the past is, however, fraught with difficulties. It is essential, if a better understanding of arthropod ecologies in the past is to be obtained, that these difficulties are clearly identified and obviated as far as is possible. They include the following:

1. A general lack of what have been termed *holotaphic biotas*
2. Problems of environmental interpretation
3. Problems of functional interpretation
4. Poorly understood high-level systematics (making the tracing of evolutionary pathways in ecology difficult)
5. A lack of body/trace fossil correlation

Despite these difficulties, arthropod ecology in the Cambrian need not stay at a "Just-So" level, for several important discoveries in the past few years have added considerable and important new data to that already accumulated.

NOTE ON TERMINOLOGY

The animals under discussion in this paper pose certain nomenclatural problems that need to be addressed in order to avoid ambiguity in subsequent discussion. In the phylogenetic scheme of Budd (1996a, 1997, 1999), animals that might broadly be described as lobopods, including the extant Onychophora and the Cambrian onychophoran-like taxa, form a paraphyletic assemblage from which—via the anomalocaridid-like taxa—the true arthropods emerge: all of the taxa together comprise the Lobopodia. Without a detailed and highly cumbersome nomenclatural scheme to resolve the nomenclatural problems caused by such grade changes (cf. Craske and Jefferies 1989), a commonsense approach is taken here, as follows: (1) *lobopod* will be used in a general way to denote a grade of organization typified by the onychophorans and the onychophoran-like taxa in the Cambrian (e.g., *Hallucigenia, Onychodictyon, Xenusion*), and it will also be applied to the tardigrades, following common but not universal practice (even if tardigrades turn out taxonomically to belong within the next grouping); (2) *Kerygmachela, Opabinia,* and the anomalocaridids will be referred to as *anomalocaridid-like taxa,* with the recognition that they possess a mix of both lobopod and arthropod characters; (3) *arthropods* will be applied to all taxa above the grade of the anomalocaridids (i.e., crown-group arthropods plus the adjacent plesions above the level of anomalocaridids); (4) *euarthropods* will be applied to the smallest clade that is inclusive of all living arthropods.

DATA SOURCES

The data for the study of the ecology of Cambrian lobopods, anomalocaridid-like taxa, and arthropods can be divided into five broad categories, each of which will be briefly examined, before taking a more detailed look at what conclusions they may lead to.

1. Burgess Shale–Type Faunas

Conway Morris (1989) and Butterfield (1995) identified 30 or so faunas from around the world, spread through the Lower and Middle Cambrian, that broadly conform in terms of preservation and faunal content to those of the Burgess Shale (Middle Cambrian, British Columbia). Their faunal coverage ranges from borehole material containing just a few taxa, through to major deposits of thousands of specimens and dozens of taxa, notably the "big three": the Burgess Shale itself, the Chengjiang fauna of South China (Hou et al. 1991), and the Sirius Passet fauna of North Greenland (Conway Morris et al. 1987). Arthropods are an important component of all these faunas.

2. Orsten and Similar Deposits

Dissolution of orsten ("stinkstone") nodules in the *Agnostus pisiformis* level of the Alum Shale of southern Sweden and northern Germany has yielded many exceptionally well preserved, phosphatized arthropods (e.g., Müller and Walossek 1985a,b,c, 1987, 1988; Walossek 1993), mostly crustacean-like in appearance (one exception being *Agnostus* itself). All of them are tiny, with the largest being less than 2 mm in length. Although many of them represent juvenile stages, it is now clear that adults are also present. Much of their anatomy has been preserved, allowing detailed suggestions about their ecology to be made. Another locality in Russia has yielded similar forms (Müller et al. 1995), and such fossils may be much more widespread than previously supposed (for similar examples from the Middle Cambrian of Australia and the Cambrian-Ordovician boundary strata of Newfoundland, see also Walossek et al. 1993, 1994, respectively).

3. Mount Cap Fragments

Of potentially equal interest are the fragments recovered from the Mount Cap Formation (Butterfield 1994). These are organic residues, again of tiny size, but with remarkable fidelity of preservation of limb structures of unidentified but cladoceran-like arthropods. Their preservation in shales, coupled with their tiny size, in some ways provides a link between the orsten and Burgess Shale–type deposits. Again, there are some indications that such preservation is widespread (e.g., Palacios and Vidal 1992: figure 7g).

4. Other Deposits

Nontrilobite arthropods and lobopods are known, rarely, from sources that cannot be readily contained within the above categories. These include, for example, the fairly widespread occurrence of aglaspidids and, toward the Upper Cambrian, so-called phyllocarid crustaceans, but also singulars such as the large lobopod *Xenusion* from Swedish Kalmarsund Sandstone erratics found in Germany (e.g., Dzik and Krumbeigel 1989). However, the conventional record is dominated by trilobites.

5. Trace Fossils

The record of trace fossils is unfortunately extremely impoverished. Well-attested arthropod trace fossils from the Cambrian, such as *Cruziana* and *Rusophycus*, are normally assigned to the trilobites (Hughes, this volume; but see also Pratt 1994; Crimes, this volume). Other traces may well also have an arthropodan origin, but evidence based on trace and trace maker co-occurrence and functional morphology is lacking. A notable exception is provided by traces from the Czech Paseky Shale, which are attributed to various nontrilobite arthropods and appear to have been made in a nonmarine environment (Chlupáč 1995; Mikuláš 1995; see also Osgood 1970 and Hesselbo 1988 for examples of aglaspidid traces). Traces that can be confidently assigned to chasmataspid chelicerates are known from the Upper Cambrian of Texas (Dunlop et al. 1996). Finally, there is a limited amount of information from the study of coprolites (e.g., those attributed to *Anomalocaris* by Conway Morris and Robison [1988]).

PREVIOUS APPROACHES TO CAMBRIAN ARTHROPOD ECOLOGY

Speculations about Cambrian arthropod ecology have naturally centered around the Burgess Shale, in connection with the reinvestigation by H. B. Whittington and coworkers (Whittington 1985; see Gould 1989 and Conway Morris 1998 for reviews). These have broadly fallen into two groupings: those about *functional morphology* of individual taxa and those about *ecologic interactions*. Although these studies have been illuminating, they both have inevitable shortcomings. Fortey (1985), dealing mostly with post-Cambrian trilobites, has carefully detailed the sorts of assumptions and re-sults possible from functional morphology, listing paradigmatic, constructional, and geologic approaches as being most important (see also e.g., Valentine 1973). Briggs and Whittington (1985) surveyed possible modes of life of Burgess Shale arthropods, placing 23 species into 6 categories (predatory and scavenging benthos; deposit-feeding benthos; scavenging and possibly predatory nektobenthos; deposit-feeding and scavenging benthos; nektonic filter feeders; and an "others" group). Analyses of this sort rely on knowledge not only of the overall morphology of the animal but also of the limbs, and even in Burgess Shale taxa this knowledge is often incomplete.

Although this cautious methodology of Fortey (1985) and Briggs and Whitting-ton (1985) has the advantage of removing from consideration effectively untestable hypotheses (for example, the several theories about agnostid ecologies such as mim-icry [Lamont 1967] or algal clinging [Pek 1977]), what one is left with can often ap-pear rather unsatisfactory. In particular, it leads to the assignment of vague, "deposit-feeding, benthos" sorts of lifestyles to large numbers of arthropods, even where their limbs are known in some detail. One question to be addressed then is whether this nebulosity comes about through lack of data or through a genuine lack of arthropod specialization; this question is discussed below.

The only full-scale investigation of interactive Burgess Shale ecology is that of Con-way Morris (1986), in which an attempt is made to identify a trophic web and to model the species distribution in terms of ecologic theory, although the Burgess Shale, like many other fossil faunas, is best modeled by a log-normal distribution rather than one more suited to a standard ecologic model (see discussion in May 1975; Conway Morris 1986). More recently, a preliminary account of the Chengjiang fauna has been given (Leslie et al. 1996), showing a very large numerical preponderance of arthro-pods in the overall distribution of taxa, although the study did not attempt to distin-guish between carcasses and molts, which would inflate the proportion of arthropods. The Sirius Passet fauna is similarly dominated by arthropods (pers. obs.).

ECOLOGY OF ARTHROPOD TAXA

I now turn to addressing in more detail the possibilities available for different groups of taxa or, where more appropriate, different ecologic realms. It should be stressed that no attempt at a comprehensive "ecology" is made here; instead, some subjects of particular interest are examined. This discussion should of course be complemented by referral to the Cambrian trilobites (Hughes, this volume), which naturally fall into the purview of Cambrian arthropod ecology. The first section focuses on three areas of recent interest: the morphologic "disparity" displayed by arthropods and its eco-logic implications, planktic filter-feeding arthropods, and predation. The second sec-tion deals with the lobopods and with *Anomalocaris* and its relatives. Finally, the evo-lution of arthropod ecology is considered as a whole.

Arthropods

Macrobenthic and Nektobenthic Arthropods:
Disparity as a Key to Ecologic Complexity

This category, although cumbersome, is nevertheless meant to identify a large and ecologically coherent group of arthropods, those of relatively large size and that interact with the sediment or other taxa living on or in it. Such taxa have been the focus of most of the studies of morphology and phylogeny in Cambrian nontrilobite arthropods, such as those previously mentioned of Briggs and Whittington (1985) and Fortey (1985). Further, and of importance to their ecology, they have also been the focus of some morphologic studies.

It is possible to examine the morphology of arthropods at more than one level. One approach is that of Wills et al. (1994), who used an overall morphology metric for assigning a concrete measure of what has rather loosely been called disparity between Cambrian and Recent arthropods. Perhaps surprisingly, they discovered that the disparity, when considered as morphospace occupancy and thus a measure of the total morphometric distance between taxa, was more or less identical between the representative groups of taxa they chose from the Cambrian and the Recent. From these results, one might make an allied claim that Cambrian arthropod ecology (in some way surely a reflection of morphology) has also remained at a similar level of complexity throughout the Phanerozoic.

Although the general approach of Wills et al. (1994) seems reasonable, it appears to contradict earlier (if rather neglected) work by Flessa et al. (1975) and Cisne (1974), which employed a remarkably novel technique for examining the change in arthropod ecology through time—that of information theory analysis. By taking a measure of the *complexity* of particular arthropod body plans, based on the permutations available of segment types, they demonstrated that during the Phanerozoic there had been a striking monotonic increase in body-plan complexity among marine arthropod orders (see also Wills et al. 1997).

I have adapted and simplified their approach here to deconvolute segmentation and segment types to demonstrate very similar patterns. Using the data of Wills et al. (1994), in terms of a morphospace defined only by segment diversity and numbers, both Cambrian and Recent arthropods have been plotted (figure 18.1). As may be seen, those of the Cambrian occupy a significantly different (and smaller) region than that of the extant ones. Cambrian arthropods—considered at the level of their tagmosis—are less complex and occupy a smaller morphospace than their Recent counterparts. However, the question may be asked, why is this analysis not rendered invalid by the more detailed and more multimetric approach of Wills et al. (1994)? To address this point, one needs to turn to the interaction between the hierarchical organization of the genome and its role in specifying body plan. Briefly, it is possible to argue that there is a fairly clear correspondence between the region of operation of

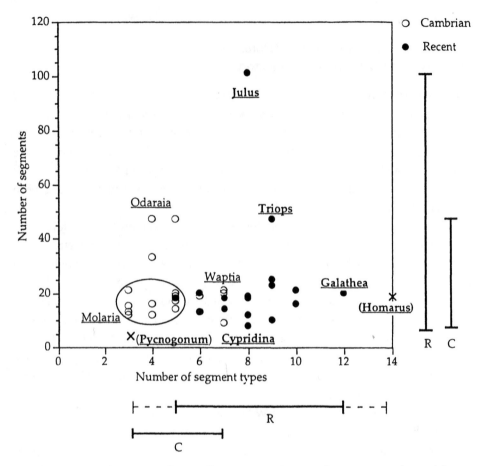

Figure 18.1 Plot of arthropods from Wills et al. (1994), showing segment diversity and number for Cambrian and extant arthropods. Two arthropods that significantly increase the range of extant morphology are also included: *Pycnogonum* and *Homarus*, an advanced decapod. Most of the Cambrian Problematica lie within the oval marked. Data from Cisne (1974), Wills et al. (1994), and personal observation.

specific and hierarchically arranged genes (segmentation and homeotic genes) and how the body plan develops at a gross level, including numbers and diversity of segments (see Akam 1995). In other words, the rather diffuse concept of a "body plan" may be broken down into hierarchical levels, which are each in principle open to analysis. By examining the body plan at these levels, one is examining a partially decoupled level of operation of the genome. If, conversely, *all* morphologic information is considered together in an undifferentiated manner, then the signal coming from specific types of morphology—in this case, tagmosis—may be obscured.

The results of this analysis confirm some rather widely held prejudices that Cambrian arthropods are in general much simpler in terms of within-body segment differentiation than arthropods of the later Phanerozoic. A view sometimes expressed, that trilobites (for example) would not be out of place in a modern benthic commu-

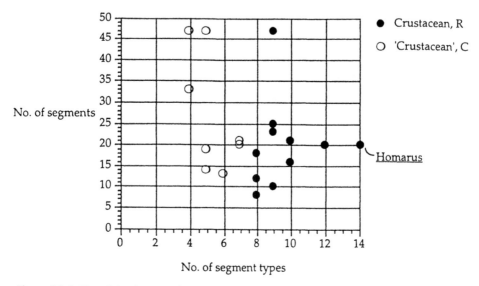

Figure 18.2 Plot of Cambrian and extant taxa falling within the "crustacean" clade of Wills et al. (1994), showing segment diversity and number.

nity, therefore seems unjustified. Trilobites, like most other Cambrian arthropods, and in particular almost all of the "problematic" arthropods (cf. Gould 1989) may be seen to have a distinctly archaic look. Only the pycnogonids of extant arthropods are as lacking in tagmosis as the trilobites (figure 18.1). By contrast, the number of segments tends to decrease from the Cambrian to the Recent, although somewhat less dramatically and with some notable exceptions, such as *Vachonisia* from the Devonian Hunsrück Shale (Stürmer and Bergström 1976), and some of the modern myriapods. One of the reasons for this change is the great rise to dominance of the crustaceans, especially after the eumalacostracan radiations of the Carboniferous. To demonstrate therefore that one is not simply seeing an effect of "clade replacement," one can plot the difference between taxa that fall into a crustacean clade (as identified by Wills et al. 1994) and their selection of extant crustaceans (figure 18.2), with *Homarus* added as an example of the most complex types of crustaceans. It should be noted that severe doubts have been expressed as to the true affinities of some of these taxa (e.g., Walossek 1999). The total morphospace occupancy is greater in the extant fauna (although not greatly so), but the most striking point is that the two areas of morphospace occupancy have no overlap: in terms of tagmosis the most highly differentiated Cambrian taxa are less complex than the least differentiated of the extant examples. Clearly, within what is allegedly the same clade, an increase in complexity is taking place.

The striking contrast between these two sets of results from the same data set suggests several interesting interpretations. First, it is clear that the Cambrian taxa look odd to our eyes partly because they have their own set of adaptations; an example is the "great appendages" possessed by taxa such as *Leanchoilia* (Bruton and Whittington 1983). Yet it is very likely that these appendages, although different in detail, are performing similar tasks to those possessed by extant arthropods. This is therefore a case of similar adaptive needs producing varied responses, although no doubt within a strong constraint of functionality. Given that (it must be repeatedly stressed) we have no particular reason to regard ancient arthropods as merely imperfect versions of more up-to-date representatives (a view perhaps partly engendered by comparison with the development of our own creations such as mechanical means of transport), there is no reason to doubt that they were as well adapted to their conditions as are modern arthropods. With this background, one might therefore expect the *detailed* complexity of limbs and so on to be equal between the Cambrian and Recent.

Nevertheless, important differences remain at the level of the tagmosis. One may have variations on themes in both the Cambrian and the Recent faunas; but the themes themselves are different. Within a regime provided by homeotic genes interacting in only a simple way, the Cambrian forms elaborate particular segments in unfamiliar ways, but their overall morphologies are strongly constrained by their lack of tagmosis. The most strikingly different region is the head, where Cambrian taxa in general have almost homonomous limbs, with the exception of a frontal pair. Most of the post-Cambrian change comes about in the reorganization and specialization of head appendages. Trilobites, for example, possess three or four pairs of postoral cephalic appendages, but the morphology hardly differs from that of thoracic ones. Cambrian crustaceans may possess a mandible, but the maxillae are hardly differentiated from the thoracic appendages, a pattern repeatedly seen in Cambrian arthropods. By contrast, an extant decapod crustacean has three highly specialized postoral cephalic appendages (mandible and two maxillae) and may also possess differentiated thoracic appendages. This contrast in tagmosis patterns between the Cambrian and the Recent has important implications for the evolution of arthropod ecology, because segment specialization lies at the heart of arthropod adaption.

The sets of specialized appendages possessed by extant crustaceans can be marshaled to perform a variety of extremely complex maneuvers. For example, extant lobsters such as *Homarus* and *Nephrops* have almost all of their appendages functionally differentiated in one way or another: for sensory purposes, feeding (chewing, crushing, shredding), swimming, copulation, grooming, and egg brooding, for example. Barker and Gibson (1977) filmed *Homarus gammarus,* the European lobster, feeding on pieces of boiled fish. The cephalic appendages are employed in a highly coordinated manner:

1. The morsel is picked up with the second pereiopod, then passed to the third maxillipeds, trapping it between the ischiopodites.

2. As the second and third maxillae move away laterally, the third maxilliped moves up toward the mandibles, which catch hold of the food particle.

3. The third maxillipeds move down again, tearing the food between them and the mandibles, while the other mouthparts move inward to assist in the tearing.

4. The food particle thus removed from the main part is released from the mandibles and pushed downward by the tips of the second maxillipeds.

5. The first and second maxillae curve around the mouth and manipulate the food particle into the mouth.

When a crustacean is faced with live prey, the procedure is likely to be more complex. Observations on the blue crab showed that prey was trapped by the thoracic limbs' forming a sort of cage, while the mouthparts and associated appendages carefully examined and manipulated the prey. In short, modern crustaceans employ a large number of feeding strategies, with often the same taxon utilizing different feeding mechanisms according to circumstance. This adaptability and utility was surely limited in most Cambrian forms. The general lack of well-differentiated cephalic mouthparts would imply, for example, that filter feeding would not even be a possibility for many taxa in the Burgess Shale (the plumose appendages of *Marrella* seem to be in the wrong position to be able to trap food particles that subsequently could be conveyed to the mouth—see Briggs and Whittington 1985 for discussion). Similarly, for the taxa listed as possible detritus feeders by Briggs and Whitington (1985), the general lack of appendage differentiation would limit the ability of the taxa to sort material prior to ingestion, making this mode of feeding rather inefficient. It thus seems likely that putatively predatory arthropods such as some *Naraoia* and *Sidneyia* (see the section "Predation in the Cambrian" below) employed a simple gnathobasic feeding technique like that of the extant *Limulus,* but that their other ecologic strategies were restricted.

At a deeper level, one might pose the question, what effect does tagmosis actually have on arthropod ecology? Even if it is true that complex tagmosis allows a greater diversity of behavior, what effect does this have on the fundamentals of ecology, for example, on the efficiency of energy transfer from one trophic level to the next? Specialization may on the one hand allow greater efficiency, although the gains from the ability to select food more efficiently may be offset to a certain extent by the greater energy involved in performing more-complicated tasks. Conversely, greater complexity may not imply anagenetic "grade improvement" but rather may be a side effect, either of "ecologic escalation" (Vermeij 1987) or of the dynamics of gene interaction (cf. Kauffman 1993 for a study of the behavior of complex systems). Hard data to study the effects of arthropod specialization are in any case hard to obtain. The only full-scale attempt at ecologic reconstruction of the Burgess Shale fauna (Conway Morris 1986) made estimations of the efficiency of transfer of energy between trophic levels and found that, considered in terms of numbers of individuals at different trophic levels, there was approximately a 7 percent efficiency of energy from primary consumers to predators/scavengers, which may be compared with the 10–20 percent efficiencies quoted for modern communities. If this difference is real and not a taphonomic artifact (predators may be less armored than their prey and thus may be less easily preserved), then Cambrian trophic webs should be correspondingly shorter than modern ones. Arthropod feeding inefficiency may be one determinant factor.

Filter Feeding: Complexity in the Microscopic Realm

The most specialized arthropods (based on tagmosis) in the Cambrian appear to be represented mainly by the tiny orsten fauna (see under "Data Sources," above), many of which were filter feeders. In addition, Briggs and Whittington (1985) suggested a nektonic filter-feeding lifestyle for *Sarotrocercus, Perspicaris,* and *Odaraia,* based on criteria such as the apparent lack of walking limbs, no sediment preserved in the gut, and large eyes.

The orsten arthropods seem to have had a wide range of ecologies, from ectoparasitism through to planktic and benthic lifestyles (for reviews, see Müller and Walossek 1985a,b; Walossek 1993). Many of the fauna as preserved, however, are interpreted to have been living in a flocculent layer near the sea floor: the absence of adults (e.g., *Rehbachiella*) or larvae (e.g., *Skara*) may give hints about migration in and out of the flocculent layer during the life cycles. For some of the taxa, such as *Skara* and *Bredocaris,* both larvae and adults are inferred to have lived on or close to the sediment-water interface. However, other taxa such as *Rehbachiella* seem to have been active swimmers and, progressively through a nauplius-metanauplius ontogeny, appear to have become better equipped filter feeders, presumably in more or less clear water. A similar mode of life has been inferred for Mount Cap arthropods, which preserve delicate filtrational setae a few micrometers wide (Butterfield 1994). These latter have been compared to extant cladocerans, although it is impossible to know from the fragments so far recovered what their overall morphology was. Nevertheless, the surmise by Butterfield (1994) that these taxa were components of the filter-feeding plankton seems reasonable, given the resemblance of recovered fragments to extant filter-feeding cladocerans (Butterfield, this volume).

The tiny arthropods of the orsten fauna, although in general possessing poorly differentiated maxillae, may represent an acme of specialization within Cambrian arthropods, at least insofar as they possess segmentation that is among the most diverse of all Cambrian arthropods. This suggestion of specialization may also be supported by consideration of two other factors: feeding mechanism and size. Walossek (1993) supports the insight of Cannon (1927) that the various crustacean filtering mechanisms are derived and do not represent the feeding mode of the last common ancestor of crown-group Crustacea. This view is supported by recent studies of the details of filtering mechanisms (e.g., Fryer 1987). As far as feeding strategy is concerned, the orsten crustaceans may represent derived states. Furthermore, although it has sometimes been suggested that small size typifies the sister groups of many extant clades, with the implication being that the last common ancestors of many extant clades were also tiny (Fortey et al. 1996), such a reconstruction does not seem to hold true for the arthropods (see Budd and Jensen 2000 for discussion). Despite the tardigrades' being of millimetric size and probably representing the *extant* closest living relatives of the euarthropods (Nielsen 1994; Budd 1996a; Dewel and Dewel 1996), in the context of a reconstructed arthropod stem group, their small size may be seen to be a derived feature (Budd 1996a). Closer relatives to the arthropods such as the anomalo-caridids *Kerygmachela* and *Opabinia* (Budd 1996a) are all of at least moderate size. If the stronger suggestion that the arthropods actually evolved from within a paraphyletic assemblage of anomalocaridid-like animals (Budd 1997) can be sustained, then crown-group arthropods, far from being primitively small, would be primitively huge (perhaps 300 mm or more in length).

It is noteworthy that all the specialized Cambrian filter feeders are demonstrably either crustaceans or crustacean-like, suggesting in turn that the later preeminence of crustaceans during the Phanerozoic may have been presaged by their complexity and, through their ability to modify their tagmosis, by their adaptability. Whether or not the linking of different ecologic systems by the evolution of arthropod filter feeding was an important factor in determining later metazoan diversification, as suggested by Butterfield (1994), the discovery of these miniature arthropods has emphasized once again how few of the routes of energy transfer in Cambrian ecosystems are directly indicated by the conventional fossil record.

Predation in the Cambrian

There has been a long debate about the presence and nature of predators in the Cambrian (see Conway Morris 1986 for review). It is now generally agreed that the activity of predators has been underemphasized, with new information such as the apparent hunting behavior of olenelloid trilobites (Jensen 1990), the description of predation-based healed injuries in trilobites (e.g., Conway Morris and Jenkins 1985), and the recognition of large, apparently predatory forms such as *Anomalocaris* (see below). Arthropods have been heavily implicated as culprits in Cambrian predation. The case rests on four lines of evidence: functional morphology (e.g., *Naraoia* [Whittington 1977] and *Sidneyia* [Bruton 1981] from the Burgess Shale possess gnathobases, and *Sanctacaris* [Briggs and Collins 1988] possesses raptorial appendages); gut contents (e.g., trilobite fragments found in the gut of *Sidneyia* [Bruton 1981]); trace fossils (see Pratt 1994 for one of the very few possible examples); and mutual cooccurrences (as has been argued for *Anomalocaris,* e.g., Vorwald 1982). Further remarks on the evolution of arthropod predatory behavior are made below in the context of the evolution of arthropod ecology.

Cambrian Lobopods

The discovery that lobopods were a significant component of Cambrian biotas (e.g., Ramsköld and Hou 1991) has been unexpected and fascinating, not least because celebrated Problematica such as *Microdictyon* and *Hallucigenia* have been recognized as such. In this respect, lobopods may be compared to the brachiopods and priapulids (Conway Morris 1977a), which seem to have been prominent in the Cambrian but are hardly represented in today's biota. Indeed, the lobopods are reduced to small numbers today. The ancestors of the Onychophora probably made the transition to land sometime before the Carboniferous (Thompson and Jones 1980) and, as far as is known from their admittedly extremely poor record, persist relatively unchanged from that time to the Recent. Significantly, no post-Cambrian marine records of onychophorans are known, and there are no extant marine forms. The report of an extant marine "pro-onychophoran" by Sundara Rajulu and Gowri (1988) seems to have been in error (S. Conway Morris, pers. comm., 1996; see also Jayaraman 1989). Other lobopod forms known from the Cambrian include the pentastomids (Walossek and Müller 1994; Walossek et al. 1994), and probably the tardigrades (Müller et al. 1995).

Assessment of Cambrian lobopod ecology is difficult for several reasons. First, virtually no work has been carried out on the functional morphology of the lobopods, and second, there are no good extant marine examples to compare them with. The only relatively common marine lobopods today are the tardigrades; it is generally agreed that the marine forms are less derived than the more familiar terrestrial ones. They are relatively poorly known (see Kinchin 1994 for a summary of their known ecology). Three broad ecologic groupings are recognized: species inhabiting organic slime or plates of barnacles, and some other ectoparasites; species occupying a psammolittoral zone; and deep-sea species, which are the most abundant. Little is known of the ecology of this last group, although they appear morphologically to be fairly diverse, with various forms of lateral cuticular extensions, or "alae," and elaborated claws. Forms adapted for digging in mud, such as the Coronarctidae, often have an elongate body form, with reduced appendages.

From the point of view of understanding the basic ecology of Cambrian lobopods, the extant marine tardigrades are unfortunately of little help. A typical Cambrian lobopod, such as *Hallucigenia,* is about 3 cm in length, whereas even the largest marine tardigrades do not exceed about 1.5 mm. The hydrodynamic regime and strictures imposed on the two are clearly of a very different order: a marine tardigrade faces a world of Reynold's numbers of less than 1; a Cambrian lobopod, of greater than 100. A marine tardigrade has been reported from the Cambrian (Müller et al. 1995). However, as miniaturized taxa, they represent a distinct fauna of their own and cannot be readily compared to the marine macrolobopods in the Cambrian.

With a lack of directly analogous extant forms, study of Cambrian lobopod ecology must fall back on functional morphology, facies association, and documented cases of species-species interactions from the fossil record. Given their only recent discovery as an important grouping, it is not surprising that virtually nothing is known of their functional morphology. At a simple level, the group of lobopods possessing spinous plates (see Ramsköld 1992; Hou and Bergström 1995; Budd 1996a) presumably used them for defensive purposes. It has also been suggested that the fleshy protrusions from the limbs of *Onychodictyon* played a role in respiration (Hou and Bergström 1995). No details of mouthparts are known, apart from the reported possibility that *Onychodictyon* possesses jaws (Ramsköld 1992; Hou and Bergström 1995), but this is not based on compelling evidence. Further, although vagrant benthic polychaete worms probably provide the closest analogs for mode of life, their Cambrian ecology is also poorly known (see, for example, Conway Morris 1979).

On the basis of claw morphology and faunal association, two specific interactions have been suggested: a relationship between the Burgess Shale *Aysheaia* and the sponge *Vauxia* (Whittington 1978), and a relationship between the Chengjiang forms of *Microdictyon* and the probable echinoderm *Eldonia* (Chen et al. 1995b). The first of these is certainly plausible, with *Aysheaia* being envisaged as climbing, and being predatory on, the sponge. Indeed, there seems to be a consistent association of *Aysheaia* with *Vauxia*. Nevertheless, this association cannot be taken as proof of a predator-prey relationship between the two animals. Although many animals are predatory on sponges today (e.g., starfish, fish, shrimps, nudibranch gastropods, and polychaetes such as *Branchiosyllis oculata* [Pawlik 1983; Chanas and Pawlik 1995], which may provide the closest extant analog to the Cambrian lobopods), many other animals are associated in a commensal or other relationship, such as inquiline polychaete species; the polynoid *Harmothoe hyalonemae* (Martin et al. 1992) is one example. Some studies on the extant fauna suggest that sponge-associated macrofauna tend not to be host specific (Koukouras et al. 1992), although *Aysheaia* appears to be. If *Aysheaia* actually ate the sponge, one might expect to see evidence such as the presence of spicules in the gut, which have not, however, been reported (see also discussion in Monge-Najera 1995). Whether or not *Aysheaia* was a true sponge predator must thus be left as an open question at present.

The putative relationship between *Microdictyon* and *Eldonia* from the Chengjiang fauna has less prima facie plausibility, if only because *Eldonia* is generally considered to be a free-floating form, probably related to pelagic holothurians (e.g., Durham 1974). Again, such a relationship is based on a consistent co-occurrence between the two forms (Chen et al. 1995b). Although a relationship between a planktic form and a macrocommensal or parasite is by no means impossible (compare, for example, gastropods living on the Mazon Creek medusoid *Essexella* [Foster 1979]), in this case it would be more reasonable to see the relationship as one of selective scavenging. A similar relationship between carcasses and scavenger might be discernible in the Burgess Shale, with at least 18 of the known *Hallucigenia* specimens being found on a single slab apparently associated with the carcass of a worm (Conway Morris 1977b:

plate 76). Finally, in this connection the so-called lobopods pentastomids (which some molecular and developmental evidence suggests to be derived branchiuran crustaceans [Abele et al. 1989]) are recorded from the Cambrian (Walossek and Müller 1994; Walossek et al. 1994) and seem to represent—as they do today—an endoparasitic lifestyle. The group of animals on which Cambrian pentastomids may have been parasitic is, however, quite obscure.

EARLY ARTHROPOD ECOLOGY

Walossek (1993) has advanced arguments that the basal euarthropod limb (leaving aside the uniramians) was biramous and may have been associated with a trunk-grappling mode of feeding rather than, for example, filter feeding, thus in effect rendering irrelevant the old dispute between disciples of Borrodaile (1926) and Cannon and Manton (1927) over whether the primitive crustacean feeding mode was by thoracic filtration (such as in the branchiopods) or cephalic appendage feeding (like the maxillary filtration of maxillopods; see Schram 1986 for discussion), respectively. Whatever the precise phylogeny of the biramous-limbed arthropods turns out to be, the widespread distribution of serially homonomous gnathobasic limbs in the Cambrian arthropods suggests that such a condition indeed characterizes a large clade of arthropods. Trunk-based predation may thus represent the primitive feeding mode for the euarthropods, from which (in, for example, the cephalocarids) crustacean thoracic feeding modes were derived, and deposit-feeding and filter-feeding modes of life may represent secondary adaptations (cf. Fortey [1994], who comes to similar conclusions for the trilobites; Walossek [1993]; Hughes, this volume). Such a conclusion is supported by the recent description of several taxa that may be regarded as lying within the stem group of the euarthropods: *Anomalocaris, Kerygmachela,* and *Opabinia.*

Study of the enigmatic Burgess Shale taxa *Anomalocaris* and *Opabinia* has received a significant boost in recent years by the description of new material from the Burgess Shale, Chengjiang, and Sirius Passet faunas (Budd 1993, 1996a, 1997, 1999; Chen et al. 1994; Hou et al. 1995; Collins 1996). Although their relationships are controversial, the anomalocaridids and their relatives may be best considered as stem-group arthropods (Budd 1993, 1996a, 1997, 1999; Dewel and Dewel 1996; Waggoner 1996): new data indicate that both *Opabinia* and some anomalocaridid-like forms were lobopodous (Budd 1996a, 1997), although taxa such as *Parapeytoia* possess arthropod-like, biramous, and gnathobasic limbs. The ecology of these forms is unclear and as yet poorly studied. Based on some functional studies of the formidable-looking anterior appendages of *Anomalocaris,* at least the anomalocaridids have been seen as the largest-known Cambrian predators, and Chen et al. (1994) extended this view to include *Opabinia* and *Kerygmachela* too. However, the frontal appendages of

Kerygmachela, for example, with their long frontal processes (Budd 1993, 1999), could hardly have been used raptorially and are more likely to have had a sensory function. The anomalocaridids themselves have usually been considered to be active predators (Whittington and Briggs 1985; Chen et al. 1994). The frontal appendages possess an impressive array of spines, similar to those seen in raptors such as the stomatopods, which use the second maxillipeds for smashing or spearing prey. However, given that the diversity of anomalocaridids is becoming known, by no means all of them are likely to have been predators. The secondary spines known from some taxa point in the wrong direction to act as barbs, as one might expect (e.g., *A. saron* [Hou et al. 1995]), and in others they form such a dense mesh that a predatory purpose seems unlikely (Nedin 1995). Without more-detailed functional analysis, the precise purposes of these structures, which must differ considerably from taxon to taxon, must remain obscure. The presence of apparently quite precise specializations such as the limb morphologies and the "*Peytoia*" mouthpart, which may also have a homolog in *Opabinia* (Hou et al. 1995), should provide a good basis for further investigation.

Despite these problems, however, more-recent data considerably strengthen the case for at least some of the anomalocaridids' being predators. The Chengjiang anomalocaridid *Parapeytoia* possesses what appear to be enormous gnathobases at the bases of biramous trunk limbs (Hou et al. 1995), suggesting that at least this form was a predator, perhaps feeding in much the same way that *Limulus* does today, with the frontal appendages playing an analogous role to the chelicerae. Such a mode of life cannot be demonstrated for the other probable stem-group arthropods *Kerygmachela* and *Opabinia.* However, *Opabinia* does appear to possess midgut digestive caecae (Budd 1996a: figure 1C), along with anomalocaridids (Budd 1997). This may suggest a shift in feeding from fairly passive or scavenging modes to more-active ones involving the rapid ingestion of live prey, thus necessitating increased digestive capacity — a possibility that, despite speculation (e.g., Buss 1995 : 201), remains to be properly tested.

EVOLUTION OF ARTHROPOD ECOLOGY

Understanding the origin and evolution of arthropod ecologic strategies must inevitably rely on some sort of phylogenetic reconstruction. Although arthropod phylogeny is highly controversial (see Budd 1996b), any one particular reconstruction will provide an accompanying ecologic scenario, particularly if character states are optimized at the internal nodes of a cladogram in order to provide hypothetical ancestral states. If the anomalocaridid taxa indeed lie within the stem group of the arthropods, then, as discussed above, it is quite possible that predation was the primitive mode of life for the euarthropod clade. In such a case, the gnathobasic limbs of *Parapeytoia* represent a primitive character state, retained by some taxa in the euarthropods such as *Sidneyia* and—primitively—the trilobites, while being lost within the crown-group crustaceans and apparently fairly basal arachnate taxa such as *Marrella* (Whittington 1971) and *Fuxianhuia* (Chen et al. 1995a; but see also Wills 1996). If so, then a broad ecologic history of the entire clade could be reconstructed. The opposite approach has been taken by Bousfield (1995), who has attempted a partial phylogenetic reconstruction based on an analysis of feeding strategies. However, the conclusions therein may need modification in light of the discovery of trunk gnathobases in the anomalocaridids (Hou et al. 1995).

The lobopodous members of the arthropod clade already display a wide range of ecologic strategies, from possible scavenging (*Hallucigenia;* Conway Morris 1977b) and miniaturization (the reported Cambrian tardigrade; Müller et al. 1995) to parasitism (*Aysheaia;* Whittington 1978). The development of large size, coupled with the beginnings of arthropodization, an increased digestive capacity as evidenced by the appearance of midgut caecae (Budd 1996a, 1997), and a ventral feeding apparatus as seen in *Parapeytoia* (Hou et al. 1995) were paralleled by a shift toward macropredation, a strategy that persisted into the arthropods. One can further speculate that the very success of this strategy may have led to high levels of competition for resources among the early predatory arthropods and thus may have been a factor in driving a reradiation of arthropods to refill lower ecologic niches. Such a scenario, in which macropredation is one end result of an important stage of "Cambrian explosion" evolution, and in which it drives the evolution more of the predator than of the prey, is in sharp contrast to so-called Garden of Ediacara hypotheses (e.g., McMenamin and McMenamin 1990), which purport to explain patterns of radiation in the Cambrian.

CAMBRIAN ECOLOGY AND ARTHROPODS

We are unfortunately a long way from a genuine understanding of the controls and processes that govern modern benthic ecology, and the prospects for the past are correspondingly worse. The general principles—the role of nutrient supply and recycling, the balance between suspension and deposit feeding, and the effect of predation—are clear enough. But there are other areas, much harder to define, in which Cambrian ecology may be profoundly different from Recent ecology. Deposit feeders seem to gain most of their actual nutrition from the fungi and bacteria that break down the largely refractory organic detritus in the sediment, and also perhaps from the protists and meiofauna that in turn feed on the bacteria (Kuipers et al. 1981). In addition, there is a large and more or less untapped store of dissolved organic matter that is highly resistant to utilization even by bacteria, although some annelids may gain some nutrition from this source (Fauchald and Jumars 1979). If any of these factors were different in the Cambrian—if, for example, many taxa were able to tap into the pool of dissolved organics—then the whole balance of energy transfers within the ecosystem might be shifted. Given the generally variable ways in which effects such as predation influence species distribution (e.g., Reise 1985), it is difficult to predict how Cambrian ecology might be affected by such considerations.

The importance of Cambrian arthropods in exceptionally preserved faunas may allow a few tentative conclusions to be applied to Cambrian ecology as a whole. First, the importance of arthropods in the *marine* realm may have been greater in the past than now. Whereas all of the Cambrian Lagerstätten are dominated by arthropods, modern-day marine assemblages, especially in deeper water, are richer in echinoderms and especially holothurians. Second, although a wide range of ecologic strategies has been documented in Cambrian arthropods, their ecologic sophistication may have been limited by their simple body plans. In addition, some strategies such as sessile filter feeding (e.g., barnacles; but see Collins and Rudkin 1981 for a possible Cambrian example), mineralization of limbs (seen in decapods), and deep burrowing may not have been employed at all. The conclusions of Conway Morris (1986) that, at least broadly, the ecologic framework of the rest of the Phanerozoic had already been established seems right, but with the important caveat that each stage of the ecologic hierarchy, at least by reference to the arthropods, may have been less efficient at transferring energy to higher trophic levels, which would have the inevitable effect of shortening trophic webs.

CONCLUSIONS

Arthropod and lobopod Cambrian ecology remains in a fairly undeveloped state at present. Newly discovered groups such as the anomalocaridids and lobopods are as yet poorly understood in terms of their functional morphology. Such understanding is essential if their role in the developing Cambrian ecosystems is to be properly assessed. However, it is already clear that (1) arthropods and lobopods played a large role in trophic webs in the Cambrian; (2) the sophistication of their ecologic strategies was restricted by their relative lack of tagmosis, providing an important limit to their efficiency; and (3) of the groups represented in the Cambrian, the most specialized may have been the crustaceans. The difficulties of building realistic ecologic models, even based on extant biotas (see critique of, for example, Polis 1991), counsels against excessive optimism that genuine understanding of Cambrian ecology will be achieved quickly.

Acknowledgments. I thank Simon Conway Morris, Stefan Bengtson, Dieter Walossek, and Sören Jensen for helpful discussions. Jason Dunlop and Simon Braddy kindly provided information on Cambrian merostomes and trace fossils. Derek Briggs, Fred Schram, Matthew Wills, and Andrey Zhuravlev provided constructive reviews. This work was supported by the Swedish Research Council (NFR).

REFERENCES

Abele, L. G., W. Kim, and B. E. Felgenhauer. 1989. Molecular evidence for the inclusion of the phylum Pentastomida in the Crustacea. *Molecular Biology and Evolution* 6:685–691.

Akam, M. 1995. Hox genes and the evolution of diverse body plans. *Philosophical Transactions of the Royal Society of London B* 349:313–319.

Barker, P. L. and R. Gibson. 1977. Observations on the feeding mechanism, structure of the gill, and digestive physiology of the European Lobster *Homarus gammarus*. *Journal of Experimental Marine Biology and Ecology* 26:297–324.

Borradaile, L. A. 1926. Notes upon crustacean limbs. *Annual Magazine of Natural History* 17:193–213.

Bousfield, E. L. 1995. A contribution to the natural classification of Lower and Middle Cambrian arthropods: food-gathering and feeding mechanisms. *Amphipacifica* 2:3–34.

Briggs, D. E. G. and D. Collins. 1988. A Middle Cambrian chelicerate from Mount Stephen, British Columbia. *Palaeontology* 31:779–798.

Briggs, D. E. G. and H. B. Whittington. 1985. Modes of life of arthropods from the Burgess Shale, British Columbia. *Transactions of the Royal Society of Edinburgh (Earth Sciences)* 76:149–160.

Bruton, D. L. 1981. The arthropod *Sidneyia inexpectans*, Middle Cambrian, Burgess Shale, British Columbia. *Philosophical Transactions of the Royal Society of London B* 295:619–656.

Bruton, D. L. and H. B. Whittington. 1983. *Emeraldella* and *Leanchoilia*, two arthropods from the Burgess Shale, Middle Cambrian, British Columbia. *Philosophical*

Transactions of the Royal Society of London B 300:553–582.

Budd, G. 1993. A Cambrian gilled lobopod from Greenland. *Nature* 364:709–711.

Budd, G. E. 1996a. The morphology of *Opabinia regalis* and the reconstruction of the arthropod stem-group. *Lethaia* 29:1–14.

Budd, G. E. 1996b. Progress and problems in arthropod phylogeny. *TREE* 11:356–358.

Budd, G. E. 1997. Stem-group arthropods from the Lower Cambrian Sirius Passet fauna of North Greenland. In R. A. Fortey and R. A. Thomas, eds., *Arthropod Relationships*, pp. 125–138. London: Chapman and Hall.

Budd, G. E. 1999. The morphology and phylogenetic significance of *Kerygmachela kierkegaardi* Budd (Buen Formation, Lower Cambrian, North Greenland). *Transactions of the Royal Society of Edinburgh (Earth Sciences)* 89:248–290.

Budd, G. E. and S. Jensen. 2000. A critical reappraisal of the fossil record of the bilaterian phyla. *Biological Reviews* 75:253–295.

Buss, L. W. 1995. A new twist on the Garstang torsion hypothesis. *Neues Jahrbuch für Geologie und Paläontologie Abhandlungen* 195:199–205.

Butterfield, N. J. 1994. Burgess Shale–type fossils from a Lower Cambrian shallow-shelf sequence in northwestern Canada. *Nature* 369:477–479.

Butterfield, N. J. 1995. Secular distribution of Burgess-Shale–type preservation. *Lethaia* 28:1–13.

Cannon, H. G. 1927. On the feeding mechanism of *Nebalia*. *Transactions of the Royal Society of Edinburgh* 55:355–369.

Cannon, H. G. and S. M. Manton. 1927. On the feeding mechanism of a mysid crustacean, *Hemimysis lamornae*. *Transactions of the Royal Society of Edinburgh* 55:219–253.

Chanas, B. and J. R. Pawlik. 1995. Defenses of Caribbean sponges against predatory reef fish. 2. Spicules, tissue toughness, and nutritional quality. *Marine Ecology Progress Series* 127:195–211.

Chen, J.-Y., L. Ramsköld, and G.-Q. Zhou. 1994. Evidence for monophyly and arthropod affinity of Cambrian giant predators. *Science* 264:1304–1308.

Chen, J.-Y., G. D. Edgecombe, L. Ramsköld, and G.-Q. Zhou. 1995a. Head segmentation in Early Cambrian *Fuxianhuia*: implications for arthropod evolution. *Science* 268:1339–1343.

Chen, J.-Y., G.-Q. Zhou, and L. Ramsköld. 1995b. The Cambrian lobopodian *Microdictyon sinicum*. *Bulletin of the National Museum of Natural Science, Taichung, Taiwan* 5:1–93.

Chlupáč, I. 1995. Lower Cambrian arthropods from the Paseky Shale (Barrandian area, Czech Republic). *Journal of the Czech Geological Society* 40:9–36.

Cisne, J. L. 1974. Evolution of the world fauna of aquatic free-living arthropods. *Evolution* 28:337–366.

Collins, D. and D. Rudkin. 1981. *Priscansermarinus barnetti*, a probable lepadomorph barnacle from the Middle Cambrian Burgess Shale. *Journal of Paleontology* 55:1006–1015.

Collins, D. H. 1996. The "evolution" of *Anomalocaris* and its classification in the arthropod class Dinocarida (nov.) and order Radiodonta (nov.). *Journal of Paleontology* 70:280–293.

Conway Morris, S. 1977a. Fossil priapulid worms. *Special Papers in Palaeontology* 20:1–95.

Conway Morris, S. 1977b. A new metazoan from the Cambrian Burgess Shale of British Columbia. *Palaeontology* 20:623–640.

Conway Morris, S. 1979. Middle Cambrian polychaetes from the Burgess Shale of British Columbia. *Philosophical Transactions of the Royal Society of London B* 285:227–274.

Conway Morris, S. 1986. The community structure of the Middle Cambrian Phyllopod Bed (Burgess Shale). *Palaeontology* 29:423–467.

Conway Morris, S. 1989. The persistence of Burgess Shale–type faunas: Implications for the evolution of deeper-water faunas. *Transactions of the Royal Society of Edinburgh (Earth Sciences)* 80:271–283.

Conway Morris, S. 1998. *The Crucible of Creation*. Oxford: Oxford University Press.

Conway Morris, S. and R. J. F. Jenkins. 1985. Healed injuries in Early Cambrian trilobites from South Australia. *Alcheringa* 9:167–177.

Conway Morris, S. and R. A. Robison. 1988. More soft-bodied animals and algae from the Middle Cambrian of Utah and British Columbia. *University of Kansas Paleontological Contributions* 122:1–48.

Conway Morris, S., J. S. Peel, A. K. Higgins, N. J. Soper, and N. C. Davis. 1987. A Burgess Shale–like fauna from the Lower Cambrian of North Greenland. *Nature* 326:181–183.

Craske, A. J. and R. P. S. Jefferies. 1989. A new mitrate from the Upper Ordovician of Norway, and a new approach to subdividing a plesion. *Palaeontology* 32:69–99.

Dewel, R. A. and W. C. Dewel. 1996. The brain of *Echiniscus viridissimus* Peterfi 1956 (Heterotardigrada): A key to understanding the phylogenetic position of tardigrades and the evolution of the arthropod head. *Zoological Journal of the Linnean Society of London* 116:35–49.

Dunlop, J. A., L. I. Anderson, and S. J. Braddy. 1996. Chasmataspids—not the ancestors of arachnids. *Palaeontology Newsletter* 32:14.

Durham, J. W. 1974. Systematic position of *Eldonia ludwigi* Walcott. *Journal of Paleontology* 48:750–755.

Dzik, J. and G. Krumbeigel. 1989. The oldest "onychophoran" *Xenusion*: A link connecting phyla? *Lethaia* 22:169–181.

Fauchald, K. and P. A. Jumars. 1979. The diet of worms: A study of polychaete feeding guilds. *Oceanography and Marine Biology Annual Review* 17:193–284.

Flessa, K. W., K. V. Powers, and J. L. Cisne. 1975. Specialisation and evolutionary longevity in the Arthropoda. *Paleobiology* 1:71–81.

Fortey, R. A. 1985. Pelagic arthropods as an example of deducing the habits of extinct arthropods. *Transactions of the Royal Society of Edinburgh (Earth Sciences)* 76:219–230.

Fortey, R. A. 1994. Adaptive deployment in feeding habits in Cambrian trilobites. *Terra Nova* 6 (Abstract Supplement 3): 3.

Fortey, R. A., D. E. G. Briggs, and M. A. Wills. 1996. The Cambrian evolutionary explosion—decoupling cladogenesis from morphological disparity. *Biological Journal of the Linnean Society* 57:13–33.

Foster, M. W. 1979. Soft-bodied coelenterates in the Pennsylvanian of Illinois. In M. H. Nitecki, ed., *Mazon Creek Fossils*, pp. 191–267. New York: Academic Press.

Fryer, G. 1987. The feeding mechanism of the Daphniidae (Crustacea: Cladocera): Recent suggestions and neglected considerations. *Journal of Plankton Research* 9:419–432.

Gould, S. J. 1989. *Wonderful Life: The Burgess Shale and the Nature of History*. New York: W. W. Norton and Co.

Hesselbo, S. P. 1988. Trace fossils of Cambrian aglaspidid arthropods. *Lethaia* 21:139–146.

Hou, X., L. Ramsköld, and J. Bergström. 1991. Composition and preservation of the Chengjiang fauna—a Lower Cambrian soft-bodied biota. *Zoologica Scripta* 20:395–411.

Hou, X.-G. and J. Bergström. 1995. Cambrian lobopodians—ancestors of extant onychophorans? *Zoological Journal of the Linnean Society of London* 114:3–19.

Hou, X.-G., J. Bergström, and P. Ahlberg. 1995. *Anomalocaris* and other large animals in the Lower Cambrian Chengjiang fauna of southwest China. *Geologiska Föreningens i Stockholm Förhandlingar* 117:163–183.

Jayaraman, K. S. 1989. Indian zoologist suspected. *Nature* 342:333.

Jensen, S. 1990. Predation by early Cambrian trilobites on infaunal worms—evidence from the Swedish *Mickwitzia* Sandstone. *Lethaia* 23:29–42.

Kauffman, S. A. 1993. *The Origins of Order: Self-Organization and Selection in Evolution*. Oxford: Oxford University Press.

Kinchin, I. M. 1994. *The Biology of Tardigrades*. London: Portland Press.

Koukouras, A., E. Russo, C. Dounas, and C. Chintiroglou. 1992. Relationship of sponge macrofauna with the morphology of their hosts in the north Aegean Sea. *Internationale Revue der Gesamten Hydrobiologie* 77:609–619.

Kuipers, B. R., P. A. W. J. de Wilder, and F. Creutzberg. 1981. Energy flow in a tidal flat ecosystem. *Marine Biology Progress Series* 5:215–221.

Lamont, A. 1967. Environmental significance of eye reduction in trilobites and recent arthropods—additional remarks. *Marine Geology* 5:377–378.

Leslie, S. A., L. E. Babcock, and W. T. Chang. 1996. Community composition and taphonomic overprint of the Chengjiang biota (early Cambrian, China). *Paleontological Society Special Publication* 8:237.

Martin, D., D. Rosell, and M. J. Uriz. 1992. *Harmothöe hyalonemae* sp. nov. (Polychaeta, Polynoidae), an exclusive inhabitant of different Atlanto-Mediterranean species of *Hylonema* (Porifera, Hexactinellida). *Ophelia* 35:169–185.

May, R. M. 1975. Patterns of species abundance and diversity. In M. L. Cody and J. M. Diamond, eds., *Ecology and Evolution of Communities*, pp. 81–120. Cambridge, Mass.: Belknap Press.

McMenamin, M. A. S. and D. L. S. McMenamin. 1990. *The Emergence of Animals: The Cambrian Breakthrough.* New York: Columbia University Press.

Mikuláš, R. 1995. Trace fossils from the Paseky Shale (Early Cambrian, Czech Republic). *Journal of the Czech Geological Society* 40:37–54.

Monge-Najera, J. 1995. Phylogeny, biogeography, and reproductive trends in the Onychophora. *Zoological Journal of the Linnean Society of London* 114:21–60.

Müller, K. J. and D. Walossek. 1985a. Arthropodal larval stages from the Upper Cambrian "Orsten" of Sweden. *Transactions of the Royal Society of Edinburgh (Earth Sciences)* 77:157–179.

Müller, K. J. and D. Walossek. 1985b. A remarkable arthropod fauna from the Upper Cambrian "Orsten" of Sweden. *Transactions of the Royal Society of Edinburgh (Earth Sciences)* 76:161–172.

Müller, K. J. and D. Walossek. 1985c. Skara-

carida, a new order of Crustacea from the Upper Cambrian of Västergötland, Sweden. *Fossils and Strata* 17:1–65.

Müller, K. J. and D. Walossek. 1987. Morphology, ontogeny, and life habit of *Agnostus pisiformis* from the Upper Cambrian of Sweden. *Fossils and Strata* 19:1–124.

Müller, K. J. and D. Walossek. 1988. External morphology and larval development of the Upper Cambrian maxillopod *Bredocaris admirabilis. Fossils and Strata* 23:1–70.

Müller, K. J., D. Walossek, and A. Zakharov. 1995. "Orsten" type phosphatized soft-integument preservation and a new record from the Middle Cambrian Kuonamka Formation in Siberia. *Neues Jahrbuch für Geologie und Paläontologie Abhandlungen* 197:101–118.

Nedin, C. 1995. The Emu Bay Shale, a Lower Cambrian fossil Lagerstätten, Kangaroo Island, South Australia. *Association of Australasian Palaeontologists, Memoir* 18:31–40.

Nielsen, 1994. *Animal Evolution: Inter-Relationships of the Living Phyla.* Oxford: Oxford University Press.

Osgood, R. G. 1970. Trace fossils of the Cincinnati area. *Palaeontographica Americana* 6:281–444.

Palacios, T. and G. Vidal. 1992. Lower Cambrian acritarchs from northern Spain: The Precambrian-Cambrian boundary and biostratigraphic implications. *Geological Magazine* 129:421–436.

Pawlik, J. R. 1983. A sponge eating worm from Bermuda, *Branchiosyllis oculata* (Polychaeta; Syllidae). *Marine Ecology Progress Series* 4:65–79.

Pek, I. 1977. Agnostid trilobites of the central Bohemian Ordovician. *Sborník geologických Věd Paleontologie* 19:7–44.

Polis, G. A. 1991. Complex trophic interactions in deserts—an empirical critique of food-web theory. *American Naturalist* 138: 123–155.

Pratt, B. R. 1994. Benthos of a Lower Cambrian sandy lagoon: *Rusophycus* burrows formed by a predatory, non-trilobite arthropod. *Terra Nova* 6 (Abstract Supplement 3):5.

Ramsköld, L. 1992. Homologies in Cambrian Onychophora. *Lethaia* 25:443–460.

Ramsköld, L. and X.-G. Hou. 1991. New Early Cambrian animal and onychophoran affinities of enigmatic metazoans. *Nature* 351:225–228.

Reise, K. 1985. *Tidal Flat Ecology: An Experimental Approach to Species Interactions*. Berlin: Springer Verlag.

Schram, F. R. 1986. *Crustacea*. Oxford: Oxford University Press.

Stürmer, W. and J. Bergström. 1976. The arthropods *Mimetaster* and *Vachonisia* from the Devonian Hunsrück Shale. *Paläontologische Zeitschrift* 50:78–111.

Sundara Rajulu, G. and N. Gowri. 1988. Discovery of pro-Onychophoran marine worms: Description of the holotype. *Indian Zoologist* 12:61–64.

Thompson, I. and D. S. Jones. 1980. A possible onychophoran from the Middle Pennsylvanian Mazon Creek Beds of northern Illinois. *Journal of Paleontology* 54:588–596.

Valentine, J. W. 1973. *Evolutionary Paleoecology of the Marine Biosphere*. Englewood Cliffs, N.J.: Prentice-Hall.

Vermeij, G. J. 1987. *Evolution and Escalation: An Ecological History of Life*. Princeton: Princeton University Press.

Vorwald, G. R. 1982. Healed injuries in trilobites—evidence for a large Cambrian predator. *Geological Society of America Abstracts with Programs* 14:639.

Waggoner, B. 1996. Phylogenetic hypotheses of the relationships of arthropods to Pre-

cambrian and Cambrian problematic fossil taxa. *Systematic Biology* 45:190–222.

Walossek, D. 1993. The Upper Cambrian *Rehbachiella* and the phylogeny of Branchiopoda and Crustacea. *Fossils and Strata* 32: 1–202.

Walossek, D. 1999. On the Cambrian diversity of Crustacea. In F. R. Schram and J. C. von Vaupel Klein, eds., *Proceedings of the Fourth International Crustacean Congress, Amsterdam, The Netherlands, July 20–24, 1998*, vol. 1, pp. 1–27.

Walossek, D. and K. J. Müller. 1994. Pentastomid parasites from the Lower Palaeozoic of Sweden. *Transactions of the Royal Society of Edinburgh (Earth Sciences)* 85:1–37.

Walossek, D., I. Hinz-Schallreuter, J. H. Shergold, and K. J. Müller. 1993. Three-dimensional preservation of arthropod integument from the Middle Cambrian of Australia. *Lethaia* 26:7–15.

Walossek, D., J. E. Repetski, and K. J. Müller, 1994. An exceptionally preserved parasitic arthropod, *Heymonsicambria taylori* n.sp. (Arthropoda—incertae sedis, Pentastomida), from Cambrian-Ordovician boundary beds of Newfoundland, Canada. *Canadian Journal of Earth Sciences* 31:1664–1671.

Whittington, H. B. 1971. Redescription of *Marrella splendens* (Trilobitoidea) from the Burgess Shale, Middle Cambrian, British Columbia. *Geological Survey of Canada, Bulletin* 209:1–24.

Whittington, H. B. 1977. The Middle Cambrian trilobite *Naraoia*, Burgess Shale, British Columbia. *Philosophical Transactions of the Royal Society of London B* 280:409–443.

Whittington, H. B. 1978. The lobopodian animal *Aysheaia pedunculata* Walcott, Middle Cambrian, Burgess Shale, British Columbia. *Philosophical Transactions of the Royal Society of London B* 284:165–197.

Whittington, H. B. 1985. *The Burgess Shale.* New Haven, Conn.: Yale University Press.

Whittington, H. B. and D. E. G. Briggs. 1985. The largest Cambrian animal, *Anomalocaris,* Burgess Shale, British Columbia. *Philosophical Transactions of the Royal Society of London B* 309:569–609.

Wills, M. A. 1996. Classification of the arthropod *Fuxianhuia. Science* 272:746–747.

Wills, M. A., D. E. G. Briggs, and R. A. Fortey. 1994. Disparity as an evolutionary index: A comparison of Cambrian and Recent arthropods. *Paleobiology* 20:93–130.

Wills, M. A., D. E. G. Briggs, and R. A. Fortey. 1997. Evolutionary correlates of arthropod tagmosis: Scrambled legs. In R. A. Fortey and R. A. Thomas, eds., *Arthropod Relationships,* pp. 57–65. Systematics Association Special Volume No. 55. London: Chapman and Hall.

CHAPTER NINETEEN

Thomas E. Guensburg and James Sprinkle

Ecologic Radiation of Cambro-Ordovician Echinoderms

Echinoderms represent a modest component of the initial metazoan radiation during the Cambrian but responded to global environmental changes across the Cambro-Ordovician boundary with rapid and prolific diversification to more varied lifestyles in expanded habitats. Many attached echinoderms were preadapted to exploit carbonate hardgrounds and other stable substrates that became abundant on shallow carbonate platforms at that time, whereas other attached—and many new free-living—echinoderms evolved structures to cope with soft substrates.

Early to Middle Cambrian echinoderms are primarily known from soft substrate environments where attached suspension-feeding eocrinoids, crinoids, and edrioasteroids clung to skeletal debris by suctorial attachment disks or were skeletally cemented by a holdfast; helicoplacoids perhaps employed other means. Vagile surface deposit-feeding echinoderms included stylophorans, homosteleans, homoiosteleans, and ctenocystoids. Echinoderms reached a diversification bottleneck in the Late Cambrian, but stemmed eocrinoids with cemented holdfasts were among the first skeletonized animals to colonize hardgrounds that became common at that time. Stylophorans, homoiosteleans, and edrioasteroids were also represented. Attached crinoids and free-living rhombiferans led the Early Ordovician radiation among suspension-feeding echinoderms and were accompanied by several other newly evolved groups with generally similar lifestyles. Vagile herbivorous echinoids and carnivorous asteroids greatly expanded echinoderm ways of life by the Middle Ordovician. This overall diversification pattern for echinoderms supports a model of two sequential evolutionary faunas in which shallow-water habitats fostered onshore origination and radiation followed by offshore expansion for many attached forms. However, the diversification pattern is not as clear among free-living echinoderm groups, and the expansion direction for several of these could have been from offshore to onshore. Bathymetry is a simplification of what must have been a complex list of controls. Most Ordovician echinoderms had regular and sturdy construction; these advanced designs were versatile and enduring by comparison with Cambrian forms, persisting through the Paleozoic and in some cases to the Recent.

DOCUMENTATION OF ECOLOGIC diversification in the fossil record provides the road map of life's temporal patterns and the context of evolutionary history. Most studies of diversification have emphasized intrinsic biotic driving factors for changes in diversification patterns and evolutionary pathways (see Sepkoski 1991 for a review), but recent field-based studies have emphasized the role of extrinsic causes. This latter approach requires extensive field observation and integration of sedimentologic, facies, and sequence stratigraphic information with paleobiologic observations (Guensburg and Sprinkle 1992; Rozhnov 1994; Droser et al. 1995). Broad-scale linkages are emerging as a result. For instance, we have previously correlated global environmental changes with the ecologic expansion and diversification of echinoderms and other metazoans during the Early Ordovician rise of the Paleozoic Evolutionary Fauna (Guensburg and Sprinkle 1992; Sprinkle and Guensburg 1995). Echinoderms of the Cambrian remained a modest component of the biota until favorable environmental shifts provided the catalyst for rapid ecologic expansion as part of the Ordovician radiation of metazoans (Sprinkle 1980). The purpose of this chapter is to review the ecologic radiation of Cambrian to Early Ordovician echinoderms and to analyze their diversification patterns. Direct associations of echinoderms and substrates are occasionally available when articulated specimens still adhere to attachment sites. In many other cases, however, life modes must be reconstructed from extensive field correlation of partial specimens and lithofacies, coupled with functional morphologic studies and extrapolation from better-preserved close relatives.

The ecologic radiation for Cambro-Ordovician echinoderms offered here differs from those suggested by Smith (1988: figure 12.3; Smith 1990) and Smith and Jell (1990: figure 53). Many Early Paleozoic echinoderms are interpreted by these authors to have rested unattached on, or had a distal structure inserted into, soft substrates. In contrast, evidence leads us to conclude that hard attachment surfaces were required and that this was an important limiting factor to the diversification of Cambrian echinoderms. This also implies that most Cambrian echinoderms were preadapted to exploit the hard substrates that became common by the Late Cambrian. These divergent functional interpretations provide an impetus for presentation of our ecologic diversification model below.

ENVIRONMENTAL CHANGES DURING THE EARLY PALEOZOIC

The time interval considered here is from the Early Cambrian (Waucoban) through the Early Ordovician (Arenig, Late Ibexian), comprising the Sauk Sequence of Sloss

(1963). The Cambrian period began long after the Varangerian glaciation and breakup of the supercontinent Rodinia. Global environmental shifts at this time can be related to the early diversification patterns of echinoderms and to the biosphere in general. Modeling of Cambrian ocean circulation patterns supports a global warming trend (Golonka et al. 1994). Sea levels rose, with interruptions, throughout the Cambrian–Early Ordovician, resulting in widespread and increasingly extensive inundation of cratons (James et al. 1989), potentially enhanced by isostatic and/or thermal subsidence of continental margins. Generally, configuration of shallow shelves changed from narrow belts with inner detrital, carbonate bank, and outer detrital zones to broad carbonate ramps that extended well into continental interiors (Cook 1989; James et al. 1989). Siliciclastic terrigenous sediments dominate Early to Middle Cambrian sequences, but carbonates compose the majority by the Early Ordovician (see Seslavinsky and Maidanskaya, this volume). This change probably resulted from gradual constriction of emergent sediment source areas by rising sea level. Evidence of slowed sedimentation during Late Cambrian time includes widespread glauconite formation; some Early Ordovician phosphatic-rich sediments have similar implications. Seawater chemistry also changed during this time. Carbonate deposition of the Early to Middle Cambrian appears to have been dominated by metastable aragonite, which later altered to calcite (Sandberg 1983). There is little evidence that encrusting organisms exploited lithified or firm sea floors at that time. In contrast, Late Cambrian to Early Ordovician carbonates were dominantly formed in a primary calcite cementation regime, fostering the formation of widespread hardgrounds or lithified substrates (Palmer and Palmer 1977; Wilson et al. 1992; Rozhnov 1994). These conditions offered ideal habitats for slow-growing (low-metabolic), calcite-secreting, epifaunal organisms such as echinoderms, and they were among the first skeletonized metazoans to exploit these habitats. The first really widespread encrinites, or echinoderm grainstones, are associated with both intraformational conglomerates and cryptalgal buildups that served as substrates for hardground formation, although a few echinoderm grainstones have been reported in association with late Early Cambrian reefs (James and Klappa 1983). Multiplated echinoderm skeletons were rapidly reduced by postmortem taphonomic processes to concentrations of durable clasts; these significantly increased the volume of sediment available for cementation (Wilson et al. 1992). Their porous construction and high-magnesium calcite composition were ideal nucleation sites for marine cements in the form of syntaxial overgrowths, thus leading to rapid lithification and formation of hardgrounds. This resulted in a self-perpetuating cycle whereby subsequent generations of echinoderms literally built upon the disarticulated remains of their ancestors.

Paleogeographic reconstructions of the Early Cambrian depict Laurentia, Baltica, and Kazakhstan (in part) separated from Gondwana and other continental masses (Golonka et al. 1994; Ruzhentsev and Mossakovsky 1995). Virtually all landmasses were concentrated in the Southern Hemisphere, with Laurentia and parts of Gondwana closest to the equator. Cratonic seas were widely distributed but covered only continental margins. These landmasses retained their identity throughout the Cambrian and Early Ordovician, and echinoderm faunas remained separate and distinctive on these continental blocks (Smith 1988; Sprinkle 1992). Baltica, Kazakhstan, and Laurentia moved slightly farther north into the tropics, then gradually converged (Golonka et al. 1994). Baltica and Laurentia collided with the closing of the Iapetus during the Middle Ordovician. The reconstructions support merging or linkage of faunal provinces for several continental blocks during the Middle to Late Ordovician, and the echinoderms reflect this greater interchange.

TEMPORAL PATTERNS IN LIFE MODES

Echinoderms constituted a small percentage of the total Cambrian biota, and the array of basic body constructions and life modes of these organisms was limited relative to younger assemblages. Eocrinoids, crinoids, edrioasteroids, and probably helicoplacoids were sessile low-to-medium-level epifaunal suspension feeders. They were either fixed or had minimal movement potential. Fossil holothurians are only rarely preserved intact, because of their slightly calcified construction. Consequently, we know little regarding their ecologic diversification, except that they were apparently present by the Middle Cambrian (*Eldonia* and relatives; undescribed fossils) and could have had both mobile benthic and planktic life modes by that time. All four classes of "carpoids"—cinctan homosteleans, solutan homoiosteleans, ctenocystoids, and cornute stylophorans—are known from the Middle Cambrian. They are generally considered to have been vagrant low-level suspension or deposit feeders, although they may constitute a polyphyletic grouping. Solutes and cornutes (later joined by mitrate stylophorans) extend into the Late Cambrian and Early Ordovician, where they constitute important groups of vagile echinoderms from this time.

Most of the life modes established by the Cambrian were carried over and expanded with a larger rapid radiation of echinoderms during the Early to Middle Ordovician. There was a dramatic increase in faunal diversity, particularly among suspension-feeding echinoderms, and a corresponding increase in fine partitioning according to substrates or attachment sites, tiering or feeding levels, and specialized food particle selection. Eocrinoids underwent considerable radiation during the Early Ordovician, giving rise to rhombiferans, diploporans, parablastoids, and paracrinoids (including rhipidocystids) (Sprinkle 1995). Blastoids were added later by the Middle Ordovician.

The most spectacular radiation during the Early Ordovician was that of the crinoids, which eclipsed blastozoans in total diversity and numbers by the Middle Ordovician. No crinoids are known from the Late Cambrian, but they had become abundant and diverse on hard substrates by the Early Ordovician and on soft substrates as well by the Middle Ordovician. Stelleroids that appeared in the Early Ordovician and echinoids and ophiocistioids that appeared in the Middle Ordovician greatly expanded the ecologic diversification of mobile benthic echinoderms to include vagrant scavengers, grazers, and carnivores. Certain edrioasteroids continued to diversify with suspension-feeding lifestyles, but only as a relatively minor faunal component. Discussions of specific morphologic changes in echinoderm systems follow.

Attachment

A wide range of habitats was exploited by Early Cambrian echinoderms, including deep slope (Poleta Formation, California) to shallow shelf detrital facies, and less common shallow carbonate bioherms and associated facies. Based upon functional morphology and taphonomy, we believe, contrary to Smith (1988), that most Early Cambrian echinoderms were attached to firm or hard substrates in life and that the limited availability of these substrates (mostly skeletal fragments) limited the distribution of the echinoderms. The fossils commonly occur in siliciclastic-dominated sequences such as fine-grained siltstones and shales that presumably formed soft substrates. Assuming that the echinoderms were not usually transported into these settings, the only available attachment sites appear to have been skeletal debris, such as trilobite molts and rare brachiopod or hyolith shells.

Specimens associated with attachment sites are rare, and the attachment mechanism in other cases is uncertain, although functional morphology and taphonomy provide important clues. No known Early to Middle Cambrian echinoderms were skeletally attached. The edrioasteroids *Stromatocystites* and *Camptostroma* had basal disks that are interpreted to have enabled clinging by suction. There is a system of radiating ridges and plate rings in the loosely plated aboral surface that was capable of being withdrawn upward, forming a partial vacuum (Smith and Jell 1990). Presumably the animals released from attachment sites following death (Guensburg and Sprinkle 1994b). Blastozoans are considered to be the sister group to edrioasteroids (Derstler 1985), and Early Cambrian examples *Kinzercystis* and *Lepidocystis* apparently retained attachment disks. Specimens of *Lepidocystis* are rarely attached to trilobite exoskeletons (Sprinkle 1973: plate 3, figures 1–4). The paleoecology of helicoplacoids is more problematic. These spindle-shaped echinoderms are most often preserved flattened parallel to bedding, but a few specimens are vertical, with a thecal pole buried in shale. Attachment sites have not been identified.

Attachment structures of Middle Cambrian edrioasteroids and eocrinoids are often modified versions of the basal disk described above (figure 19.1). For the most part, these fossils occur in fine-grained siliciclastic and mixed siliciclastic to carbonate (micritic) sequences of the outer detrital belt (Sprinkle 1976). The diverse and widespread eocrinoid *Gogia* and close relatives were the most common echinoderms during this time. Specimens occasionally occur attached to skeletal fragments (Sprinkle 1973: plate 23, figures 1–6) by a small multiplated button-shaped holdfast at the end

Figure 19.1 Reconstruction of soft-substrate echinoderm community from the Middle Cambrian Burgess Shale (British Columbia, Canada). Community is reconstructed at the base of a carbonate bank in about 150 m of water and includes short- and long-stalked eocrinoids (*Gogia,* left, and *G. radiata,* left center), the crinoid *Echmatocrinus* (right center), the edrioasteroid *Walcottidiscus* (right rear), and tiny mobile *Ctenocystis* (left front). Echinoderms, which make up less than 5 percent of the fauna, are shown with other components of the fauna, including trilobites, sponges, *Marrella,* a hyolith, and the priapulid *Ottoia.* Front width of block diagram about 0.5 m. *Source:* Modified from Sprinkle and Guensburg (1997) by James Sprinkle and Jennifer Logothetti.

of a short-to-long multiplated stalk (figure 19.1). The lower holdfast surfaces are not well known, so it is uncertain whether suction was still used for adherence or if there was actually skeletal cementation to the attachment surface. *Lichenoides* is a *Gogia* relative whose thickened plates of the lower theca as an adult possibly anchored the animal. *Cymbionites* is a problematic Middle Cambrian taxon known by a greatly thickened basal plate that must have enabled anchoring in a similar manner. Edrioasteroids such as *Totiglobus* and *Edriodiscus* had basal disks functionally similar to those of earlier relatives (Guensburg and Sprinkle 1994b). A *Totiglobus* from southern Idaho is attached to a trilobite free-cheek. The earliest probable crinoid *Echmatocrinus* occurs attached to worm tubes (figure 19.1), hyoliths, and possible stalks of other *Echmatocrinus* specimens using a medium-length stalk tipped by a low conical holdfast that appears to have been cemented to the attachment surface (Sprinkle 1973; Sprinkle and Collins 1998).

Late Cambrian echinoderms are poorly known, but based upon skeletal debris, they were locally common in shallow shelf environments of cratonic seas, and echinoderms were among the first metazoans to attach to widespread hardgrounds devel-

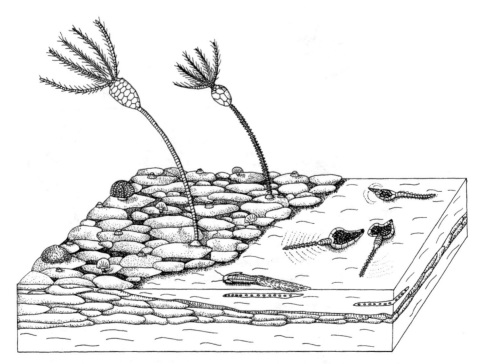

Figure 19.2 Reconstruction of hardground and soft-substrate echinoderm communities from the Upper Cambrian Snowy Range Formation (Wyoming, USA). A flat-pebble conglomerate bed is slowly being covered by soft muddy substrate (right), but thicker parts of the bed (left) have become pitted and corroded during a long period of exposure on the shallow sea floor. Two genera of stemmed trachelocrinid eocrinoids (left center), along with many additional holdfasts, a biscuit-shaped edrioasteroid (upper left), a sponge (lower left), and several *Billingsella* calciate brachiopods, are attached to this lithified surface, while two cornute stylophorans (right front), a solute homoiostelean (right rear), and a trilobite feed in the soft muddy sediment. Front width of block diagram about 0.5 m. *Source:* Modified from Sprinkle and Guensburg (1997) by James Sprinkle and Jennifer Logothetti.

oped on grainstones, intraformational limestones, and cryptalgal biohermal mounds. Some hardground surfaces are encrusted by numerous subconical massive cemented holdfasts (figure 19.2), which, based on association with disklike columnals having trilobate lumens and distinctive thecal plates, we assign to eocrinoids. No Late Cambrian crinoids are known. This is curious because they commonly encrusted Ordovician hardgrounds (Palmer and Palmer 1977; Brett and Brookfield 1984; Guensburg 1992; Guensburg and Sprinkle 1992; Sprinkle and Guensburg 1995). The eocrinoid *Ridersia* may represent a sister taxon to later rhombiferans (Jell et al. 1985) and has a strongly heteromorphic stem that may indicate a free-living adult life mode. Edrioasteroids continued to attach with a basal disk, but there were modifications that presumably increased efficiency by adding a well-developed peripheral rim that sealed the thecal margin (Smith and Jell 1990). Undescribed edrioasteroids from the Late Cambrian of Missouri have long conical aboral surfaces that could have been inserted into firm but plastic siliciclastic substrates or attached to skeletal fragments. Early and Middle Ordovician attached echinoderms continued encrusting hardgrounds and other solid surfaces. Eocrinoids, paracrinoids, and crinoids all exploited these surfaces in great numbers. Rootlike and radicular holdfasts among crinoids first appeared during the Middle Ordovician, corresponding to the rapid ecologic radiation of this group.

Locomotion

"Carpoids" were flattened, more or less bilaterally symmetric, benthic vagrant organisms. Among these, homosteleans, or cinctans, had a single biserial appendage that perhaps facilitated limited movement. Homoiosteleans, or solutes, used the larger of their two appendages in a similar manner. Ctenocystoids lacked appendages and presumably moved by means of water pulses channeled through the alimentary canal. Cornute stylophorans often have highly asymmetrical thecae, and the nature of locomotion is difficult to discern. A highly flexible appendage, the aulacophore, presumably propelled these animals with a wriggling or sculling motion. Mitrate stylophorans that appeared in the Early Ordovician were bilaterally symmetrical and may have been more active. Many rhombiferan cystoids are thought to have broken free or autotomized from a holdfast as juveniles and been essentially free-living as adults. A short flexible proximal stem and a long relatively stiff distal stem perhaps enabled these animals to move across the sea floor. Edrioasteroids are rarely skeletally attached, and some may have been capable of limited movement.

Stem Development and Tiering

Elevation of the feeding appendages and oral surface among early eocrinoids was accomplished by a stalk or stem. This stalk is generally short in Cambrian species, one or two times the thecal length, but longer in a few taxa (see figure 19.1). Stalks are covered with small irregular plates and terminate at a holdfast below. By the late Middle Cambrian, the eocrinoid *Akadocrinus* had a stem with polymeric columnals. Among eocrinoids, the transition to a fully formed stem with holomeric columnals permitting effective elevation of the theca and feeding structures was accomplished by the latest Middle Cambrian (Sprinkle 1973). Late Cambrian trachelocrinid eocrinoids had stem lengths several times the thecal height, allowing feeding to intermediate or high levels (see figure 19.2), perhaps as much as 0.5 m above the substrate. Other echinoderms appear to have followed a similar pattern, but the record is not as good and the timing was apparently different. The earliest fossil we believe to be a crinoid, *Echmatocrinus,* has a medium-length stalk that tapers gradually to a thin zone immediately above a small encrusting holdfast (see figure 19.1). The next record of crinoids is not until the Early Ordovician, and by then well-developed meric stems more than 0.5 m in length and attached to large calyces are known, enabling them to reach high feeding tiers. Stems tend to be polymeric, usually pentameric, and flexible to allow advantageous feeding strategies. A few crinoids developed stems up to 0.9 m long by the Middle Ordovician (Guensburg 1992; Brower 1994), earlier than proposed by Ausich and Bottjer (1982). The edrioblastoid (edrioasteroid) *Cambroblastus* from the Late Cambrian has a short stalklike structure generally similar to an eocrinoid stalk, and later edrioblastoids such as *Lampteroblastus* and *Astrocystites* from the Ordovician had short stems with columnal-like plates.

Feeding

The majority of Cambrian to Early Ordovician echinoderms were suspension feeders. Edrioasteroids lacked feeding appendages and probably fed using cilia or tube foot–generated currents, combined with mucous chains along ambulacral food grooves exposed by opening cover plate flaps. Beginning in the Late Cambrian, advanced isorophid edrioasteroids apparently had modified or lost the tube feet and perhaps gathered food by cilia-driven mucous on the epithelial lining of the food grooves. Blastozoans such as eocrinoids and rhombiferans had thin biserial erect feeding appendages called brachioles arising from short ambulacra on the thecal summit or upper sides. Most Early to Middle Cambrian species had relatively few brachioles, and those formed an open uncoordinated array. Brachioles are usually extremely thin and are thought to have lacked tube feet (Sprinkle 1973); feeding is assumed to have been accomplished by the ciliated mucous style. Food grooves are narrow, limiting these organisms to small food particles.

Many later blastozoans, such as rhombiferans, retained that basic construction, but in other cases there was modification and elaboration. Trachelocrinid eocrinoids of the Late Cambrian have thick erect biserial arms with widely spaced brachioles branching off both sides (see figure 19.2) in a pattern that is functionally similar to and convergent with crinoids that have a loose filtration fan. This basic pattern continued in eumorphocystids, hemicosmitids, and some paracrinoids, but failed to achieve the success of crinoids. Early Ordovician cylindrical rhombiferans have brachioles arising from the top of the theca, similar to those of *Gogia* and as such probably represent a continuation of the initial blastozoan feeding strategy. Pleurocystitids were convergent with many carpoids in their feeding style. They have a specialized and reduced ambulacral system consisting of two large brachioles with the food grooves usually facing the substrate, allowing exploitation of presumably nutrient-rich larger particles at the sediment-water interface. Rhipidocystids and some paracrinoids have small filter-feeding systems. *Echmatocrinus* had short thick nonbranching arms with wide food grooves and large tube feet (see figure 19.1), indicating specialization toward capture of large food particles. In general, Early Ordovician crinoids retained relatively larger food grooves than most blastozoans, indicating that feeding strategies of blastozoans and crinoids remained separate well into the Early

Ordovician. Thereafter, early pinnulate crinoids, particularly camerates, were presumably microplankton feeders and competitors to blastozoans.

Feeding styles of carpoids apparently differed from those of the suspension feeders described above. Ctenocystoids presumably strained particulate matter from the sediment-water interface using the ctenoid apparatus surrounding the mouth. Cinctans apparently had soft tissues protruding from an aperture along the thecal margin, allowing low-level suspension or deposit feeding. Mitrate and cornute stylophorans used the aulacophore for surface deposit feeding and/or extended upward, allowing low-level suspension feeding (see figure 19.2). Homoiosteleans had a short feeding arm extending from the anterior thecal margin and presumably used for surface deposit feeding. The first echinoderm carnivores (asteroids) and herbivores (asteroids, echinoids, and ophiocystoids) did not appear until the Early to Middle Ordovician, respectively.

Respiration

Cambrian echinoderms have two types of respiratory structures: widespread pores, called epispires, between the thecal plates; and tube feet in the ambulacra that connected to the water vascular system and an external hydropore, or madreporite, near the mouth. Epispires are found in many Early and Middle Cambrian eocrinoids (Sprinkle 1973), in some Early and Middle Cambrian edrioasteroids (Jell et al. 1985; Smith 1985), and in most Middle Cambrian homosteleans (Friedrich 1993). Epispires occur at the sutures of thick tessellate plates and were apparently occupied by thin outpouchings of epidermis (podia) across which gas exchange could take place. Epispires were perhaps vulnerable and were either lost by many of these groups after the Middle Cambrian and replaced by taxa with thin tessellate thecal plates, or they evolved into diplopores, which are paired pores within the thecal plates, allowing efficient water flow and better protection on the thecal exterior.

Thin-plated echinoderms that respired through the entire plate surface were especially common in the Late Cambrian (see examples in figure 19.2), where only a few epispire-bearing echinoderms have been found, and continued into the Early Ordovician. Thecae with this design were easily disarticulated, contributing to a poor fossil record for echinoderms during this interval (Sprinkle 1973; Smith 1988). Thin-plated eocrinoids were mostly replaced in the Early Ordovician by new groups of blastozoans that developed thicker and stronger thecal plates with specialized respiratory structures, such as pectinirhombs and diplopores. Other echinoderms that retained thin thecal plates (especially early crinoids and rhombiferans) had stellate plates with one or more strengthening ridges radiating either to the plate sides or less commonly to the plate corners (Paul 1972; Dzik and Orłowski 1993). Several early crinoids that appeared during the Early Ordovician had an anal sac or tube with pore-bearing plates that may also have augmented respiration.

Tube feet or podia in the ambulacra were probably important respiratory structures of echinoderm groups that had moderate to thick thecal plates but lacked epispires or other accessory respiratory structures. These would include edrioasteroids, some of which preserve podial pores into the thecal interior (Bell 1976; Guensburg and Sprinkle 1994b); helicoplacoids (Durham 1967; Derstler 1985); cornute and mitrate stylophorans, which have tube foot podial basins on the aulacophore (Ubaghs 1968); crinoids that had tube feet on their medium to long arms (Sprinkle 1973); and early asteroids that had tube foot pores in their wide ambulacra (Blake 1994). Blastozoan groups such as later eocrinoids, early rhombiferans, and early diploporans that may have lacked tube feet in their ambulacra had elaborate accessory thecal respiratory structures (pores or thin folds) (Sprinkle 1973).

Protection

Almost all echinoderms were fully encased in a multiplated calcite skeleton that provided protection and support for soft tissues. The earliest echinoderms all had thin imbricate plating, but thicker tessellate-plated taxa were common by the Middle Cambrian. High visibility, large size, and passive behavior would seemingly have made early echinoderms attractive prey, but by analogy with their modern relatives, thecal interiors were mostly fluid-filled with low nutritional value. Suitable predators may have been less common in the Cambrian, but large animals such as anomalocaridids are possibilities; cephalopods became abundant only in the Ordovician, and jawed fish appeared much later in the Paleozoic. We have not seen examples of unsuccessful predation, but we have identified plate-filled coprolites, or regurgitation wads, in the Middle Cambrian, evidence that predation or scavenging did occur (Sprinkle 1973:100).

Spine-bearing plates first occurred in a few Late Cambrian solute homoiosteleans (see figure 19.2), and spines mounted on sockets appeared in Early Ordovician pyrgocystid edrioasteroids and asteroids. Spinose echinoids first appeared in the Middle Ordovician. Ordovician echinoderms typically have fewer plates, arranged in better-defined patterns (circlets) than those of their Cambrian ancestors. Variations in patterns are useful taxonomically at various levels.

Plate-thickening ridges across sutures, forming a trusswork, and fewer but larger plates, are all common thecal strengthening trends. Protection from predators was also increased by the development of mobility in many Cambrian and Ordovician echinoderms, including the ability to swim short distances that is inferred for Middle Cambrian ctenocystoids, and the development of a shallow infaunal way of life in many homalozoans and perhaps some asterozoans. Modern echinoderms have biochemical defenses that make them distasteful, but this feature may have been developed much later in the fossil record when predation increased. Protection from storm activity, including high currents and rapid sediment deposition, would include such factors as better attachment structures, developing a columnal-bearing stem from a multiplated stalk (Middle Cambrian), streamlining the theca in high-level suspension feeders, furling or enrolling the feeding appendages over the summit (Middle Cambrian; Sprinkle and Collins 1998), developing an expandable theca that could be retracted into a blister or low-domal shape under high currents (Middle Cambrian?), and the ability to raise the theca or burrow back up to the surface if covered by a layer of soft sediment.

EVOLUTIONARY FAUNAS AND ECHINODERMS

Evolutionary faunas are groupings of marine shelf metazoans that reached their maximum diversity (and ecologic dominance) at different times during the Phanerozoic (Sepkoski 1981, 1984, 1991). Three Phanerozoic evolutionary faunas are recognized: the small Cambrian Evolutionary Fauna (CEF), which dominated that period; the larger Paleozoic Evolutionary Fauna (PEF), which radiated in the Ordovician and dominated the rest of the Paleozoic; and the Modern Evolutionary Fauna (MEF), which radiated in the early Mesozoic and still dominates today. Analysis of evolutionary faunas has led to the proposal that new metazoan groups originated in shallow onshore environments, then expanded offshore to middle and outer shelf areas, and finally were gradually reduced from the shore to outer shelf and slope environments, as elements from the next evolutionary fauna repeated this onshore-offshore expansion pattern (Jablonski et al. 1983; Sepkoski and Sheehan 1983).

Early echinoderms seem to agree with the evolutionary fauna model, producing a small radiation of taxa with 8 classes and about 35 genera in the Cambrian, followed by a larger radiation with 17 classes and several hundred genera in the Ordovician (Sprinkle 1980, 1992; Sprinkle and Guensburg 1995) (figure 19.3). Many of the Cambrian classes are small and short-lived and have unusual morphology, such as helicoplacoids (Durham 1967; Derstler 1985) and homosteleans (Friedrich 1993). Echinoderm diversity dropped to low levels after the Middle Cambrian, and only eocrinoids, cornute stylophorans, homoiosteleans, and edrioasteroids are known from the Late Cambrian, although crinoids and holothurians must also have survived this interval. We term this echinoderm component of the CEF as the eocrinoid-stylophoran fauna (see Sumrall et al. 1997).

The Early and Middle Ordovician marked the continuance and modest to rapid expansion of these Cambrian groups and the first appearance and rapid expansion of many new echinoderm groups belonging to the Paleozoic Evolutionary Fauna, such as crinoids (?continuance); many groups of blastozoans, especially rhombiferans and diploporans (new); asteroids and ophiuroids (new); edrioasteroids (continuance); echinoids (new); and stylophorans (continuance) (figure 19.3). Crinoids became the dominant echinoderms by the Middle Ordovician, a position they held throughout the rest of the Paleozoic, followed in diversity by rhombiferans in the Ordovician and Silurian and blastoids in the Devonian through Permian (Sprinkle 1980). This initial

Early Radiation of the Echinoderms

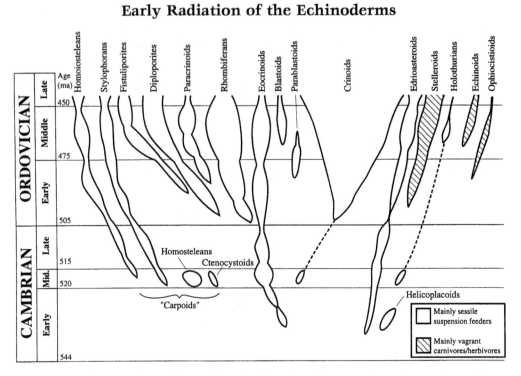

Figure 19.3 Diversification diagram for echinoderms based upon numbers of genera (approximate only). A modest Middle Cambrian radiation is followed by a Late Cambrian decline and a rapid Early Ordovician expansion.

new echinoderm component of the PEF in the Early and Middle Ordovician is termed the crinoid-rhombiferan fauna (Guensburg and Sprinkle 1994a; Sprinkle and Guensburg 1997).

Although several groups of Cambrian echinoderms, including lepidocystids, *Camptostroma, Gogia,* and *Echmatocrinus,* first appeared or developed their maximum abundance in offshore outer detrital belt settings, we are not sure that these limited data indicate an offshore-to-onshore expansion pattern for Cambrian echinoderms in general, which is opposite to the prevailing model for abundant groups such as trilobites in the CEF. Evidence is better for echinoderm groups in the Early Ordovician at the beginning of the PEF. Here, we find strong evidence that some echinoderm groups, such as crinoids and edrioasteroids, were much more common in shallow onshore settings, where hard substrates were present, than in deeper-water offshore settings that mostly lacked extensive hard substrates. These groups appear to have radiated in shallow onshore areas in the Early Ordovician, then spread offshore later in the Ordovician (Guensburg and Sprinkle 1992; Sprinkle and Guensburg 1995). This agrees with the onshore-to-offshore expansion model (Sepkoski and Sheehan 1983; Sepkoski 1991), but the pattern was apparently produced by an extrinsic environmental cause (the availability of hard substrates onshore) and not by an intrinsic biotic cause favored by some workers. We also found some evidence of the opposite pattern in other echinoderm groups, such as rhombiferans and mitrate stylophorans, that are more common in deeper-water offshore areas (Sprinkle and Guensburg 1995). This distribution hints at a possible offshore-to-onshore expansion pattern for these two echinoderm groups, which is again opposite to the prevailing model.

CONCLUSIONS

1. Cambrian echinoderms diversified modestly into two lifestyles: attached epifaunal suspension feeders and vagile surface deposit feeders. The majority of suspension feeders attached to skeletal fragments by means of a suctorial attachment disk or later a cemented holdfast. Availability of attachment surfaces could have been a significant limiting factor for their distribution.

2. Vagile carnivorous asteroids and herbivorous echinoids were late, appearing in the Early to Middle Ordovician.

3. The Cambrian echinoderm radiation was followed in the Early Ordovician with a rapid expansion led by suspension-feeding blastozoans and crinoids. Initially, crinoids flourished on hard substrates, and blastozoans dominated in soft substrates. Both the distribution of fossils within lithofacies and the functional morphologic evidence support this conclusion. Surface detritus feeders continued in modest numbers.

4. The widespread availability of suitable habitats appears to have fostered the onshore (shallow-water) diversification of Ordovician echinoderms of the Paleozoic Evolutionary Fauna. A few groups, such as rhombiferans, may have diversified in the reverse direction.

Acknowledgments. We thank Jennifer Logothetti (Rockford, IL), for helping with illustration preparation, and Kathy Baker (Rock Valley College), for assisting in manuscript preparation. J. F. Bockelie and an anonymous reviewer provided insightful contributions to a draft version of the manuscript, and Andrey Zhuravlev offered important editorial suggestions. This paper is based in part on fieldwork between 1989 and 1994 supported by the National Science Foundation under grants BSR-8906568 (JS) and EAR-9304253 (JS and TEG) and by a Petroleum Research Fund, American Chemical Society Grant to Mark Wilson, College of Wooster (TEG). The Geology Foundation, University of Texas at Austin, provided additional funds for field and publication expenses. This is a contribution to IGCP Project 366, Ecological Aspects of the Cambrian Radiation.

REFERENCES

Ausich, W. I. and D. L. Bottjer. 1982. Tiering in suspension-feeding communities on soft substrata throughout the Phanerozoic. *Science* 216:173–174.

Bell, B. M. 1976. *A Study of North American Edrioasteroidea*. New York State Museum and Science Service, Memoir 21.

Blake, D. B. 1994. Reevaluation of the Palasteriscidae Gregory, 1900, and the early phylogeny of the Asteroidea (Echinodermata). *Journal of Paleontology* 68:123–134.

Brett, C. E. and M. E. Brookfield. 1984. Morphology, faunas, and genesis of Ordovician hardgrounds from southern Ontario, Canada. *Palaeogeography, Palaeoclimatology, Palaeoecology* 46:233–290.

Brower, J. C. 1994. Camerate crinoids from the Middle Ordovician (Galena Group, Dunleith Formation) of northern Iowa and southern Minnesota. *Journal of Paleontology* 68:510–599.

Cook, H. E. 1989. Geology of the Basin and Range Province, western United States: An overview. In M. E. Taylor, ed., *Cambrian and Early Ordovician Stratigraphy and Paleontology of the Basin and Range Province, Western United States, Field Trip Guidebook T125*, pp. 6–13. Washington, D.C.: American Geophysical Union.

Derstler, K. 1985. *Studies on the Morphological Evolution of Echinoderms*. Ph.D. Dissertation, University of California, Davis.

Droser, M. L., P. M. Sheehan, R. A. Fortey, and X. Li. 1995. The nature of diversification and paleoecology of the Ordovician Radiation with evidence from the Great Basin. In J. D. Cooper, M. L. Droser, and S. C. Finney, eds., *Ordovician Odyssey: Short Papers for the Seventh International Symposium on the Ordovician System*, pp. 405–408. Pacific Section SEPM, Book 77.

Durham, J. W. 1967. Notes on the Helicoplacoidea and early echinoderms. *Journal of Paleontology* 41:97–102.

Dzik, J. and S. Orłowski. 1993. Late Cambrian eocrinoid *Cambrocrinus*. *Acta Palaeontologica Polonica* 138:21–34.

Friedrich, W. P. 1993. Systematik und Funktionsmorphologie Mittelkambrischer Cincta (Carpoidea, Echinodermata). *Beringeria* 7:1–190.

Golonka, J., M. L. Ross, and C. R. Scotese. 1994. Phanerozoic paleogeographic and paleoclimatic modeling maps. In A. F. Embry, B. Beauchamp, and A. J. Glass, eds., *Pangea: Global Environments and Resources*, pp. 1–47. Canadian Society of Petroleum Geologists Memoir 17.

Guensburg, T. E. 1992. Paleoecology of hardground encrusting and commensal crinoids, Middle Ordovician, Tennessee. *Journal of Paleontology* 66:129–147.

Guensburg, T. E. and J. Sprinkle. 1992. Rise of echinoderms in the Paleozoic Evolutionary Fauna: Significance of paleoenvironmental controls. *Geology* 20:407–410.

Guensburg, T. E. and J. Sprinkle. 1994a. Echinoderm rapid diversification across the Cambro-Ordovician boundary. *Geological Society of America Abstracts with Programs* 26:A427.

Guensburg, T. E. and J. Sprinkle. 1994b. Revised phylogeny of the Edrioasteroidea based on new taxa from the Early and Middle Ordovician of western Utah. *Fieldiana (Geology)*, n.s., 29:1–43.

Jablonski, D., J. J. Sepkoski, Jr., D. J. Bottjer, and P. M. Sheehan. 1983. Onshore-offshore patterns in the evolution of Phanerozoic shelf communities. *Science* 222:1123–1125.

James, N. P. and C. F. Klappa. 1983. Petrogenesis of Early Cambrian reef limestones, Labrador, Canada. *Journal of Sedimentary Petrology* 53:1051–1096.

James, N. P., P. K. Stevens, C. R. Barnes, and I. Knight. 1989. Evolution of a lower Paleozoic continental-margin carbonate platform, northern Canadian Appalachians. In P. D. Cravello, J. L. Wilson, J. F. Sarg, and J. F. Read, eds., *Controls on Carbonate Plat-*

form and Basin Development, pp. 123–146. Society of Economic Paleontologists and Mineralogists Special Publication 44.

Jell, P. A., C. F. Burrett, and M. R. Banks. 1985. Cambrian and Ordovician echinoderms from eastern Australia. Alcheringa 9:183–208.

Palmer, T. J. and C. D. Palmer. 1977. Faunal distribution and colonization strategy in a Middle Ordovician hardground community. Lethaia 10:179–199.

Paul, C. R. C. 1972. Cheirocystella antiqua gen. et sp. nov. from the Lower Ordovician of western Utah, and its bearing on the evolution of the Cheirocrinidae (Rhombifera: Glyptocystitida). Brigham Young University Geology Studies 19:15–63.

Rozhnov, S. V. 1994. Changes in the hardground at the Cambrian-Ordovician Boundary. Paleontological Journal 28:84–91.

Ruzhentsev, S. V. and A. A. Mossakovsky. 1995. Geodinamika i tektonicheskoe razvitie paleozoid Tsentral'noy Azii kak rezul'tat vzaimodeystviya Tikhookeanskogo i Indo-Atlanticheskogo segmentov Zemli [Geodynamics and tectonic evolution of the Central Asia paleozoids as a resultant of interaction between the Pacific and Indo-Atlantic segments of Earth]. Geotektonika 1995 (4):29–47.

Sandberg, P. A. 1983. An oscillating trend in Phanerozoic nonskeletal carbonate mineralogy. Nature 305:19–22.

Sepkoski, J. J., Jr. 1981. A factor analytic description of the Phanerozoic marine fossil record. Paleobiology 7:36–53.

Sepkoski, J. J., Jr. 1984. A kinetic model of Phanerozoic taxonomic diversity. III. Post-Paleozoic families and mass extinctions. Paleobiology 10:246–267.

Sepkoski, J. J., Jr. 1991. A model of onshore-offshore change in faunal diversity. Paleobiology 17:58–77.

Sepkoski, J. J., Jr. and P. M. Sheehan. 1983. Diversification, faunal change, and community replacement during the Ordovician radiations. In M. J. S. Tevesz and P. L. McCall, eds., Biotic Interactions in Recent and Fossil Benthic Communities, pp. 673–717. New York: Plenum Press.

Sloss, L. L. 1963. Sequences in the cratonic interior of North America. Geological Society of America Bulletin 74:93–114.

Smith, A. B. 1985. Cambrian eleutherozoan echinoderms and the early diversification of edrioasteroids. Palaeontology 28:715–56.

Smith, A. B. 1988. Patterns of diversification and extinction in Early Palaeozoic echinoderms. Palaeontology 31:799–828.

Smith, A. B. 1990. Evolutionary diversification of echinoderms during the Early Palaeozoic. In P. D. Taylor and G. P. Larwood, eds., Major Evolutionary Radiations, pp. 265–286. Oxford: Clarendon Press.

Smith, A. B. and P. A. Jell. 1990. Cambrian edrioasteroids from Australia and the origin of starfishes. Memoirs of the Queensland Museum 28:715–778.

Sprinkle, J. 1973. Morphology and Evolution of Blastozoan Echinoderms. Museum of Comparative Zoology Special Publication.

Sprinkle, J. 1976. Biostratigraphy and paleoecology of Cambrian echinoderms from the Rocky Mountains. Brigham Young University Geology Studies 23:61–73.

Sprinkle, J. 1980. An overview of the fossil record. In T. W. Broadhead and J. A. Waters, eds., Echinoderms: Notes for a Short Course, pp. 15–26. University of Tennessee Department of Geological Sciences, Studies in Geology 3.

Sprinkle, J. 1992. Radiation of Echinodermata. In J. H. Lipps and P. W. Signor, eds., Origin and Early Evolution of the Metazoa, pp. 375–398. New York: Plenum Press.

Sprinkle, J. 1995. Do eocrinoids belong to the Cambrian or to the Paleozoic Evolutionary Fauna? In J. D. Cooper, M. L. Droser, and S. C. Finney, eds., *Ordovician Odyssey: Short Papers for the Seventh International Symposium on the Ordovician System,* pp. 397–400. Pacific Section SEPM, Book 77.

Sprinkle, J. and D. Collins. 1998. Revision of *Echmatocrinus* from the Middle Cambrian Burgess Shale of British Columbia. *Lethaia* 31:269–282.

Sprinkle, J. and T. E. Guensburg. 1995. Origin of echinoderms in the Paleozoic Evolutionary Fauna: The role of substrates. *Palaios* 10:437–453.

Sprinkle, J. and T. E. Guensberg. 1997. Early radiation of echinoderms. In J. A. Waters and C. G. Maples, eds., *Geobiology of Echinoderms,* pp. 205–224. Paleontological Society Papers 3.

Sumrall, C. D., J. Sprinkle, and T. E. Guensburg. 1997. Systematics and paleoecology of Late Cambrian echinoderms from the western United States. *Journal of Paleontology* 71:1091–1109.

Ubaghs, G. 1968. Stylophora. In R. C. Moore, ed., *Treatise on Invertebrate Paleontology, Part S, Echinodermata 1,* pp. S495–S565. New York: Geological Society of America; Lawrence: University of Kansas.

Wilson, M. A., T. J. Palmer, T. E. Guensburg, C. D. Finton, and L. E. Kaufman. 1992. The development of an Early Ordovician hardground community in response to rapid sea-floor calcite precipitation. *Lethaia* 25:19–34.

Calcified Algae and Bacteria

Calcified microbes expanded rapidly in abundance and diversity from Nemakit-Daldynian to Tommotian. This rapid diversification near the base of the Cambrian reflects a burst of cyanobacterial evolution, and commencement of an environmentally facilitated Cyanobacterial Calcification Episode that continued into the Ordovician. No new genera appeared during the Middle-Late Cambrian, and apparent diversity declined. Correlation between generic diversity and number of studies suggests that this decline might be a monographic artifact. Calcified microbes remained important components of shallow marine carbonates throughout the Cambrian. Most groups represent cyanobacteria (Angusticellularia, Botomaella, Girvanella, and Obruchevella groups), or probable cyanobacteria (Epiphyton, Proaulopora, and Renalcis groups). Chabakovia, Nuia, and Wetheredella are Microproblematica. Calcified microbes created rigid, compact reef frameworks. During the Early Cambrian they were commonly associated with archaeocyaths, but they continued their successful reef-building role into the Middle-Late Cambrian in the absence of a significant metazoan contribution. Distribution patterns suggest that filamentous and dendritic forms (Angusticellularia, Epiphyton, and Girvanella groups) preferred high-energy conditions and formed reefs in grainy locations; whereas botryoidal forms (Renalcis Group) formed mudstone-associated reefs in shelf and midramp environments. There is no evidence that calcified microbes were affected by metazoan grazing, disturbance, or competition during the Cambrian. Conversely, these microbes may have inhibited metazoan larval settlement and growth. Cambrian calcified algae are very scarce and are much less diverse than cyanobacteria. Amgaella, Mejerella, and Seletonella may be dasycladaleans. They are known only from the Middle (Amgaella) and Late (Mejerella and Seletonella) Cambrian of Russia and adjacent regions.

THE LONG-TERM HISTORY of microbes and metazoans has been seen as a displacement of prokaryotes by eukaryotes (Garrett 1970). In the Cambrian, it is tempting to emphasize invertebrate newcomers and to expect that microbial fossils should be scarce and in decline. Yet calcified microbial fossils are common in the Cambrian, and they appeared rapidly, early in the period, as if switched on by some event (Riding 1984). In part, this biota represents continuation of the old Proterozoic order, but in many respects it was a new development, with few earlier counterparts. Marine calcified microbes had never been so abundant and diverse before and were never to be so abundant in subsequent periods. In the Cambrian, calcified microbes are major reef builders (Pratt et al., this volume). In comparison, calcified algae are of minor importance (Chuvashov and Riding 1984), and their major radiation was in the Ordovician. The abundance and diversity of calcified cyanobacteria—or at least of microfossils that appear to be cyanobacteria—during the Cambrian reflect both suitable conditions for calcification and an evolutionary radiation that parallels that seen in many invertebrate groups.

TAXONOMIC GROUPS

Research on Cambrian calcified microbes began with the discovery of *Epiphyton*, by Bornemann (1886) and was given tremendous impetus by K. B. Korde, V. P. Maslov, A. G. Vologdin, and colleagues in the USSR between 1930 and 1980 (Riding 1991a). Cambrian calcified algae and cyanobacteria are here grouped into cyanobacteria (*Angusticellularia, Botomaella, Girvanella,* and *Obruchevella* groups), possible cyanobacteria (*Epiphyton, Proaulopora,* and *Renalcis* groups), Microproblematica (*Chabakovia, Nuia,* and *Wetheredella*), possible dasycladalean algae (*Amgaella*), and Problematica that have at times been assigned to these groups and to possible red algae (*Cambroporella, Edelsteinia,* and *Lenaella*). Recognition of 21 genera in 7 groups, together with Microproblematica, possible dasycladaleans, and Problematica, provides an outline classification (table 20.1) that omits numerous junior synonyms and minor and misidentified genera. Riding (1991b: table 1, figure 1) listed 74 of the most widely known of these, all but 5 of which were created by researchers in the USSR during the period 1930–1980. The total number of genera involved probably approaches 125.

The most striking general feature of the calcified Cambrian flora is the scarcity of algae. This understanding has emerged relatively recently. During the 1960s and early 1970s, many of the Cambrian calcified microbes were regarded as algae (Riding 1991a: tables 2 and 4). Vologdin (1962), for example, regarded members of the *Angusticellularia, Renalcis, Epiphyton,* and *Botomaella* groups as red algae, and Korde (1973) considered that the Cambrian flora was dominated by red algae. This opinion began to change after the suggestion of Luchinina (1975) that most of these genera represent cyanobacteria was supported by studies of modern analogs (Riding and Voronova 1982a,b). The only Cambrian fossils that have continued to be generally regarded as heavily calcified algae are much larger and include genera such as *Sele-*

Table 20.1 Classification of Cambrian Calcified Algae and Bacteria:
Groups, Principal Genera, and Affinities

Angusticellularia Group *Angusticellularia*	CYANOBACTERIA
Botomaella Group *Bajanophyton, Bija, Botomaella, Kordephyton*	
Girvanella Group *Batinevia, Cladogirvanella, Girvanella, Razumovskia, Subtifloria*	
Obruchevella Group *Obruchevella*	
Epiphyton Group *Acanthina, Epiphyton, Gordonophyton, Korilophyton, Sajania,* *Tubomorphophyton*	?CYANOBACTERIA
Proaulopora Group *Proaulopora*	
Renalcis Group *Gemma, Renalcis, Tarthinia*	
Chabakovia *Nuia* *Wetheredella*	MICROPROBLEMATICA
Amgaella Group *Amgaella, Mejerella, Seletonella*	?DASYCLADALEANS
Cambroporella *Edelsteinia* *Lenaella*	PROBLEMATICA

Source: Modified from Riding 1991a.

tonella and *Amgaella*, which may be dasycladalean green algae (Korde 1950, 1957). Their known distribution is very limited; *Seletonella*, for example, is known only from its type-locality.

Of the 30 principal genera (table 20.1), 11 can confidently be regarded as cyanobacteria (*Angusticellularia, Botomaella, Girvanella,* and *Obruchevella* groups), a further 10 (*Epiphyton, Proaulopora,* and *Renalcis* groups) are possible cyanobacteria, 3 (*Chabakovia, Nuia,* and *Wetheredella*) are Microproblematica, 3 (*Amgaella* group) may be dasycladalean algae, and 3 are Problematica that have been thought to be algae. Members of the *Angusticellularia, Botomaella, Girvanella, Epiphyton,* and *Renalcis* groups (figure 20.1) overwhelmingly dominate the flora through much of the Cambrian and make a major contribution to the construction of domes, reefs, and oncoids. These may all represent cyanobacteria, but for important groups such as *Epiphyton* and *Renalcis,* this interpretation, although likely, has yet to be confirmed from modern analogs. Consequently, collective names have been applied to these calcified microfossils

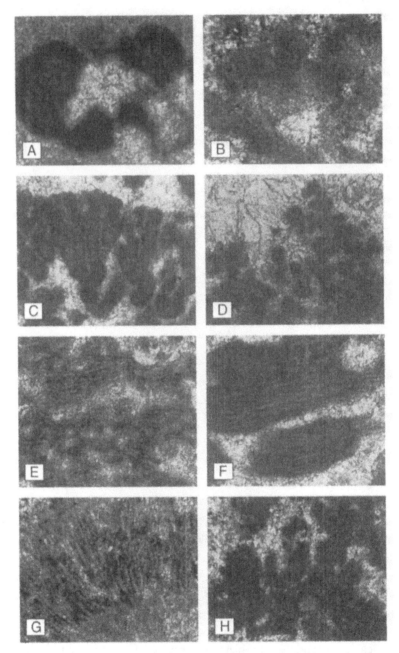

Figure 20.1 Common Cambrian calcified cyanobacteria and possible cyanobacteria. A, *Renalcis*, Salaany Gol, western Mongolia, ?Atdabanian; B, *Tarthinia* (*Renalcis* Group), Olenek River, Siberia, Tommotian; C, *Tubomor-phophyton* (*Epiphyton* Group), Oi-Muraan, Lena River, Siberia, Atdabanian; D, *Korilophyton* (*Epiphyton* Group), Fomich River, Anabar, Siberia, Nemakit-Daldynian; E, *Girvanella*, Tyuser River, Lena River, Siberia, Atdabanian; F, *Subtifloria* (*Girvanella* Group), Salaany Gol, western Mongolia, Tommotian; G, *Botomaella*, Olenek River, Siberia, Tommotian; H, *Angusti-cellularia* (=*Angulocellularia*), Olenek River, Siberia, Tommotian. Magnification for all ×70.

in order to distinguish them as a group, even though their collective affinities are not altogether certain. These names include calcibionts (Luchinina 1991, 1998) and calcimicrobes (=calcified microbial microfossils; James and Gravestock 1990:460).

DIVERSITY

Taxonomic Treatment

The many taxa described among these fossils have not been widely recorded outside northern Asia, reflecting the dominance of Soviet systematic work. In contrast, many sedimentological studies of limestones containing these fossils have been done outside the former Soviet Union, by workers often unfamiliar with and unsupportive of the complex taxonomies formulated by paleontologists (see Mankiewicz 1992). The significant contribution to the study of these fossils by K. B. Korde has been limited by the following tendencies: (1) to split taxa (Gudymovich 1967; Luchinina 1975; Pratt 1984)—e.g., Korde (1961) created 62 species for *Epiphyton;* (2) to incorporate diagenetically altered (Mankiewicz 1992) and inorganic (Riding 1991a) material; and (3) to discern cellular and sporangial detail in obscure microstructures (Riding and Voronova 1982a). As a result, assessment of the biodiversity represented by these fossils must take account of a variety of intricate systematic problems whose resolution is under way but not yet complete.

Ecophenotypic Variation

To what extent do these fossils represent biologically distinct taxa? Cyanobacterial calcification is a sheath-related character influenced, but not controlled, by the organism (Golubic 1973; Pentecost and Riding 1986). Could similar-appearing calcified forms be created by different organisms? The answer appears to be positive, as is likely in the case of *Girvanella* (Riding 1977a). To add to this complication, one organism may produce different morphotypes. Maslov (1956) suggested that *Renalcis* shows ecophenotypic variation, and Riding (1991a) reported that *Botomaella* and *Hedstroemia,* which appear morphologically distinct, both resemble extant rivulariaceans, although not necessarily the same strain. Saltovskaya (1975) went much further and suggested that some genera, including *Epiphyton, Renalcis,* and *Chabakovia,* were identical because they show intergradation. She placed them in synonymy and believed them all to be filamentous. Pratt (1984) also suggested that *Renalcis* and *Epiphyton* might not be genetically distinct, but proposed that they were both coccoid cyanobacteria.

It is likely that ecophenotypic variation does exist within some of these groups. However, several lines of evidence suggest that distinct taxa nonetheless are present. Despite the presence of morphologic series, there are some clear differences between major groups. For example, botryoidal fossils such as *Renalcis* are quite different in organization and construction from dendritic forms such as *Epiphyton* (Riding and Voronova 1985). In addition, although precise analogs of these fossils have yet to be reported, available evidence also indicates significant differences (see the section Cyanobacteria, below). Furthermore, most of the morphotypes intimately coexist, sometimes being mutually attached, while retaining distinct differences in morphology, including chamber size, wall thickness and structure, and filament shape and size. The absence of complete intergradation strengthens the view that they are not simply morphologic variants of one form. Moreover, these taxa exhibit changes in morphology and occurrence through time. This can be seen by comparing Cambrian *Epiphyton* and *Renalcis* with Devonian specimens. These observations suggest that the morphologic similarities reflect parallelism in structurally simple but biologically distinct organisms.

AFFINITIES

Cyanobacteria

Important questions concerning the calcified microbial fossils that dominate the Cambrian flora include not only their affinity but also their mutual distinctness and the timing of their calcification. Cyanobacterial affinity applies particularly to members of the *Angusticellularia*, *Botomaella*, *Girvanella*, and *Obruchevella* groups and is based on similarities in size and shape among these fossils and extant examples (Riding 1991a). Precise modern analogs are still required for the *Epiphyton*, *Proaulopora*, and *Renalcis* groups. *Epiphyton* group fossils can be compared to stigonemataleans such as *Loriella* (Riding and Voronova 1982a). Korde (1958) regarded *Renalcis* as a cyanobacterium, and Hofmann (1975) suggested that it could represent coccoid colonies. *Proaulopora*, too, can be compared to extant cyanobacteria such as *Calothrix* (Luchinina in Chuvashov et al. 1987) but lacks a precise modern analog.

Differences between taxa can be complicated by apparent intergradation. In particular, the *Epiphyton* and *Renalcis* groups, together with *Angusticellularia*, constitute a morphologic series (Pratt 1984) involving at least 5 genera: *Epiphyton*, *Angusticellularia*, *Tarthinia*, *Renalcis*, and *Chabakovia* (Riding and Voronova 1985). Pratt (1984) suggested that *Epiphyton*, *Renalcis*, and their intermediates formed by calcification of dead and degrading colonies of coccoid cyanobacteria. This interpretation therefore involves both biologic affinity and the timing of calcification. So far as calcification is concerned, no postmortem, subaqueous, preburial calcification mechanism is known to account for the quality and quantity of preservation seen in *Epiphyton* and *Renalcis*. In contrast, in vivo calcification, as seen in extant cyanobacteria, can result in intense impregnation that preserves sheath morphology in detail. This mechanism would account for the delicate morphologic details exhibited by Cambrian calcified microbes where they are well preserved. These details include internal spaces, ranging from

Table 20.2 Cambrian Ranges of Calcified Microproblematica and Possible Algae

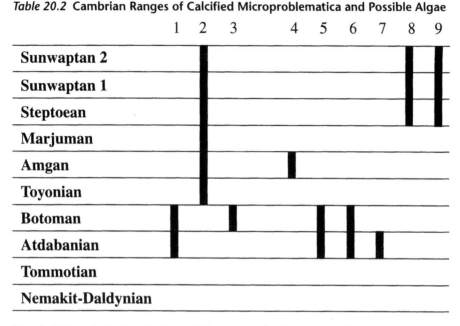

Note: *1, Chabakovia; 2, Nuia; 3, Wetheredella; 4, Amgaella; 5, Cambroporella; 6, Edelsteinia; 7, Lenaella; 8, Mejerella; 9, Seletonella.* TG = 5 total genera; OR = 5 number of originations. Ranges of *Mejerella* and *Seletonella* lack stage resolution.

tubes to inflated and irregular chambers, and the micritic, delicately fibrous, or—in some cases—peloidal structure of the wall (Riding and Voronova 1985).

At the same time, consistency of appearance of these details for particular taxa support evidence from extant analogs that they represent genetically distinct organisms. Furthermore, despite recognition of morphologic series (Pratt 1984; Riding and Voronova 1985), it can be seen that in most cases intergradation is not complete and taxa are disjunct. Even superficially, *Epiphyton* and *Renalcis* are distinctly different, and they most likely represent filamentous cyanobacteria (Riding and Voronova 1982a; Luchinina in Chuvashov et al. 1987) and coccoid cyanobacteria (Hofmann 1975; Luchinina in Chuvashov et al. 1987), respectively. Nonetheless, anomalies remain, as in the case of *Angusticellularia*, which has a filamentous extant analog (Riding and Voronova 1982b), but grades as a fossil toward *Tarthinia*. In this respect, it has to be remembered that morphologic parallelism is common among algae and cyanobacteria.

Microproblematica

Although more abundant in the Ordovician, the Problematica *Nuia* and *Wetheredella* are known in the Cambrian (table 20.2). *Wetheredella* is very rare in the Cambrian and has been recorded only from the Botoman (Kobluk and James 1979, figure 8). *Nuia* was first described from the Late Cambrian of Siberia (Maslov 1954). Its oldest occurrence is Toyonian (Ross et al. 1988). Its radial fibrous structure can, in transverse section, resemble ooids, but it is characteristically elongate and multilayered. Maslov (1954) considered *Nuia* a green alga, but its affinity continues to defy explanation (Ross et al. 1988). The affinities of *Chabakovia* are also uncertain. It is comparable to some members of the *Renalcis* group and also shows resemblance to foraminifers (Elias 1950; Loeblich and Tappan 1964).

Algae

Dasycladaleans

The earliest representatives of calcified dasycladaleans have been sought in a heterogeneous group of rare and poorly understood genera that include *Amgaella, Cambroporella, Edelsteinia, Lenaella, Mejerella,* and *Seletonella* (Maslov 1956:82; Bassoullet et al. 1979). These are mostly centimetric in size, hollow, and cup- or pear-shaped and have been described as having pores or branches. In these respects, they do broadly resemble dasycladaleans. *Seletonella* has had its name given to a major dasycladalean family (Seletonellaceae; Korde 1973:239), yet the affinities of these Cambrian fossils are not at all certain. In particular, those recorded from the Early Cambrian (*Cambroporella, Edelsteinia,* and *Lenaella*) are unlikely to be algae (Debrenne and Reitner, this volume).

Cambroporella (Atdabanian-Botoman) has been regarded as the oldest calcified dasycladalean (Bassoullet et al. 1979), but it has also been compared with bryozoans (Elias 1954) and hydroconozoans (Sayutina 1985:73). *Edelsteinia,* also from the Early Cambrian, was regarded as a possible green alga by Maslov (1956:82), but Webby (1986) suggested a relationship with stromatoporoid sponges. A smaller conical Atdabanian fossil, *Lenaella,* originally thought to be a hydrozoan, sponge, or alga (Korde 1959:626), is of uncertain affinity.

The three younger genera—*Amgaella* (Middle Cambrian), *Mejerella,* and *Seletonella* (Late Cambrian)—show more resemblances to algae and have been regarded as dasycladaleans. *Amgaella,* from the Amgan of the Amga River, Siberian Platform (Korde 1957), has a thick wall, pierced by numerous pores and surrounding a hollow interior. At its type-locality *Amgaella* is reef-building (Hamdi et al. 1995). Both *Mejerella* and *Seletonella* are known only from a single Late Cambrian locality in Kazakhstan (Korde 1950). They differ from *Amgaella* in having thinner walls and numerous external branches that superficially resemble those of dasycladaleans in life but are atypical of dasycladalean skeletons, in which the branches are uncalcified and normally preserved as pores that pierce the calcareous wall. Like *Amgaella, Seletonella* appears to be reef-building.

Of these genera, *Amgaella* is perhaps the most likely alga, and it may be the oldest calcified dasycladalean. *Palaeoporella,* described originally from the Late Ordovician by Stolley (1893) as a dasycladalean but now thought to be a udoteacean (codiacean;

Pia 1927), has been reported from the Late Cambrian of Texas (Johnson 1954, 1961, 1966), but the unit in which it occurs (Ellenburger Group) is of Early Ordovician age. The oldest certain record of a calcified dasycladalean is *Rhabdoporella* of probable Late Ordovician age (Høeg 1932). The radiation of calcified algae was apparently more an Ordovician than a Cambrian event (Riding 1994).

Rhodophytes

The oldest bona fide calcified red alga is *Petrophyton* from the Middle-Late Ordovician (Edwards et al. 1993; Riding 1994). In the Cambrian, *Solenopora* Dybowski has been confused with *Epiphyton* (Priestley and David 1912 : 768) and with *Bija* (Maslov (1937: plate 1, figures 3–6). *Bija,* first described from the Toyonian, is also known from the Atdabanian and the Botoman and has been placed by Luchinina (1975) in the cyanobacteria (Riding 1991a). It is regarded here as a member of the *Botomaella* Group (see table 20.1). Solenoporaceans are a heterogeneous group that includes metazoans (e.g., *Solenopora spongioides* Dybowski 1877, the type species), red algae (e.g., *Solenopora gotlandica* Rothpletz 1908), and cyanobacteria (e.g., *Solenopora compacta* Billings 1865) (Riding 1977b; Brooke and Riding 1987). None of these is definitely known from the Cambrian.

RADIATION

Knowledge of the distribution of Cambrian calcified cyanobacteria and associated groups would be better if, in spite of its faults, the detailed taxonomy developed in the USSR had been more widely applied elsewhere. Stratigraphic distribution plots (Riding and Voronova 1984; Riding 1991a: figure 6; Mankiewicz 1992; Zhuravlev 1996: figure 4; Zhuravlev, this volume) show highest diversity in the Early Cambrian, particularly Atdabanian-Botoman (table 20.3). However, the pattern shown corresponds proportionally with the number of areas from which calcified cyanobacteria have been reported: Nemakit-Daldynian, 9 taxa, 2 areas; Early Cambrian, 68 taxa, 28 areas; Middle Cambrian, 23 taxa, 13 areas; Late Cambrian, 18 taxa, 9 areas. The pattern may thus reflect monographic bias, due to concentration of detailed studies in the Early Cambrian of Siberia and adjacent areas (cf. Zhuravlev [1996], who attributes diversity decline to reduction in reef spatial heterogeneity). Future studies of the Middle-Late Cambrian may reveal diversity similar to that of the Early Cambrian.

Nonetheless, some of the patterns presently observed may be real. The appearance in the Nemakit-Daldynian of a number of calcified cyanobacteria that are unknown in the Proterozoic was an evolutionary event for these microbes (Riding 1994 : 433), just as it was for metazoans. Of the 7 genera recorded in the Nemakit-Daldynian, 5 are first appearances. This flora diversified during the Tommotian-Botoman, the problematic *Wetheredella* was added during the Botoman, and *Nuia* was added during the

Table 20.3 Cambrian Ranges of Calcified Cyanobacteria and Possible Cyanobacteria

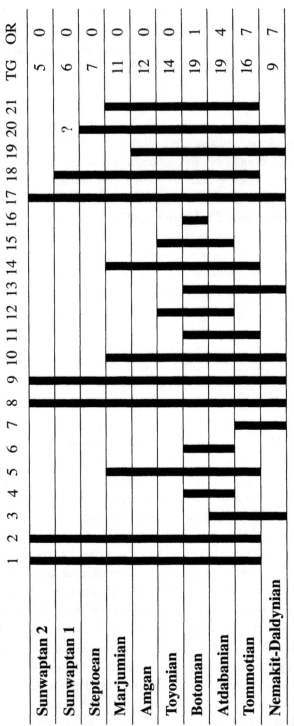

Note: 1, *Epiphyton*; 2, *Gordomophyton*; 3, *Korilophyton*; 4, *Sajania*; 5, *Tubomorphophyton*; 6, *Acanthina*; 7, *Gemma*; 8, *Renalcis*; 9, *Tarthinia*; 10, *Angusticellularia*; 11, *Bajanophyton*; 12, *Bija*; 13, *Botomaella*; 14, *Kordephyton*; 15, *Batinevia*; 16, *Cladogirvanella*; 17, *Girvanella*; 18, *Razumovskia*; 19, *Subtifloria*; 20, *Obruchevella*; 21, *Proaulopora*. TG = total genera; OR = number of originations. Originations in the Nemakit-Daldynian do not count *Angusticellularia* and *Girvanella*, because these genera occur in the Proterozoic.

Toyonian. There were no subsequent originations for cyanobacteria-like calcified taxa during the remainder of the Cambrian, and change is limited to extinctions. The only post–Early Cambrian originations are among the possible algae *Amgaella, Mejerella,* and *Seletonella.*

Cyanobacteria

Proterozoic Antecedents

Calcified cyanobacteria-like microfossils in the Proterozoic may show a patterned abundance distribution through time (Riding 1994: table 1) and include forms similar to *Girvanella* and *Angusticellularia* (Raaben 1969; Hofmann and Grotzinger 1985; Turner et al. 1993; Pratt 1995). Records (e.g., Kolosov 1970, 1975; Green et al. 1989; Fairchild 1991) suggest that they are generally scarce and of low diversity. Subsequent to about 700 Ma, this may have been due to global low temperatures (Riding 1994).

Cambrian Radiation

It is not yet known precisely when the radiation of calcified cyanobacteria-like fossils "of Paleozoic type" (Voronova 1979:868) first significantly developed. Future work may push back this event earlier into the late Neoproterozoic. Reefal associations are common in the Nemakit-Daldynian of the Siberian Platform (Voronova in Voronova and Radionova 1976; Kolosov 1977; Zhuravleva et al. 1982; Luchinina 1985, 1990, 1999), Altay Sayan Foldbelt (Zadorozhnaya 1974), Mongolia (Drozdova 1980; Kruse et al. 1996), and probably Oman (Mattes and Conway Morris 1990) in which *Angusticellularia, Gemma, Korilophyton, Renalcis,* and *Tarthinia* are prominent, and *Botomaella, Girvanella, Obruchevella,* and *Subtifloria* also occur. Most of these genera are long-ranging (table 20.3), but *Gemma* and *Korilophyton* appear restricted to the lower part of the Early Cambrian. At present, therefore, the Nemakit-Daldynian marks the appearance of the "Cambrian flora" (Chuvashov and Riding 1984), and all major groups, with the exception of *Proaulopora,* are represented. This flora represents a marked departure from Proterozoic calcified microfossils, both in abundance and diversity. *Renalcis* may have noncalcified analogs in silicified palmelloid coccoid colonies of Proterozoic age (Hofmann 1975), and *Obruchevella* possesses silicified (Reitlinger 1959) and phosphatized (Peel 1988) analogs, but I am not aware of *Botomaella*-like or *Epiphyton*-like organization in Proterozoic microfossils.

Diversity increased sharply in the Tommotian, and peaked in the Atdabanian and Botoman, before progressively declining during the Middle-Late Cambrian (table 20.3). Apart from northern Asia, Cambrian calcified cyanobacteria have been recorded widely, although notably not in South America, sub-Saharan Africa, or southern Asia. Early Cambrian records are numerous: from northern Asia (Siberian Platform, Altay Sayan Foldbelt, the South Urals, Eastern and Western Transbaikalia, the Russian Far East, Tuva, Mongolia, Kazakhstan, Uzbekistan), China (North and South China Platforms), Europe (Germany, Normandy, Sardinia, Spain), Morocco, Australia, Antarctica, the Appalachians (Virginia, Newfoundland, Labrador), western North America (Sonora, Mexico; Nevada; British Columbia, Yukon Territory, Northwestern Territories), Ellesmere Island in Canada, and Greenland. Areas from which Middle–Late Cambrian cyanobacteria have been recorded are notably fewer: Siberian Platform, Eastern and Western Sayan, Kazakhstan, North China Platform, Newfoundland, Quebec, Mackenzie Mountains, Northwestern Territories, British Columbia, Alberta, Virginia, Wyoming, Nevada, Texas.

Similarity between generic diversity and number of regions studied complicates assessment of this apparent decline, from 19 genera in the Atdabanian to 5 in the latest Cambrian. Patterns of Cambrian diversification will remain uncertain until there have been more studies of the Middle and Upper Cambrian. Apart from probable pre-Ediacaran occurrences of *Angusticellularia* and *Girvanella,* all originations are Lower Cambrian, and most are pre-Botoman.

Post-Cambrian

Some of these genera (*Angusticellularia, Cladogirvanella, Epiphyton, Girvanella,* and *Renalcis*) continue to occur during the Paleozoic, but more than 75 percent of this flora has not been recorded after the Cambrian. It remains to be seen to what extent this pattern will be confirmed by future studies. At present, the Paleozoic distribution of calcified cyanobacteria-like fossils is markedly episodic (see the section "Calcification: Patterns" below). The Cambrian flora continues its decline into the Early Ordovician and largely disappears, apart from a weak resurgence in the Silurian, for much of the middle Paleozoic (Riding 1991b). By the time that it reappears in the Late Devonian, it has changed considerably; only members of the *Epiphyton, Girvanella,* and *Renalcis* groups are conspicuous (Wray 1967; Riding 1979), and the component genera appear different: *Paraepiphyton*—which somewhat resembles *Korilophyton*—is the single representative of the *Epiphyton* Group, and *Izhella* and *Shuguria* occur in place of *Renalcis.* Possibly they are synonyms of *Renalcis* (Saltovskaya 1975; Pratt 1984), but some features emphasized in them are not typical of the Cambrian. Notably absent are important reef builders such as *Angusticellularia, Gordonophyton,* and *Tarthinia.*

Algae

Proterozoic Antecedents

A variety of benthic algae, including chlorophytes (Han and Runnegar 1992), rhodophytes (Butterfield et al. 1990), and carbonaceous films (Hofmann 1992) that could represent phaeophytes and other algae, have been reported from the Proterozoic, but the only records of calcified algae are based on two tentative reports. Horodyski and Mankiewicz (1990) suggested that *Tenuocharta* (600–700 Ma) might be a red alga or cyanobacterium. Grant et al. (1991) compare a possible red alga (530–650 Ma) with phylloid algae.

Cambrian

Realization that many calcified microfossils previously thought to be algae are most likely cyanobacteria (Luchinina 1975) indicates that calcified algae are scarce in the Cambrian, as well as in the Proterozoic. The dasycladalean affinities of the rare Cambrian fossils *Amgaella*, *Mejerella*, and *Seletonella* still require confirmation.

Post-Cambrian

The first confirmed records of both heavily calcified dasycladaleans and rhodophytes are *Rhabdoporella* and *Petrophyton*, respectively, from the Middle-Late Ordovician (Høeg 1932).

ENVIRONMENTAL ECOLOGY

The following statements regarding the environmental ecology of these calcified microbes are advanced here as working hypotheses for future evaluation.

Models

The distributions of calcified microbes on Cambrian carbonate and mixed siliciclastic-carbonate platforms reflect their abundance and mutual associations with respect to energy and turbidity/water clarity.

Abundance Model

Calcified microbes were more abundant and diverse in high-energy shallow-water environments. Calcified microbes required (1) firm substrates in conditions that favored (2) photosynthesis and (3) calcification. These requirements were provided most readily in shallow high-energy outer-shelf environments, for the following reasons:

(1) Because of their small size, calcified microbes had very low tolerance of particulate sediment. They grew most abundantly in particulate sediment-free conditions on external and internal (cryptic) reef surfaces and on other hard substrates such as microbial domes and oncoids. In many cases they were the principal builders of these substrates. (2) Low turbidity facilitated photosynthesis, although light flux requirements of cyanobacteria were not high. This requirement of low turbidity was fa-

vored by high-energy conditions that removed sediment. In turn, high energy enhanced (3) calcification.

Calcified microbes therefore grew best in shallow shelf margin locations. Nonetheless, they could also colonize firm substrates in many shallow-water carbonate and mixed siliciclastic-carbonate environments, provided turbidity was not too high and water circulation too low.

Association Model

Filamentous microbes (e.g., *Epiphyton* Group, *Girvanella* Group) preferred high-energy and less turbid conditions than botryoidal (coccoid; e.g., *Renalcis* Group) microbes. These requirements were provided most readily in outer shelf (*Epiphyton* and *Girvanella* groups) and inner platform (*Renalcis* Group) environments, respectively.

This contention regarding preferences of cyanobacteria is supported not only by lithologic observations but also by data from present-day calcifying freshwater environments, where coccoid cyanobacteria (e.g., *Gloeocapsa*) are more common in quieter water (lakes), and filamentous forms (e.g., rivulariaceans) are more common in fast-flowing streams (Riding, pers. obs.).

Distribution Patterns

These dual models of abundance and association patterns are supported by reports of the distribution of calcified microbes from a variety of locations and ages during the Cambrian.

1. Low energy/inner shelf and midramp. Small shale-mudstone-enveloped inner platform reefs and domes are characterized by *Renalcis* Group fossils (e.g., James and Gravestock 1990; Latham and Riding 1990; Kruse et al. 1995; Riding and Zhuravlev 1995).

2. High energy/shelf margin and inner ramp. Grainy and shelf edge locations are characterized by *Epiphyton* and *Girvanella* group framestones, together with *Tarthinia,* often forming biostromes (e.g., Zadorozhnaya 1974; McIlraeth 1977; James 1981; Read and Pfeil 1983; Coniglio and James 1985; Rees et al. 1989; Bao et al. 1991; Debrenne et al. 1991; Wood et al. 1993) locally seen in downslope transported blocks.

Discussion

As a generalization, Cambrian calcified microbes most commonly occur either in shale enclosed domes and bioherms or in biostromes and domes in current swept environments (see the section "Sedimentologic Roles: Reefs: Categories" below). Coniglio and James (1985:752) drew attention to the "apparent preference of *Epiphyton* for platform-edge lithofacies." James and Gravestock (1990) and Debrenne and Zhuravlev (1996) note a similar distribution for *Girvanella* and also that calcified microbes increase in abundance toward shelf margins. However, these associations and patterns are not exclusive: apparently turbid inner-platform settings also have fossils of the *Epiphyton* group (e.g., Rowland and Gangloff 1988: figure 7; Debrenne and Zhuravlev 1996: figure 1a; Pratt et al., this volume: figure 12.3) and the *Angusticellularia* group (common in the Labrador reefs described in James and Kobluk 1978). Conversely, *Renalcis* has been reported from oolite-associated Nevadan reefs (Rowland and Gangloff 1988), and *Tarthinia* is common in high-energy biostromes in China (Bao et al. 1991) and Mongolia (Wood et al. 1993). At present, it seems that calcified microbes that could occupy turbid low-energy settings also found habitats in shelf-edge reefs, but not necessarily vice versa. This could partly account for increased diversity toward platform margins. There may also be a significant time component operating on microbial associations. Middle Cambrian reefs in China (Bao et al. 1991) have abundant representatives of both *Epiphyton* and *Renalcis* groups but appear to lack both *Epiphyton* and *Renalcis* themselves. The models offered here need to be refined by future work documenting both sedimentary environments and taxonomy more precisely through time.

Reef Succession

Although reef succession has been described in which archaeocyaths increase in abundance upward (Rowland and Gangloff 1988: figure 18), there is little information regarding within-reef variation in calcified microbes. Rowland and Gangloff (1988:119) describe *Renalcis* in the lower part and *Epiphyton* in the upper part of Siberian reefs. This appears to be consistent with the association model outlined above.

Light, Depth, and Cryptic Habitats

Initially, uncertainties concerning the affinities of Cambrian calcified microbes confused their paleoecologic application as depth indicators (Zhuravleva 1960; cf. Riding 1975). Cyanobacteria have low light flux requirements. This enables them to have a wide depth range and to occupy cryptic habitats. Pratt (1989) described a deepwater Middle Cambrian reef dominated by *Girvanella* and *Epiphyton*, but he did not estimate its depth. *Epiphyton* and *Renalcis* could grow upward on outer reef surfaces and downward in cavities (Rowland and Gangloff 1988), forming dendritic masses in both cases. Kobluk and James (1979) figure cryptic *Angusticellularia, Renalcis,* and *Wetheredella,* and Zhuravlev and Wood (1995:453) note *Angulocellularia* (=*Angusticellularia*), *Chabakovia, Epiphyton, Gordonophyton,* and *Renalcis* among cryptic biotas.

However, it is also possible that in some cases interpretation of cryptic habitats from apparently downward growth of calcified microbes may have been mistaken: double-geopetals suggest that apparent downward growth of specimens (probably *Gordono-phyton*) identified as *Epiphyton* (James 1981) in allochthonous blocks may actually have been upward. Nonetheless, whereas some reefs consist of tight vertically erect *Epiphyton* group genera with prostrate *Razumovskia*, creating a reticulate frame, others appear to have pendant *Renalcis*, *Gordonophyton*, and other genera growing down into large cavities (Pratt et al., this volume: figures 12.1A and 12.2B).

COMMUNITY ECOLOGY

Calcified microbes conceivably offered a potential source of food for grazers and a substrate for reefal and encrusting organisms and competed for substrate with other benthic organisms.

Grazing

It has been suggested that metazoan interference, including grazing, caused stromato-lite decline (Garrett 1970; Awramik 1971) and that grazers were active in Early Cambrian stromatolites (Edhorn 1977). However, there are few reports of effects on calcified microbes that can confidently be attributed to grazers. Microburrows are locally common in reefal muds (Kruse et al. 1995), but have rarely been reported to have affected calcified microbes (Kobluk 1985). The grazers inferred by Edhorn (1977) are in situ sessile orthothecimorph hyoliths, which probably were suspension feeders (Landing 1993; Zhuravlev 1996). Potential grazers, including halkieriids, tommoti-ids, and mollusks (Kruse et al. 1995: table 1), were small and probably ineffective. Modern grazers of microbial mats are mainly insects and crustaceans (Farmer 1992), which were not present during the Proterozoic and Early Paleozoic.

Substrate Provision

Calcified microbes provided extensive potential firm substrate sites for attachment, especially in the form of microbial domes and reefs and also as oncoids. Locally these were utilized by metazoans such as archaeocyaths and radiocyaths. Reefs generally provided habitats for mobile organisms such as trilobites.

Competition for Substrate

It is possible that microbes may actively have inhibited sessile metazoans. In mixed reefal communities of calcified microbes and metazoans, calcified microbes were often epiphytic on archaeocyaths and radiocyaths (e.g., James and Gravestock 1990:

figure 6a). There are also examples of both archaeocyath-dominated and microbe-dominated reefs. It has been suggested that filter-feeding archaeocyaths were more tolerant of muddy environments, whereas cyanobacteria preferred clearer water (Riding and Zhuravlev 1995). In archaeocyath-dominated reef environments, low tolerance shown by calcified microbes for particulate sediment, and their need for firm substrate, may have forced them to be epiphytic, often in cryptic locations. Conversely, in microbe-dominated reefs, archaeocyaths are often equally scarce. The reason for this is less clear, since microbe reefs provided extensive attachment sites. A possible explanation is that microbial surfaces and/or growth rates inhibited archaeocyath attachment and growth (Riding and Zhuravlev 1994). Kruse et al. (1995, 1996) and Gravestock and Shergold (this volume) report dwarf archaeocyaths in microbial reefs.

SEDIMENTOLOGIC ROLES

Reefs

Cambrian calcified microbes assumed unrivaled importance as reef builders (see Pratt et al., this volume). Most well-studied Early Cambrian reefs are biohermal lenses surrounded by mudrocks and consist of varying mixtures of calcified microbes, among which *Epiphyton, Renalcis,* and *Angusticellularia* are prominent, together with archaeocyaths. Overall, there is little doubt that calcified microbes were more abundant than archaeocyaths in Early Cambrian reefs (Copper 1974). Middle-Late Cambrian reefs are less well known but include extensive biostromes in high-energy ooid grainstone shoal environments. During this interval archaeocyaths are virtually absent, and with few exceptions (Hamdi et al. 1995), these reefs are almost wholly microbial in character and contain conspicuous *Gordonophyton, Razumovskia, Tarthinia,* and *Tubomorphophyton,* although these have often been misidentified (Ahr 1971; McIlreath 1977; James 1981; Markello and Read 1981; Astashkin et al. 1984; Coniglio and James 1985; Waters 1989). These Middle–Late Cambrian examples in particular show that these calcified microbes were capable of constructing strong, early lithified reefs in grainy mobile shoal environments, independently of metazoans.

Categories

There may be two basic categories of Cambrian reef: (1) high-energy, biostromal, grain-shoal–associated, macrocavity-poor, calcified microbe–dominated microframes (exemplified by the Middle Cambrian Zhangxia Formation of North China; Bao et al. 1991); and (2) low-energy, biohermal, shale-silt–enclosed, cavernous, cluster, and frame reefs with conspicuous sponge biotas (exemplified by the oldest-known archaeocyath reef; Riding and Zhuravlev 1995). However, intermediates between these end members are certainly widespread.

Structure

Many of the characteristics of calcified microbe reefs—particularly their strength and tight structure—derive from the small size, intense calcification, and tendency for mutual attachment of genera such as *Epiphyton*, *Renalcis*, and *Angusticellularia*. Two distinct but intergrading common reef fabrics are (1) dendrolite (Riding 1988), vertically oriented, mutually attached frameworks with dendritic mesofabric, and (2) thrombolite (Aitken 1967), unoriented skeletal microclusters mutually separated by early lithified micrite with clotted mesofabric. Botryoidal microfossils (e.g., *Renalcis* Group) created clots, whereas dendritic forms (e.g., *Epiphyton* Group) made dendrolites. *Razumovskia* crusts formed reticulate gridworks with *Gordonophyton*. These created lithified masses and turflike layers on upper sediment surfaces, and cryptic hangers on lower surfaces of reefal cavities of centimeter to decimeter scale. More-restricted crusts enveloped archaeocyath external and internal surfaces.

Overall Construction

In the Cambrian, both high-energy and low-energy reefs can consist of dendrolite and thrombolite. High-energy reefs are typically layered, dendrolitic and/or thrombolitic, laterally extensive biostromes. Low-energy reefs are typically thrombolitic and/or dendrolitic meter-scale domes (kalyptrae of Soviet workers; see Rowland and Gangloff 1988:120, figure 10) that can amalgamate into bioherms up to 100 m thick. All appear to have had relatively low (meter-scale) relief (see Fagerstrom 1987:327).

Calcified Microbe Domination

Cambrian reefs demonstrate the ability of calcified microbes to construct strong, early lithified reefs in grainy mobile shoal environments independently of metazoans. However, calcified microbes and archaeocyaths were neither necessarily rivals in reef formation (but see the section "Competition for Substrate," above), nor were archaeocyaths dependent on calcified microbes for reefal success.

Early Cementation and Cyanobacterial Calcification

The ability of calcified microbes to build strong, rapidly accreting reefs depended upon their early lithification. Microbial calcification and marine cementation went hand in hand to create these deposits, forming dendritic and clotted crusts on most hard surfaces—even in shaley environments (see the section "Calcification," below).

Postscript

Reefal association of calcified microbes with sponges was renewed in the Early Ordovician (e.g., Riding and Toomey 1972; Pratt and James 1982). Calcified microbe abundance appears to have waned in the late Early Ordovician (Riding 1992), but none-theless locally recurred—e.g., in the Early Silurian (Riding and Watts 1983). The de-cline of calcified microbes could have been due to metazoan and/or algal competition and also to reduced lithification (Riding 1997). Abundance—although not diver-sity—of calcified microbes similar to that seen in the Cambrian reappears, possibly for the last time in the Late Devonian (Riding 1992).

Fragments and Oncoids

Calcified microbes all were attached, except possibly for isolated semiplanktic flocs of *Girvanella* and *Obruchevella*. Most are found in situ, although they also formed on-coids through encrustation of allochthonous nuclei. Early cementation and mutual attachment of many calcified microbes may have provided sufficient strength to with-stand significant breakage even in high-energy environments. In addition, these were small organisms that may be difficult to recognize when synsedimentarily broken through abrasion or bioerosion. Coniglio and James (1985) suggested that *Epiphyton* Group fossils could have contributed silt-size peloids whose origin would be difficult to attribute. They report sand-size *Epiphyton, Nuia,* and also *Girvanella* fragments and scarce *Girvanella* oncoids. More recently, micrite-coated calcified microbe grains have been found to be locally abundant in grainstones (Wood et al. 1993).

CALCIFICATION

Processes and Timing

Calcification of these microbes took place in apparently normal marine subtidal con-ditions. Cyanobacterial affinities have been established for many of these taxa and are plausible for most of the remainder (see the section "Affinities," above). Calcification in extant cyanobacteria has two characteristics: it requires conditions in which car-bonate saturation is high, and it is localized on or in the external protective muci-laginous sheath that surrounds the cells (Riding 1991b). Calcification occurs during the life of the organisms and is encouraged both by the properties of the sheath, which provide suitable sites for crystal nucleation, and by the creation of alkalinity gradients due to photosynthetic metabolism of the cells enclosed by the sheath. Calcification in *Renalcis* might have been partly postmortem, as Hofmann (1975) and Pratt (1984) suggested, but the replicate and often detailed preservation of most of these fossils suggests that calcification occurred during life. Furthermore, these fos-sils calcified in the subtidal environments in which they lived, and prior to burial (Pratt 1984). If this environment facilitated intense calcification shortly after death, then it is difficult to see how it would not also have favored in vivo calcification. Ex-tant cyanobacteria in freshwater exhibit in vivo calcification that is promoted by both biological and environmental factors and would account for the preservation observed in these fossils. Changes through time within the Cambrian, and also through the Paleozoic, show these fossils undergoing changes that can be attributed to evolution and extinction but not to ecophenotypic variation and differential preservation.

Temporal Patterns

Cyanobacterial calcification is rare in modern marine environments. It is widespread in freshwater streams in limestone areas, but in this environment precipitation is so rapid that it often results in encrustation rather than impregnation of the sheath. The analogs of cyanobacterial calcification that most closely resemble ancient marine examples have been found in calcareous pools and lakes (e.g., Riding 1977a; Riding and Voronova 1982b), where sheath material is heavily impregnated but not externally encrusted.

Factors

The scarcity of other reef builders emphasizes the Cambrian prominence of calcified microbes, but the main reason for their importance at this time probably lies in the factors that led to their heavy calcification. The occurrence of intense calcification in microbes in marine environments in the Cambrian indicates not only that suitable microbes were present but also that environmental conditions suitable for calcification were maintained. It suggests that carbonate saturation levels may have been substantially raised relative to those of the present-day (Merz-Prei and Riding 1995).

Patterns

Calcified cyanobacteria were episodically abundant in marine environments during the Phanerozoic (Riding 1991b, 1992) and probably also during the Proterozoic (Riding 1994; see Turner et al. 1993). These extended episodes of marine calcification that affected cyanobacteria and similar microbes have been termed Cyanobacterial Calcification Episodes (CCEs; Riding 1992). It has been suggested that the environmental factors that resulted in elevated saturation levels with respect to calcium carbonate minerals during these calcification events promoted enhanced marine cementation and ooid formation, as well as microbial calcification (Riding 1992, 2000). These would have been further stimulated by local water turbulence and climatic conditions. Later in the Paleozoic, calcified microbes similar to those of the Cambrian reappear as Lazarus taxa, particularly in the Silurian and Late Devonian.

CONCLUSIONS

1. Calcified microbes are prominent in Cambrian shallow marine carbonates and have worldwide distribution. Those whose affinities are known (*Angusticellularia, Bo-*

tomaella, Girvanella, and Obruchevella groups) are cyanobacteria. The Epiphyton, Proaulopora, and Renalcis groups probably also represent cyanobacteria. Chabakovia, Nuia, and Wetheredella remain Problematica. Further elucidation of affinities requires identification of modern analogs.

2. Calcified algae are much scarcer and less diverse than calcified microbes. Their precise affinities remain in doubt. Amgaella, Mejerella, and Seletonella may be dasycladalean chlorophytes. They are known only from the Middle Cambrian (Amgaella) and Late Cambrian (Mejerella and Seletonella) of Russia and Kazakhstan.

3. Calcified microbes diversified rapidly during the Nemakit-Daldynian and Tommotian. Apparent diversity was highest in the Early Cambrian. There are no records of new genera during the Middle-Late Cambrian. Positive correlation between taxonomic diversity and number of studies suggests that Early Cambrian high diversity could be monographic. Details of the space-time distribution of calcified algae and cyanobacteria are scant, largely because of insufficient application of precise taxonomy.

4. Abrupt appearance of calcified microbes in the Nemakit-Daldynian may be revised when there is more information from the late Neoproterozoic. At present, rapid diversification of these fossils suggests both commencement of a Cyanobacterial Calcification Episode (CCE) and a burst of cyanobacterial evolution close to the Neoproterozoic-Cambrian boundary.

5. This Calcification Episode determined the importance of calcified cyanobacteria during the Cambrian. It was facilitated not only by the presence of microbes capable of calcification but also by environmental factors that elevated carbonate saturation.

6. Calcified microbes, particularly members of the Epiphyton, Renalcis, and Angusticellularia groups, dominated reefs in nonturbid environments. They created rigid, strong, compact frameworks both before and after archaeocyath demise. There is no evidence that calcified microbes were affected by metazoan grazing, disturbance, or competition, and although calcified microbes provided extensive hard substrates, these were underutilized by metazoans such as archaeocyaths during the Early Cambrian. Calcified microbes may have inhibited metazoan larval settlement and growth.

7. Filamentous forms (e.g., Epiphyton and Girvanella groups) preferred high-energy and less turbid conditions than botryoidal forms (e.g., Renalcis Group). High- and low-energy associations can be discerned: Epiphyton and Girvanella groups formed biostromes in grainy and shelf-edge locations, whereas Renalcis Group genera formed mudstone-enveloped bioherms in lower-energy inner shelf and midramp environments. However, there was a good deal of overlap in these preferences.

Acknowledgments. I am indebted to Brian R. Pratt for helpful and stimulating review, Larisa G. Voronova for providing the specimens illustrated in figure 20.1, and Andrey Yu. Zhuravlev for discussion of stratigraphic distributions and comments on the manuscript. This paper is a contribution to IGCP Project 366.

REFERENCES

Ahr, W. M. 1971. Paleoenvironment, algal structures, and fossil algae in the Upper Cambrian of Central Texas. *Journal of Sedimentary Petrology* 41:205–216.

Aitken, J.D. 1967. Classification and environmental significance of cryptalgal limestones and dolomites with illustrations from the Cambrian and Ordovician of southwestern Alberta. *Journal of Sedimentary Petrology* 37:1163–1178.

Astashkin, V. A., A. I. Varlamov, N. K. Gubina, A. E. Ekhanin, V. S. Pereladov, V. I. Romenko, S. S. Sukhov, N. V. Umperovich, A. B. Fedorov, B. B. Shishkin, and E. I. Khobnya. 1984. *Geologiya i perspektivy neftegazonosnosti kembriyskikh rifovykh sistem Sibirskoy platformy* [Geology and prospects of oil-gas bearing of Cambrian reef systems of the Siberian Platform]. Moscow: Nedra.

Awramik, S. M. 1971. Precambrian columnar stromatolite diversity: Reflection of metazoan appearance. *Science* 174:825–827.

Bao, H., X. Mu, and R. Riding. 1991. Middle Cambrian dendrolite biostromes, Jinan, East China. *Fifth International Symposium on Fossil Algae, Capri, Italy, April 1991, Abstracts,* pp. 7–8.

Bassoullet, J.-P., P. Bernier, R. Deloffre, P. Génot, M. Jaffrezo, and D. Vachard. 1979. Essai de classification des Dasycladales en tribus. *Bulletin des Centres de Recherches Exploration-Production Elf-Aquitaine* 3:429–442.

Billings, E. 1865. *Palaeozoic Fossils,* vol. 1, pp. 169–426. Ottawa: Canada Geological Survey.

Bornemann, J. 1886. Die Versteinerungen des cambrischen Schichtensystems der Insel Sardinien nebst vergleichenden Untersuchungen über analoge Vorkommnisse aus anderen Ländern 1. *Nova Acta der Kaiserslichen Leopoldinisch-Carolinischen Deutschen Akademie der Naturforscher* 51 (1):1–147.

Brooke, C. and R. Riding. 1987. A new look at the Solenoporaceae. *Fourth International Symposium on Fossil Algae, Cardiff, July 1987, Abstracts,* p. 7.

Brooke, C. and R. Riding. 1998. Ordovician and Silurian coralline red algae. *Lethaia* 31:185–195.

Butterfield, N. J., A. H. Knoll, and K. Swett. 1990. A bangiophyte red alga from the Proterozoic of Arctic Canada. *Science* 250:104–107.

Chuvashov, B. and R. Riding. 1984. Principal floras of Palaeozoic marine calcareous algae. *Palaeontology* 27:487–500.

Chuvashov, B. I., V. A. Luchinina, V. P. Shuysky, I. M. Shaykin, O. I. Berchenko, A. A. Ishchenko, V. D. Saltovskaya, and D. I. Shirshova. 1987. *Iskopaemye izvestkovye vodorosli (morfologiya, sistematika, metody izucheniya)* [Fossil calcareous algae (morphology, systematics, methods of study)]. *Trudy, Institut geologii i geofiziki, Sibirskoe otdelenie, Akademiya nauk SSSR* 674:1–225.

Coniglio, M. and N. P. James. 1985. Calcified algae as sediment contributors to early Paleozoic limestones: Evidence from deep-water sediment of the Cow Head Group, western Newfoundland. *Journal of Sedimentary Petrology* 55:746–754.

Copper, P. 1974. Structure and development of Early Paleozoic reefs. *Proceedings of the Second International Coral Reef Symposium* 1:365–386.

Debrenne, F. and A. Yu. Zhuravlev. 1996. Archaeocyatha, palaeoecology: A Cambrian sessile fauna. Bollettino della Società Paleontologica Italiana, Special Volume 3:77–85.

Debrenne, F., A. Gandin, and A. Zhuravlev. 1991. Palaeoecological and sedimentological remarks on some Lower Cambrian sediments of the Yangtze platform (China).

Bulletin de la Société géologique de France 162:575–584.

Drozdova, N. A. 1980. *Vodorosli v organogennykh postroykakh nizhnego kembriya Zapadnoy Mongolii* [Algae in Lower Cambrian organogenous buildups of western Mongolia]. *Trudy, Sovmestnaya Sovetsko-Mongol'skaya paleontologicheskaya ekspeditsiya* 10:1–140.

Dybowski, W. 1877. Die Chaetetiden der ostbaltischen Silur-Formation. *Russisch-Kaiserliche Mineralogische Gesellschaft zu St. Petersburg Verhandlungen,* 2d ser., 1878(14):1–134 [1879].

Edhorn, A.-S. 1977. Early Cambrian algae croppers. *Canadian Journal of Earth Sciences* 14:1014–1020.

Edwards, D., J. G. Baldauf, P. R. Bown, K. J. Dorning, M. Feist, L. T. Gallagher, N. Grambast-Fessard, M. B. Hart, A. J. Powell, and R. Riding. 1993. Algae. In M. J. Benton, ed., *The Fossil Record 2*, pp. 15–40. London: Chapman and Hall.

Elias, M. K. 1950. Paleozoic Ptychocladia and related Foraminifera. *Journal of Paleontology* 24:287–306.

Elias, M. K. 1954. *Cambroporella* and *Coeloclema,* Lower Cambrian and Ordovician bryozoans. *Journal of Paleontology* 28:52–58.

Fagerstrom, J. A. 1987. *The Evolution of Reef Communities.* New York: John Wiley and Sons.

Fairchild, I. 1991. Origins of carbonate in Neoproterozoic stromatolites and the identification of modern analogues. *Precambrian Research* 53:281–299.

Farmer, J. D. 1992. Grazing and bioturbation in modern microbial mats. In J. W. Schopf and C. Klein, eds., *The Proterozoic Biosphere: A Multidisciplinary Study,* pp. 295–297. Cambridge: Cambridge University Press.

Garrett, P. 1970. Phanerozoic stromatolites: Noncompetitive ecologic restriction by grazing and burrowing animals. *Science* 169:171–173.

Golubic, S. 1973. The relationship between blue-green algae and carbonate deposits. In G. Carr and B. A. Whitton, eds., *The Biology of Blue-Green Algae,* pp. 434–472. Botanical Monographs 19. Oxford: Blackwell.

Grant, S. W. F., A. H. Knoll, and G. J. B. Germs. 1991. Probable calcified metaphytes in the latest Proterozoic Nama Group, Namibia: Origin, diagenesis, and implications. *Journal of Paleontology* 65:1–18.

Green, J. W., A. H. Knoll, and K. Swett. 1989. Microfossils from silicified stromatolitic carbonates of the Upper Proterozoic Limestone-Dolomite "Series," central East Greenland. *Geological Magazine* 126:567–585.

Gudymovich, S. S. 1967. Izvestkovye vodorosli anastas'inskoy i ungutskoy svit pozdnego dokembriya (?)—nizhnego kembriya severo-zapadnoy chasti Vostochnogo Sayana [Calcareous algae from the Cambrian Anastas'ino and Ungut formations in the northwestern part of eastern Sayan]. In A. P. Zhuze, ed., *Iskopaemye vodorosli SSSR* [Fossil algae of the USSR], pp. 134–138. Moscow: Nauka.

Hamdi, B., A. Yu. Rozanov, and A. Yu. Zhuravlev. 1995. Latest Middle Cambrian metazoan reef from northern Iran. *Geological Magazine* 132:367–373.

Han, T.-M. and B. Runnegar. 1992. Megascopic eukaryotic algae from the 2.1 billion-year-old Negaunee Iron-Formation, Michigan. *Science* 257:232–235.

Høeg, O. A. 1932. Ordovician algae from the Trondheim area. *Skrifter utgitt av Det Norske Videnskaps-Akademi i Oslo I, Mat.-Naturv. Klasse* 4:63–96.

Hofmann, H. J. 1975. Stratiform Precambrian stromatolites, Belcher Islands, Canada: Relations between silicified microfossils and microstructure. *American Journal of Science* 275:1121–1132.

Hofmann, H. J. 1992. Proterozoic carbonaceous films. In J. W. Schopf and C. Klein, eds., *The Proterozoic Biosphere: A Multidisciplinary Study*, pp. 349–358. Cambridge: Cambridge University Press.

Hofmann, H. J. and J. P. Grotzinger. 1985. Shelf-facies microbiotas from the Odjick and Rocknest formations (Epworth Group; 1.89 Ga), northwestern Canada. *Canadian Journal of Earth Sciences* 22:1781–1792.

Horodyski, R. J. and C. Mankiewicz. 1990. Possible late Proterozoic skeletal algae from the Pahrump Group, Kingston Range, southeastern California. *American Journal of Science* 290-A:149–169.

James, N. P. 1981. Megablocks of calcified algae in the Cow Head Breccia, western Newfoundland: Vestiges of a Cambro-Ordovician platform margin. *Geological Society of America Bulletin* 92:799–811.

James, N. P. and D. I. Gravestock. 1990. Lower Cambrian shelf and shelf-margin buildups, Flinders Ranges, South Australia. *Sedimentology* 37:455–480.

James, N. P. and D. R. Kobluk. 1978. Lower Cambrian patch reefs and associated sediments: Southern Labrador, Canada. *Sedimentology* 25:1–35.

Johnson, J.H. 1954. An introduction to the study of rock-building algae and algal limestones. *Quarterly of the Colorado School of Mines* 49:1–117.

Johnson, J. H. 1961. *Limestone-building algae and algal limestone*. Boulder: Colorado School of Mines, 297 pp.

Johnson, J. H. 1966. A review of the Cambrian algae. *Quarterly of the Colorado School of Mines* 61:1–162.

Kobluk, D. R. 1985. Biota preserved within cavities in Cambrian *Epiphyton* mounds, Upper Shady Dolomite, southwestern Vir-

ginia. *Journal of Paleontology* 59:1158–1172.

Kobluk, D. R. and N. P. James. 1979. Cavity-dwelling organisms in Lower Cambrian patch reefs from southern Labrador. *Lethaia* 12:193–218.

Kolosov, P. N. 1970. Organicheskie ostatki verkhnego dokembriya yuga Yakutii [Upper Precambrian organic remains from the south of Yakutia]. In A. K. Bobrov, ed., *Stratigrafiya i paleontologiya proterozoya i kembriya vostoka Sibirskoy platformy* [Proterozoic and Cambrian stratigraphy and paleontology on the east of the Siberian Platform], pp. 57–70. Yakutsk: Yakutsk Publishing House.

Kolosov, P. N. 1975. *Stratigrafiya verkhnego dokembriya yuga Yakutii* [Upper Precambrian Stratigraphy on the South of Yakutia]. Yakutsk: Institute of Geology, Yakutsk Branch, Siberian Division, USSR Academy of Sciences.

Kolosov, P. N. 1977. *Drevnie neftegazonosnye tolshchi yugo-vostoka Sibirskoy platformy (stratigrafiya, vodorosli, mikrofitolity i organogennye obrazovaniya)* [Ancient oil-gas–bearing strata on the southeast of the Siberian Platform (stratigraphy, algae, microphytolites, and organogenous buildups)]. Novosibirsk: Nauka.

Korde, K. B. 1950. Ostatki vodorosli iz kembriya Kazakhstana [Algal remains from the Cambrian of Kazakhstan]. *Doklady Akademii nauk SSSR* 73:809–812.

Korde, K. B. 1957. Novye predstaviteli sifonnikovykh vodorosli [New representatives of siphoneous algae]. In K. B. Korde, ed., *Materialy k "Osnovam paleontologii" 1* [Materials to the "Principals of Paleontology" 1], pp. 67–75. Moscow: USSR Academy of Sciences Publishing House.

Korde, K. B. 1958. K sistematike iskopaemykh Cyanophycea [To the systematics of fossil Cyanophycea]. In A. G. Sharov, ed., *Materialy k "Osnovam paleontologii" 2* [Materials to the "Principles of Paleontology" 2], pp. 99–111. Moscow: USSR Academy of Sciences Publishing House.

Korde, K. B. 1959. Problematicheskie ostatki iz kembriyskikh otlozheniy yugo-vostoka Sibirskoy platformy [Problematic remains from the Cambrian strata on the southeast of the Siberian Platform]. *Doklady Akademii nauk SSSR* 125:625–627.

Korde, K. B. 1961. *Vodorosli kembriya yugo-vostoka Sibirskoy platformy* [Cambrian algae from the southeast of the Siberian Platform]. *Trudy, Paleontologicheskiy institut, Akademiya nauk SSSR* 89:1–148.

Korde, K. B. 1973. *Vodorosli kembriya* [Cambrian algae]. *Trudy, Paleontologicheskiy institut, Akademiya nauk SSSR* 139:1–350.

Kruse, P. D., A. Yu. Zhuravlev, and N. P. James. 1995. Primordial metazoan-calcimicrobial reefs: Tommotian (Early Cambrian) of the Siberian Platform. *Palaios* 10:291–321.

Kruse, P. D., A. Gandin, F. Debrenne, and R. Wood. 1996. Early Cambrian bioconstructions in the Zavkhan Basin of western Mongolia. *Geological Magazine* 133:429–444.

Landing, E. 1993. In situ earliest Cambrian tube worms and the oldest metazoan-constructed biostrome (Placentian Series, southeastern Newfoundland). *Journal of Paleontology* 67:333–342.

Latham, A. and R. Riding. 1990. Fossil evidence for the location of the Precambrian/Cambrian boundary in Morocco. *Nature* 344:752–754.

Loeblich, A. R., Jr. and H. Tappan. 1964. Part C, Protista 2, Sarcodina, chiefly "thecamoebians" and Foraminiferida. In R. C. Moore, ed., *Treatise on Invertebrate Paleontology*, pp. C1–C900. Boulder, Colo.: Geological Society of America.

Luchinina, V. A. 1975. *Paleoal'gologhicheskaya kharakteristika rannego kembriya Sibirskoy platformy (yugo-vostok)* [Paleoalgo-logical characteristics of the Early Cambrian on the Siberian Platform (southeast)]. *Trudy, Institut geologii i geofiziki, Sibirskoe otdelenie, Akademiya nauk SSSR* 216:1–100.

Luchinina, V. A. 1985. Vodoroslevye postroyki rannego paleozoya severa Sibirskoy platformy [Early Paleozoic algal buildups on the North of the Siberian Platform]. *Trudy, Institut geologii i geofiziki, Sibirskoe otdelenie, Akademiya nauk SSSR* 628:45–49.

Luchinina, V. A. 1990. Raschlenenie i korrelyatsiya pogranichnykh otlozheniy venda i kembriya Sibirskoy platformy po izvestkovym vodoroslyam [Subdivision and correlation of the Vendian and Cambrian strata of the Siberian Platform by calcareous algae]. *Trudy, Institut geologii i geofiziki, Sibirskoe otdelenie, Akademiya nauk SSSR* 765:32–43.

Luchinina, V. A. 1991. Calcibionta—calcareous algae of Vendian-Phanerozoic. *Fifth International Symposium on Fossil Algae, Capri, Italy, April 1991, Abstracts*, p. 59.

Luchinina, V. A. 1998. Nekotorye osobennosti razvitiya izvestkovykh vodorosley Calcibionta na granitse venda i kembriya. [Some peculiarities of the evolution of the calcareous algae Calcibionta at the Vendian-Cambrian boundary]. *Geologiya i geofizika* 39:568–574.

Luchinina, V. A. 1999. Organogennye postroyki v pogranichnom intervaliye venda i kembriya Sibirskoy platformy [Buildups at the Vendian-Cambrian boundary on the Siberian Platform]. *Geologiya i geofizika* 40:1785–1794.

Mankiewicz, C. 1992. Proterozoic and Early Cambrian calcareous algae. In J. W. Schopf and C. Klein, eds., *The Proterozoic Biosphere: A Multidisciplinary Study*, pp. 359–367. Cambridge: Cambridge University Press.

Markello, J. R., and J. F. Read. 1981. Carbonate ramp-to-deeper shale transition of an Upper Cambrian intrashelf basin, Nolichucky Formation, southwest Virginia Appalachians. *Sedimentology* 28:573–597.

Maslov, V. P. 1937. Nizhnepaleozoyskie porodoobrazuyushchie vodorosli Vostochnoy Sibiri [Lower Paleozoic rock-forming algae of eastern Siberia]. *Problemy paleontologii* 2/3:249–325. Moscow: Moscow University Publishing House.

Maslov, V. P. 1954. O nizhnem silure Vostochnoy Sibiri [On the Lower Silurian of eastern Siberia]. In N. S. Shatskiy, ed., *Voprosy geologii Azii 1* [Problems of the geology of Asia 1], pp. 495–531. Moscow: USSR Academy of Sciences Publishing House.

Maslov, V. P. 1956. *Iskopaemye izvestkovye vodorosli SSSR* [Fossil Calcareous Algae of the USSR]. *Trudy, Institut geologicheskikh nauk, Akademiya nauk SSSR* 160:1–302.

Mattes, B. W. and S. Conway Morris. 1990. Carbonate/evaporite deposition in the Late Precambrian–Early Cambrian Ara Formation of southern Oman. In A. H. F. Robertson, M. P. Searle, and A. C. Ries, eds., *The Geology and Tectonics of the Oman Region*, pp. 617–636. Geological Society Special Publication 49.

McIlreath, I. 1977. Accumulation of a Middle Cambrian deep water limestone debris apron adjacent to a vertical, submarine carbonate escarpment, southern Rocky Mountains, Canada. In H. E. Cook and M. E. Taylor, eds., *Deep-Water Carbonate Environments*, pp. 113–124. *Society of Economic Paleontologists and Mineralogists Special Publication* 25. Tulsa, Okla.

Merz-Prei, M. and R. Riding. 1995. Supersaturation in Recent freshwater tufa streams: Calibration for marine environmentally controlled $CaCO_3$ precipitation in the past? *10th Bathurst Meeting of Carbonate Sedimentologists, Royal Holloway, London, 2–5 July 1995, Abstracts of Talks*, pp. 38, 44.

Peel, J. S. 1988. *Spirellus* and related helically

coiled microfossils (cyanobacteria) from the Lower Cambrian of North Greenland. *Rapport Grønlands Geologiske Undersøgelse* 137:5–32.

Pentecost, A. and R. Riding. 1986. Calcification in cyanobacteria. In B. S. C. Leadbeater and R. Riding, eds., *Biomineralization in Lower Plants and Animals*, pp. 73–90. Systematics Association Special Volume 30. Oxford: Clarendon Press.

Pia, J. 1927. Thallophyta. In M. Hirmer, ed., *Handbuch der Paläobotanik 1*, pp. 31–136. München: Oldenburg.

Pratt, B. R. 1984. *Epiphyton* and *Renalcis*—diagenetic microfossils from calcification of coccoid blue-green algae. *Journal of Sedimentary Petrology* 54:948–971.

Pratt, B. R. 1989. Deep-water *Girvanella-Epiphyton* reef on a mid-Cambrian continental slope, Rockslide Formation, Mackenzie Mountains, Northwest Territories. In H. H. J. Geldsetzer, N. P. James, and E. Tebbutt, eds., *Reefs, Canada and Adjacent Area*, pp. 161–164. Canadian Society of Petroleum Geologists, Memoir 13.

Pratt, B. R. 1995. The origin, biota, and evolution of deep-water mud-mounds. In C. L. V. Monty, D. W. J. Bosence, P. H. Bridges, and B. R. Pratt, eds., *Carbonate Mud-Mounds*, pp. 49–123. International Association of Sedimentology, Special Publication 23. Oxford: Blackwell Science,

Pratt, B. R. and N. P. James. 1982. *Epiphyton* and *Renalcis*—diagenetic microfossils from calcification of coccoid blue-green algae. *American Association of Petroleum Geologists, Bulletin* 66:619.

Priestley, R. E. and T. W. E. David. 1912. Geological notes of the British Antarctic Expedition. *11th International Geological Congress, Stockholm 1910, Compte Rendu*, pp. 767–811.

Raaben, M. E. 1969. *Stromatolity verkhnego rifeya (gimnosolenidy)* [Stromatolites of the

Upper Riphean (gymnosolenids)]. *Trudy, Geologicheskiy institut, Akademiya nauk SSSR* 203:1–100.

Read, J. F. and R. W. Pfeil. 1983. Fabrics of allochthonous reefal blocks, Shady Dolomite (Lower to Middle Cambrian), Virginia Appalachians. *Journal of Sedimentary Petrology* 53:761–778.

Rees, M. N., B. R. Pratt, and A. J. Rowell. 1989. Early Cambrian reefs, reef complexes, and associated lithofacies of the Shackleton Limestone, Transantarctic Mountains. *Sedimentology* 36:341–361.

Reitlinger, E. A. 1959. *Atlas mikroskopicheskikh organicheskikh ostatkov i problematiki drevnikh tolshch Sibiri* [Atlas of the Microscopic Organic Remains and Problematica from the ancient strata of Siberia]. *Trudy, Institut geologicheskikh nauk, Akademiya nauk SSSR* 25:1–62.

Riding, R. 1975. *Girvanella* and other algae as depth indicators. *Lethaia* 8:173–179.

Riding, R. 1977a. Calcified *Plectonema* (blue-green algae): A recent example of *Girvanella* from Aldabra Atoll. *Palaeontology* 20:33–46.

Riding, R. 1977b. Problems of affinity in Palaeozoic calcareous algae. In E. Flügel, ed., *Fossil Algae: Recent Results and Developments*, pp. 202–211. Berlin: Springer Verlag.

Riding, R. 1979. Devonian calcareous algae. In M. R. House, C. T. Scrutton, and M. G. Bassett, eds., *The Devonian System*, pp. 141–144. *Special Papers in Palaeontology* 23.

Riding, R. 1984. Sea-level changes and the evolution of benthic marine calcareous algae during the Palaeozoic. *Journal of the Geological Society, London* 141:547–553.

Riding, R. 1988. Classification of microbial carbonates. In *Sixth International Coral Reef Symposium, Townsville, August 1988. Discussion Sessions: Benthic Microbes and Reefs. Abstracts*, p. 5.

Riding, R. 1991a. Calcified cyanobacteria. In R. Riding, ed., *Calcareous Algae and Stromatolites*, pp. 55–87. New York: Springer Verlag.

Riding, R. 1991b. Cambrian calcareous cyanobacteria and algae. In R. Riding, ed., *Calcareous Algae and Stromatolites*, pp. 305–334. New York: Springer Verlag.

Riding, R. 1992. Temporal variation in calcification in marine cyanobacteria. *Journal of the Geological Society, London* 149:979–989.

Riding, R. 1994. Evolution of algal and cyanobacterial calcification. In S. Bengtson, ed., *Early Life on Earth: Nobel Symposium 84*, pp. 426–438. New York: Columbia University Press.

Riding, R. 1997. Stromatolite decline: A brief reassessment. In F. Neuweiler, J. Reitner, and C. Monty, eds., *Biosedimentology of Microbial Buildups*, IGCP Project No. 380, Proceedings of 2nd Meeting, Göttingen, Germany. *Facies* 36:227–230.

Riding, R. 2000. Microbial carbonates: The geological record of calcified bacterial-algal mats and biofilms. *Sedimentology* 47 (Supplement 1, Millennium Reviews): 179–214.

Riding, R. and Toomey, D. F. 1972. The sedimentological role of *Epiphyton* and *Renalcis* in Lower Ordovician mounds, southern Oklahoma. *Journal of Paleontology* 46:509–519.

Riding, R. and L. Voronova. 1982a. Affinity of the Cambrian alga *Tubomorphophyton* and its significance for the Epiphytaceae. *Palaeontology* 25:869–878.

Riding, R. and L. Voronova. 1982b. Recent freshwater oscillatoriacean analogue of the Lower Palaeozoic calcareous algae *Angulocellularia*. *Lethaia* 15:105–114.

Riding, R. and L. Voronova. 1984. Assemblages of calcareous algae near the Precambrian/Cambrian boundary in Siberia and Mongolia. *Geological Magazine* 121: 205–210.

Riding, R. and L. Voronova. 1985. Morphological groups and series in Cambrian calcareous algae. In D. F. Toomey and M. H. Nitecki, eds., *Paleoalgology: Contemporary Research and Applications*, pp. 56–78. New York: Springer Verlag.

Riding, R. and N. Watts. 1983. Silurian *Renalcis* (?cyanophyte) from reef facies in Gotland (Sweden). *Neues Jahrbuch für Geologie und Paläontologie Monatshefte* 4: 242–248.

Riding, R. and A. Yu. Zhuravlev. 1994. Cambrian reef builders: Calcimicrobes and archaeocyaths. *Terra Nova* 6 (Abstract Supplement 3): 7.

Riding, R. and A. Yu. Zhuravlev. 1995. Structure and diversity of oldest sponge-microbe reefs: Lower Cambrian, Aldan River, Siberia. *Geology* 23: 649–652.

Ross, R. J., J. Valusek, and N. P. James. 1988. *Nuia* and its environmental significance. *New Mexico Bureau of Mines and Mineral Resources, Memoir* 44: 115–121.

Rothpletz, A. 1908. Ueber Algen und Hydrozoen im Silur von Gotland und Oesel. *Kungl. Svenska Vetenskapsakademiens Handlingar* 43 (5).

Rowland, S. M. and R. A. Gangloff. 1988. Structure and paleoecology of Lower Cambrian reefs. *Palaios* 3: 111–135.

Saltovskaya, V. D. 1975. Rod *Epiphyton* Bornemann (ego veroyatnye sinonimy i stratigraficheskoe znachenie) [Genus *Epiphyton* Bornemann (its probable synonyms and stratigraphic significance)]. In *Voprosy paleontologii Tadzhikistana* [Problems of the paleontology of Tajikistan], pp. 70–88. Dushanbe.

Sayutina, T. A. 1985. K revizii roda *Yakovle-*

vites Korde, 1979. In B. S. Sokolov and I. T. Zhuravleva, eds., *Problematiki pozdnego dokembriya i paleozoya, Trudy, Instituta geologii i geofiziki, Sibirskoe otdelenie, Akademiya nauk SSSR* 632: 70–74. Moscow: Nauka.

Stolley, E. 1893. Über silurische Siphoneen. *Neues Jahrbuch für Mineralogie, Geologie, und Paläontologie* 1893: 135–146.

Turner, E. C., G. M. Narbonne, and N. P. James. 1993. Neoproterozoic reef microstructures from the Little Dal Group, northwestern Canada. *Geology* 21: 259–262.

Vologdin, A. G. 1962. *Drevneyshie vodorosli SSSR* [The oldest algae of the USSR]. Moscow: USSR Academy of Sciences Publishing House.

Voronova, L. G. 1979. Calcified algae of the Precambrian and Early Cambrian. *Bulletin des Centres de Recherches Exploration-Production Elf-Aquitaine* 3: 867–871.

Voronova, L. G. and E. P. Radionova. 1976. *Vodorosli i mikrofitolity paleozoya* [Palaeozoic algae and microphytolites]. *Trudy, Geologicheskiy institut, Akademiya nauk SSSR* 294: 1–220.

Waters, B. B. 1989. Upper Cambrian *Renalcis-Girvanella* framestone mounds, Alberta. In H. H. J. Geldsetzer, N. P. James, and E. Tebbutt, eds., *Reefs: Canada and Adjacent Area*, pp. 165–169. Canadian Society of Petroleum Geologists, Memoir 13.

Webby, B. D. 1986. Early stromatoporoids. In A. Hoffman and M. H. Nitecki, eds., *Problematic Fossil Taxa*, pp. 148–166. New York: Oxford University Press.

Wood, R., A. Yu. Zhuravlev, and A. Chimed Tseren. 1993. The ecology of Lower Cambrian buildups from Zuune Arts, Mongolia: Implications for early metazoan reef evolution. *Sedimentology* 40: 829–858.

Wray, J. L. 1967. Upper Devonian calcareous

algae from the Canning Basin, Western Australia. *Professional Contribution of the Colorado School of Mines* 3:1–76.

Zadorozhnaya, N. M. 1974. Rannekembriyskie organogennye postroyki vostochnoy chasti Altae-Sayanskoy skladchatoy oblasti [Early Cambrian organogenous buildups from the eastern part of the Altay Sayan Foldbelt]. *Trudy, Institut geologii i geofiziki, Sibirskoe otdelenie, Akademiya nauk SSSR* 84:159–186.

Zhuravlev, A. Yu. 1996. Reef system recovery in the Early Cambrian extinction. In M. B. Hart, ed., *Biotic Recovery from Mass Extinction Events*, pp. 79–96. Geological Society of London Special Publication 102.

Zhuravlev, A. Yu. and R. Wood. 1995. Lower Cambrian reefal cryptic communities. *Palaeontology* 38:443–470.

Zhuravleva, I. T. 1960. *Arkheotsiaty Sibirskoy platformy* [Archaeocyaths of the Siberian Platform]. Moscow: USSR Academy of Sciences Publishing House.

Zhuravleva, I. T., N. P. Meshkova, V. A. Luchinina, and L. N. Kashina. 1982. Biofatsii Anabarskogo morya v pozdnem dokembrii i rannem kembrii [Biofacies of the Anabar Sea in the late Precambrian and Early Cambrian]. *Trudy, Institut geologii i geofiziki, Sibirskoe otdelenie, Akademiya nauk SSSR* 510:74–103.

CHAPTER TWENTY-ONE

J. Michael Moldowan, Stephen R. Jacobson, Jeremy Dahl,
Adnan Al-Hajji, Bradley J. Huizinga, and Frederick J. Fago

Molecular Fossils Demonstrate Precambrian Origin of Dinoflagellates

The natural product chemistry of modern organisms shows that dinosterols are concentrated in, and are nearly exclusive to, dinoflagellates. Saturated dinosteroid (dinosteranes) and triaromatic dinosteroid hydrocarbons found in rock extracts and petroleum are molecular fossils of dinosterols. We observed a virtually continuous dinosterane record in Precambrian to Cenozoic organic-rich marine rocks. Ratios of dinosterane concentrations to those of steranes with affinities to other taxa are uneven, with relatively high ratios in some Vendian to Devonian extracts, low ratios in Carboniferous to Permian extracts, and high ratios in Upper Triassic through Cretaceous extracts. A similar record was found for triaromatic dinosteroids, which were absent (undetected) in the Carboniferous to Permian extracts. These results show a parallel trend between fossil dinosteroids and the combined cyst records of acritarchs and dinoflagellates. This record reflects the high abundance and diversity of Cambrian to Devonian acritarchs, the relatively low abundance and diversity of Carboniferous to Permian acritarchs, and emergence, diversification, and increasing biomass of dinoflagellates in Triassic to Cretaceous rocks. The dinosteroid hydrocarbon record supplements morphologic and ultrastructural arguments that either modern dinoflagellates evolved from ancient (Precambrian) acritarchs or early dinoflagellates did not commonly encyst. In either case the chemical lineage shown by the dinosteroid hydrocarbons indicates a heritage that dates at least from the Riphean.

IN A SURVEY OF marine rocks of various geological ages, Moldowan et al. (1996) reported triaromatic dinosteroids (*1*—numbers in this style refer to figure 21.1) in Precambrian to Devonian organic-rich sedimentary rocks. Also, in an earlier report Summons et al. (1992) noted dinosterane (2) occurrences in extracted organic matter from Precambrian rocks. These data appear to provide the long sought-for evidence that dinoflagellates (or closely related protists) have an ancient origin, a hypothesis

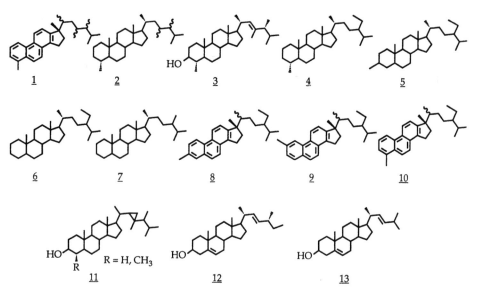

Figure 21.1 Steroid structures mentioned in text: *1*, triaromatic dinosteroid; *2*, dinosterane; *3*, dinosterol (most abundant structure); *4*, 4α-methylstigmastane; *5*, 3β-methylstigmastane; *6*, stigmastane; *7*, ergostane; *8*, triaromatic 3-methyl-24-ethylcholesteroid; *9*, triaromatic 2-methyl-24-ethylcholesteroid; *10*, triaromatic 4-methyl-24-ethylcholesteroid; *11*, gorgosterol and 4α-methylgorgosterol; *12*, 27-norcholest-5,22-en-3β-ol; *13*, 24-norcholest-5,22-en-3β-ol.

previously suggested by evolutionary biologists. Dinoflagellates are to a large extent primary producers, and rRNA and ultrastructure studies suggest their primitive nature (Margulis 1970; Wainright et al. 1993). Proof of a pre-Mesozoic diagnostic signature could provide pivotal information for understanding and reconstructing ancient food webs and the presence of environmentally important zooxanthellae-driven carbonate precipitation (Fensome et al. 1993), despite the apparent absence of definitive examples of dinoflagellate cysts in the Paleozoic and Precambrian record.

Dinoflagellates are the nearly exclusive producers of dinosterols (3). An exception was noted by Volkman et al. (1993) in a marine diatom. Dinosterols, in turn, are the biochemical precursors for the geologically preserved dinosteroid hydrocarbons (dinosteranes, triaromatic dinosteroids). These dinosteroid compounds are part of a large group of geologically preserved hydrocarbons known as biomarkers. Biomarkers are defined as "complex organic compounds composed of carbon, hydrogen, and other elements which are found in oil, bitumen, rocks and sediments and show little or no change in structure from their parent organic molecules in living organisms" (Peters and Moldowan 1993).

Organic geochemical data on dinosteranes (2) are presented here, in addition to the detailed triaromatic dinosteroid hydrocarbon (1) information omitted from Moldowan et al. (1996) in the abbreviated journal format. This information supports the concept of a pre-Mesozoic chemical record of dinoflagellates or closely related forms.

Dinoflagellates' probable endosymbiotic origin is based on ultrastructure studies (Margulis 1970). Such a symbiotic incorporation of organelles representing discrete taxonomic entities as organelles, and therefore multiple sources of discrete genetic material merged in a single cell, has presented conundrums to those evolutionary biologists and paleontologists trying to use the conventional "tree" model for describing evolution of single-celled taxa. Statistical treatment of rRNA data (Wainright et al. 1993: figure 1) shows dinoflagellates (included as alveolates) diverging relatively early. Based primarily on ultrastructure, Taylor (1994) has provided a "tree" that also shows dinoflagellates with an ancient origin. He cites the triple-membraned envelopes of dinoflagellates and their multiple nuclei as evidence of likely endosymbiotic origin. Ambiguities in microbiological evolution introduced by endosymbiosis, rather than by mutation, also cloud chemocladistic evolutionary interpretations. Therefore, we believe our empirical data from the fossil record adds new information for addressing the ancient origin for the dinoflagellate lineage.

METHODS

We analyzed extracts from 129 organic-rich stratigraphically dated samples from Proterozoic to Cretaceous age cores, side-wall cores, and outcrop samples for triaromatic dinosteroids and dinosteranes (1 and 2; table 21.1). The rock samples were screened for percentage of total organic carbon (>1 percent) and Rock-Eval pyrolysis parameters useful for discriminating contamination by migrated oil (Peters 1986).

Dinosteranes (2) were identified by gas chromatography–mass spectrometry–mass spectrometry (GC-MS-MS) coelution experiments, using synthetic standards of four (20R)-5α-dinosterane diastereomers having the 20R,23S,24R (RSR), RRR, RSS, and RRS stereochemistries (Stoilov et al. 1993). Other methyl steranes, namely (20R)-4α-methylstigmastane (4) (Stoilov et al. 1993) and (20R)-3β-methylstigmastane (5) (Summons and Capon 1988), were identified by similar means. Dinosteranes were detected and measured using m/z 414 → 98 and 414 → 231 GC-MS-MS transitions on a Hewlett-Packard 5890 Series II GC coupled to a VG Micromass Autospec Q hybrid mass spectrometer system at Stanford University. The m/z 414 → 98 transition is highly selective for dinosteranes over other methylsteranes and was used for identification. The more intense m/z 414 → 231 transition was used for quantification of dinosteranes and methylstigmastanes. Because of partial coelution interferences from other steranes in the m/z 414 → 231 transition, it was necessary to measure dinosteranes in the m/z 414 → 98 transition in some cases and to apply a correction factor to obtain the amount consistent with m/z 414 → 231 measurements.

Because of the great effect of thermal maturation on absolute biomarker concentrations, we followed an approach suggested by Peters and Moldowan (1993), using ratios of biomarkers of similar thermal stability, that is, ratios of dinosteranes with structurally similar steranes rather than absolute concentrations of dinosterane. Concentration effects caused by thermal maturity cancel out in such ratios and yield results indicating one biomarker input compared with another, which reflects the relative contributions of two taxa to a given sample.

RESULTS

Dinosterane (2) data, presented in figure 21.2, were recorded in ratios to (20R)-4α-methylstigmastane (4), (20R)-3β-methylstigmastane (5), (20R)-5α-stigmastane (6), and (20R)-5α-ergostane (7). Precursors for the denominator compounds appear to be less restricted in modern organisms and environments than dinosterols (3). The 4α-methylstigmastanes are related to 4α-methyl-24-ethylcholesterols found in both dinoflagellates and haptophytes (synonym "prymnesiophytes," sensu Siesser 1993) (Volkman et al. 1990). However, the occurrence of 4α-methylstigmastane precursors is much more restricted than those of stigmastane itself. Two possible stigmastane precursors, fucosterol and isofucosterol, have been found in Dinophyceae but, if present, are generally minor sterols. These and other stigmastane precursors are widespread major or minor sterols in some members of virtually all the algal families (Volkman 1986). The same can be said for ergostane precursors, except they have not been found in the Dinophyceae. The precursors of 3-methylstigmastanes are unknown in modern organisms. These modified steranes are ubiquitous in rock extracts and oils and are thought to be derived by microbial alkylation of 2-sterenes produced during diagenesis from common desmethylsterols (Summons and Capon 1988, 1991; Dahl et al. 1992). Therefore, 3-methylstigmastanes are likely to have the same widespread algal origins as stigmastanes. The (desmethyl) stigmastane precursors stigmasterol and sitosterol are common to algae and vascular plants, which were important contributors to some paleodepositional environments.

Triaromatic dinosteroids were compared (figure 21.3) in ratios to triaromatic 3-methyl-24-ethylcholesteroids (8) plus triaromatic 2-methyl-24-ethylcholesteroids (9) and to triaromatic 4-methyl-24-ethylcholesteroids (10) (Dahl et al. 1995). The key triaromatic dinosteroids were analyzed in rock-extract aromatic fractions prepared by high-performance liquid chromatography (Peters and Moldowan 1993), using gas chromatography–mass spectrometry for monitoring the m/z 245 ion (loss of side chain), and identified by coelution with authentic standards (Ludwig et al. 1981; Lichtfouse et al. 1990; Shetty et al. 1994; Stoilov et al. 1994). The precursors for compounds selected as denominators in these ratios appear to be widely distributed in modern organisms or environments. The precursors of triaromatic 3-methyl-24-ethylcholesteroids (8) and triaromatic 2-methyl-24-ethylcholesteroids (9), like their saturated analogs, appear to be formed by diagenesis of 4-desmethylsterols. The triaromatic 4-methyl-24-ethylcholesteroids (10) are related to 4-methyl-24-ethylcholesterols, which are abundant both in dinoflagellates and prymnesiophytes (Volkman et al. 1990).

Table 21.1 **Identification of Samples and Measurements of Their Dinosteroid Ratios**

SAMPLE ID	COUNTRY-REGION	BASIN	FORMATION	DEPTH
A565	Australia	McArthur	Velkerri	181.3 m
A564	Australia	McArthur	McMinn	64.0 m
A468	US/Arizona	Colorado Plateau	Kwagunt, Chuar	Surface
A466	US/Arizona	Colorado Plateau	Kwagunt, Chuar	Surface
S2	Middle East	Central Arabia	Huqf	>1,000 m
A476	Middle East	Central Arabia	Huqf	4,447 m
C9	US/Missouri	?	Bonneterre	2,337 ft
333	US/North Dakota	Williston	?	?
S3	Middle East	Central Arabia	Huqf	>1,000 m
C151	E. Siberia, Russia	Yudoma-Olenek	Kuonamka	Surface
S1	Middle East	Central Arabia	Huqf	>2,000 m
C152	E. Siberia, Russia	Yudoma-Olenek	Kuonamka	Surface
A316	E. Siberia, Russia	Yudoma-Olenek	Kuonamka	Surface?
C156	E. Siberia	?	?	Surface
A221	Australia	Georgina	Inca	107.6 m
A220	Australia	Georgina	Currant Bush	22.18 m
A197	Sweden	?	Alum Shale	44.88 m
A198	Sweden	?	Alum Shale	24.21 m
A200	Sweden	?	Alum Shale	22 m
A199	Sweden	?	Alum Shale	9 m
315	USSR	?	Dictyonema	<1,000 m
A569	Australia	Amadeus	Horn Valley	227.9 m
A568	Australia	Amadeus	Horn Valley	217.5 m
A436	Australia	Amadeus	Horn Valley	216.5 m
A382	Middle East	Central Arabia	Hanadir	14,829 ft
A229	Australia	Canning	Goldwyer	936 m
A333	US/Iowa	?	Glenwood	752.1 ft
A030	US/Iowa		Decorah-Guttenberg	953.3 ft
A330	US/Nevada	Basin & Range Province	Vinini	8 ft
A437	US/Iowa	?	St. Peter	761 ft
A900	Jordan	?	?	2,576 m
C25	US/Nevada		Vinini Cr	Surface
A242	Libya	Cryenacian Platform		?
A291	Algeria	Ghadames	Rhazziane	3,214 m
A289	Algeria	Illizi	?	2,083 m
A332	US/Iowa		Maquoketa Shale	461.6 ft
A241	Libya	Sirte		?
A383	Middle East	Central Arabia	Qusaiba	13,505 ft
A288	Algeria	Ghadames	Oued Ali	?
A449	Bolivia	South Sub-Andean FTB	?	2,191.5 m
A448	Bolivia	South Sub-Andean FTB	?	2,188.9 m
A447	Bolivia	Chaco		?
A376	Middle East	Central Arabia	Jauf	14,379 ft
A239	Russia	Timan-Pechora	?	4,028–4,034 m
A237	Russia	Timan-Pechora	?	3,700–3,708 m

PERIOD	AGE	BIOMARKER RATIOS*					
		2/(2+5)	2/(2+7)	2/(2+6)	2/(2+4)	1/(1+8+9)	1/(1+10)
M. Pro.	Riphean	0.131	0.104	0.713	0.496	0.000	0.000
M. Pro.	Riphean	0.104	0.064	0.915	0.365	0.727	0.842
E. L. Pro.	Sturtian	0.213	0.263	0.766	0.539	0.000	0.000
E. L. Pro.	Sturtian	0.130	0.055	0.648	0.360	0.224	0.437
L. L. Pro.	L.V.	0.000	0.000	0.000	0.000	n/a	n/a
L. Pro.	Vendian	n/a	n/a	n/a	n/a	0.741	0.785
Cam.		0.047	0.032	0.373	0.152	n/a	n/a
Cam.	?	0.077	0.051	0.336	0.207	0.000	0.000
E. Cam.		0.000	0.000	0.000	0.000	n/a	n/a
E. Cam.		0.039	0.007	0.321	0.051	n/a	n/a
E. Cam.		0.000	0.000	0.000	0.000	n/a	n/a
E. Cam.		0.028	0.006	0.243	0.042	n/a	n/a
E. M. Cam.	?	0.067	0.045	0.565	0.777	0.000	0.000
M. Cam.		0.228	0.079	0.242	0.396	n/a	n/a
M. Cam.	L. Templeton–Floran	0.088	0.050	0.522	0.481	0.000	0.000
M. Cam.	Floran-Undillan	0.042	0.000	0.329	0.102	0.000	0.000
M. Cam.		0.051	0.012	0.619	0.140	0.000	0.000
L. Cam.		0.095	0.054	0.548	0.498	L	L
E. Ord.	Tremadocian	0.000	0.000	0.000	0.000	L	L
E. Ord.	Tremadocian	0.086	0.042	0.480	0.250	0.000	0.000
E. Ord.	?	0.110	0.080	0.468	0.441	0.000	0.000
E. Ord.	Arenigian	0.169	0.100	0.476	0.387	L	L
E. Ord.	Arenigian	0.062	0.048	0.607	0.280	L	L
E. Ord.	Arenigian	0.000	0.000	0.000	0.000	L	L
M. Ord.	?	0.050	0.070	0.604	0.505	0.777	0.836
M. Ord.	Llanvirnian	0.000	0.000	0.000	0.000	0.000	0.000
M. Ord.	E. Caradocian	0.141	0.080	0.336	0.394	0.000	0.000
M. Ord.	E. Caradocian	0.115	0.030	0.426	0.153	0.000	0.000
M. Ord.	E. Caradocian	0.277	0.190	0.495	0.592	0.769	0.787
M. Ord.	E. Caradocian	0.003	0.006	0.142	0.077	0.000	0.000
M. Ord.	E. Caradocian	0.000	0.000	0.000	0.000	0.838	0.850
M. L. Ord.		0.039	0.016	0.311	0.114	n/a	n/a
L. Ord.	Caradocian	0.260	0.147	0.524	0.464	0.611	0.767
L. Ord.	Caradocian	0.312	0.190	0.520	0.542	0.378	0.616
L. Ord.	Caradocian	0.287	0.113	0.584	0.453	0.000	0.000
L. Ord.	M. Ashgillian	0.085	0.019	0.347	0.159	0.000	0.000
E. Sil.	Llandoverian	0.047	0.074	0.464	0.311	0.647	0.467
E. Sil.	?	0.000	0.000	0.000	0.000	0.373	0.589
Sil.	?	0.057	0.027	0.348	0.133	0.000	0.000
L. Sil.	Ludlovian	0.061	0.035	0.569	0.189	0.117	0.313
L. Sil.	Ludlovian	0.008	0.005	0.296	0.101	0.060	0.145
E. Dev.	?	0.030	0.012	0.130	0.135	0.000	0.000
E. Dev.	?	0.000	0.000	0.000	0.000	L	L
E. Dev.	Emsian	0.000	0.000	0.000	0.000	0.000	0.000
E. Dev.	Emsian	0.115	0.055	0.333	0.260	0.000	0.000

(continues)

Table 21.1 (Continued)

SAMPLE ID	COUNTRY-REGION	BASIN	FORMATION	DEPTH
A446	Bolivia	?	Tita	?
A605	Bolivia	Sub-Andean	?	?
C73-d93	Bolivia	?	?	?
A606	Bolivia	Sub-Andean	?	?
A604	Bolivia	Sub-Andean	?	?
A602	Bolivia	Sub-Andean	?	?
A949	Bolivia	?	?	?
A940	Bolivia	?	?	?
A445	Bolivia	?	?	?
A227	USA	?	Woodford Shale	8,943 ft
654	US/North Dakota	Williston	Bakken (Upper Member)	10,002 ft
A948	UK/Manchester	?	?	Surface
A226	US/Oklahoma	Anadarko	Big Lime	4,022.2 ft
A225	US/Utah	Paradox	Hermosa	2,782 ft
A190	US/Texas	Palo Duro	Motley Co., Texas	5,461–5,471 ft
A391	Middle East	Central Arabia	Unayzah	6,196 ft
C73-d3	Bolivia	?	Bolivia	?
A603	Bolivia	Sub-Andean	?	?
A601	Bolivia	Sub-Andean	?	?
A559P	Brazil	?	?	?
A367	Brazil	Parana	Irati	?
A337	US/Montana	Rocky Mountain	Phosphoria (Mead Peak Member)	408.5 ft
A342	US/Montana	Rocky Mountain	Phosphoria (Mead Peak Member)	431.5 ft
A336	US/Montana	Rocky Mountain	Phosphoria (Mead Peak Member)	207 ft
A335	US/Montana	Rocky Mountain	Phosphoria (Mead Peak Member)	205 ft
C124	Svalbard		Sticky Keep	?
C125	Svalbard		Sticky Keep	?
A576	US/Alaska			Surface
A416	US/Wyoming	?	Dinwoody	34.1 ft
A415	US/Wyoming	?	Dinwoody	32.1 ft
A414	US/Wyoming	?	Dinwoody	30.9 ft
C123	Svalbard		Botneheia	?
393	Switzerland	Southern Alps FTB	Meride Shale	327–332 ft
898	Syria	Northwest Arabian	?	675–825 m
280	US/Idaho		Thaynes Limestone	1,941 ft
208	US/Alaska	North Slope	Shublik	8,891–8,903 ft
207	US/Alaska	North Slope	Shublik	8,825–8,830 ft
797	Italy	Central Apennines	Dolomia Principale	Surface
A793	Papua New Guinea	South Papuan FTB	Koi-Iange	?
213	US/Alaska	North Slope	Kingak Shale	9,390 ft
210	US/Alaska	North Slope	Kingak Shale	8,399–8,405 ft

		BIOMARKER RATIOS*					
PERIOD	AGE	2/(2+5)	2/(2+7)	2/(2+6)	2/(2+4)	1/(1+8+9)	1/(1+10)
Dev.	?	0.067	0.026	0.250	0.212	0.000	.000
M. Dev.	L. Givetian	0.091	0.040	0.539	0.247	0.000	.000
L. Dev.	?	0.057	0.026	0.473	0.127	n/a	n/a
L. Dev.	Frasnian	0.111	0.035	0.384	0.279	0.351	0.585
L. Dev.	Frasnian	0.091	0.031	0.494	0.177	0.000	0.000
L. Dev.	Frasnian	0.076	0.023	0.604	0.228	0.114	0.259
L. Dev.	?	0.078	0.024	0.267	0.175	0.000	0.000
L. Dev.	?	0.103	0.034	0.576	0.166	0.000	0.000
L. Dev.	?	0.036	0.019	0.401	0.140	0.000	0.000
Car.-Dev.		0.049	0.014	0.280	0.077	0.076	0.236
E. Car.	Kinderhookian	0.074	0.026	0.320	0.190	0.069	0.157
E. Car.	E. Namurian	0.000	0.000	0.000	0.000	L	L
M. Car.	Desmoinesian	0.075	0.047	0.430	0.241	0.121	0.399
M. Car.	Desmoinesian	0.020	0.013	0.347	0.098	0.000	0.000
M.–L. Car.	Penn.	0.018	0.007	0.177	0.189	0.000	0.000
Per.-Car.		0.013	0.013	0.647	0.075	0.000	0.000
Per.		0.070	0.012	0.343	0.080	n/a	n/a
E. Per.	Aktas.-Leonard.	0.016	0.015	0.502	0.194	0.000	0.000
E. Per.	Aktas.-Leonard.	0.019	0.018	0.571	0.202	0.000	0.000
E. Per.	Art.-Kungurian	0.022	0.005	0.614	0.069	0.000	0.000
E. Per.	Art.-Kungurian	0.027	0.005	0.345	0.106	0.000	0.000
E. Per.	Leonardian	0.037	0.027	0.648	0.268	0.000	0.000
E. Per.	Leonardian	0.096	0.064	0.378	0.285	0.000	0.000
E. Per.	Guadalupian	0.050	0.025	0.569	0.177	0.000	0.000
L. Per.	Guadalupian	0.039	0.021	0.537	0.154	0.000	0.000
E. Tri.		0.043	0.018	0.350	0.147	n/a	n/a
E. Tri.		0.053	0.024	0.336	0.185	n/a	n/a
E. Tri.		n/a	n/a	n/a	n/a	0.808	0.649
E. Tri.	E. Scythian	0.351	0.174	0.521	0.518	0.615	0.824
E. Tri.	E. Scythian	0.088	0.066	0.799	0.437	0.154	0.332
E. Tri.	E. Scythian	0.073	0.046	0.627	0.291	0.300	0.467
		0.038	0.016	0.260	0.126	n/a	n/a
M. Tri.	Anisian-Landinian	0.479	0.221	0.815	0.782	0.915	0.814
Tri.	?	0.151	0.043	0.683	0.646	0.633	0.616
Tri.	?	0.254	0.200	0.725	0.524	0.662	0.756
Tri.	?	0.261	0.154	0.631	0.425	0.632	0.714
Tri.	?	0.275	0.169	0.633	0.531	0.593	0.701
L. Tri.	Norian	0.463	0.116	0.648	0.873	0.920	0.783
M. Jur.	E. M. Callovian	0.391	0.113	0.951	0.408	n/a	n/a
L. Jur.	?	0.254	0.150	0.754	0.536	0.813	0.808
L. Jur.	?	0.257	0.127	0.596	0.532	0.801	0.833

(continues)

Table 21.1 (*Continued*)

SAMPLE ID	COUNTRY-REGION	BASIN	FORMATION	DEPTH
313	Australia	Barrow Dampier	Dingo Claystone	2,835 ft
314	Australia	Barrow Dampier	Dingo Claystone	2,545 ft
997	UK	North Sea		6,888 ft
993	Scotland	Celtic Sea	Cullaidh	
A793	Papua New Guinea	South Papuan	Koi-Iange	
413	Middle East	Central Arabia	Tuwaiq Mountain	6,815 ft
412	Middle East	Central Arabia	Hanifa	6,540 ft
A789P	UK	North Sea	Kimmeridge Clay	5,320 ft
410	Middle East	Central Arabia	Hanifa	1,009 ft
A790	Papua New Guinea	South Papuan FTB	Imburu	Surface
A788P	UK	North Sea	Kimmeridge Clay	5,320 ft
A799	Papua New Guinea	South Papuan FTB	Imburu	9,380 ft
A791	Papua New Guinea	South Papuan FTB	Imburu	Surface
A142	Russia	West Siberia	Bazhenov	2,773 m
A798	Papua New Guinea	South Papuan FTB	Leru	7,880 ft
A503	Argentina	Neuquen	Vaca Muerta	Surface
312	Australia	Barrow Dampier	Muderong Shale	1,291 m
209	US/Alaska	North Slope	?	11,568 ft
A619	Colombia	Upper Magdalena	Paja	Surface
A631P	Italy	South Alps FTB	?	Surface
A627P	Ecuador	Oriente	Napo	295 m
A617	Colombia	Upper Magdalena		Surface
A629P	Australia	Eromanga	Toolebuc	?
A616	Colombia	Upper Magdalena	Similti	Surface
A539P	Angola	Lower Congo	Vermelha	9,500 ft
A630P	Italy	South Alps FTB	?	Surface
A533	Angola	Lower Congo	Vermelha	7,088 ft
A600	Bolivia	Sub-Andean Basin	?	Surface
C34	US/Colorado	?		?
A554P	Angola	Lower Congo	Vermelha	8,233 ft
A411	Ecuador	Oriente	Napo	144 m
A547P	Angola	Lower Congo	Vermelha	6,100 ft
226	US/Alaska	North Slope	Seabee Shale	11,161 ft
A552	Angola	Lower Congo	Vermelha	7,536 ft
A545	Angola	Lower Congo	Vermelha	5,366 ft
583	US/Wyoming		Hilliard (Upper Member)	70–80 ft
A623	US/California	San Joaquin	?	14,112 ft
A543	Angola/Cabinda	Lower Congo	Vermelha	4,545 ft
A613	Colombia	Upper Magdalena	LaLuna	Surface
A608P	Colombia	Upper Magdalena	LaLuna	Surface
A537	Angola	Lower Congo	Labe	9,952 ft
A386	Trinidad	Onshore Trinidad	Naparima Hill	Surface
A385	Trinidad	Onshore Trinidad	Naparima Hill	Surface

*Refer to Figure 21.1 and text for compound structures and names. n/a = measurement not taken. L = low biomarker concentrations.

PERIOD	AGE	BIOMARKER RATIOS*					
		2/(2+5)	2/(2+7)	2/(2+6)	2/(2+4)	1/(1+8+9)	1/(1+10)
L. Jur.	?	0.409	0.197	0.673	0.477	0.616	0.632
L. Jur.	?	0.082	0.043	0.565	0.304	0.516	0.776
M. Jur.		n/a	n/a	n/a	n/a	0.353	0.615
M. Jur.	Bathonian	n/a	n/a	n/a	n/a	0.955	0.920
M. Jur.	E.–M. Callovian	n/a	n/a	n/a	n/a	.492	0.719
L. Jur.	Oxfordian	0.296	0.139	0.597	0.429	0.892	0.811
L. Jur.	L. Oxfordian	0.421	0.147	0.487	0.408	0.830	0.746
L. Jur.	Oxford.–E. Kimm.	0.139	0.078	0.645	0.532	n/a	n/a
L. Jur.	L. Oxford.–Kimm.	0.298	0.132	0.422	0.334	0.683	0.769
L. Jur.	E. Kimmeridgian	0.454	0.154	0.645	0.514	0.788	0.809
L. Jur.	Kimmeridgian	0.346	0.182	0.800	0.618	L	L
L. Jur.	L. Kimmeridgian	0.175	0.080	0.683	0.328	0.684	0.785
L. Jur.	E. Tithonian	0.522	0.284	0.668	0.673	0.840	0.803
L. Jur.	Volgian	0.189	0.159	0.627	0.428	0.827	0.863
E. Cret.	L. Berriasian	0.252	0.145	0.587	0.479	0.815	0.826
E. Cret.	Berriasian L. Valang.	0.366	0.328	0.744	0.583	0.627	0.730
E. Cret.	Hauterivian	0.099	0.092	0.460	0.579	0.807	0.528
E. Cret.	Barremian-Albian	0.099	0.104	0.493	0.395	0.876	0.826
E. Cret.	L. Haut.–L. Berri.	n/a	n/a	n/a	n/a	0.842	0.807
E. Cret.	Aptian	0.080	0.035	0.588	0.246	0.311	0.502
E. Cret.	Aptian	n/a	n/a	n/a	n/a	0.821	0.851
E. Cret.	L. M. Albian	0.164	0.171	0.694	0.443	0.794	0.863
E. Cret.	L. Albian	0.079	0.112	0.614	0.492	0.676	0.825
L.–E. Cret.	E. Cenomanian	0.248	0.266	0.762	0.734	0.768	0.810
L. Cret.	E. Cenomanian	0.184	0.151	0.632	0.455	0.638	0.771
L. Cret.	Cenomanian	0.073	0.041	0.580	0.264	0.330	0.605
L. Cret.	L. Cenomanian	0.079	0.115	0.640	0.711	0.853	0.860
L. Cret.	Cenomanian-Turonian	0.174	0.122	0.655	0.605	0.527	0.735
L. Cret.	L. Turonian	0.035	0.029	0.358	0.002	n/a	n/a
L. Cret.	Turonian	0.134	0.121	0.552	0.616	0.590	0.720
L. Cret.	Turonian	0.241	0.246	0.674	0.627	0.877	0.893
L. Cret.	Coniacian	0.175	0.130	0.562	0.512	0.688	0.764
L. Cret.	?	0.033	0.062	0.535	0.522	0.775	0.823
L. Cret.	Santonian	0.099	0.130	0.480	0.536	0.762	0.835
L. Cret.	Santonian	0.165	0.189	0.614	0.669	0.791	0.862
L. Cret.	Santonian	0.072	0.078	0.548	0.283	0.848	0.719
L. Cret.	Campanian	0.051	0.012	0.534	0.480	0.763	0.835
L. Cret.	Campanian	0.035	0.150	0.580	0.599	0.752	0.809
L. Cret.	L. Campanian	0.158	0.205	0.615	0.607	0.932	0.932
L. Cret.	L. Campanian	0.283	0.316	0.682	0.744	0.733	0.819
L. Cret.	L. Campanian	0.223	0.221	0.663	0.392	0.692	0.755
L. Cret.	L. Campanian–Maast.	0.226	0.236	0.691	0.617	0.870	0.898
L. Cret.	Maastrichtian	0.235	0.215	n/a	n/a	0.845	0.891

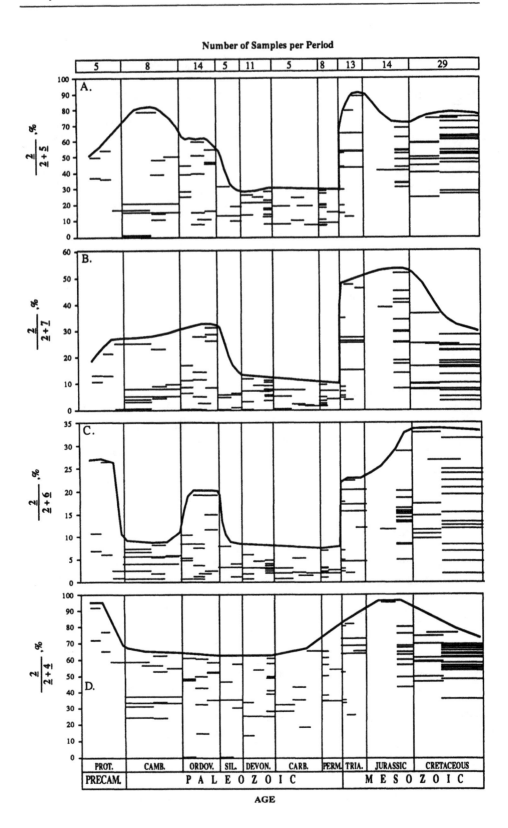

Figure 21.2 Distribution of dinosteranes versus other steranes and methylsteranes through geologic time. Horizontal lines indicate dinosterane amounts in 122 rock extracts expressed as a percentage of 4 dinosterane stereoisomers (20R,23S,24S + 20R,23S,24R + 20R,23R,23S + 20R,23R,24R) in the sum of the 4 dinosterane stereoisomers, plus the following: in A, 3β-methylstigmastane 20R (5 in figure 21.1); in B, ergostane 20R (7); in C, stigmastane 20R (6); in D, 4α-methylstigmastane (4).

Lichtfouse et al. (1990) proposed that positions of methyl groups in triaromatic hydrocarbons are altered by methyl group migrations, and 4-methylsteroid precursors could be responsible for 2- and 3-methyl-substituted triaromatics (or the reverse). Furthermore, a methyl shift (from C-10 to C-1 or C-4) in the aromatization of sterols can result in a methyl at the 4-position in ring-A monoaromatic steroids derived from 4-desmethylsterols found in thermally immature sediments. These compounds could also form 4-methyl triaromatic steroids upon further diagenesis (Hussler et al. 1981). Therefore, diagenetic rearrangements could also explain the similarities in triaromatic dinosteroids to triaromatic 2- + 3-methyl-24-ethylcholesteroids ratios (figure 21.3A) and to triaromatic 4-methyl-24-ethylcholesteroids ratios (figure 21.3B). However, it is not certain that these rearrangement mechanisms are active here. Opposing these mechanisms is the fact that significant amounts of only the 4-methyl and not 1-, 2-, or 3-methyl isomers of triaromatic dinosteroids are present in these samples. They would require, then, a selective methyl rearrangement active in the triaromatic 24-ethylcholesteroid series and not active in the triaromatic dinosteroid series. Thus, it appears unlikely that methyl rearrangements contribute in any significant way to the formation of the analyzed compounds.

DISCUSSION

Dinoflagellate Evidence in the Fossil Record

For many years dinoflagellates have been considered primitive and, therefore, ancient organisms, on the basis of their morphology, ultrastructure, and biochemistry (e.g., Margulis 1970; Evitt 1985; Taylor 1987; Withers 1987; Knoll and Lipps 1993). In the fossil record they are recognized by their acid-resistant organic-walled cysts in marine sediments. Unfortunately, the geologic record bears no undisputed fossil dinoflagellate cysts older than Middle Triassic (Goodman 1987; Helby et al. 1987), although there are 22 specimens of the enigmatic and controversial, thermally altered, organic-walled Late Silurian microfossil *Arpylorus antiquus* (Calandra) Sarjeant, from Tunisia (Calandra 1964; Sarjeant 1978; Bujak and Williams 1981; Evitt 1985; Goodman 1987), and the Devonian *Palaeodinophysis altaica*, which requires confirmation by further studies (Fensome et al. 1993). The inability of the fossil record to provide a thorough dinoflagellate history has been explained by noting that only 6 of 15 orders of living dinoflagellates produce fossilizable cysts (Goodman 1987). Head (1996)

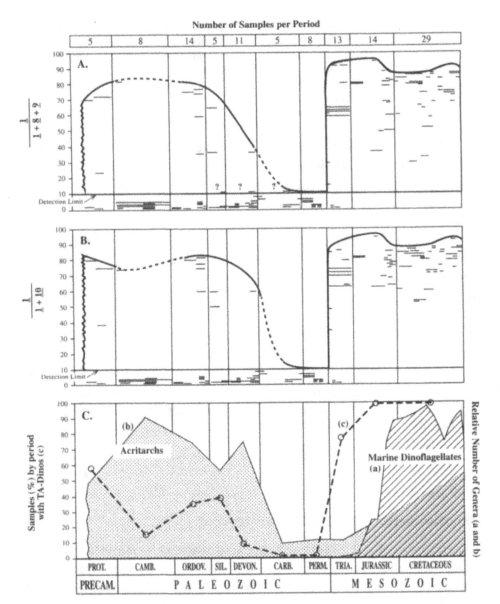

Figure 21.3 Comparison of diversity of dino-flagellate and acritarch cysts with abundance and frequency of occurrence of triaromatic di-nosteroids over geologic time. A and B, Horizontal lines indicate triaromatic dinosteroid (1 in figure 21.1) amounts in 129 rock extracts in which methyltriaromatic steroids were detected. Values are expressed as percentage of triaromatic dinosteroids in the sum of triaromatic dinosteroids + 2- + 3-methyl-24-ethyl-cholesteroids ($1/[1 + 8 + 9]$ in A) and triaromatic dinosteroids + 4-methyl-24-ethylcholes-teroids ($1/[1 + 10]$ in B). Lower detection limit is ~10 percent for triaromatic dinosteroids, and samples with ~10 percent are indicated by horizontal marks stacked below the detection limit line. Nonzero amounts below 10 percent are not implied. C, Schematic representations of numbers of (a) dinoflagellate cyst genera (adapted from MacRae et al. 1996) and (b) acritarch genera (adapted from Strother 1996). Circles and dashed lines (c) give frequency of occurrence of detectable triaromatic dinosteroids in samples from each geologic time period.

noted that only 13–16 percent of living species produce preservable cysts. Therefore, absence of dinoflagellate cysts cannot prove that dinoflagellates did not exist (Evitt 1985). Some modern dinoflagellates form acritarchous cysts lacking morphologic features diagnostic of dinoflagellate cysts (e.g., Anderson and Wall 1978; Dale 1978). These taxonomically undiagnostic cysts reinforce the long and widely held hypothesis that at least some acritarchs (incertae sedis organic-walled microfossils that originate in the Precambrian) record the heritage of dinoflagellates.

Chemical Fossil Evidence for Dinoflagellates

An unusual suite of sterols, including dinosterols (3), are associated almost exclusively with dinoflagellates. Dinosterol is typically the major sterol in dinoflagellates (Shimuzu et al. 1976; Withers 1987), although it is absent in one genus, *Amphidinium* (Kokke et al. 1981). Minor amounts of dinosterol (2.0–3.6 percent of total sterols) have been detected in a cultured marine diatom (Volkman et al. 1993), but based on this single occurrence (among at least 25 other analyzed diatom species that lack dinosterols; Volkman 1986) and the low concentration levels, diatoms do not appear to be an important sedimentary source for dinosterols and their diagenetic dinosteroids. The function of these sterols is poorly understood. It has been presumed that they are membrane constituents, but dinosterol esters have been isolated from extraplastid lipid globules of the dinoflagellate eyespot (Withers and Nevenzel 1977; Withers and Haxo 1978). Therefore, the function and localization of these sterols remains unknown (Withers 1987).

The unusual structures of several dinoflagellate sterols reinforce the suggestion that dinoflagellates are a highly specialized group (Withers 1987). They exist as both free-living algae and symbiotic zooxanthellae that inhabit various host invertebrates. The structure of dinosterol (3) is unusual in three ways: (1) by having 4,23,24-trimethylation, (2) by the lack of a double bond in the sterol ring system, and (3) by the presence of the 4-methyl group. The last feature, in particular, is considered a vestige of primitive biochemistry. In most (modern) eukaryotes, sterols consist exclusively of 4-desmethyl structures (e.g., 13). This demethylation can be seen as a biosynthetic distancing from the squalene cyclization step that occurs in all sterol biosyntheses and produces an intermediate with geminal methyls at the 4-position. Similarly, sterols synthesized by the few bacteria, such as *Methylococcus capsulatus*, that are capable of sterol synthesis display only 4-methyl and 4,4-dimethyl but not 4-desmethyl structures (Bloch 1976). This primitive chemistry can be taken along with morphologic arguments as circumstantial evidence, but not proof, that dinoflagellates have a primitive and therefore ancient evolutionary history. Other dinoflagellate sterols include several that contain a cyclopropyl ring (22,23-methylene), such as gorgosterol and 4-methylgorgosterol (11), that are presumably synthesized in the organism from dinosterol (Withers et al. 1979). These cyclopropyl-containing sterols are widely found in invertebrates that contain zooxanthellae and apparently are synthesized by both free-living and zooxanthellae dinoflagellate strains (Withers et al. 1979, 1982). Other very unusual compounds identified in dinoflagellates include 27-nor- (12) and 24-nor- (13) sterols (Goad and Withers 1982).

We systematically searched marine rocks through the geologic column for the taxonomically restricted dinosteroids (dinosteranes and triaromatic dinosteroids, sensu Mackenzie et al. 1982) and recorded them by geologic age (figures 21.2 and 21.3). Assuming that these steroids are derived exclusively from the dinoflagellate-restricted sterols, our results confirm that the lineage of dinoflagellates is rooted in the Precambrian.

The occurrence of triaromatic dinosteroids (1) in all of our Upper Triassic to Cretaceous marine rock extracts, and their absence in the Carboniferous-Permian ones, is similar to the fossil record for dinoflagellate cysts and acritarchs. (A small number of Tertiary marine rock samples, such as Miocene [Monterey Formation], California, USA, and Miocene [Malembo Formation] and Eocene [Landana Formation], Angola, show dinosteroid concentrations of similar magnitude to the Cretaceous data set.) Some minor differences may be significant. Triaromatic dinosteroids are abundant throughout most of the Mesozoic, suggesting that dinoflagellates were quantitatively important in Mesozoic marine environments where organic matter is preserved.

Triaromatic dinosteroids are abundant in some Middle to Upper Triassic rocks, even though relatively few Triassic dinoflagellate species are known (about 10; Fensome et al. 1996). This abundance suggests that dinoflagellates already thrived in certain paleoenvironmental niches. Three samples from the Lower Triassic (Scythian) Dinwoody Formation, Wyoming, show significant triaromatic dinosteroid concentrations. However, in three Lower to Middle Triassic rock extracts (two samples from the Sticky Keep Member of the Tvillingodden Formation, upper Scythian, and one from the Botneheia Member of the Bravaisberget Formation, Middle Triassic, Svalbard), triaromatic dinosteroids were not detected. Also, Scythian and Ladinian age marine and paralic extracts from three locations in Australia showed only the "probable presence of trace amounts" of dinosterane (2) (Summons et al. 1992). These sporadic or low-level dinosteroid occurrences in the Early to Middle Triassic suggest that dinoflagellate populations were not globally significant during that time span. However, consistently strong dinosteroid concentrations from the Late Triassic onward suggest increased dinoflagellate populations. These data supplement the major evolutionary radiation of dinoflagellates in the early Mesozoic (e.g., Fensome et al. 1996). Thus, to a first approximation, increased dinoflagellate species diversity, biomass, and dinosteroid abundance appear to correlate in the early Mesozoic.

Earlier occurrences of triaromatic dinosteroids (1) are observed in some extracts from Proterozoic to Devonian rocks (figures 21.3A,B). A hiatus during the Middle Cambrian to Lower Ordovician could be due to our sample selection, although the species diversity of acritarchs also shows a significant drop during this time interval (Zhuravlev, this volume: figure 8.1C). The evidence for dinoflagellates during the Paleozoic consists of the unproven affinity of *Arpylorus antiquus*, the questionable occurrence of *Palaeodinophysis altaica,* and acritarchs that lack the diagnostic morpho-logic features of dinoflagellates (Anderson and Wall 1978; Dale 1978). The Paleozoic parts of the triaromatic dinosteroid abundance curves correlate with acritarch species diversity (*b* in figure 21.3C), although the frequency of triaromatic dinosteroid oc-currence (*c* in figure 21.3C) in the Proterozoic to Devonian samples is low (~9–60 percent) relative to Mesozoic samples (100 percent). Overall, high relative abun-dances of triaromatic dinosteroids for organically rich rocks reflect geologic times when either acritarchs and/or dinoflagellates flourished (figure 21.3).

The ratio of dinosteranes to (20*R*)-3β-methylstigmastane (figure 21.2A) shows a similar pattern to the analogous ratio of triaromatic steroids (triaromatic dinosteroids to triaromatic stigmasteroids, *1* and *8,* respectively, figure 21.3A), *with one important difference.* Nearly all of our marine rock extracts show detectable levels of dinosteranes, irrespective of age. (The oldest samples analyzed were from the McMinn Formation [1,429 + 31 Ma; Jackson et al. 1986], the slightly older Velkerri Formation, McArthur Basin, Australia, and the Kwagunt Formation [Sturtian], Colorado Plateau, USA. These all have detectable dinosteranes [table 21.1]. The McMinn sample shows an espe-cially strong abundance of triaromatic dinosteroids, which were also detected in one of two Kwagunt samples, and not the Velkerri sample.) This is due mainly to analyti-cal technology, namely, the use of the highly dinosterane-selective and diagnostic GC-MS-MS transition m/z 414 → 98, which does not have an analog for the triaro-matic dinosteroids. These analyses show no break in dinosterane occurrence from the Precambrian through the Cretaceous. This suggests that the hiatus in triaromatic di-nosteroid detection in Carboniferous to Permian rocks (figure 21.3) reflects relatively low dinosteroid concentrations rather than their total absence. Nevertheless, these data support the aromatic dinosteroid trend showing high relative abundances in Tri-assic to Cretaceous extracts, and low relative abundances throughout the Paleozoic, with some Devonian and older extracts showing high levels comparable to those of the Mesozoic. Both the radiation of cyst-forming dinoflagellates (Fensome et al. 1996) during the Mesozoic and the related increase in preserved dinoflagellate cyst biomass in the fossil record are still supported by these data, which suggest that the late Pa-leozoic dinoflagellates or their precursors were much less abundant than their Meso-zoic descendants.

Dinosterane distributions versus ergostane (*7*) and stigmastane (*6*), whose pre-cursors are produced by a wide variety of eukaryotes, also show a trend similar to the triaromatic dinosteroid trend (figures 21.2B and 21.2C, respectively). Low Pa-leozoic dinosterane abundances and high Mesozoic ones are shown by the dinoster-ane ratio with stigmastane (figure 21.2C). This implies that certain algae that pro-duced stigmastane precursors were dominant during the Paleozoic. They accounted for a smaller part of the Mesozoic biomass relative to dinoflagellates. Notably variant samples with relatively high stigmastane occur in the Precambrian and Ordovician. Possible C_{29}-sterol precursors for stigmastane occur in a wide variety of algae, and these are often the predominant sterols in various green algae (e.g., Prasinophyceae),

Prymnesiophyceae, and cyanobacteria (Volkman 1986). They are also the predominant sterols in many vascular plants. However, such C_{29}-sterols are not generally abundant in dinoflagellates.

The dinosterane (2) distribution versus (20R)-4α-methylstigmastane (4), on the other hand, shows no particular trend (figure 21.2D). Because (20R)-4α-methylstigmastane (4) is derived from precursor sterols that are also common in dinoflagellates, a ratio of the two compounds might not be expected to track dinoflagellate abundance. This contrasts with the results for the analogous triaromatic dinosteroids versus triaromatic 4-methyl-24-ethylcholesteroids (10) (figure 21.3B). A possible explanation for this difference is that (20R)-4α-methylstigmastane and triaromatic 4-methyl-24-ethylcholesteroid probably have chemical precursors that differ in functionality and are, therefore, likely to be derived from different organisms. Precursors that are likely to aromatize require an appropriate double bond that can migrate into the C-ring of the molecule to initiate the aromatization process (Riolo et al. 1986; Moldowan and Fago 1986). Precursors that are likely to become fully saturated steranes would tend to have fewer double bonds or lack them entirely. Modern dinoflagellates have been found to carry either or both kinds of sterol (Withers 1987), unusually functionalized 4-methyl-24-ethylcholest-8(14)-en-3-ols (containing a C-ring double bond) that are ideally set up for aromatization, and 4-methyl-24-ethylcholestan-3-ols that are fully saturated.

CONCLUSIONS

Our geologically well-dated, chemostratigraphic data show that lower Paleozoic rocks contain dinosteroid hydrocarbons (Moldowan et al. 1996), predating the oldest unequivocal dinoflagellate fossil by nearly 300 million years and even the first equivocal ones by more than 100 million years. The documentation of this stratigraphic set of biomarker data does not substantiate the presence of dinoflagellates in Precambrian rocks, but it does show that a dinosteroid precursor was in existence during these times. Combined with other data that argue for an ancient appearance (e.g., rRNA data [Wainright et al. 1993] and ultrastructural features [Margulis 1970; Taylor 1994]), this information supports an earlier evolution than the cyst record currently supports.

REFERENCES

Anderson, D. M. and D. Wall. 1978. Potential importance of benthic cysts of *Gonyaulax tamarensis* and *G. excavata* in initiating toxic dinoflagellate blooms. *Journal of Phycology* 14:224–234.

Bloch, K. 1976. On the evolution of a biosynthetic pathway. In A. Kornberg, B. L. Honecker, L. Cornudella, and L. Oro, eds., *Reflections on Biochemistry,* pp. 143–150. New York: Pergamon.

Bujak, J. P. and G. L. Williams. 1981. The evolution of dinoflagellates. *Canadian Journal of Botany* 59:2077–2087.

Calandra, F. 1964. Sur un présumé dinoflagellé, *Arpylorus* nov. genre du Gothlandien de Tunisie. *Compte rendus sommaires de l'Académie des Sciences, Paris,* ser. D, 258:4112–4114.

Dahl, J., J. M. Moldowan, M. A. McCaffrey, and P. A. Lipton. 1992. A new class of natural products revealed by 3-alkyl steranes in petroleum. *Nature* 355:154–157.

Dahl, J., J. M. Moldowan, R. E. Summons, M. A. McCaffrey, P. Lipton, D. S. Watt, and J. M. Hope. 1995. Extended 3-alkyl steranes and 3-alkyl triaromatic steroids in crude oils and rock extracts. *Geochimica et Cosmochimica Acta* 59:3717–3729.

Dale, B. 1978. Acritarchous cysts of *Pterdinium faeroensis* Paulsen: Implications for dinoflagellate systematics. *Palynology* 2:187–193.

Evitt, W. R. 1985. *Sporopollenin Dinoflagellate Cysts: Their Morphology and Interpretation.* Austin, Tex.: American Association of Stratigraphic Palynologists Foundation.

Fensome, R. A., F. J. R. Taylor, G. Norris, W. A. S. Sarjeant, D. I. Wharton, and G. L. Williams. 1993. *A Classification of Fossil and Living Dinoflagellates.* Micropaleontology Press Special Paper 7.

Fensome, R. A., R. A. MacRae, J. M. Moldowan, F. J. R. Taylor, and G. L. Williams. 1996. The early Mesozoic radiation of dinoflagellates. *Paleobiology* 22:329–238.

Goad, L. J. and N. W. Withers. 1982. The identification of 27-nor (24R)-24-methyl-cholesta-5,22-dien-3β-ol and brassicasterol as the major sterols of the marine dinoflagellate *Gymnodinium simplex. Lipids* 17:853–858.

Goodman, D. K. 1987. Dinoflagellate cysts in ancient and modern sediments. In F. J. R.

Taylor, ed., *The Biology of Dinoflagellates,* pp. 649–722. Oxford: Blackwell Scientific.

Head, M. J. 1996. Modern dinoflagellate cysts and their biological affinities. In J. Jansonius and D. C. McGregor, eds., *Palynology: Principles and Applications,* pp. 1197–1248. American Association of Stratigraphic Palynologists 13. Dallas.

Helby, R., R. Morgan, and A. D. Partridge. 1987. A palynological zonation of the Australian Mesozoic. In P. A. Jell, ed., *Studies in Australian Mesozoic Palynology,* pp. 1–94. Sydney: Association of Australasian Palynologists.

Hussler, G., B. Chappe, P. Wehrung, and P. Albrecht. 1981. C_{27}-C_{29} ring A monoaromatic steroids in Cretaceous black shales. *Nature* 294:556–558.

Jackson, M. J., T. G. Powell, R. E. Summons, and I. P. Sweet. 1986. Hydrocarbon shows and petroleum source rocks in sediments as old as 1.7×10^9 years. *Nature* 322:727–729.

Knoll, A. H. and J. H. Lipps. 1993. Evolutionary history of prokaryotes and protists. In J. H. Lipps, ed., *Fossil Prokaryotes and Protists,* pp. 19–29. Boston: Blackwell Scientific.

Kokke, W. C. M. C., W. Fenical, and C. Djerassi. 1981. Sterols with unusual nuclear unsaturation in three cultured marine dinoflagellates. *Phytochemistry* 20:127–134.

Lichtfouse, E., J. Riolo, and P. Albrecht. 1990. Occurrence of 2-methyl-, 3-methyl-, and 6-methyltriaromatic steroid hydrocarbons in geological samples. *Tetrahedron Letters* 31:3937–3940.

Ludwig, B., G. Hussler, P. Wehrung, and P. Albrecht. 1981. C_{26}-C_{29} triaromatic steroid derivatives in sediments and petroleums. *Tetrahedron Letters* 22:3313–3316.

Mackenzie, A. S., N. A. Lamb, and J. R. Maxwell. 1982. Steroid hydrocarbons and the thermal history of sediments. *Nature* 295:223–226.

MacRae, R. A., R. A. Fensome, and Gl. L. Williams. 1996. Fossil dinoflagellate diversity, originations, and extinctions and their significance. *Canadian Journal of Botany* 74:1687–1694.

Margulis, L. 1970. *Origin of eukaryotic cells.* New Haven, Conn.: Yale University Press.

Moldowan, J. M. and F. J. Fago. 1986. Structure and significance of a novel rearranged monoaromatic steroid hydrocarbon in petroleum. *Geochimica et Cosmochimica Acta* 50:343–351.

Moldowan, J. M., J. Dahl, S. R. Jacobson, B. J. Huizinga, F. J. Fago, R. Shetty, D. S. Watt, and K. E. Peters. 1996. Chemostratigraphic reconstruction of biofacies: Molecular evidence linking cyst-forming dinoflagellates with pre-Triassic ancestors. *Geology* 24:159–162.

Peters, K. E. 1986. Guidelines for evaluating petroleum source rock using programmed pyrolysis. *American Association of Petroleum Geologists, Bulletin* 70:318–329.

Peters, K. E. and J. M. Moldowan. 1993. *The Biomarker Guide: Interpreting Molecular Fossils in Petroleum and Ancient Sediments.* Englewood Cliffs, N.J.: Prentice-Hall.

Riolo, J., G. Hussler, P. Albrecht, and J. Connan. 1986. Distributions of aromatic steroids in geological samples: Their evaluation as geochemical parameters. *Organic Geochemistry* 10:981–990.

Sarjeant, W. A. S. 1978. *Arpylorus antiquus* Calandra, emend., a dinoflagellate cyst from the Upper Silurian. *Palynology* 2:167–179.

Shetty, R., I. Stoilov, D. S. Watt, R. M. K. Carlson, F. J. Fago, and J. M. Moldowan. 1994. Synthesis of biomarkers in fossil fuels: C-23 and C-24 diastereomers of (20R)-4,17,23,24-tetramethyl-18,19-dinorcho-

lesta-1,3,5,7,9,11,13-heptaene. *Journal of Organic Chemistry* 59:8203–8208.

Shimuzu, Y., M. Alam, and A. Kobayashi. 1976. Dinosterol, the major sterol with a unique side chain in the toxic dinoflagellate, *Gonyaulax tamarensis. Journal of the American Chemical Society* 98:1059–1060.

Siesser, W. G. 1993. Calcareous nannoplankton. In J. H. Lipps, ed., *Fossil Prokaryotes and Protists*, pp. 169–201. Boston: Blackwell Scientific.

Stoilov, I., E. Kolaczkowska, D. S. Watt, R. M. K. Carlson, F. J. Fago, and J. M. Moldowan. 1993. Synthesis of biological markers in fossil fuels. 7. Selected diastereomers of 4-methyl-5-stigmastane and 5-dinosterane. *Journal of Organic Chemistry* 58:3444–3454.

Stoilov, I., R. Shetty, J. S. Pyrek, S. L. Smith, W. J. Layton, D. S. Watt, R. M. K. Carlson, and J. M. Moldowan. 1994. A synthesis of a triaromatic steroid biomarker, (20R,24R)-4,17-dimethyl-18,19-dinorstigmasta-1,3,5,7,9,11,13-heptaene, from stigmasterol. *Journal of Organic Chemistry* 59:926–928.

Strother, P. K. 1996. Acritarchs. In J. Jansonius and D. C. McGregor, eds., *Palynology: Principles and Applications*, pp. 81–106. Dallas: American Association of Stratigraphic Palynologists Foundation.

Summons, R. E. and R. J. Capon. 1988. Fossil steranes with unprecedented methylation in ring-A. *Geochimica et Cosmochimica Acta* 52:2733–2736.

Summons, R. E. and R. J. Capon. 1991. Identification and significance of 3-ethyl steranes in sediments and petroleum. *Geochimica et Cosmochimica Acta* 55:2391–2395.

Summons, R. E., J. Thomas, J. R. Maxwell, and C. J. Boreham. 1992. Secular and environmental constraints on the occurrence of dinosterane in sediments. *Geochimica et Cosmochimica Acta* 56:2437–2444.

Taylor, F. J. R. 1987. General group characteristics; special features of interest; short history of dinoflagellate study. In F. J. R. Taylor, ed., *The Biology of Dinoflagellates*, pp. 1–23. Oxford: Blackwell Scientific.

Taylor, F. J. R. 1994. The role of phenotypic comparisons of living protists in the determination of probable eukaryotic phylogeny. In S. Bengston, ed., *Early Life on Earth, Nobel Symposium 84*, pp. 312–326. New York: Columbia University Press.

Volkman, J. K. 1986. A review of sterol markers for marine and terrigenous organic matter. *Organic Geochemistry* 9:83–99.

Volkman, J. K., P. Kearney, and S. W. Jeffrey. 1990. A new source of 4-methyl sterols and 5α(H)-stanols in sediments: Prymnesiophyte microalgae of the genus *Pavlova*. *Organic Geochemistry* 15:489–497.

Volkman, J. K., S. M. Barrett, G. A. Dunstan, and S. W. Jeffrey. 1993. Geochemical significance of the occurrence of dinosterol and other 4-methyl sterols in a marine diatom. *Organic Geochemistry* 20:7–15.

Wainright, P. O., G. Hinkle, M. L. Sogin, and S. S. Stickel. 1993. Monophyletic origins of the Metazoa: An evolutionary link with Fungi. *Science* 260:340–342.

Withers, N. 1987. Dinoflagellate sterols. In F. J. R. Taylor, ed., *The Biology of Dinoflagellates*, pp. 316–359. Oxford: Blackwell Scientific.

Withers, N. W. and F. T. Haxo. 1978. Isolation and characterization of carotenoid-rich lipid globules from *Peridinium foliaceum*. *Plant Physiology* 62:36–39.

Withers, N. W. and J. C. Nevenzel. 1977. Phytyl esters in a marine dinoflagellate. *Lipids* 12:989–993.

Withers, N. W., W. C. M. C. Kokke, M. Rohmer, W. H. Fenical, and C. Djerassi. 1979. Isolation of sterols with cyclopropyl-containing side chains from the cultured marine alga *Peridinium foliaceum*. *Tetrahedron Letters* 385:3605–3608.

Withers, N. W., W. C. M. C. Kokke, W. H. Fenical, and C. Djerassi. 1982. Sterol patterns of cultured zooxanthellae isolated from marine invertebrates. Synthesis of gorgosterol and 23-desmethylgorgosterol by aposymbiotic algae. *Proceedings of the National Academy of Sciences, USA* 79: 3764–3768.

Contributors

Adnan Al-Hajji
Saudi Aramco
P.O. Box 8745
Dhahran
Saudi Arabia

Martin D. Brasier
Department of Earth Sciences
Oxford University
Parks Road
Oxford OX1 3PR
United Kingdom

Graham E. Budd
Department of Earth Sciences
University of Uppsala
Norbyvägen 22
SE-752 36 Uppsala
Sweden

Mikhail B. Burzin
Paleontological Institute
Russian Academy of Sciences
Profsoyuznaya ulitsa 123
Moscow 117868
Russia

Nicholas J. Butterfield
Department of Earth Sciences
University of Cambridge
Downing Street
Cambridge CB2 3EQ
United Kingdom

T. Peter Crimes
Department of Earth Sciences
University of Liverpool
P.O. Box 147
Liverpool L69 3BX
United Kingdom

Jeremy Dahl
Department of Geological and
    Environmental Sciences
Stanford University
Stanford CA 94305-2115
USA

Françoise Debrenne
UMR 8569
Laboratoire de Paléontologie
Muséum National d'Histoire Naturelle
8, rue Buffon
F-75005 Paris
France

Mary L. Droser
Department of Earth Sciences
University of California
Riverside CA 92521
USA

Toni T. Eerola
Department of Geology and Mineralogy
University of Helsinki
P.O. Box 11
FIN-00014 Helsinki
Finland

Frederick J. Fago
Department of Geological and
    Environmental Sciences
Stanford University
Stanford CA 94305-2115
USA

David I. Gravestock (deceased)
Mines and Energy Resources, South
    Australia
P.O. Box 151
Eastwood
South Australia 5063
Australia

Thomas E. Guensburg
Physical Science Division
Rock Valley College
Rockford IL 61114
USA

Nigel C. Hughes
Department of Earth Sciences
University of California
Riverside CA 92521
USA

Bradley J. Huizinga
ARCO International Oil and Gas
    Company
2300 West Plano Parkway
Plano TX 75075-8499
USA

Stephen R. Jacobson
Department of Geological Sciences
Ohio State University
Columbus OH 43210-1397
USA

Artem V. Kouchinsky
Geological Institute
Russian Academy of Sciences
Pyzhevskiy pereulok 7
Moscow 109017
Russia

Xing Li
Department of Earth Sciences
University of California
Riverside CA 92521
USA

John F. Lindsay
Australian Geological Survey Organisation
GPO Box 378
Canberra ACT 2601
Australia

Irina D. Maidanskaya
United Institute of Physics of the Earth
Russian Academy of Sciences
Bol'shaya Gruzinskaya ulitsa 10
Moscow 123810
Russia

J. Michael Moldowan
Department of Geological and
    Environmental Sciences
Stanford University
Stanford CA 94305-2115
USA

Brian R. Pratt
Department of Geological Sciences
University of Saskatchewan
Saskatoon, Saskatchewan S7N 5E2
Canada

Joachim Reitner
Institut und Museum für Geologie und
    Paläontologie
Georg-August-Universität
Goldschmidtstrasse 3
D-37077 Göttingen
Germany

Robert Riding
Department of Earth Sciences
Cardiff University
Cardiff CF10 3YE
United Kingdom

Sergei V. Rozhnov
Paleontological Institute
Russian Academy of Sciences
Profsoyuznaya ulitsa 123
Moscow 117868
Russia

Kirill B. Seslavinsky (deceased)
United Institute of Physics of the Earth
Russian Academy of Sciences
Bol'shaya Gruzinskaya ulitsa 10
Moscow 123810
Russia

John H. Shergold
Laboratoire de Paléontologie
Institut des Sciences de l'Evolution (URA
    327)
Université de Montpellier II
Place Eugène Bataillon
F-34095 Montpellier Cedex 05
France
(formerly Australian Geological Survey
    Organisation)

Alan G. Smith
Department of Earth Sciences
University of Cambridge
Downing Street
Cambridge CB2 3EQ
United Kingdom

Ben R. Spincer
Department of Earth Sciences
University of Cambridge
Downing Street
Cambridge CB2 3EQ
United Kingdom

James Sprinkle
Department of Geological Sciences
University of Texas
Austin TX 78712
USA

Galina T. Ushatinskaya
Paleontological Institute
Russian Academy of Sciences
Profsoyuznaya ulitsa 123
Moscow 117868
Russia

Rachel A. Wood
Department of Earth Sciences
University of Cambridge
Downing Street
Cambridge CB2 3EQ
United Kingdom

Andrey Yu. Zhuravlev
Paleontological Institute
Russian Academy of Sciences
Profsoyuznaya ulitsa 123
Moscow 117868
Russia

Index

For complete coverage, see both scientific and common names of flora and fauna.

Printed in the USA
CPSIA information can be obtained
at www.ICGtesting.com
JSHW051451221024
72172JS00010B/81

9 780231 106139